QCD
SPECTRAL SUM RULES

World Scientific Lecture Notes in Physics

ISSN: 1793-1436

---

*Published titles**

*For the complete list of published titles, please visit
http://www.worldscientific.com/series/wslnp

World Scientific Lecture Notes in Physics — Vol. 26

QCD SPECTRAL SUM RULES

STEPHAN NARISON

Laboratoire de Physique Mathématique UA768
USTL — Montpellier (F)
and
Theory Division CERN — Geneva (CH)

World Scientific

Singapore • New Jersey • London • Hong Kong

Published by

World Scientific Publishing Co. Pte. Ltd.
5 Toh Tuck Link, Singapore 596224
USA office: 27 Warren Street, Suite 401-402, Hackensack, NJ 07601
UK office: 57 Shelton Street, Covent Garden, London WC2H 9HE

Library of Congress Cataloging-in-Publication Data
Narison, S.
QCD spectral sum rules / Stephan Narison.
(World Scientific lecture notes in physics; vol. 26)
ISBN-13 978-9971-5-0653-7 -- ISBN-10 9971-5-0653-X
1. Quantum chromodynamics. 2. Sum rules (Physics) I. Title. II. Series.
QC793.3.Q35N35 1989
539.7'548--dc20 89-29138
CIP

British Library Cataloguing-in-Publication Data
A catalogue record for this book is available from the British Library.

PREFACE

The aim of this book is to provide :

1) an elementary introduction to the method of QCD Spectral Sum Rules (QSSR) for non-specialists and experimental physicists ;

2) a specialized review of QSSR developments over the last ten years of activity in this field.

The book is intended as an extension of and complement to my previous report in Ref. 34). In the first two chapters, I give a short general discussion of chiral symmetry, perturbative QCD and different QCD-like non-perturbative approaches other than QSSR. Chapters 3 and 4 are devoted to the theoretical foundations of QSSR and to different methods of evaluating the Wilson coefficients of the QCD condensates. The remaining chapters are concerned with different phenomenological applications of QSSR in QCD. Finally, in Chapter 18, I discuss the extension of QSSR to the composite models of electroweak interactions.

However, owing to the space-time limitation, I have been obliged to be selective in my discussion of QSSR methods and the hadronic channels. The former discussion is focused on the Laplace Sum Rule (LSR), the Finite Energy Sum Rule (FESR) and the Moments Sum Rules which are complementary methods. For the latter, I have selected the hadronic channels relevant for an understanding of chiral symmetry and the dynamics of light- and heavy-quark systems, for a non-perturbative test of QCD such as gluonia and for subsequent experimental research, e.g. the search for exotic states.

Most of our discussion of the phenomenology of QSSR is based on the optimal estimates extracted within the sum rule variable stability criterion and the value of the onset of the QCD continuum fixed either from FESR and/or from a stability criterion. Thus, in the course of this book it will be noticed that real progress has been made in improving the previous QSSR results by freeing them from handwaving and ad hoc phenomenological arguments. As a consequence, a coherent picture of hadron dynamics emerges from QSSR and it is impressive to see how *QSSR fix the properties of the lowest ground state* ***plus*** *the correlated* value of the onset of the QCD continuum by the few perturbative and lowest dimension non-perturbative parameters of the naive and the somewhat disappointing QCD Lagrangian.

Some other interesting problems naturally remain to be tackled before the picture is complete. However, in the present state of the art where our control of non-perturbative effects is very poor and while we await a new generation of Lattice Gauge Theory results, QSSR appear to be a powerful and competitive semi-phenomenological approach to the complex dynamics of hadrons.

Little further initial comment is required. R.P. Feynman's remark about gauge theories (Omni magazine 1979) sums up my feeling about QSSR :

> *"... A few years ago, I was very skeptical..*
> *I was expecting mist and now it looks like ridges and valleys after all..."*

to which I would add a Malagasy saying :

> *"Vary iray no nafafy, ka vary zato no miakatra"*

which means : with one grain of rice sowed, one can gather by the thousand.

CONTENTS

QCD SPECTRAL SUM RULES

CHAPTER 1

CHIRAL SYMMETRY - SPECTRAL SUM RULES - MS-SCHEME AND PERTURBATIVE QCD

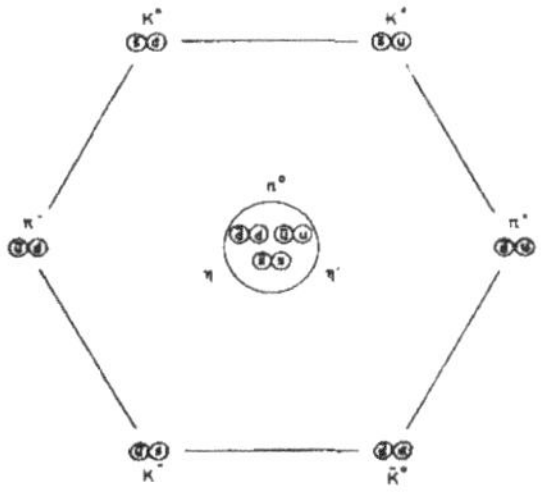

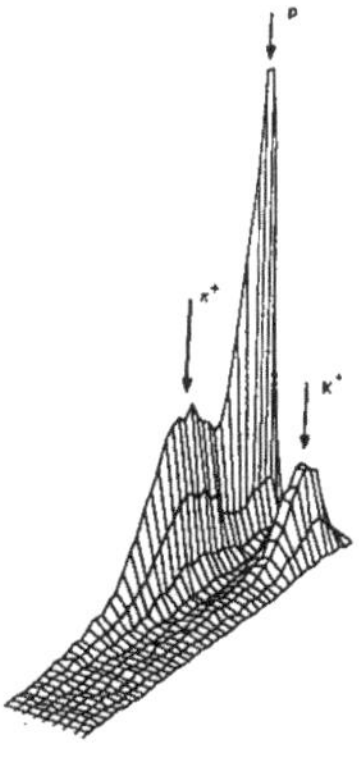

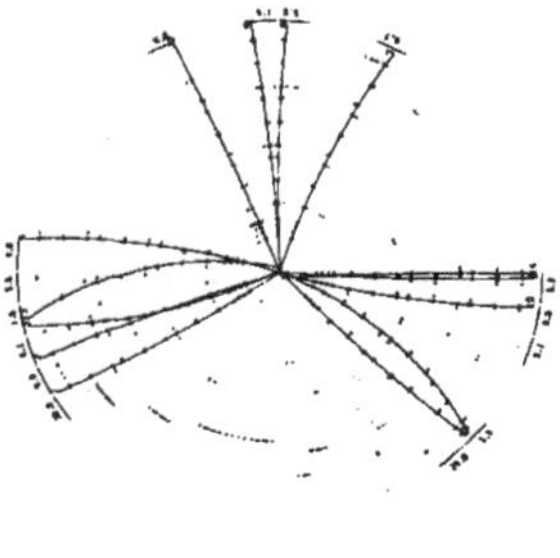

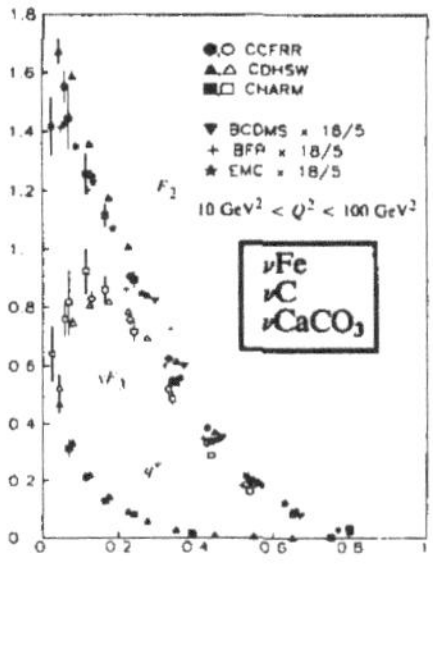

Prof. Murray Gell-Mann contemplating his meson classification, some meson peaks, the QCD jets and the nucleon structure function

1. PERTURBATIVE QCD AND CHIRAL SYMMETRY

Since the pioneering work of Gell-Mann, Fritzsch and Leutwyler[1] and after the discovery of the asymptotic freedom properties of QCD[2], there are various reasons for believing that QCD is the best candidate theory of strong interactions though the confinement problem is still unsolved owing to the peculiar infrared behaviour of the theory. The asymptotic freedom property of QCD at high momentum allows perturbative calculations in a series-expansion of the strong interaction coupling constant to give a nice description of various hard processes[3] (deep inelastic scattering, Drell-Yan, jets, high P_T...)

a) The QCD Lagrangian and its Symmetry

The QCD Lagragian density is :

$$\mathcal{L}_{QCD}(x) = -\frac{1}{4} G^a_{\mu\nu} G^{\mu\nu}_a + i \sum_{j=1}^{n} \bar{\psi}^\alpha_j \gamma^\mu (D_\mu)_{\alpha\beta} \psi^\beta_j - \sum_{j=1}^{n} m_j \bar{\psi}^\alpha_j \psi_j$$

$$- \frac{1}{2\alpha_G} \partial_\mu A^\mu_a \partial^\mu A_\mu - \partial_\mu \bar{\varphi}_a D^\mu \varphi_a , \qquad (1.1)$$

where $G^a_{\mu\nu} \equiv \partial_\mu A^a_\nu - \partial_\nu A^a_\mu + g f_{abc} A^b_\mu A^c_\nu$ $(a \equiv 1,2,\ldots,8)$ are Yang-Mills[4] field strengths constructed from the gluon fields $A^a_\mu(x)$. $(D_\mu)_{\alpha\beta} \equiv \delta_{\alpha\beta} \partial_\mu - ig \sum_a \frac{1}{2} \lambda^a_{\alpha\beta} A^a_\mu$ are covariant derivatives acting on the quark colour component $\alpha,\beta \equiv$ red, blue and yellow. $\lambda^a_{\alpha\beta}$ are the eight 3×3 colour matrices and f_{abc} the structure constants which close the SU(3) Lie algebra :

$$[T_a, T_b] = i f_{abc} T_c \quad , \qquad (1.2)$$

where $(T^a)_{\alpha\beta} = \frac{1}{2} \lambda^a_{\alpha\beta}$ in the fundamental colour $\underline{3}$ representation, whilst in the adjoint $\underline{8}$ representation of gluon basis $(T^a)_{bc} = -i f^a_{bc}$. The last two terms in (1.1) are respectively the gauge-fixing term necessary for a covariant quantization in the gluon sector [$\alpha_G = O(1)$ in the Landau (Feynman) gauge] and the Fadeev-Popov[5)] ghost term necessary to eliminate unphysical particles from the theory ($\varphi^a(x)$ are eight anticommuting scalar fields in the $\underline{8}$ of SU(3)).

$\mathcal{L}_{QCD}(x)$ is locally invariant under the BRS[6)] transformations:

$$
\begin{aligned}
A_\mu(x) &\longrightarrow A_\mu(x) + \omega\, D_\mu\, \varphi \quad , \\
\psi_i(x) &\longrightarrow \exp(-\, ig\omega\, \vec{T}.\vec{\varphi})\, \psi_i \quad , \\
\overline{\varphi} &\longrightarrow \overline{\varphi} + \frac{\omega}{\alpha_G}\, \partial_\mu\, A^\mu \quad , \\
\varphi &\longrightarrow \varphi - \frac{1}{2}\, g\, \omega\, \vec{\varphi} \times \vec{\varphi} \quad ,
\end{aligned}
\qquad (1.3)
$$

where $\omega(x)$ is an arbitrary parameter.

$\mathcal{L}_{QCD}(x)$ is invariant under the $U(1)_B$ global transformation:

$$\psi_i(x) \longrightarrow \exp(-\, i\, \theta\, \mathbb{1})\, \psi_i(x) \quad , \qquad (1.4)$$

to which corresponds the conserved baryonic current :

$$J^\mu(x) = \sum_i \overline{\psi}_i\, \gamma^\mu\, \psi_i \qquad (1.5)$$

and the baryonic charge generator of the $U(1)_B$ group :

$$B = \int d^3x\; J^0(\vec{x}, t) \quad . \qquad (1.6)$$

For massless quarks $\mathcal{L}_{QCD}$ is also invariant under the axial $U(1)_A$ transformation :

$$\psi_i \longrightarrow (-i\, \theta\, \mathbb{1}\, \gamma_5)\, \psi_i \quad , \qquad (1.7)$$

acting on quark-flavour components. The corresponding current

$$J_5^\mu(x) = \sum_i \bar{\psi}_i \gamma^\mu \gamma^5 \psi_i \tag{1.8}$$

has an anomalous divergence

$$\partial_\mu J_5^\mu(x) = \frac{g^2}{4\pi^2} \frac{n}{8} \epsilon_{\mu\nu\rho\sigma} G_a^{\mu\nu} G_a^{\rho\sigma} , \tag{1.9}$$

where the rate of the change of the associated axial charge

$$\dot{Q}_5 = \int d^3x \, \partial_0 J_5^0 (\vec{x}, t) , \tag{1.10}$$

is zero in the absence of instanton-type solutions[7].

In the massless quark limit ($m_j = 0$), $\mathcal{L}_{QCD}$ also possesses a $SU(n)_L \times SU(n)_R$ global chiral symmetry and is invariant under the global transformation :

$$\psi_i \longrightarrow \exp\left(-i\ \theta^A T_A\right) \psi_i ,$$
$$\psi_i \longrightarrow \exp\left(-i\ \theta^A T_A \gamma_5\right) \psi_i , \tag{1.11}$$

where T^A $A \equiv 1,\ldots,n^2-1$ are the infinitesimal generators of the SU(n) group acting on the quark-flavour components. The associated Noether currents are the vector and axial-vector currents :

$$V_\mu^A(x) = \bar{\psi}_i \gamma_\mu T_{ij}^A \psi_j ,$$
$$A_\mu^A(x) = \bar{\psi}_i \gamma_\mu \gamma_5 T_{ij}^A \psi_j , \tag{1.12}$$

which are the currents of the algebra of the currents of Gell-Mann[1,8].

The corresponding charges which are the generators of $SU(n)_L \times SU(n)_R$

are :

$$Q_L^A = \int d^3x \left(V_o^A - A_o^A\right) \quad ,$$

$$Q_R^A = \int d^3x \left(V_o^A + A_o^A\right) \quad . \qquad (1.13)$$

b) Chiral Symmetry Breaking and PCAC

The charges in 1.13 are conserved in the massless quark limit. In the Nambu-Goldstone[9] realization of chiral symmetry, the axial charge does not annihilate the vacuum. This is the basis of the successes of current algebra and pion PCAC [8]. In this scheme, the chiral flavour group $G \equiv SU(n)_L \times SU(n)_R$ is broken spontaneously by the light (u,d,s) quark vacuum condensates down to a subgroup $H \equiv SU(n)_{L+R}$ where the vacuums are symmetrical :

$$\left\langle \bar\psi_u \psi_u \right\rangle = \left\langle \bar\psi_d \psi_d \right\rangle = \left\langle \bar\psi_s \psi_s \right\rangle \quad . \qquad (1.14)$$

This spontaneous breaking mechanism is accompanied by n^2-1 massless Goldstone P (pion-like) bosons which are associated with each unbroken generator of the coset space G/H. On the other hand, the vector charge is assumed to annihilate the vacuum and the corresponding symmetry is achieved à la Wigner-Weyl[10]. In this case, the particles are classified in irreducible representations of $SU(n)_{L+R}$ and form parity doublets. In addition to the electromagnetic mass which the Goldstone boson can aquire[11], they get a mass mainly from an explicit breaking ($m_i \neq 0$) of the $SU(n)_L \times SU(n)_R$ global symmetry. In this case, the divergence of the axial-vector current reads :

$$\partial_\mu A^\mu(x)_j^i = (m_i + m_j)\, \bar\psi_i (i\, \gamma_5)\, \psi_j \quad , \qquad (1.15a)$$

to which are associated the quasi-Goldstone parameters defined as :

$$\left\langle 0 \left| \partial_\mu A^\mu \right| \pi \right\rangle = \sqrt{2}\, f_\pi\, m_\pi^2\, \vec{\pi} \quad , \tag{1.15b}$$

where $\vec{\pi}$ is the pion field and f_π = 93.3 MeV controls the $\pi \to \mu\nu$ decay. Current algebra also tells us that the two-point correlator associated with (1.15) is related to the axial-current one via a Ward identity[8] (up to equal-time commutators) :

$$q^\mu q^\nu \Pi_5^{\mu\nu} = \psi_5(q^2) - \int d^4 x\, e^{iqx} \delta(x_0)\, q^\nu \left\langle 0 \left| [A^0(x), A^{\nu+}(0)] \right| 0 \right\rangle$$

$$+ i \int d^4x\, e^{iqx}\, \delta(x_0) \,.\, \left\langle 0 \left| \left[\partial_\mu A^\mu, (A^0(x))^+ \right] \right| 0 \right\rangle \quad , \tag{1.16a}$$

with :

$$\psi_5(q^2) = i \int d^4x\, e^{iqx} \left\langle 0 \left| \mathrm{T}\, \partial_\mu A^\mu(x) \left(\partial_\nu A^\nu(0) \right)^+ \right| 0 \right\rangle \quad ,$$

$$\Pi_5^{\mu\nu}(q^2) = i \int d^4 x\, e^{iqx} \left\langle 0 \left| \mathrm{T}\, A^\mu(x) \, (A^\nu(0))^+ \right| 0 \right\rangle \quad . \tag{1.16b}$$

At $q = 0$, the identity (1.16) reduces to :

$$\psi_5(0) = -i\, (m_u + m_d) \left\langle 0 \left| \left[\bar{\psi}_d(0)\, i\, \gamma_5\, \psi_u(0)\, Q_5^+ \right] \right| 0 \right\rangle , \tag{1.17a}$$

where Q_5 is the axial-charge generator. In the Nambu-Goldstone realization of chiral symmetry $Q_5 \;|\, 0 \,\rangle \neq 0$. Then we get

$$\psi_5(0) = -(m_u + m_d) \left\langle \bar{\psi}_d\, \psi_d + \bar{\psi}_u\, \psi_u \right\rangle . \tag{1.17b}$$

Using (1.15b) in the definition of $\psi_5(q^2)$ and equating this with (1.17b), we have the pion PCAC relation at $q = 0$,

$$2\, m_\pi^2\, f_\pi^2 = -(m_u+m_d) \left\langle \bar{\psi}_u\, \psi_u + \bar{\psi}_d\, \psi_d \right\rangle \quad . \tag{1.18}$$

We also know from experiments that the spectrum of the pseudo-scalar bosons octet (π, K, η) does not show any degeneracy in the masses. This suggests a large explicit breaking of the $SU(3)_L \times SU(3)_R$ chiral flavour group à la Gell-Mann, Oakes and Renner[12] :

$$\mathcal{L}_{GOR} = -\, \epsilon_0\, U_0(x) - \epsilon_8\, U_8(x) - \epsilon_3\, U_3(x) \quad , \tag{1.19a}$$

where the Hermitian scalar densities $U_a(x)$ expressed in terms of quark bilinears read :

$$U_a(x) = \mathrm{Tr}\, \bar{\psi}\, \lambda_a \psi \qquad , \qquad a = 0,3,8$$

with (1.19b)

$$\lambda_0 = \sqrt{\frac{2}{3}} \begin{pmatrix} 1 & & 0 \\ & 1 & \\ 0 & & 1 \end{pmatrix} \qquad \lambda_3 = \begin{pmatrix} 1 & & 0 \\ & -1 & \\ 0 & & 0 \end{pmatrix} \qquad \lambda_8 = \frac{1}{\sqrt{3}} \begin{pmatrix} 1 & & 0 \\ & 1 & \\ 0 & & -2 \end{pmatrix} ,$$

and where the symmetry-breaking parameters ϵ_a are combinations of the quark masses :

$$\epsilon_0 = \frac{1}{\sqrt{2}} (m_u + m_d + m_s) \quad ,$$

$$\epsilon_3 = \frac{1}{2} (m_u - m_d) \quad ,$$

$$\epsilon_8 = \frac{1}{\sqrt{3}} \cdot \frac{1}{2} (m_u + m_d - 2\, m_s) \; . \tag{1.19c}$$

Therefore, it is essential to have good control of the quark mass values and a deviation from the SU(3) symmetrical relation in (1.14). In fact, there are estimates of the quark-mass ratios from current algebra [13,14], which are reliable (with perhaps the exception of the up quark) as the mass ratio is not renormalized. On the other hand, it is much more difficult to estimate the absolute values of the quark masses and similarly of the vacuum condensate which is correlated to it via the PCAC relation in (1.18). This reliability needs a consistent renormalization framework rendered possible only with the advent of QCD as we shall see later on.

2. PRE-QCD CURRENT ALGEBRA WEINBERG SUM RULES

Spectral function sum rules were used long before the advent of QCD. They are usually known as dispersion sum rules in current algebra[8] and most of them are based on the assumed asymptotic behaviour of the absorptive amplitudes. Let us discuss in detail two types of superconvergent sum rules :

a) Asymptotic realizations of $SU(2)_L \times SU(2)_R$ chiral symmetry

Weinberg has proposed two sum rules[15] (WSR) based on the belief that the $SU(2)_L \times SU(2)_R$ chiral symmetry is realized asymptotically in nature. In modern QCD language, his discussion is based on the fact that the axial-vector current has two QCD realizations: the first one is the short-distance realization in terms of the quark fields :

$$\left\langle 0 \left| \bar{d} \, \gamma^\mu \, \gamma^5 \, u \right| \pi \right\rangle = \sqrt{2} \, f_\pi \, p^\mu \quad , \tag{1.20}$$

where $f_\pi = 93.3$ MeV is the pion decay constant and p^μ is its momentum. The long-distance realization of the axial current is obtained from the chiral Lagrangian of the non-linear σ model :

$$\mathcal{L}_\sigma = -\frac{1}{4} f_\pi^2 \, \mathrm{Tr} \left\{ \partial_\mu \, U \, \partial^\mu \, U^+ \right\} \quad , \tag{1.21}$$

where $U = \exp\left(i \, \vec{\tau}.\vec{\pi} \, / f_\pi\right)$ is the pion rotation matrix, $\vec{\pi}$ the pion field and τ the Pauli matrices. Now one can show how the WSR connect these two realizations. It is appropriate to study the two-point correlator :

$$W^{\mu\nu}_{LR} = i \int d^4 x\, e^{iqx} \left\langle 0 \,\middle|\, \mathbb{T}\, J^{\mu}_{L}(x) \left(J^{\nu}_{R}(o)\right)^{+} \,\middle|\, 0\right\rangle$$

$$= -\left(g^{\mu\nu} q^2 - q^{\mu} q^{\nu}\right) \Pi^{(1)}_{LR}(q^2) + q^{\mu} q^{\nu}\, \Pi^{(o)}_{LR}(q^2)\,, \qquad (1.22)$$

where J^{μ}_{L} and J^{ν}_{R} are left- and right-handed currents which read in terms of the quark fields :

$$J^{\mu}_{L} \equiv \bar{u}\, \gamma^{\mu}(1-\gamma_5)\, d \quad , \quad J^{\mu}_{R} \equiv \bar{u}\, \gamma^{\mu}(1+\gamma_5)\, d \quad . \qquad (1.23)$$

$\Pi^{(1)}$ and $\Pi^{(o)}$ are the transverse and longitudinal parts of the correlator. In the asymptotic ($q^2 \longrightarrow \infty$) or chiral limit ($m_{u,d} = 0$) where $SU(2)_L \times SU(2)_R$ chiral symmetry is realized, the asymptotic expression of $W^{\mu\nu}_{LR}$ is zero. This vanishing of $W^{\mu\nu}_{LR}$ can be expressed in terms of two WSR of the absorptive parts simply using the well-known Hilbert representation issued from the analyticity properties of Green's function :

$$\mathrm{Re}\, \Pi_{LR}(q^2) = \frac{1}{\pi} \int_0^{\infty} \frac{dt}{t-q^2-i\epsilon}\, \frac{1}{\pi}\, \mathrm{Im}\, \Pi_{LR}(t) + \text{"subtraction..."} \,. \qquad (1.24)$$

Then, the two famous WSR read :

$$\int_0^{\infty} dt\ \mathrm{Im} \left(\Pi^{(1)}_{LR} + \Pi^{(o)}_{LR}\right)(t) = 0 \quad , \qquad (1.25)$$

$$\int_0^{\infty} dt\ t\ \mathrm{Im}\, \Pi^{(1)}_{LR}(t) = 0 \quad , \qquad (1.26)$$

where the first is the $q^{\mu}q^{\nu}$ component of $W^{\mu\nu}_{LR}$ and the second its $g^{\mu\nu}$ parts. Eqs (1.25) and (1.26) express a duality between the long-range (spectral function) and high-energy (theory) parts of hadrons. The spectral function appearing in the sum rules can be studied using the long-distance behaviour of the axial and vector currents :

$$A^{\mu}(x) = -\sqrt{2}\, f_{\pi}\, \partial_{\mu} \vec{\pi} + \left(\sqrt{\frac{2}{3}}\right) [\vec{\pi}, \vec{\pi}\, \partial^{\mu} \vec{\pi}] \Big/ f_{\pi} + \dots ,$$

$$V^{\mu}(x) = i\, \vec{\pi}\, \partial_{\mu} \vec{\pi} \quad . \qquad (1.27)$$

Possible final-state interactions between pseudoscalar particles can lead to the formation of resonances having the quantum numbers 1^{--}, 1^{++}, 0^{-+} and 0^{++}. Using a narrow-width approximation and assuming that the π, A_1 and ρ dominate the spectral functions, Weinberg has derived from (1.25) and (1.26) the constraints :

$$\frac{M_{\rho}^2}{2\gamma_{\rho}^2} - \frac{M_{A_1}^2}{2\gamma_{A_1}^2} - 2\, f_{\pi}^2 = 0 \quad ,$$

$$\frac{M_{\rho}^4}{2\gamma_{\rho}^2} - \frac{M_{A_1}^4}{2\gamma_{A_1}^2} = 0 \quad , \qquad (1.28)$$

where γ_V is the V-meson coupling to the corresponding current :

$$\left\langle 0 \,|V^{\mu}|\, \rho \right\rangle = \sqrt{2}\, \frac{M_{\rho}^2}{2\gamma_{\rho}}\, \epsilon^{\mu} \quad , \qquad (1.29a)$$

with the normalization :

$$\Gamma_{\rho \to e^+ e^-} \simeq \frac{2}{3}\, \pi\, \alpha^2\, \frac{M_{\rho}}{2\gamma_{\rho}^2} \quad . \qquad (1.29b)$$

From the above crude assumptions, one can already deduce from (1.28) a prediction of the A_1 mass by giving $M_{\rho} = 0.77$ GeV, $\gamma_{\rho} \simeq 2.55$ and f_{π} :

$$M_{A_1} \simeq 1.1 \text{ GeV} \quad . \tag{1.30}$$

However, if one adds to (1.28) a relation between f_π, γ_ρ and M_ρ deduced from the use of soft-pion techniques plus ρ-universality for the ρ into $\pi\pi$ decay (the approximate KSFR relation[16]) :

$$f_\pi^2 \simeq \frac{M_\rho^2}{16\gamma_\rho^2} \quad , \tag{1.31}$$

one arrives at the Weinberg mass formula for the A_1 meson :

$$M_{A_1} \simeq \sqrt{2}\, M_\rho \quad , \tag{1.32}$$

which, within the crude approximation used, is very successful compared to the data. Possible QCD improvements of the WSR will be discussed later on.

b) Asymptotic Realizations of $SU(3)_F$ Symmetry

Weinberg-inspired sum rules have been also derived from the asymptotic realization of flavour symmetry. These are the so-called Das-Mathur-Okubo (DMO) sum rules[17]. The DMO sum rules can be studied from the two-point correlator :

$$\Pi_i^{\mu\nu}(q) = i \int d^4x\, e^{iqx} \left\langle 0 \left| \mathbb{T}\, V_i^\mu(x) \left(V_i^\nu(0)\right)^+ \right| 0 \right\rangle$$

$$\equiv -(g^{\mu\nu} q^2 - q^\mu q^\nu)\, \Pi_i(q^2) \quad , \tag{1.33a}$$

where $V_i^\mu \equiv \bar\psi_i \gamma^\mu \psi_i$ $(i \equiv u,d,s \ldots)$ are the flavour components of the electromagnetic current :

$$J_{EM}^\mu(x) = \frac{2}{3} V_u^\mu - \frac{1}{3} V_d^\mu + \frac{2}{3} V_c^\mu - \frac{1}{3} V_s^\mu + \ldots \quad . \tag{1.34b}$$

Within the asymptotic ($q^2 \longrightarrow \infty$) or massless ($m_i = 0$) limit, we can derive the DMO sum rule[17]:

$$\int_0^\infty dt\ (\text{Im}\ \Pi_3(t) - \text{Im}\ \Pi_8(t)) \equiv \int_0^\infty dt\ \text{Im}\ (\Pi_u + \Pi_d - 2\ \Pi_s) = 0, \qquad (1.35)$$

which corresponds to the difference between the isovector and isoscalar spectral functions associated with $SU(3)_F$. Saturating (1.35) by the lowest resonance masses, we obtain the well-known successful phenomenological relation among vector mesons :

$$M_\rho\ \Gamma_{\rho \to e^+e^-} - 3\left(M_\omega\ \Gamma_{\omega \to e^+e^-} + M_\varphi\ \Gamma_{\varphi \to e^+e^-}\right) \simeq 0. \qquad (1.36)$$

An alternative way of writing the sum rules in (1.35) is in terms of the $e^+e^- \longrightarrow$ Hadrons total cross-section which follows from the optical theorem :

$$\sigma_H(t) = \frac{4\pi^2\alpha}{t}\ e^2\ \frac{1}{\pi}\ \text{Im}\ \Pi(t) \qquad (1.37)$$

and which is useful as we have complete data for the total cross-section. Eqs (1.25),(1.26) and (1.35) have shown constraints for low-energy data which follow from the asymptotic behaviour of the spectral functions. These are the prototype sum rules which will be refined and extended within QCD.

3. A SURVEY OF QCD SPECTRAL SUM RULES

Spectral sum rules are different versions and/or improvements of the Hilbert representation in (1.24). For the purposes of more general discussion, let us forget QCD for the moment, i.e. the theoretical side Re $\Pi(q^2)$, and we shall concentrate on the RHS spectral integral.

a) **Laplace Transform Sum Rule**

This type of sum rule is derived from (1.24) by applying to both sides the inverse Laplace operator[18]: ($Q^2 \equiv -q^2 > 0$)

$$\hat{\mathbb{L}} \equiv \lim_{\substack{Q^2, N \to \infty \\ N/Q^2 \equiv \tau \text{ fixed}}} (-1)^N \frac{(Q^2)^N}{(N-1)!} \frac{\partial^N}{(\partial Q^2)^N} . \tag{1.38}$$

In this case, one gets the exponential form :

$$\hat{\mathbb{L}}\,\Pi = \tau \int_0^\infty dt\, e^{-t\tau} \frac{1}{\pi} \text{Im}\,\Pi(t) , \tag{1.39}$$

from which one can derive the ratio of moments[19]:

$$R(\tau) = -\frac{d}{d\tau} \log \int_0^\infty dt\, e^{-t\tau} \frac{1}{\pi} \text{Im}\,\Pi(t) \tag{1.40}$$

or the finite energy like[18]:

$$R_c(\tau) = \frac{\int_0^{t_c} t\, dt\, e^{-t\tau} \frac{1}{\pi} \text{Im}\,\Pi(t)}{\int_0^{t_c} dt\, e^{-t\tau} \frac{1}{\pi} \text{Im}\,\Pi(t)} . \tag{1.41}$$

As can be seen in the derivation of the Laplace sum rule, one has to assume that various derivatives exist. For an approximate truncated series like in QCD[20], this existence is satisfied. The advantages of $\hat{\mathbb{L}}\Pi$ are two-fold. Firstly, the use of various derivatives helps to eliminate the subtraction terms in (1.24) which are often polynomials in q^2. Secondly, the exponential factor increases the role of the ground state into the spectral integral if the QSSR variable τ is not too small. This fact is welcome for low-energy physics. The advantage of (1.40) and (1.41) can explicitly be seen if one uses the simple duality ansatz "one resonance" plus "continuum" for parametrizing the spectral function. One can see in this parametrization that these two sum rules give an expression of the mass squared of the ground state. The accuracy of this simple duality ansatz will be tested later on. One can illustrate the sum rule by taking the example of the three-dimensional harmonic oscillator in quantum mechanics[19]. In this case, the RHS of the sum rule in (1.39) reads :

$$F(\tau) = \sum_{n=0,2,4...} (R_n)^2 e^{-E_n \tau} , \quad (1.42)$$

where R_n is the radial wave function for zero angular momentum and E_n the corresponding eigenvalue. τ is the parameter which regulates the energy resolution of the sum rule and plays the role of an "imaginary time" variable. As one has an exact solution of the LHS for the harmonic oscillator potential $V(r) = \frac{1}{2} m \omega^2 r^2$, one can see that in the limit $\tau \to \infty$ the exact expression $R(\tau \to \infty) \equiv -\frac{d}{d\tau} \operatorname{Log} F(\tau)$ tends to the lowest eigenvalues $E_o = \frac{3}{2}\omega$. At finite τ and for a truncated series in τ, one can observe (see Fig. 1.1) that $R_{approx}(\tau)$ stays above the eigenvalue E_o as a consequence of the positivity of R. The agreement between R_{approx} and R_{exact} increases if one adds more and more terms in the τ-expansion. The minimum of R_{approx} provides an upper bound to the value of E_o while the distance between R_{approx} and E_o controls the strength of the continuum to the sum rule. However, by working with a truncated series as in QCD, we do not often have a nice minimum for R_{approx}. This minimum is replaced in some cases by an inflexion point where the optimal information on the resonance properties is obtained. We shall see later on that this previous example mimics the case of QCD quite well.

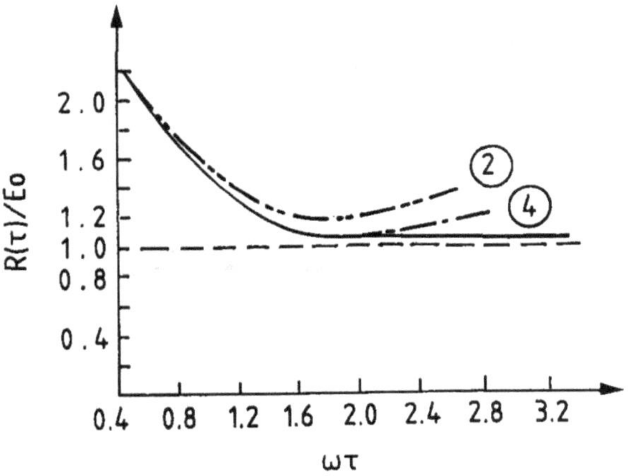

Fig. 1.1 : The ratio of moments normalized to the ground-state energy versus the imaginary time variable for the case of the harmonic oscillator potential. (2) and (4) : approximate series including the second and fourth order terms; ___ exact solution.

b) Finite Energy Sum Rule (FESR)

Another version of QSSR is the FESR :

$$\int_0^{Q^2} dt\ t^n\ \frac{1}{\pi}\ \mathrm{Im}\ \Pi_{Theor}(t) \simeq \int_0^{Q^2} dt\ t^n\ \frac{1}{\pi}\ \mathrm{Im}\ \Pi_{EXP}\ (t) \quad n = 0,1,.. \quad , \qquad (1.43)$$

which was known a long time before QCD[21]. Eq.(1.43) can be derived in many ways. A use of the Cauchy theorem (Fig. 1.2) on a finite radius contour in the complex q^2 plane is one way[22] :

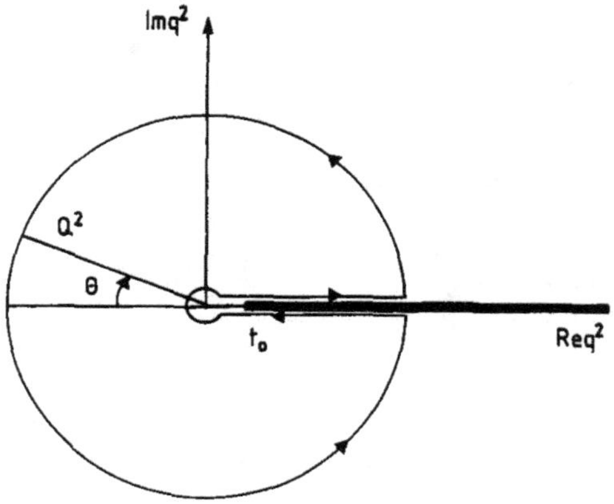

Fig. 1.2 : Cauchy contour in the complex q^2 plane.

$$\frac{1}{2\pi i}\oint dz\ z^n\ \Pi(z) = 0 \quad . \qquad (1.44)$$

If one neglects the contribution of the little circle around the origin which is safer if $\Pi(0) = 0$, one deduces the moments :

$$\mathcal{M}^{(n)}(Q^2) = \int_0^{Q^2} dt\ t^n\ \frac{1}{\pi}\ \mathrm{Im}\ \Pi(t) = (-1)^{n+1}\ \frac{(Q^2)^{n-1}}{2\pi}\ .$$

$$.\int_{-\pi}^{+\pi} d\theta\ e^{i(n+1)\theta}\ \Pi(Q^2\ e^{i\theta})\ , \qquad (1.45)$$

where the *LHS* is *known from* the data and the *RHS* from the theory. However, as the FESR diverges for increasing n, the real axis is dominated by the high Q^2 region. For the RHS to reproduce this correctly, more information on the behaviour of the two-point correlator in the region of the big circle near the cut is needed. This means that more and more non-leading terms in the series expansion become important at large n and can destroy the convergences of the series.

Another way of deriving the FESR which casts light upon the meaning of local duality is the Gaussian sum rule[23] which reads:

$$G(\hat{t},\sigma) = \frac{1}{\sqrt{4\pi\sigma}} \int_0^\infty dt\, e^{-\frac{(t+\hat{t})^2}{4\sigma}} \frac{1}{\pi} \operatorname{Im} \Pi(t) , \qquad (1.46)$$

for a Gaussian centred at $\hat{t}$ with a finite width resolution $\sqrt{4\pi\sigma}$. Eq.(1.46) can be derived by applying the inverse Laplace operator :

$$\hat{\mathbb{L}} \equiv \lim_{\substack{N,\tau^2\to\infty \\ \frac{N}{\tau^2}\equiv\sigma}} \frac{(-\tau^2)^N}{(N-1)!} \frac{d^N}{(d\tau^2)^N} , \qquad (1.47)$$

to the already Laplace-transformed quantity :

$$F(\tau) = e^{-\hat{t}\tau}\, \tau^{-1} \int_0^\infty dt\, e^{-t\tau} \frac{1}{\pi} \operatorname{Im} \Pi(t) . \qquad (1.48)$$

One can already note from (1.46) that in the limit $\sigma = 0$, one has the strict local duality :

$$G(\hat{t},0) = \frac{1}{\pi} \operatorname{Im} \Pi(\hat{t}) . \qquad (1.49)$$

Also, (1.46) obeys the heat-evolution equation :

$$\left(\frac{\partial^2}{\partial \hat{t}^2} - \frac{\partial}{\partial\sigma}\right) G(\hat{t}, \sigma) = 0 , \qquad (1.50a)$$

with the initial condition in (1.49), where now $\hat{t}$ is the position, σ the time evolution and $\frac{1}{\pi} \operatorname{Im} \Pi(t)$ the temperature distribution in the region $0 \leq \hat{t} \leq \infty$. The two boundary conditions for $\sigma > 0$:

$$G(\hat{t} = 0, \sigma) = 0 \quad ,$$
$$\frac{\partial G}{\partial \hat{t}}(\hat{t}, \sigma)\Big|_{\hat{t}=0} = 0 \quad , \tag{1.50b}$$

lead to two independent solutions $U^-(\hat{t}, \sigma)$ and $U^+(\hat{t}, \sigma)$ where $G(\hat{t}, \sigma) = \frac{1}{2}(U^+ + U^-)(\hat{t}, \sigma)$. These solutions can be expressed in terms of Hermite polynomials. The conservation of the total heat implies the duality relation :

$$\int_{-\infty}^{+\infty} d\hat{t}\ G(\hat{t}, \sigma) = \int_0^{\infty} dt\ \frac{1}{\pi}\ \mathrm{Im}\ \Pi(t) = \int_0^{\infty} d\hat{t}\ U^+(\hat{t}, \sigma) \quad , \tag{1.51}$$

where the last equality comes from the symmetry properties of $U^+(\hat{t}, \sigma)$. A relation involving higher moments of the spectral function can also be deduced using the generating function of Hermite polynomials and leads to the sum rules :

$$\sigma^n \int_0^{\infty} d\hat{t}\ H_{2n}\left(\hat{t}/_{2\sqrt{\sigma}}\right) U^+(\hat{t}, \sigma) = \int_0^{\infty} dt\ t^{2n}\ \frac{1}{\pi}\ \mathrm{Im}\ \Pi(t) \quad , \tag{1.52a}$$

$$\sigma^{n+1/2} \int_0^{\infty} d\hat{t}\ H_{2n+1}\left(\hat{t}/_{2\sqrt{\sigma}}\right) U^-(\hat{t},\sigma) = \int_0^{\infty} dt\ t^{2n+1}\ \frac{1}{\pi}\ \mathrm{Im}\ \Pi(t) \quad , \tag{1.52b}$$

which only become useful once statements about the restriction to finite intervals can be made. In this case, (1.52) leads to the FESR in (1.43). Finally, the last (but not the least) way of deriving (1.43) is simply to take the coefficient of the τ variable in the two sides of Laplace sum rule in (1.39) [24]. This latter method can be formalized by using the zeta function prescription inspired from the non-relativistic approach[25]. In fact, if H is a Hamilton operator, the associated zeta-function can be written as :

$$\zeta(n) = \frac{1}{\Gamma(n)} \int_0^\infty dt\ \tau^{n-1}\ \mathrm{tr}\ e^{-Ht} \quad , \qquad (1.53a)$$

i.e. in field theory :

$$\zeta(n) = \frac{1}{\Gamma(n)} \int_0^\infty d\tau\ \tau^{n-1} \int_0^\infty dt\ e^{-t\tau}\ \frac{1}{\pi}\ \mathrm{Im}\ \Pi(t) \quad , \qquad (1.53b)$$

where the last integral is the familiar Laplace transform of Im $\Pi(t)$. If this Laplace-transform and its successive derivatives are a series in τ, then, one can easily derive (1.43) by comparing the exact expression of $\zeta(n = 0)$ with its approximate form.

Now, let us return to the FESR in (1.43). Contrary to the Laplace transform (1.39), where the role of the lowest ground state is important, the FESR is governed by the effects of high-mass resonances, i.e. it needs good control of the continuum contributions to the sum rule. In some cases, where a stability in t_c (continuum threshold) does not occur, this is a great disadvantage.

c) Analytical Continuation

Various versions of this method have been discussed in the literature [26]. In most cases, the problem is formulated in terms of norm problems for the input errors and is quite similar to the standard χ^2- minimization used in numerical analysis. More explicitly let us take a simple example. A polynomial in t is used for approximating the $\frac{1}{t-q^2}$ term of (1.24) in the real axis[26a]. Then, applying the Cauchy theorem to the finite Q^2 contour in the complex Q^2 plane, one arrives at the sum rule :

$$\Pi(q^2) = \frac{1}{2i\pi} \oint_C dt \left(\frac{1}{t-q^2} - \sum a_n\ t^n \right) \Pi(t) +$$

$$+\left[\Delta_n \equiv \frac{1}{\pi}\int_0^{Q^2} dt \left(\frac{1}{t-q^2} - \sum_n a_n t^n\right) \mathrm{Im}\, \Pi(t)\right], \tag{1.54}$$

where Δ_n is the "fit error" which should tend to zero, if the result is optimal. An important difference with previous sum rules is that in the RHS the data enters only in Δ_n whilst the main part of $\Pi(q^2)$ is given by its theoretical side. However, it is difficult to appreciate the reliability of the results coming from the method due to the ad hoc uses of the polynomial parametrization (or in general of the kernels in the integrals) and to the strong dependence of the results on the values of the input errors. Moreover, the sum rule in its form (1.54) might also depend on the arbitrary subtraction scale and would be less appropriate for the study of the resonance parameters. Moreover, the way of extrapolating the QCD information up to small q^2 is doubtful. From these weak points all mathematical bagages used to formulate the sum rules might loose their efficiency in the physical uses of the sum rule. More refinements and phenomenological tests of this approach are needed before a definite claim about its advantage can be made.

d) **Moment Sum Rules**

Sum rules of the type :

$$\mathcal{M}^{(N)} \equiv \frac{(-1)^N}{N!} \frac{d^N}{(dQ^2)^N} \Pi(Q^2)\Big|_{Q^2=Q_0^2} = \int_0^\infty \frac{dt}{\left(t+Q_0^2\right)^{N+1}} \frac{1}{\pi} \mathrm{Im}\, \Pi(t) \tag{1.55}$$

for finite N are often used in the literature. To our knowledge, these sum rules were first discussed by Yndurain[27] in connection with the study of $e^+e^- \to$ Hadrons data and used later for heavy-quark systems [28,29]. As in the case of the Laplace sum rule, one needs to assume that various derivatives of $\Pi(Q^2)$ exist. Also, one can see that for high moments, the role of the ground state is enhanced in the sum rule. Therefore (1.55) is a good candidate for studying the low-energy properties of hadrons as we shall see later on.

We have given a brief general survey of spectral function sum-rule methods which we believe can be applied for a general class of QCD-like theories. As one can see all the methods presented here have their own advantages and disadvantages. For the particular case of QCD where the theory is not yet solved exactly, some questions, though important, such as the existence of high derivatives at high Q^2 as well as a correct and convincing way of estimating the true theoretical systematic errors in the sum rules analysis remain academic. We have checked in a QCD-like model*) such as the non-linear σ model in two dimensions that the high derivatives for a two-point correlator exist unambiguously. Also, one can always test a posteriori whether the assumptions used for the analysis make sense.

In this review, we shall mainly concentrate on the use of the Laplace-transform (1.39-41) and moments (1.55) owing to their sensitivity to the low-energy behaviour of the spectral functions. However, in some cases, we shall also discuss for comparison constraints from FESR (1.43) which complement in many cases the Laplace Sum Rule (LSR) results.

Now, let us come to a study of the perturbative and non-perturbative aspects of QCD, which are the basis of the sum rules approach discussed in this book.

4. $\overline{MS}$ SCHEME FOR PERTURBATIVE QCD**)

a) Renormalization Constants

As in QED, the evaluation of QCD Feynman diagrams leads (in many cases) to divergent results. Finite physical answers need a renormalization of the QCD parameters (vertices, coupling, masses...). However, the renormalization programme of QED[30] cannot be extended trivially to QCD. Here quarks are off-shell and the standard on-shell

*) I wish to thank G. Veneziano for this suggestion.

**)This discussion is mainly based on the Physics Report quoted in Ref. 34).

renormalization and a Pauli-Villars[31] regularization, which are successful in QED, cannot often be used. In QCD, one uses instead the method of dimensional regularization and renormalization (so-called $\overline{MS}$-scheme[32 to 35]) which is proven to preserve gauge invariance to all orders of perturbation theory. Its most important feature is the concept of analytical continuation of the dimension of space-time to complex n (n=4 for low-energy space-time). In this approach, the infrared and ultraviolet divergences are transformed into poles in $\epsilon \equiv n-4$. They are of the form :

$$\sum_{p=1} \frac{Z^{(p)}}{\epsilon^{p}} \tag{1.56}$$

and will appear as counterterms in the initial Lagrangian constrained by the Slavnov-Taylor identities[36]. For renormalizable theories the $Z^{(p)}$ are constants or polynomials in inverse of the square of some momentum. In QCD, the counterterms of the Lagrangian are :

$$\begin{aligned}\Delta \mathcal{L}_{QCD} = {} & \Delta_{3YM} \frac{1}{4} (\partial_\mu A_\nu - \partial_\nu A_\mu) (\partial^\mu A^\nu - \partial^\mu A^\nu) + \\ & \Delta_{1YM} \frac{1}{2} (\partial_\mu A_\nu - \partial_\nu A_\mu) g \vec{A}^\nu \times \vec{A}^\mu + \\ & \Delta_5 \frac{1}{4} g^2 \left(\vec{A}_\mu \times \vec{A}_\nu\right) (\vec{A}^\mu \times \vec{A}^\nu) - \\ & \Delta_{2F}\, i \sum_j \bar{\psi}_j \gamma^\mu \partial_\mu \psi_j + \Delta_4 \sum_j m_j \bar{\psi}_j \psi_j \\ & - \Delta_{1F}\, g \bar{\psi} \frac{\lambda}{2} \gamma^\mu \psi \vec{A}_\mu + \\ & \Delta_6 \frac{1}{2\alpha_G} \left(\partial_\mu \vec{A}^\mu\right)^2 + \tilde{\Delta}_3 \left(\partial_\mu \bar{\varphi}\right)^2 + \tilde{\Delta}_1\, g\, \partial_\mu \vec{\varphi} A^\mu \times \psi\end{aligned} \tag{1.57}$$

where $\vec{A}_\mu \times \vec{A}_\nu \equiv f_{abc} A^b_\mu A^c_\nu$. It is possible to rescale the fields in such a way that $\mathcal{L}_{QCD} + \Delta \mathcal{L}_{QCD}$ has the form in (1.1) but in terms of "bare" quantities. This manipulation is correlated to the introduction of renormalization constants.

Table 1.1

Dimensions of Couplings and fields

Name	Notation	Dimension
gauge coupling	g	$\frac{1}{2}$ (4-n)
quark mass	m_i	1
covariant gauge parameter	α_G	0
fermion field	$\psi_j(x)$	$\frac{1}{2}$ (n-1)
gluon field	A^a_μ (x)	$\frac{1}{2}$ (n-2)
Fadeev-Popov field	$\varphi^a(x)$	$\frac{1}{2}$ (n-2)

Taking into account the dimension obtained in the 4-ϵ world (see Table 1.1) via the mass scale ν, one has relations between renormalized and bare parameters :

$$g^R = \nu^{-\epsilon/2} g^B Z_\alpha^{-1/2} \quad : \quad g^2/4\pi \equiv \alpha_s \ ,$$
$$m^R_j = m^B_j Z_m^{-1} \ ,$$
$$\alpha^R_G = \alpha^B_G Z_G^{-1} \ ,$$
$$\left(\psi^\alpha_j\right)^R = \nu^{\epsilon/2} \left(\psi^j_\alpha\right)^B (Z_{2F})^{-1/2} \ ,$$
$$A^\mu_R = \nu^{\epsilon/2} \left(A^a_\mu\right)_B (Z_{3YM})^{-1/2} \ ,$$

$$(\varphi^a)_R = \nu^{\varepsilon/2} (\varphi^a)_B \left(\tilde{Z}_3\right)^{-1/2} , \tag{1.58}$$

where $Z_i \equiv 1-\Delta_i$. Introducing renormalization constants for the quark-gluon-quark vertex as

$$\left(g \bar{\psi} A \psi\right)_R = \left(g_B \bar{\psi}_B A_B \psi_B\right) \nu^{\varepsilon} Z_{1F}^{-1} , \tag{1.59}$$

and analogously for the three gluon (Z_{1YM}), ghost-gluon-ghost $\left(\tilde{Z}_1\right)$ and four-gluon (Z_5) vertices one can deduce :

$$\begin{aligned} g_B^{YM} &= Z_{1YM} Z_{3YM}^{-3/2} g_R , \\ \tilde{g}_B &= \tilde{Z}_1^{-1} Z_{3YM}^{-1/2} g_R , \\ g_B^F &= Z_{1F} Z_{3YM}^{-1/2} Z_{2F}^{-1} g_R , \\ \left(g_B^{(5)}\right)^2 &= Z_5 Z_{3YM}^{-2} g_R^2 , \end{aligned} \tag{1.60}$$

which are related to each over by BRS[6] invariance :

$$g_{YM}^B = \ldots\ldots = g_B^{(5)} , \tag{1.61a}$$

leading to the Slavnov-Taylor[36] identities :

$$Z_{3YM}/Z_{1YM} = \tilde{Z}_3/\tilde{Z}_1 = Z_{2F}/Z_{1F} , \tag{1.61b}$$

and

$$Z_5 = Z_{1YM}^2/Z_{3YM} . \tag{1.62}$$

The mass renormalization constant is :

$$m_B = \left(Z_m \equiv Z_4 Z_{2F}^{-1}\right) m_R , \tag{1.63}$$

and the gauge one is :

$$\alpha_G^B = \alpha_G^R \, Z_G^{-1} \, Z_{3YM} \quad . \tag{1.64}$$

More generally, for a Green's function with n_{YM}, $\tilde{n}$ and n_F external gluons, ghost and fermion fields, one can associate the renormalization constants :

$$Z_\Gamma = \left(Z_{3YM}^{1/2}\right)^{-n_{YM}} \left(Z_3^{1/2}\right)^{-\tilde{n}} \left(Z_{2F}^{1/2}\right)^{-n_F} \quad . \tag{1.65}$$

Expressions of these renormalization constants and the corresponding anomalous dimensions are known from standard diagram techniques (see Table 1.2).

Table 1.2 : Anomalous dimension $\gamma_i = \dfrac{\nu}{Z_i}\dfrac{dZ_i}{d\nu}$ *in the t'Hooft scheme for* $SU(N)_c \times SU(n)_F$

Fermion field	$\gamma_{2F} = \left(\dfrac{\alpha_s}{\pi}\right) \dfrac{N^2-1}{2N} \dfrac{\alpha_G}{2} + \mathcal{O}\left(\dfrac{\alpha_s}{\pi}\right)^2$
Gluon field	$\gamma_{3YM} = -\left(\dfrac{\alpha_s}{\pi}\right)\left\{\dfrac{N}{4}\left(\dfrac{13}{3} - \alpha_G\right) - \dfrac{2}{3}\left(\dfrac{1}{2}\right)n\right\}$
Ghost field	$\tilde{\gamma}_3 = -\left(\dfrac{\alpha_s}{\pi}\right)\dfrac{N}{8}(3 - \alpha_G)$
Mass	$\gamma_m = (\gamma_1 \equiv 2)\left(\dfrac{\alpha_s}{\pi}\right) + \left(\gamma_2 \equiv \dfrac{1}{6}\left(\dfrac{101}{2} - \dfrac{5n}{3}\right)\right)\left(\dfrac{\alpha_s}{\pi}\right)^2$ $+ \left(\gamma_3 \equiv \dfrac{1}{128}\left[1249 + \left(\dfrac{160}{3}\xi(3) - \dfrac{2216}{27}\right)n - \dfrac{140}{81}n^2\right]\right)\left(\dfrac{\alpha_s}{\pi}\right)^3$ for N=3

$$\text{Coupling constant } \beta(\alpha_s) = \frac{\nu}{\alpha_s}\frac{d\alpha_s}{d\nu} = -\frac{\nu}{Z_\alpha}\frac{dZ_\alpha}{d\nu}$$

$$= \left[\beta_1 \equiv -\frac{1}{2}b_1 = \frac{1}{2}\left(-11 + \frac{2}{3}n\right)\right]\left(\frac{\alpha_s}{\pi}\right) + \left[\beta_2 \equiv -\frac{1}{8}b_2 = -\frac{1}{4}\left(51 - \frac{19}{3}n\right)\right]\left(\frac{\alpha_s}{\pi}\right)^2$$

$$+ \left[\beta_3 = -\frac{1}{32}\left(\frac{2857}{2} + \frac{325}{54}n^2 - \frac{5033}{18}n\right)\right]\left(\frac{\alpha_s}{\pi}\right)^3 \quad \text{for } N=3$$

$$\text{Gauge} \qquad \beta_G = \nu\frac{d\alpha_G}{d\nu} = -\alpha_G\gamma_{3YM}$$

$$\text{Three-gluon :} \qquad \gamma_{1YM} = -\left(\frac{\alpha_s}{\pi}\right)\left\{\left(\frac{17}{6} - \frac{3}{2}\alpha_G\right)\frac{N}{4} - \frac{2}{3}\left(\frac{1}{2}\right)n\right\}$$

$$\text{Ghost-gluon-ghost :} \quad \tilde{\gamma}_1 = \left(\frac{\alpha_s}{\pi}\right)\alpha_G\frac{N}{4}$$

$$\text{Fermion-gluon-fermion :} \quad \gamma_{1F} = \left(\frac{\alpha_s}{\pi}\right)\frac{1}{2}\left\{\frac{N}{4}(3 + \alpha_G) - \alpha_G\frac{N^2-1}{2N}\right\}$$

---

i) Z_{2F} and Z_m come from the evaluation of the quark self-energy diagram parametrized as :

$$\Sigma = m_B\,\Sigma_1 + \left(\hat{p} - m_B\right)\Sigma_2 \equiv P \qquad + \ldots, \qquad (1.66a)$$

where the full unrenormalized propagator reads :

$$S_F = \left(\frac{1}{1-\Sigma_2}\right) \frac{1}{\hat{p} - m_B(1 + \Sigma_1/(1-\Sigma_2))} . \qquad (1.66b)$$

Then

$$Z_{2F} = \frac{1}{(1-\Sigma_2)\text{pole}} \qquad Z_m = 1 - \Sigma_1 \Big|_{\text{pole}} . \qquad (1.66c)$$

Expressions of Σ_1^B and Σ_2^B derived from the rules and properties in appendix A read :

$$\Sigma_1^B = \left(g_B . \nu^{-\epsilon/2}\right)^2 \frac{N^2-1}{2N} \frac{1}{(16\pi^2)^{1-\epsilon/4}} \int_0^1 dx \left\{\Gamma\left(\frac{\epsilon}{2}\right)\left(\frac{\mathbb{R}^2}{\nu^2}\right)^{-\epsilon/2}\right.$$

$$\left. \times \left[2(2-x) - \epsilon(1-x) + (1-\alpha_G)(1-2x)\right] + (1-\alpha_G)\ 2x(1-x) \frac{p^2}{m_B^2-p^2x}\right\} . (1.67a)$$

with $\mathbb{R}^2 = (1-x)\left(m_B^2 - p^2x\right) - i\epsilon'$.

Then, (see e.g. Ref. 33) :

$$\Sigma_1^B = \left(\frac{\alpha_s}{\pi}\right) \frac{N^2-1}{2N} \frac{1}{2} \left\{ \frac{3}{2}\left(\frac{2}{\epsilon}\right) + \frac{3}{2}(\log 4\pi-\gamma) + \frac{5}{2}\right.$$

$$+ \frac{3}{2} \log \frac{\nu^2}{m_B^2 - p^2} + \left(\frac{1}{2}\right)\frac{m_B^2}{-p^2} - \left(\frac{1}{2}\right)\frac{m_B^2}{-p^2}\left(4 + \frac{m_B^2}{-p^2}\right) \log\left(1 - \frac{p^2}{m_B^2}\right)$$

$$\left. + (1-\alpha_G)\left[-\frac{1}{2} - \left(\frac{1}{2}\right)\frac{m_B^2}{-p^2} + \left(\frac{1}{2}\right)\frac{m_B^2}{-p^2}\left(1 + \frac{m_B^2}{-p^2}\right)\log\left(1-\frac{p^2}{m_B^2}\right)\right]\right\}, \qquad (1.67b)$$

where here and in the following $\gamma \equiv \gamma_E = 0.5772...$ denotes the Euler constant.

Also

$$\Sigma_2^B = \left(g_B \cdot \nu^{-\epsilon/2}\right)^2 \frac{N^2-1}{2N} \frac{1}{(16\pi^2)^{1-\epsilon/4}} \int_0^1 dx \left\{ \Gamma\left(\frac{\epsilon}{2}\right) \left(\frac{R^2}{\nu^2}\right)^{-\epsilon/2} \right.$$

$$\left. \times \left[-2x + \epsilon x + (1-\alpha_G)\, 2(1-x)\right] + (1-\alpha_G)\, 2x(1-x) \frac{p^2}{m_B^2 - p^2 x} \right\} \quad (1.67c)$$

and then

$$\Sigma_2^B = \left(\frac{\alpha_s}{\pi}\right) \frac{N^2-1}{2N} \left(\frac{1}{4}\right) [-1 + (1-\alpha_G)] \left\{ \frac{2}{\epsilon} + \log 4\pi - \gamma + 1 - \log \frac{m_B^2 - p^2}{\nu^2} \right.$$

$$\left. + \left(\frac{m_B^2}{-p^2}\right)^2 \log\left(1 - \frac{p^2}{m_B^2}\right) - \frac{m_B^2}{-p^2} \right\}, \quad (1.67d)$$

which shows that Σ_2 vanishes, in the Landau gauge ($\alpha_G = 0$), at order α_s/π. For completeness, we give the asymptotic expressions of Σ_1^B and Σ_2^B.

For Σ_1, one has

$$\Sigma_1^B \Big|_{-p^2 \gg m^2} = \left(\frac{\alpha_s}{\pi}\right) \frac{N^2-1}{2N} \left(\frac{1}{2}\right) \left\{ \frac{3}{\epsilon} + \frac{3}{2} (\log 4\pi - \gamma) + \frac{5}{2} - \frac{3}{2} \log \frac{-p^2}{\nu^2} \right.$$

$$+ (1 - \alpha_G)\left(-\frac{1}{2}\right) + \mathcal{O}\left(\frac{m^2}{-p^2}\log\frac{-p^2}{m^1}\right)\Bigg\} , \tag{1.67e}$$

$$\Sigma_1^B\Big|_{-p^2 \ll m^2} = \left(\frac{\alpha_s}{\pi}\right)\frac{N^2-1}{2N}\left(\frac{1}{2}\right)\Bigg\{\frac{3}{\epsilon} + \frac{3}{2}(\log 4\pi - \gamma) + \frac{3}{2}\log\frac{\nu^2}{m_B^2} + \frac{3}{4}$$

$$+ \frac{5}{6}\left(\frac{-p^2}{m_B^2}\right) + (1-\alpha_G)\left(-\frac{1}{4} - \frac{1}{12}\left(\frac{-p^2}{m_B^2}\right)\right)\Bigg\}. \tag{1.67f}$$

In the time-like region, one has :

$$\Sigma_1^B\Big|_{p^2 = m^2 = \nu^2} = \left(\frac{\alpha_s}{\pi}\right)\frac{N^2-1}{2N}\left(\frac{1}{2}\right)\left\{\frac{3}{\epsilon} + \frac{3}{2}(\log 4\pi - \gamma) + 2\right\} , \tag{1.67g}$$

where $\Sigma_1^B\Big|_{p^2 = m^2 = \nu^2}$ is gauge-independent and is related to the mass defined at the pole of the fermion propagator (see section 2). For Σ_2 one gets

$$\Sigma_2^B\Big|_{-p^2 \gg m^2} = \left(\frac{\alpha_s}{\pi}\right)\frac{N^2-1}{2N}\left(\frac{1}{4}\right)[-1+(1-\alpha_G)]\Bigg\{\frac{2}{\epsilon} + \log 4\pi - \gamma + 1 - \log\frac{-p^2}{\nu^2}$$

$$+ \mathcal{O}\left(\frac{m^2}{p^2}\right)^2 \log\frac{-p^2}{m^2}\Bigg\} , \tag{1.67h}$$

$$\Sigma_2^B \bigg|_{-p^2 \ll m^2} = \left(\frac{\alpha_s}{\pi}\right) \frac{N^2-1}{2N} \left(\frac{1}{4}\right) [-1 + (1-\alpha_G)] \left\{ \frac{2}{\epsilon} + \log 4\pi - \gamma + \log \frac{\nu^2}{m^2} \right.$$

$$\left. + \frac{1}{2} - \frac{2}{3}\left(\frac{-p^2}{m_B^2}\right) + \mathcal{O}\left(\frac{-p^2}{m_B^2}\right)^2 \right\} . \qquad (1.67i)$$

Note that the pole in $1/\epsilon$ is constant according to the theorem given in Ref.32) for a renormalizable theory.

ii) Z_{3YM} comes from the gluon propagator :

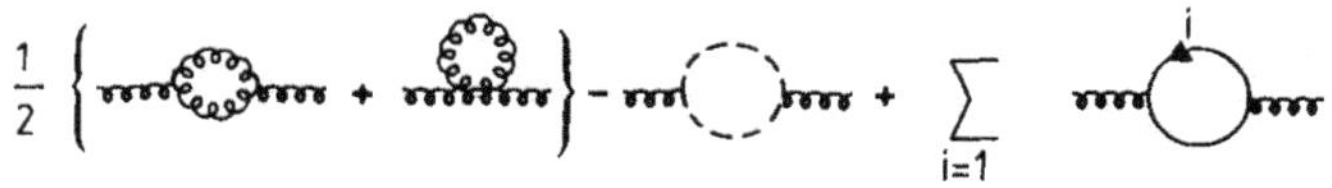

(1.68)

iii) Z_{1YM} comes from the three-gluon vertex :

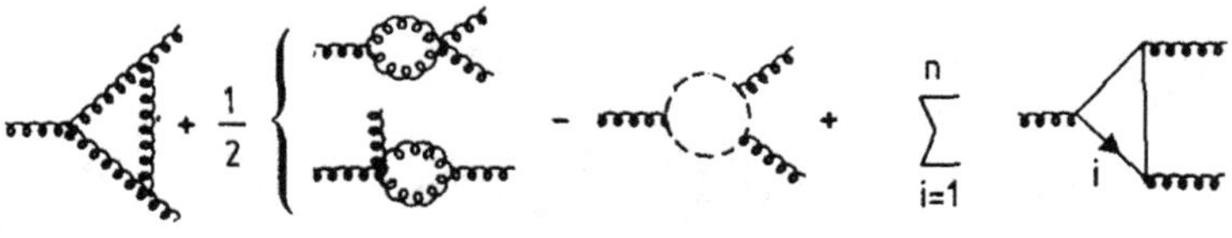

(1.69)

iv) $\tilde{Z}_1$ and $\tilde{Z}_3$ come respectively from the ghost self-energy and ghost--gluon-ghost vertex

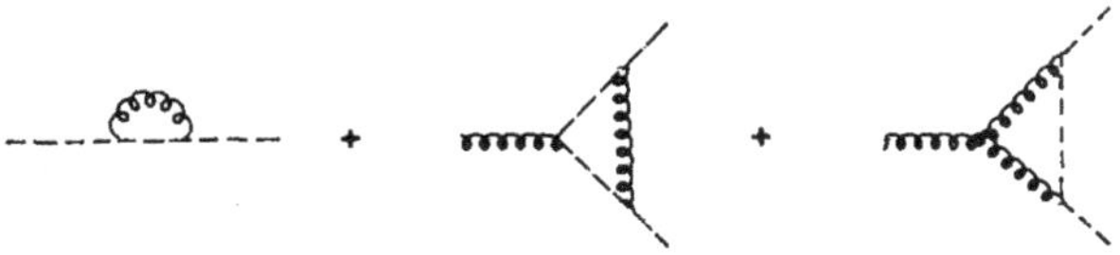

(1.70)

v) Z_{1F} is obtained from the fermion-gluon-fermion vertex

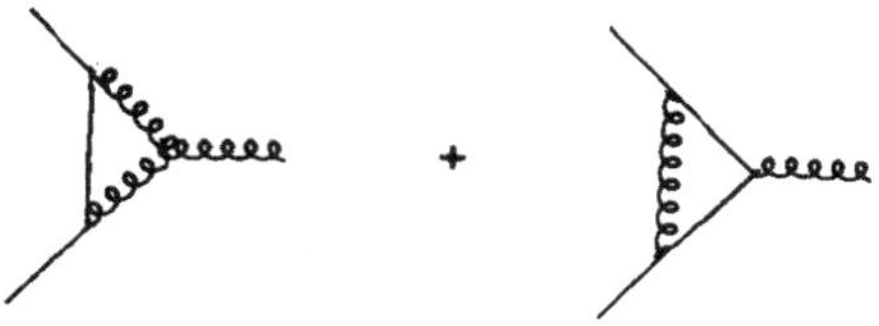

(1.71)

Once we have these previous renormalization constants, we can for instance deduce that of the coupling constant :

$$Z_\alpha = Z^2_{1YM}\, Z^{-3}_{3YM} = 1 - \left(\frac{\alpha_s}{\pi}\right)\left(\frac{11}{3}\frac{N}{2} - \frac{2}{3}\frac{n}{2}\right)\frac{1}{\epsilon} + \ldots \qquad (1.72)$$

b) **Renormalization Group Equation (RGE) :**

Now, we are ready to study the renormalization group equation introduced by Stueckelberg and Peterman in the contex of QED[37]. Let the renormalized Green's function be:

$$\Gamma_R(\nu, p_1 \ldots p_N, g, \alpha_G, m_j) = Z_\Gamma\, \Gamma_B(p_1 \ldots p_N, g, \alpha_G, m_j) \quad . \qquad (1.73)$$

The fact that Γ_B is independent of ν implies the disappearance of the total derivative $\nu \dfrac{d\Gamma_B}{d\nu} = 0$ which is equivalent to :

$$\left\{\nu \frac{\partial}{\partial \nu} + \nu \frac{d\alpha_s}{d\nu}\frac{\partial}{\partial \alpha_s} + \sum_j \frac{\nu}{m_j}\frac{dm_j}{d\nu} m_j \frac{\partial}{\partial m_j} + \nu \frac{d\alpha_s}{d\nu}\frac{\partial}{\partial \alpha_G} - \right.$$

$$\left. - \frac{1}{Z_\Gamma} \nu \frac{dZ_\Gamma}{d\nu} \right\} \Gamma_R (\nu, p_1 \ldots p_N; g \ldots) = 0. \tag{1.74}$$

One can introduce the universal parameters β function β_i and anomalous dimension γ_i defined as :

$$\alpha_s \beta(\alpha_s) = \nu \left.\frac{d\alpha_s}{d\nu}\right|_{g_B, m_B \text{ fixed}} ; \qquad \beta_G(\alpha_s) = \nu \left.\frac{d\alpha_G}{d\nu}\right|_{g_B, m_B \text{ fixed}}$$

$$\gamma_m = - \frac{\nu}{m^R} \left.\frac{dm^R}{d\nu}\right|_{g_B, m_B \text{ fixed}} ; \qquad \gamma_{2F} = \frac{\nu}{Z_{2F}} \frac{dZ_{2F}}{d\nu} ,$$

$$\gamma_{3YM} = \frac{\nu}{Z_{3YM}} \frac{d\, Z_{3YM}}{d\nu} ; \qquad \tilde{\gamma}_3 = \frac{\nu}{\tilde{Z}_3} \frac{d\tilde{Z}_3}{d\nu} ,$$

$$\gamma_\Gamma = \frac{\nu}{Z_\Gamma} \frac{dZ_\Gamma}{d\nu} = - \frac{1}{2} \left(n_{YM} \gamma_{3YM} + n_F \gamma_{2F} + \tilde{n}\, \tilde{\gamma}_3 \right) , \tag{1.75}$$

which transforms (1.74) into the renormalization group equation (RGE) :

$$\left\{ \nu \frac{\partial}{\partial \nu} + \beta(\alpha_s)\, \alpha_s \frac{\partial}{\partial \alpha_s} - \sum_j \gamma_m(\alpha_s)\, m_j \frac{\partial}{\partial m_j} + \beta_G \frac{\partial}{\partial \alpha_G} - \gamma_\Gamma \right\} \Gamma^R = 0 \quad . \tag{1.76}$$

The expressions of the above universal parameters can be easily obtained from (1.75) and the values of Z_i in Table 1.2. However, noting that $\beta(\alpha_s)$ is independent of the fermion mass in the $\overline{MS}$-scheme, one can write :

$$\alpha_s\ \beta(\alpha_s, \epsilon) = \nu \frac{d\, \alpha_s^R}{d\nu} = \nu \frac{d}{d\nu} \left(\alpha_s^B\ \nu^{-\epsilon}\ Z_\alpha^{-1} \right)$$

$$= - \epsilon \alpha_s - \alpha_s^R \frac{1}{Z_\alpha} \nu \frac{d Z_\alpha}{d\nu} . \qquad (1.77)$$

The fact that Z_α in (1.72) is ν-independent allows one to write

$$\left\{\alpha_s^R \beta(\alpha_s, \epsilon) + \epsilon \alpha_s^R + \left(\alpha_s^R\right)^2 \beta(\alpha_s, \epsilon) \frac{\partial}{\partial \alpha_s^R}\right\} Z_\alpha = 0 \quad , \qquad (1.78)$$

which inserted into the expression of Z_α in terms of $\frac{1}{\epsilon}$ poles (Eq. 1.56) gives the differential equation :

$$\alpha_s^R \beta(\alpha_s, \epsilon) = - \epsilon \alpha_s^R + \left(\text{finite term} \equiv \alpha_s^R \beta(\alpha_s)\right) \qquad (1.79)$$

and then :

$$\beta(\alpha_s) = \alpha_s \frac{\partial Z_\alpha^{(1)}}{\partial \alpha_s} , \qquad (1.80)$$

i.e. $\beta(\alpha_s)$ is just the coefficient of the $\frac{1}{\epsilon}$ term of Z_α.
The same reasoning applies to the anomalous dimensions, i.e with the sign convention used in (1.75), they are the opposite of the $\frac{1}{\epsilon}$ coefficient of the corresponding renormalization constant. Gauge and scheme invariance of the β and γ functions can also be proven [33)34)].

Let us now solve the RGE in (1.76). If D is the dimension of Γ in units of mass and we scale the momenta $p_1 \ldots p_N$ by a dimensionless factor λ, the Euler theorem on homogeneous function gives :

$$\left\{\lambda \frac{\partial}{\partial \lambda} + \sum_j m_j \frac{\partial}{\partial m_j} + \nu \frac{\partial}{\partial \nu} - D\right\} \Gamma_R \ (\lambda p_1 \ldots \lambda p_N \ ; \ \alpha_s, \alpha_G, \ m_{j}, \nu) = 0. \qquad (1.81a)$$

Introducing for convenience the dimensionless variables

$$t \equiv \log \lambda \qquad x_j \equiv m_j/\nu \quad , \tag{1.81b}$$

we arrive at the desired form of the RGE :

$$\left\{ -\frac{\partial}{\partial t} + \beta(\alpha_s)\, \alpha_s \frac{\partial}{\partial \alpha_s} + \beta_G \frac{\partial}{\partial \alpha_G} - \sum_j (1+\gamma_m)\, x_j \frac{\partial}{\partial x_j} + D - \gamma_\Gamma \right\} \cdot$$

$$\cdot\, \Gamma_R \left(e^t p_1, \ldots, e^t p_B \,;\, \alpha_s, \alpha_G, x_j, \nu \right) = 0 \quad , \tag{1.82}$$

with the solution :

$$\Gamma^R \left(e^t p_1 \ldots, e^t p_N \,;\, \alpha_s, \alpha_G, x_j, \nu \right) = \lambda^D \cdot \Gamma\, (p_1, \ldots, p_N ; t = 0,$$

$$\bar{\alpha}_s, \bar{\alpha}_G, \bar{x}_j)\, \exp \left\{ -\int_0^t dt'\, \gamma_\Gamma \left(\bar{\alpha}_s\, (t', \alpha_s) \right) \right\} . \tag{1.83}$$

One should note that the Green's function has acquired an extra dimension induced by the exponential factor. This is why it is called anomalous dimension.

c) **Running Parameters**

$\bar{\alpha}_s(t)$, $\bar{x}_i(t)$ and $\bar{\alpha}_G(t)$ of (1.83) are respectively the running-coupling, mass and gauge solutions of the differential equations :

$$\frac{d\bar{\alpha}_s}{dt} = \bar{\alpha}_s\, \beta\left(\bar{\alpha}_s\right) \quad : \quad \bar{\alpha}_s(0, \alpha_s) = \alpha_s^R(\nu) \quad , \tag{1.84}$$

$$\frac{d\bar{x}_i}{dt} = -\left[1 + \gamma_m\left(\bar{\alpha}_s\right) \right] \bar{x}_i(t) \quad : \quad \bar{x}_i(0, \alpha_s) = x_i^R(\nu) \quad , \tag{1.85}$$

$$\frac{d\bar{\alpha}_G}{dt} = \beta_G\left(\bar{\alpha}_s\right) \quad : \quad \bar{\alpha}_G(0,\ \alpha_s) = \alpha_G(\nu) \ . \tag{1.86}$$

The solutions of the two first equations read to two loops[34]:

$$\bar{\alpha}_s = \bar{\alpha}_s^{(1)} \left\{1 - \frac{\bar{\alpha}_s^{(1)}}{\pi} \frac{\beta_2}{\beta_1} \log\log \frac{-q^2}{\Lambda^2}\right\} \ , \tag{1.87}$$

$$\bar{m}_1 = \bar{m}_1^{(1)} \left\{1 + \left(\gamma_1 \frac{\beta_2}{\beta_1^2} \log\log \frac{-q^2}{\Lambda^2} - \frac{1}{\beta_1}\left(\gamma_2 - \gamma_1 \frac{\beta_2}{\beta_1}\right)\right) \frac{\bar{\alpha}_s^{(1)}}{\pi}\right\} \ , \tag{1.88}$$

with : $\bar{\alpha}_s^{(1)} = \pi/-\left(\beta_1 \log \sqrt{-q^2}/\Lambda\right)$ the one loop solution of (1.84) ;
$\gamma_m(\alpha_s) \equiv \gamma_1 \frac{\alpha_s}{\pi} + \gamma_2 \left(\frac{\alpha_s}{\pi}\right)^2$ and $\beta(\alpha_s) \equiv \beta_1 \left(\frac{\alpha_s}{\pi}\right) + \beta_2 \left(\frac{\alpha_s}{\pi}\right)^2$. Λ is a renormalization group invariant (RGI) but scheme-dependent defined as :

$$\frac{1}{2}\log \nu^2 + \frac{\pi}{\beta_1 \alpha_\sigma(\nu)} - \frac{\beta_2}{\beta_1^2} \log \left(\frac{1 + \beta_2/\beta_1 \ \frac{\alpha_s(\nu)}{\pi}}{\alpha_s(\nu)}\right)$$

$$= \frac{1}{2}\log \Lambda^2 - \frac{\beta_2}{\beta_1^2} \log \left(- \frac{\beta_1}{2\pi}\right) \ . \tag{1.89}$$

Analogously to Λ, one can also introduce the RGI mass $\hat{m}_1$ defined to one loop as[22b) :

$$\hat{m}_1 = m_1(\nu)\left(\frac{\pi}{-\beta_1\ \alpha_s(\nu)}\right)^{\gamma_1/-\beta_1} , \qquad (1.90)$$

which is related to the one-loop running mass as :

$$\bar{m}_1^{(1)} = \hat{m}_1 \Big/ \left(\log \sqrt{-q^2}/\Lambda\right)^{\gamma_1/-\beta_1} . \qquad (1.91)$$

One can also introduce a RGI spontaneous mass μ_1 associated with the quark-vacuum condensate using the fact that the product $m\langle\bar{\psi}\psi\rangle$ entering into the PCAC relation (1.19) does not get renormalized. Then :

$$\langle\bar{\psi}_1\psi_1\rangle = -\mu_1^3\left(\log\sqrt{-q^2}/\Lambda\right)^{\gamma_1/-\beta_1} . \qquad (1.92)$$

For completeness, we also give $\bar{\alpha}_s(t)$ and $\bar{m}_1(t)$ to three loops and for $SU(3)_F$ where γ_3 and β_3 have been calculated by Tarasov. This calculation is quoted in Ref. 38) :

$$\frac{\bar{\alpha}_s}{\pi} = \frac{4}{9L}\left\{1 - 0.79\,\frac{\log L}{L} + 0.62\,\frac{\log^2 L}{L^2} - 0.62\,\frac{\log L}{L^2}\right\} ,$$

$$\bar{m}_1 = \hat{m}_1\left(\frac{9}{2}\frac{\bar{\alpha}_s}{\pi}\right)^{4/9}\left\{1 + 0.895\,\frac{\bar{\alpha}_s}{\pi} + 2.707\left(\frac{\bar{\alpha}_s}{\pi}\right)^2\right\} , \qquad (1.93)$$

with $L \equiv \log - q^2/\Lambda^2$.

Within the previous generalities, we are now ready to study the specific example of the two-point correlator useful for the QSSR analysis.

d) The RGE for the Two-Point Correlator

Let $\Pi\left(q^2, \alpha_s, m_i, \nu\right)$ be a generic notation for the two-point correlator :

$$\Pi(q^2) = i \int d^4x\, e^{iqx} \left\langle 0 \mid \mathbb{T}\, J_H(x)\, (J_H(0))^\dagger \mid 0 \right\rangle , \qquad (1.94)$$

built from the hadronic current of quarks and/or gluon bilinear fields. In $n = 4-\epsilon$ dimension, $\Pi(q^2)$ gets an extra $\nu^{-\epsilon}$ dimension. The renormalized two-point correlator is[22b,33,34]:

$$\Pi_R\left(q^2, \alpha_s, m_i, \nu\right) \equiv \Pi_B\left(q^2, \alpha_s^B, m_i^B, \epsilon\right) - \nu^{-\epsilon}\, C\left(q^2, \alpha_s^B, m_i^B, \epsilon\right) , \qquad (1.95)$$

where in the $\overline{MS}$ scheme, C are the ϵ-pole terms :

$$C\left(q^2, \alpha_s^B, m_i^B, \epsilon\right) = \sum_K \frac{1}{\epsilon^k}\, C_k\left(q^2, \alpha_s, m_i\right) , \qquad (1.96)$$

where as usual C_k are constants or polynomials in m^2/q^2. The fact that Π_B is independent of ν, implies the relation :

$$\left\{\nu \frac{\partial}{\partial \nu} + \beta(\alpha_s)\, \alpha_s \frac{\partial}{\partial \alpha_s} - \sum \gamma_m\, m_i \frac{\partial}{\partial m_i}\right\} \Pi_R\left(q^2, \alpha_s, m_i, \nu\right)$$

$$= -\nu \frac{d}{d\nu}\left(\nu^{-\epsilon} \sum_K \frac{1}{\epsilon^k}\, C_k\right) . \qquad (1.97)$$

Rewriting :

$$\nu \frac{d}{d\nu}\, \nu^{-\epsilon} \sum_K \frac{1}{\epsilon^k}\, C_k = \left\{\nu \frac{\partial}{\partial \nu} + \nu \frac{d\alpha_s}{d\nu} \frac{\partial}{\partial \alpha_s} - \sum \gamma_m\, m_i \frac{\partial}{\partial m_i}\right\} . \ \nu^{-\epsilon} \sum_K \frac{1}{\epsilon^k}\, C_k \qquad (1.98)$$

and using :

$$\nu \frac{d\alpha_s}{d\nu} = - \epsilon\, \alpha_s + \alpha_s\, \beta(\alpha_s) \quad , \tag{1.99}$$

and the fact that the equation is finite for $\epsilon \to 0$, one gets :

$$\lim_{\epsilon\to 0} \nu \frac{d}{d\nu} \left(\nu^{-\epsilon} \sum_k \frac{1}{\epsilon^k} C_k \right) = - \frac{\partial}{\partial\alpha_s} (\alpha_s\, C_1) \quad , \tag{1.101a}$$

and the set of recursive equations for $k > 1$:

$$\left(\alpha_s\, \beta(\alpha_s) - \sum_i \gamma_m\, m_i \frac{\partial}{\partial m_i} \right) C_k = \frac{\partial}{\partial\alpha_s} (\alpha_s\, C_{k+1}) \quad . \tag{1.101b}$$

The dimensionless condition of Π reads :

$$\left\{ \nu \frac{\partial}{\partial\nu} + \lambda \frac{\partial}{\partial\lambda} + \sum_i m_i \frac{\partial}{\partial m_i} \right\} \Pi \left(\lambda^2 \nu^2 ,\ \alpha_s , m , \nu \right) = 0 \quad , \tag{1.102}$$

where $t \equiv \log \lambda$. Therefore, one arrives at the RGE for the two-point correlator :

$$\left\{ - \frac{\partial}{\partial t} + \beta(\alpha_s)\, \alpha_s \frac{\partial}{\partial\alpha_s} - \sum_i (1+\gamma_m)\, x_i \frac{\partial}{\partial x_i} \right\} \Pi(t,\ \alpha_s , x_i) =$$

$$\frac{\partial}{\partial\alpha_s} (\alpha_s\, C_1) \equiv D_s \quad , \tag{1.103}$$

with the solution :

$$\Pi(t, \alpha_s ,\ x_i) = \Pi \left(t{=}0,\ \bar{\alpha}_s(t),\ \bar{x}_i(t) \right) -$$

$$\int_0^t dt'\ D_s \left[t-t',\ \bar{\alpha}_s(t'),\ \bar{x}(t') \right] \quad . \tag{1.104}$$

e) **Renormalization of composite operators**

We have seen from the example of WSR in paragraph 2) that we are dealing with two-point correlators associated with the local current colourless operators $J_B(x)$ built from quarks and/or gluon fields. So, before any QSSR analysis, one should control the renormalization of such operators. This problem can be conveniently studied using background field techniques[39]. Operators can be classified into three classes : class I are gauge invariants which do not vanish after use of the classical equation of motions. Class II are gauge-invariant but vanish after use of the classical equation of motion. Class III are gauge-dependent operators. Therefore any composite renormalized operators can be written as :

$$\mathcal{O} = Z_I\ O_I^B + Z_{II}\ O_{II}^B + Z_{III}\ O_{III}^B \quad . \tag{1.105a}$$

The great advantage of the background field techniques is that for graphs with external quark and background fields, one only needs gauge invariant counterterms, i.e :

$$Z_{III} = 0 \quad , \tag{1.105b}$$

which is a consequence of the background field gauge invariance under quantization and renormalization. Let us illustrate the approach by studying the renormalization of the $G_{\mu\nu}\ G^{\mu\nu}$ gluon operator[40] in the presence of massive quarks. We have three types of dimension-four operators :

$$O_1 = -\frac{1}{4}\ i\ GG \quad ; \quad O_2 = -\ \bar{\psi}\left(\hat{D}+im\right)\psi \quad ; \quad O_3 = im\ \bar{\psi}\psi \quad . \tag{1.106}$$

The renormalized O_1 operator is in general a combination of these three "bare" operators :

$$O_1 = Z_{11} O_1^B + Z_{12} O_2^B + Z_{13} O_3^B \tag{1.107}$$

The renormalization constants Z_{ij} are mass-independent in the $\overline{MS}$ scheme. In particular, one can already get Z_{11} and Z_{12} in the massless limit. In any case, we insert at zero momentum the O_1 operator into the gluon and quark propagators :

$$\left\langle A_a^\mu \; O_1 \; A_b^\nu \right\rangle = Z_\alpha^{-1} \; Z_{11} \left\langle A_a^\mu \; O_1^B \; A_b^\nu \right\rangle + Z_{12} \left\langle A_a^\mu \; O_2^B \; A_b^\nu \right\rangle \; ,$$

$$\left\langle \bar\psi \; O_1 \; \psi \right\rangle = Z_{11} \left\langle \bar\psi \; O_1^B \; \psi \right\rangle + Z_{2F} \; Z_{12} \left\langle \bar\psi \; O_2^B \; \psi \right\rangle + Z_{2F} Z_{13} \left\langle \bar\psi \; O_3^B \; \psi \right\rangle. \tag{1.108}$$

The insertion of O_1^B into the gluon propagator corresponds to the Feynman rule $-i \; \delta_{ab} \left(p^2 g_{\mu\nu} - p_\mu p_\nu \right)$ and one has to evaluate :

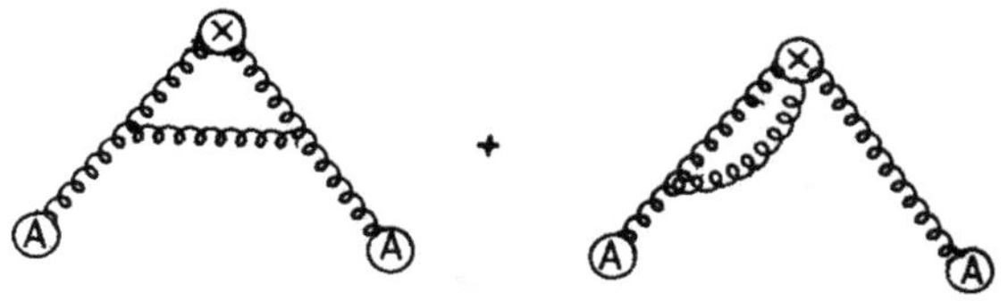

(1.109)

The zero momentum insertion of O_2^B into the gluon propagator corresponds to the rule $i\hat{p}$ and we have to calculate :

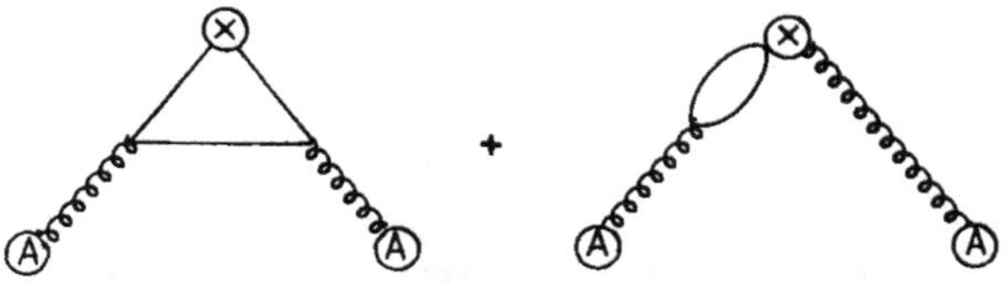

(1.110)

The insertion into the quark propagator corresponds respectively to $ig\frac{\lambda a}{2}\gamma^\mu$, $i\left(\hat{p} - m_B\right)$ and im_B for the O_1, O_2 and O_3 operators.

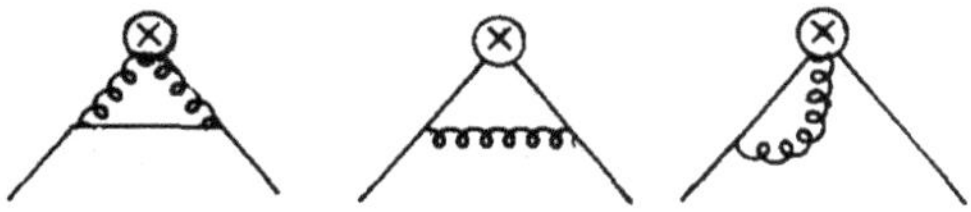

(1.111)

Evaluations of the above diagrams give in the Landau gauge[40]:

$$Z_{11}^{(2)} = Z_\alpha^{(2)} \quad ; \quad Z_{12}^{(2)} = 0 \quad ; \quad Z_{13}^{(2)} = -\frac{\gamma_1}{\epsilon}\frac{\alpha_s}{\pi}, \tag{1.112}$$

as $Z_{2F}^{(2)} = 1$ in the Landau gauge. The index (2) means second order in (α_s/π). Therefore one has :

$$GG = \left(1 + \frac{\alpha_s}{\pi}\frac{B_1}{\epsilon}\right)(GG)_B + 4\frac{\gamma_1}{\epsilon}\frac{\alpha_s}{\pi}m_B\left(\bar{\psi}\psi\right)_B, \tag{1.113}$$

i.e., GG is not multiplicatively renormalizable. In fact, one can deduce from (1.113) the renormalization group invariant combination:

$$\frac{1}{4}\beta(\alpha_s)\ GG + \gamma_m\ m\ \bar{\psi}\psi, \tag{1.114a}$$

which appears in the trace of the energy-momentum tensor :

$$\Theta_\mu^\mu = \frac{1}{4}\beta(\alpha_s)\ GG + (1+\gamma_m)\ \bar{\psi}\psi, \tag{1.114b}$$

where $\beta(\alpha_s)$ and $\gamma_m(\alpha_s)$ are the β function and the mass anomalous dimension respectively. Proofs of the RGI of quantities like the vector current or the (pseudo) scalar currents are straightforward. Much more involved operators can be found in Refs 41), 42) and 48).

f) **Pseudoscalar Correlator and Current Algebra Ward Identity**

In order to illustrate our discussion of this section, let us show explicitly the evaluation within dimensional regularization of the two-point correlator built from the pseudoscalar current defined in (1.25).

i) To lowest order of QCD, we shall compute the following diagram :

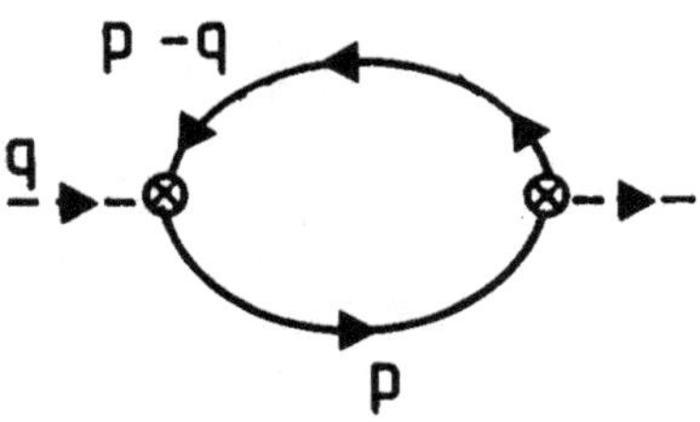

Using usual Feynman rules, it reads :

$$\nu^{\epsilon}\,\psi_5(q^2) = (m_i+m_j)^2\,(-i)\,N\int\frac{d^n p}{(2\pi)^n}$$

$$\mathrm{Tr}\left\{(i\gamma_5)\,\frac{1}{\hat{p}-m+i\epsilon'}\,(i\gamma_5)\,\frac{1}{\hat{p}-\hat{q}-m_j+i\epsilon'}\right\}. \tag{1.115}$$

Parametrizing the quark propagators and using the properties of Dirac matrices in n dimensions[34] (see Appendix A), we obtain :

$$\psi_5^B(q^2) = (m_i+m_j)^2\,\frac{N}{4\pi^2}\int_0^1 dx\left(\frac{2}{\epsilon}+\log 4\pi-\gamma\right)\left(\frac{R^2-i\epsilon'}{\nu^2}\right)^{-\epsilon/2}.$$

$$\cdot \left\{ \left(3 + \frac{\epsilon}{2} \right) q^2 x(1-x) - 2\left(1 + \frac{\epsilon}{4} \right) \left(m_i^2 x + m_j^2 (1-x) + m_i m_j \right) \right\} , \quad (1.116)$$

where $\mathbb{R}^2 \equiv -q^2 x\ (1-x) + m_i^2 x + m_j^2\ (1-x)$ and $\gamma = 0.5772...$ the Euler constant.

Two limits of (1.116) are particularly interesting :

$$\psi_5^B \left(-q^2 \gg m_{i,j}^2 \right) \simeq \frac{3}{8\pi^2} (m_i + m_j)^2 \ q^2 \left\{ \left(\frac{2}{\epsilon} + \log 4\pi - \gamma - \log - \frac{q^2}{\nu^2} \right) \cdot \right.$$

$$\cdot \left(1 + \frac{2\left(m_i^2 + m_j^2 - m_i m_j \right)}{-q^2} \right) + 2 + \frac{\epsilon}{4} \log^2 \left(- \frac{q^2}{\nu^2} \right) -$$

$$\left. \frac{\epsilon}{2} (\log 4\pi - \gamma + 2) \log - \frac{q^2}{\nu^2} \right\} , \quad (1.117)$$

$$\psi_5^B (q^2 \to 0) \simeq \frac{3}{4\pi^2} (m_i + m_j) \left\{ \left(m_i^3 \log \frac{m_i^2}{\nu^2} + m_j^3 \log \frac{m_j^2}{\nu^2} \right) \right.$$

$$\left. - \left(\frac{2}{\epsilon} + \log 4\pi - \gamma - 1 \right) \left(m_i^3 + m_j^3 \right) \right\} . \quad (1.118)$$

For $q \to 0$, we know from the current algebra Ward identity in (1.16) that $\psi_5 (q^2)$ is related to $\Pi_5^{\mu\nu}(q^2)$. A perturbative evaluation of the axial two-point correlator gives :

$$q_\mu q_\nu \, \Pi_5^{\mu\nu} = \frac{N}{8\pi^2} \int_0^1 dx \, \frac{m_i^2 x + m_j^2 (1-x) + m_i m_j}{q^2} \left(\frac{\mathbb{R}^2 - i\epsilon'}{\nu^2} \right)^{-\epsilon/2} \Gamma(\epsilon/2) . \quad (1.119)$$

Using $\log \mathbb{R}^2 = \log|R^2| - i\pi\, \theta\,(-R^2)$, one gets from (1.116) and (1.117) the well-known result :

$$\mathrm{Im}\, \psi_5(t) = \mathrm{Im} \left(q_\mu q_\nu \, \Pi_5^{\mu\nu}(t) \right) =$$

$$= \frac{3}{8\pi^2} (m_i + m_j)^2 \left(1 - \frac{(m_i - m_j)^2}{t} \right) \lambda^{1/2} \left(1, \frac{m_i^2}{t}, \frac{m_j^2}{t} \right) .$$

$$. \, \theta\left(t - (m_i + m_j)^2 \right) \quad (1.120)$$

$$q_\mu q_\nu \, \Pi_5^{\mu\nu} = \psi_5(q^2) - \frac{3}{4\pi^2} (m_i + m_j) \left(m_i^3 \log \frac{m_i^2}{\nu^2} + m_j^3 \log \frac{m_j^2}{\nu^2} \right) . \quad (1.121)$$

(1.121) suggests that at zero momentum $\psi_5(0)$ possesses a small perturbative contribution or alternatively the "true" quark condensate $\langle \bar\psi \psi \rangle$ can be a combination of the one in (1.17b) and the perturbative piece in (1.121). We shall come back to this point later.

ii) $\psi_5(q^2)$ to two loops in the $\overline{MS}$ scheme

In order to simplify life for the reader, we shall ignore the quark mass in the internal loops and quote the bare result of Becchi et al to two loops[43]:

$$\psi_5^B(q^2) = \nu^{-\epsilon} \left(m_i^B + m_j^B \right)^2 . \left(\frac{3q^2}{8\pi^2} \right) \left\{ \frac{2}{\epsilon} + \log 4\pi - \gamma + 2 - \log \frac{-q^2}{\nu^2} \right.$$

$$- \frac{\epsilon}{2}\,(\log 4\pi - \gamma + 2)\,\log\frac{-q^2}{\nu^2} + \frac{\epsilon}{4}\,\log^2\frac{-q^2}{\nu^2} +$$

$$\left.\left(\frac{g^B\nu^{-\epsilon/2}}{4\pi^2}\right)^2 \left[\frac{4}{\epsilon^2} + \frac{4}{\epsilon}\,(\log 4\pi - \gamma) + \frac{29}{3\epsilon} + \mathcal{O}(1)\right]\left(\frac{-q^2}{\nu^2}\right)^{-\epsilon}\right\} \quad . \qquad (1.122)$$

We introduce the renormalized parameter :

$$g^B\ \nu^{-\epsilon/2} = g\left\{1 + \mathcal{O}\left(\frac{\alpha_s}{\pi}\right)\right\} \quad ,$$

$$m_i^B = m_i\ \left(1 - \frac{2}{\epsilon}\,\frac{\alpha_s}{\pi}\right) \quad , \qquad (1.123)$$

and we obtain :

$$\psi_5^B(q^2) = \frac{3}{8\pi^2}(m_i + m_j)^2\ q^2\ \left\{\frac{2}{\epsilon} + \log 4\pi - \gamma + 2 - \log\frac{-q^2}{\nu^2} + \left(\frac{\alpha_s}{\pi}\right)\right.$$

$$\left.\left[-\frac{4}{\epsilon^2} + \frac{5}{3\epsilon} + \log^2\frac{-q^2}{\nu^2} - \left(\frac{17}{3} + 2\,(\log 4\pi - \gamma)\right)\log\frac{-q^2}{\nu^2}\right]\right\} . \qquad (1.124)$$

First we learn that the lowest order terms which vanish for $\epsilon \to 0$ induce terms in the two-loop results via the mass renormalization. We also see that the non-local $\frac{1}{\epsilon}\log\frac{-q^2}{\nu^2}$ pole has disappeared. This consists of a double check of our results. A last check can be done with the RGE[34) . $\psi_5(q^2)$ obeys the RGE of the type in (1.103) where D_s is the coefficient of the $\frac{1}{\epsilon}$ term :

$$D_s \equiv D_0 + D_1 \left(\frac{\alpha_s}{\pi}\right) \quad , \qquad (1.125a)$$

with :

$$D_0 = -\frac{3}{8\pi^2}(x_i + x_j)^2 \; 2 \; e^{-2t} \quad ,$$

$$D_1 - \frac{3}{8\pi^2}(x_i + x_j)^2 \; \frac{10}{3} \; e^{-2t} \quad . \qquad (1.125b)$$

Let us now write $\psi_5(q^2)$ in terms of dimensionless variables :

$$\psi_5(t, \alpha_s, x_i) = -\frac{3q^4}{8\pi^2}(x_i + x_j)^2 \; e^{-2t} \left\{ -2t + \log 4\pi - \gamma + 2 + \right.$$

$$\left. + \left(\frac{\alpha_s}{\pi}\right)(4at^2 + 2bt + c) \right\} \quad , \qquad (1.126)$$

where a,b,c have to be determined. Using the RGE, one obtains the constraint :

$$D_0 = -\left(\frac{3}{8\pi^2}\right)(x_i + x_j)^2 \; e^{-2t} \; . \; 2 \quad ,$$

$$D_1 = -\left(\frac{3}{8\pi^2}\right)(x_i + x_j)^2 e^{-2t} \left\{ -8at - 2b - 2\gamma_1(\log 4\pi - \gamma + 2) + 2\gamma_1 . 2t \right\} \quad , \qquad (1.127)$$

where $\gamma_1 = 2$ is the mass anomalous dimension. The fact that D_1 cannot depend on t implies :

$$- 4a + 2\,\gamma_1 = 0 \quad \Longrightarrow \quad a = 1 \ . \tag{1.128}$$

The relation between C_1 and D in (1.103) implies :

$$C_1^{(0)} = D_0 \ . \tag{1.129}$$

$C_1^{(1)}$ is not fixed by the RGE but we know it from the calculation in (1.124). It is :

$$C_1^{(1)} = \left(\frac{5}{3}\right) \left(- \frac{3}{8\pi^2}\right) (x_i + x_j)^2 \; e^{-2t} \ , \tag{1.130}$$

We can now deduce from (1.103)

$$2\; C_1^{(1)} = D_1 \tag{1.131a}$$

and the recursive relation implies :

$$C_2^{(1)} = - \frac{3}{8\pi^2} (x_i + x_j)^2 \; e^{-2t} \; 2\gamma_1 \ . \tag{1.131b}$$

Eq. (1.131) inserted into (1.127) implies :

$$- 2b - 2\,\gamma_1 (\log 4\pi - \gamma + 2) = \frac{10}{3} \ . \tag{1.132}$$

As one can see, the RGE and the explicit calculation of the $\frac{1}{\epsilon}$ coefficient at $\left(\frac{\alpha_s}{\pi}\right)$ allows one to fix the coefficients of $\frac{1}{\epsilon^2}$, $\log^2 - q^2/\nu^2$ and $\log - q^2/\nu^2$ at that order. This is a really impressive result !

A complete two-loop expression of $\psi_5(q^2)$ including quark masses has been obtained by Broadhurst[44]. The use of his result at $q = 0$ implies :

$$\psi_5^R(0) = \frac{3}{4\pi^2}(m_i + m_j)\left(m_i^3 Z_i + m_j^3 Z_j\right) \quad , \tag{1.133a}$$

with :

$$Z_i = 1 - \log\frac{m_i^2}{\nu^2} + \left(\frac{2\alpha_s}{3\pi}\right)\left(5 - 5\log\frac{m_i^2}{\nu^2} + 3\log^2\frac{m_i^2}{\nu^2}\right) \quad , \tag{1.133b}$$

and improves the Ward identity in (1.121).
Three-loop expressions of $\psi_5(q^2)$ also exist in the chiral limit $m_i=0$.
In the $\overline{MS}$ scheme and for $SU(n)_F$, it reads[24c] :

$$\psi_5(q^2) = \frac{3}{8\pi^2}(m_i + m_j)^2 \Bigg\{ -q^2 \log -\frac{q^2}{\nu^2}\Bigg[1 + \frac{17}{3}\frac{\alpha_s}{\pi} + \left(\frac{\alpha_s}{\pi}\right)^2$$

$$\left(\frac{11089}{144} - \frac{611}{24}\zeta(3) + n\left(\frac{2}{3}\zeta(3) - \frac{65}{24}\right)\right)\Bigg]$$

$$+ \left(\frac{\alpha_s}{\pi}\right)\left(q^2 \log^2 -\frac{q^2}{\nu^2}\right)\left[1 - \frac{\alpha_s}{\pi}\left(-\frac{53}{3} + \frac{11}{18}n\right)\right]$$

$$- \left(\frac{\alpha_s}{\pi}\right)^2 q^2 \log^3 -\frac{q^2}{\nu^2}\left(\frac{19}{12} - \frac{n}{18}\right)\Bigg\} \quad , \tag{1.134}$$

where $\zeta(3) = 1.202$ is the Riemann's zeta function.
The complete two-loop result for the spectral function is[46]:

$$\frac{1}{\pi}\,\mathrm{Im}\,\psi_5(t) = \frac{3(m_i+m_j)^2}{8\pi^2\, q^2}\,\bar{q}^4\, v\left\{1 + \frac{4\alpha_s}{3\pi}\left\{\frac{3}{8}(7-v^2)\right.\right.$$

$$+ \sum_i (v+v^{-1})\left(\mathrm{Li}_2(\alpha_1\alpha_2) - \mathrm{Li}_2(-\alpha_i) - \log\alpha_i \log\beta_i\right)$$

$$\left.\left. + A_i \log\alpha_i + B_i \log\beta_i\right\} \cdot \mathcal{O}\left(\alpha_s^2\right)\right\}\ , \qquad (1.135a)$$

where $\mathrm{Li}_2(x) = -\int_0^x dx/x \log(1-x)$ and :

$$A_i = \frac{3}{4}\left(\frac{3m_i+m_j}{m_i+m_j}\right) - \frac{19+2v^2+3v^4}{32v} - \frac{m_i(m_i-m_j)}{\bar{q}^2\, v(1+v)}\left(1+v+\frac{2v}{1+\alpha_i}\right)\ ;$$

$$B_i = 2 + 2\left(m_i^2 - m_j^2\right)\Big/\bar{q}^2\, v\ ;$$

$$\alpha_i = \frac{m_i}{m_j}\left(\frac{1-v}{1+v}\right)\ ;\quad \beta_i = (1+\alpha_i)^{1/2}\,(1+v)^2\,/4v$$

$$\bar{q}^2 = q^2 - (m_i - m_j)^2\ ;\quad v = \left(1 - 4\,\frac{m_i m_j}{\bar{q}^2}\right)^{1/2}\ . \qquad (1.135b)$$

In this case where $m_j = 0$ this expression simplifies as :

$$\frac{1}{\pi}\,\mathrm{Im}\,\psi_5(t) = \frac{3}{8\pi^2}\, xt^2(1-x)^2\left\{1 + \frac{4}{3}\frac{\alpha_s}{\pi}\left[\frac{9}{4} + 2\,\mathrm{Li}_2(x)\right.\right.$$

$$+ \log x\log(1-x) - \frac{3}{2}\log\left(\frac{1}{x}-1\right) - \log(1-x)$$

$$\left.\left. + x\log\left(\frac{1}{x}-1\right) - (x/(1-x))\log x\right\}\ ,$$

where $$x \equiv m_i^2 / t. \tag{1.135c}$$

The expression of the two-point correlator of the scalar current $\partial_\mu \left(\bar{\Psi}_i \gamma_\mu \psi_j\right)$ can be obtained from (1.135) by changing m_j into $-m_j$ in all terms.

g) **Vector Correlator to Two Loops**

We shall be concerned with :

$$\Pi_V^{\mu\nu} = -(g^{\mu\nu} q^2 - q^\mu q^\nu)\, \Pi_V^{(1)}(q^2) + q^\mu q^\nu\, \Pi^{(0)}(q^2)$$

$$\equiv i \int d^4x\, e^{iqx} \left\langle 0 \mid \mathbb{T}\, V_{ij}^\mu(x) \left(V_{ij}^\nu(0)\right)^+ \mid 0\right\rangle, \tag{1.136}$$

with $V_{ij}^\mu = \bar{\Psi}_i \gamma^\mu \psi_j$. The longitudinal part is related to $\psi(q^2)$ from the Ward identity of the type of (1.121) which reads to two loops (see (1.133)) :

$$(q^2)^2\, \Pi_V^{(0)} = \psi(q^2) - \frac{3}{4\pi^2}(m_i - m_j)\left(m_i^3 Z_i - m_j^3 Z_j\right), \tag{1.137}$$

where, as we have already said, $\psi(q^2)$ can be deduced from $\psi_5(q^2)$ by interchanging m_j into $-m_j$ in (1.134) and (1.135). The lowest order contribution to $\Pi_V^{(1)}(q^2)$ reads[34]:

$$\Pi_V^{(1)} = \left(\frac{3}{4\pi^2}\right) \nu^{-\epsilon}\, \Gamma(\epsilon/2) \int_0^1 dx \,.\left[2x(1-x) + \frac{m_i^2 x + m_j^2(1-x) - m_i m_j}{-q^2}\right].$$

$$.\left(\frac{R^2}{\nu^2}\right)^{-\epsilon/2}. \tag{1.138}$$

The complete two-loop expression in the $\overline{MS}$ scheme is[47]:

$$\Pi^{(1)}(q^2) = \nu^{-\epsilon}\,\frac{1}{4\pi^2}\left\{\frac{2}{\epsilon}\left(1+\frac{\alpha_s}{\pi}\right) - \frac{1}{3}\left(1+\frac{\alpha_s}{\pi}\,\frac{15}{4}\right)\right.$$

$$+ \left(1-\alpha-\beta - 2(\alpha-\beta)^2\right)K + \alpha\ell_i + \beta\ell_j + 2(\alpha-\beta)\,(\alpha Z_i - \beta Z_j)$$

$$+ \left(\frac{2\alpha_s}{3\pi}\right)\left\{\frac{1}{2}\,(1-\alpha-\beta - 2(\alpha-\beta)^2)\,L + \alpha\ell_i\,(1+2\ell_i)\right.$$

$$+ \beta\ell_j(1+2\ell_j) - \frac{1}{4}\,(1+2N_A + 2N_B)\left(1+6\,\frac{(m_i-m_j)^2}{q^2}\right)$$

$$+ x_i\,f_i^2 + x_j\,f_j^2 + \frac{1}{2}\,(N_A - N_B)^2$$

$$-\,(\alpha-\beta)\,\left(G(x_i) - G(x_j)\right) - \left(3-2\,(\alpha+\beta)\right)\,K^2$$

$$\left.\left. - \frac{1}{2} + \alpha\,(2+\ell_i) + \beta\,(2+\ell_j)\,K\right\}\right\}\,, \qquad (1.139a)$$

with Z_i defined (1.133b) and :

$$\alpha \equiv -\,m_i^2/q^2 \quad ; \quad \beta \equiv -\,m_j^2/q^2 \quad ; \quad \ell_i \equiv -\log\frac{m_i^2}{\nu^2} \quad ;$$

$$K \equiv 1 + \frac{\ell_i}{2} + \frac{1}{2}\,(1+x_i)\,f_i \quad ; \quad f_i \equiv \log x_i/(1-x_i) \quad ;$$

$$x_i \equiv m_i^2/E\,\left[1+\sqrt{1-(m_i m_j/E)^2}\right] \;;$$

$$E \equiv \frac{1}{2}\left(m_i^2 + m_j^2 - q^2\right) \quad ;$$

$$N_A = \alpha(1 + f_i)(1 + x_j f_j) \quad ; \quad N_B = \beta(1+f_j)(1 + x_i f_i) \; ;$$

$$L \equiv 3K^2 + 2K + 6 - 6(1+\alpha+\beta)\, I - 10\, x_i f_i^2 + m_i\left[(3K-2)\frac{\partial K}{\partial m_i} - (1+\alpha+3\beta)\frac{\partial I}{\partial m_i}\right] + (i \leftrightarrow j) \; ;$$

$$G(x) = \int_0^x dy \left(\frac{\log y}{1-y}\right)^2 = \sum_{n=1}^{\infty} \frac{x^n}{n^2}\left(1 + (1-n \log x)^2\right) \; ;$$

$$F(x) = \int_0^x dy \left(\frac{\log y}{1-y}\right)^2 \log\left(\frac{x}{y}\right) = \sum_{x=1}^{\infty} \frac{x^n}{n^3}\left(2 + (2 - n \log x)^2\right) \quad ;$$

$$I \equiv F(1) + F(x_i x_j) - F(x_i) - F(x_j) \quad . \tag{1.139b}$$

We have not checked these horrible expressions ! It appears that in the limit $m_j = 0$ the result becomes less horrible. In this case, one can deduce from (1.139) the spectral function :

$$\mathrm{Im}\, \Pi_V^{(1)}(m_j = 0) = (2+x)\frac{\mathrm{Im}\, \psi_5(t)}{3\, m^2 t} - \left(\frac{\alpha_s}{6\pi^2}\right)\left\{(3+x)(1-x)^3 \log \frac{x}{1-x} + 2\, x \log x + (3-x^2)(1-x)\right\} , \tag{1.140}$$

with $x \equiv m^2/t$.

For the equal mass case, the QCD expression of $Q^2 \frac{d}{dQ^2}\Pi(Q^2)$ which is related to the $e^+e^- \to$ Hadrons total cross-section is also known to four loops for $m_i = 0$ and to three loops up to $\frac{m^2}{Q^2}$ terms. After renormalization group improvement in the $\overline{MS}$ scheme[38] it reads:

$$\frac{Q^2}{dQ^2}\,\Pi_V^{(1)}(Q^2) = -\frac{3}{12\pi^2}\left\{1 + \frac{\bar{\alpha}_s}{\pi} + \left((1.986 - 0.115n) + \sum_{j=1}^{n}\frac{\bar{m}_j^2}{Q^2}\,1.05\right)\left(\frac{\bar{\alpha}_s}{\pi}\right)^2\right.$$

$$-\frac{\bar{m}^2}{Q^2}\left[6 + 28\,\frac{\bar{\alpha}_s}{\pi} + (269.15 - 12.25n)\left(\frac{\bar{\alpha}_s}{\pi}\right)^2\right] + \left[70.985 - 1.200n - 0.005\,n^2\right.$$

$$\left.\left. - 1.679\left(\sum_n Q_i\right)^2\right]\left(\frac{\bar{\alpha}_s}{\pi}\right)^3 \right. , \qquad (1.141)$$

where here $\bar{\alpha}_s$ and $\bar{m}$ are the three-loop running parameters defined in (1.93). The m^4 terms of $\Pi_V^{(1)}$ have also been obtained to two loops. In the $\overline{MS}$ scheme[47] it reads :

$$\Pi_V^{(1)}\Big|_{m^4} = \left(\frac{3}{2\pi^2}\right)\frac{\bar{m}^4}{Q^4}\left\{1 + \log\frac{Q^2}{\bar{m}^2} + 2\left(\frac{\bar{\alpha}_s}{3\pi}\right)\left(5 + 5\log\frac{Q^2}{\bar{m}^2} + 3\log^2\frac{Q^2}{\bar{m}^2}\right)\right\}.$$

$$\cdot\left(1 + \frac{\bar{\alpha}_s}{3\pi} + \mathcal{O}\left(\frac{\bar{\alpha}_s}{\pi}\right)^2\right) - \frac{3}{4\pi^2}\,\frac{\bar{m}^4}{Q^4}\left(1 + \mathcal{O}(\bar{\alpha}_s)\right)\Bigg\} , \qquad (1.142a)$$

where one should note that the terms appearing in the first { } of (1.142a) are involved in the current algebra Ward identity obeyed by the quark $\langle\bar\psi\psi\rangle$ condensate in (1.133b). After a resummation of the mass-singular logarithms in this { } , one can rewrite the improved expression of the $\bar m^4$ terms for $SU(n)_F$ as :

$$\Pi_V^{(1)}\Bigg|_{\bar m^4} \simeq \frac{3}{12\pi^2}\left\{\frac{\bar m^4}{Q^4}\cdot\frac{1}{(15+2n)}\left[-\frac{72\pi}{\bar\alpha_s}+\frac{357-23n}{4}\right]+\frac{8\pi^2}{Q^4}\,m\langle\bar\psi\psi\rangle\left(1+\frac{\bar\alpha_s}{3\pi}\right)\right\} \quad (1.142b)$$

h) **Weinberg Sum Rules to Two Loops**

We are now in a good position to check the validity of the Weinberg superconvergent assumptions done in (1.25) and (1.26). Let us, for instance, analyze the spectral functions to the lowest order. They can be deduced from (1.135) and (1.138) using the usual change m_i into $-m_i$ in order to deduce the expressions of the scalar and axial spectral functions. It is easy to see that for the combination involved in the WSR (1.25) and (1.26), the first one is still zero while the second one reads :

$$\mathrm{Im}\,\Pi_{LR}^{(1)} = -\,\mathrm{Im}\,\Pi_{LR}^{(o)} = \frac{3}{2\pi}\cdot\frac{m_i m_j}{t}\,\lambda^{1/2}\left(1,\frac{m_i^2}{t},\frac{m_j^2}{t}\right) \quad . \quad (1.143)$$

One can compute the next corrections for the first sum rule, which reads[22b)]:

$$\Pi_{LR}^{(1)} + \Pi_{LR}^{(o)}\ (-q^2 \gg m^2) \simeq \left(\frac{\alpha_s}{\pi}\right)\frac{1}{\pi^2}\left\{\frac{m_i m_j}{q^2}+\ldots\right\} \quad . \quad (1.144)$$

For the second to be convergent, one has to look for some other combinations like[22b)] :

$$R_{ijk} = \Pi^{(1)}_{LR}\Big|_{ij} - \frac{m_j}{m_k}\Pi^{(1)}_{LR}\Big|_{ik} , \tag{1.145}$$

or to improve the sum rule with, for example, the Laplace transform, as we shall see later on.

i) Tensor Current Correlator to Two Loops

We shall be concerned with the current :

$$J_{\mu\nu}(x) = i\,\bar\psi\left(\gamma_\mu \overleftrightarrow{D}_\nu + \gamma_\nu \overleftrightarrow{D}_\mu\right)\psi \tag{1.146a}$$

$$\hat{J}_{\mu\nu}(x) = :\, - G_{\mu\alpha}G^{\alpha}_{\nu} + \frac{1}{4}\, g_{\mu\nu}G^{\alpha\beta}G_{\alpha\beta}: \tag{1.146b}$$

where $\overleftrightarrow{D}_\mu \equiv \overrightarrow{D}_\mu - \overleftarrow{D}_\mu$ is the covariant derivative.
We now study the renormalization of the two currents according to Ref. 48) :

$$\begin{aligned} J^{\mu\nu} &= Z_{11}\, J^{B}_{\mu\nu} + Z_{12}\, \hat{J}^{B}_{\mu\nu} \\ \hat{J}^{\mu\nu} &= Z_{21}\, J^{B}_{\mu\nu} + Z_{22}\, \hat{J}^{B}_{\mu\nu} \end{aligned} \tag{1.147a}$$

where in the $\overline{MS}$ scheme :

$$Z_{11} = 1 + \left(\frac{\alpha_s}{\pi}\right)\frac{4}{3}\,C_2(R)\,\frac{1}{\epsilon} \quad ,$$

$$Z_{12} = -\frac{8}{3}\,T(R)\left(\frac{\alpha_s}{\pi}\right)\frac{1}{\epsilon} \quad ,$$

$$Z_{21} = -\frac{1}{3} C_2(R) \left(\frac{\alpha_s}{\pi}\right) \frac{1}{\epsilon} ,$$

$$Z_{22} = 1 + \frac{n}{3} \left(\frac{\alpha_s}{\pi}\right) \frac{1}{\epsilon} . \qquad (1.147b)$$

The associated two-point correlator can be written as :

$$\Pi_{\mu\nu,\rho\sigma} = \frac{1}{2} \left\{ \eta_{\mu\rho} \eta_{\nu\sigma} + \eta_{\mu\sigma} \eta_{\nu\rho} + \frac{2}{1-d} \eta_{\mu\nu} \eta_{\rho\sigma} \right\} \Pi(q^2) , \qquad (1.148a)$$

where $d \equiv 4-\epsilon$ is the space-time dimension and $\eta_{\mu\nu} \equiv \left(g_{\mu\nu} - q_\mu q_\nu / q^2\right)$.
$\Pi(q^2)$ can be extracted using the projector.

$$R_{\mu\nu\rho\sigma} = \frac{2}{(d+1)(d-1)(d-2)} \left\{ (n-2) q_\mu q_\nu q_\rho q_\sigma / q^4 \right.$$

$$+ \left(q_\mu q_\nu g_{\rho\sigma} + g_{\mu\nu} q_\rho q_\sigma\right) / q^2 - g_{\mu\nu} g_{\rho\sigma}$$

$$\left. - (d-1) \left(2 q_\mu q_\rho g_{\nu\sigma} / q^2 - g_{\mu\rho} g_{\nu\sigma}\right) \right\} , \qquad (1.148b)$$

which gives

$$\Pi(q^2) = R_{\mu\nu\rho\sigma} \Pi^{\mu\nu\rho\sigma} . \qquad (1.148c)$$

To the lowest order, the spectral function reads[48b] :

$$\text{Im}\, \Pi(t) = \frac{t^2}{10\pi} (5-2v^2) v^3 \theta(t-4m^2) \qquad (1.148d)$$

where $v \equiv \left(1 - 4 \frac{m^2}{t}\right)^{1/2}$.

The evaluation of the two-loop quark diagram gives for $m_1 = 0$ [48b]:

$$\Pi_B(q^2) = -\frac{3q^4}{10\,\pi^2}\left\{-\frac{2}{\epsilon} + \gamma - \frac{12}{5} + \log\frac{-q^2}{\nu^2} + \left(\frac{\alpha_s}{9\pi}\right)\left[\frac{32}{\epsilon^2} + \frac{1}{\epsilon}\left(\frac{1049}{15} - 32\gamma - 32\log - \frac{q^2}{4\pi\nu^2}\right) + \left(32\gamma - \frac{1049}{15}\right)\log - \frac{q^2}{4\pi\nu^2} + 16\log^2 - \frac{q^2}{4\pi\nu^2}\right]\right\}, \qquad (1.149a)$$

where one should notice the non-local pole $\frac{1}{\epsilon}\log - \frac{q^2}{\nu^2}$. The renormalized correlator is

$$\Pi^R = Z_{11}^2\ \Pi^B \qquad (1.149b)$$

Then one gets in the $\overline{MS}$ scheme[48b]:

$$\Pi^R = -\frac{3}{10\pi^2}\left(q^4\log - \frac{q^2}{4\pi\nu^2}\right)\left\{1 + \left(\frac{\alpha_s}{\pi}\right)\left[-\frac{473}{135} + \frac{8}{9}\left(2\gamma + \log - \frac{q^2}{4\pi\nu^2}\right)\right]\right\}. \qquad (1.149c)$$

This result is free of the $\frac{1}{\epsilon}\log - \frac{q^2}{\nu^2}$ non-local pole. This serves as a check of the partial validity of the expression. The α_s coefficient differs from that given in Ref. 29) both in the $\log^2$ and in the log terms. In fact, renormalization of the current has not been done carefully there. The above examples have really shown how much care should be taken in dealing with two-loop corrections and renormalizations. Let us now come to one example of renormalization and regularization scheme dependences of the QCD parameters.

j) $\overline{MS}$ and on-shell schemes : Scheme Invariance of the Quark Pole Mass

Let us call the on-shell scheme the QED-like scheme where Green's functions are Pauli-Villars regularized and the renormalization is done on shell :

$$\Pi^B_{OS} = \Pi^B_{PV}(q^2) - \Pi^B_{PV}(q^2 = 0) \quad . \qquad (1.150)$$

The relevance of this question for the sum rules is that one uses (mainly in the heavy-quark case[18,19,29]) the expression obtained in QED for the spectral function but there is often confusion regarding definition of the quark mass to be used in the analysis. For definiteness, let us concentrate on the vector spectral function.

In a QED-like on-shell scheme, the vector spectral function is known from the calculation of Källen and Sabry[49] which is accurately approximated by the Schwinger[50] expression :

$$\mathrm{Im}\,\Pi^{(1)}_V(t)\ (m_i = m_j) = \frac{3}{12\pi}\,\theta(t - 4m^2)\ .\ v\left(\frac{3-v^2}{2}\right)\left\{1 + \frac{4}{3}\,\alpha_s(t)\ f(v)\right\},$$

$$\text{where : } \quad v = \left(1 - 4\,\frac{m^2}{t}\right)^{1/2} \quad \text{and} \quad f(v) = \frac{\pi}{2v} - \frac{3+v}{4}\left(\frac{\pi}{2} - \frac{3}{4\pi}\right) \quad . \qquad (1.151)$$

In QED, the mass appearing in (1.151) is well defined as the electron is observed, i.e it is the mass at $p^2 = M^2 = m^2$, the pole of the electron propagator $S_F(p)$. Radiative corrections do not affect this mass and this is achieved via the mass renormalization constant[51]:

$$Z_m^{QED} = 1 - \left(\frac{\alpha_s}{\pi}\right)\left[\log\frac{\Lambda^2_{UV}}{M^2} + \frac{1}{2}\right] . \qquad (1.152)$$

This pole mass can be related to the Euclidian mass $m(p^2 = -M^2)$. Using the expression of Σ_1 in (1.67), one gets :

$$m(p^2) = M(p^2=M^2)\left\{1 - \frac{\alpha_s}{\pi}\left(1 - \frac{M^2}{p^2}\right)\left\{\log\left(1 - \frac{p^2}{M^2}\right) - \frac{1}{3}\alpha_G\right.\right. .$$

$$\left.\left. \cdot \left[1 + \frac{M^2}{p^2}\log\left(1 - \frac{p^2}{M^2}\right)\right]\right\}\right\} . \qquad (1.153)$$

Now the question is to know how to connect these QED-like masses to the masses used in the $\overline{MS}$ off-shell schemes. This question has been studied in Ref. 47) by calculating the vector correlator in two ways. Firstly, they calculate Re $\Pi(q^2)$ from the dispersive integral using the spectral function in (1.151) where the QED-like pole mass should be used. Secondly, they calculate directly Re $\Pi(q^2)$ in the Euclidian region by using the "light"-quark expansion m^2/q^2. In this case, one should use the quark mass in (1.91) after a leading-log resummation. Using the RGI of $\Pi(q^2)$ which is related to the physical observable e^+e^- into hadrons total cross-section, one obtains after requiring the same radiative corrections :

$$M_{QED}(p^2 = M^2) = \bar{m}_{\overline{MS}}(M)\left\{1 + \frac{4}{3}\frac{\alpha_s}{\pi} + \mathcal{O}\left(\alpha_s^2\right)\right\} . \qquad (1.154)$$

The same procedure has also been applied to other channels[46] and leads to (1.154). Let us now compare Eq. (1.154) with the relation between the pole and running mass in the $\overline{MS}$ scheme. This relation is[34,52]:

$$M_{\overline{MS}}(p^2=M^2) = \bar{m}_{\overline{MS}}(M)\left\{1 + \Sigma_1^{FP}(M^2 = \nu^2)\right\} , \qquad (1.155a)$$

where Σ_1^{FP} is the finite part of the quark self-energy decomposed as in (1.66). In the $\overline{MS}$ scheme, one can easily deduce from (1.67) :

$$\Sigma_1^{FP}(M^2 = \nu^2) = \frac{4}{3}\left(\frac{\alpha_s}{\pi}\right) , \qquad (1.155b)$$

which states that the pole mass in QED and $\overline{MS}$ schemes are equal. Then, we have the <u>important</u> conclusion[52]:

"*The pole mass is regularization <u>and</u> renormalization schemes invariant*"

These relations in (1.153) to (1.155) are sufficient for a consistent use of the mass definitions appearing in the sum rules analysis as we shall see later. Let us now consider the non-perturbative QCD effects on the correlators which we have discussed previously.

CHAPTER 2

A MINI-REVIEW OF SOME NON PERTURBATIVE ASPECTS OF QCD

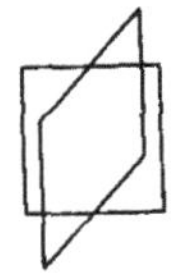

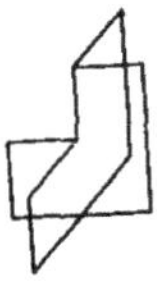

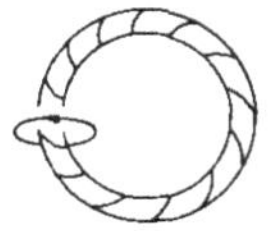

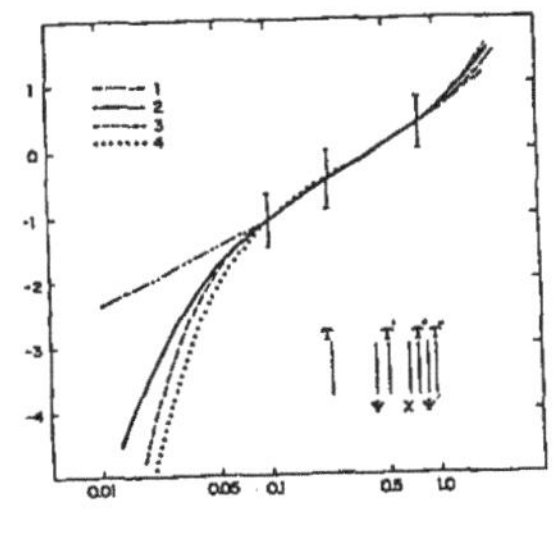

Some lattice operators, geometrical mapping of the baryons and potentials for the quarkonia systems.

We have already mentioned that the long distance properties of QCD are far from being understood owing to the peculiar infrared behaviour of the theory, although there have been several attempts to tackle this problem. Various approximation schemes are therefore needed for the study of bound-state problems.

1. LATTICE GAUGE THEORIES

This approach based on Monte-Carlo numerical simulations seems to be "very promising for demonstrating QCD phenomena with increasing precision but not for proving it". However, at present, lattice results should be viewed with caution as we need to understand the contamination of lattice size effects, the subtraction of continuum physics and the method of introducing dynamical fermions[53]. There has been progress in these directions together with an increase in computer time of runs and in hypercube volumes and the introduction of some special lattice dedicated computers. A simple example to show how this method works is the evaluation of a two-point correlator in pure $SU(3)_c$ Yang-Mills.

We start from a two-point correlator built with local time operators and perform a rotation in Euclidian space time (see e.g. Teper, Ref. 53).

$$\langle \phi(t)\ \phi(o) \rangle = \langle \phi(o) e^{-iHt}\ \phi(o) \rangle \underset{\tau\equiv it}{=} \langle \phi(o) e^{-H\tau}\ \phi(o) \rangle . \tag{2.1}$$

We insert a complete set of energy eigenstates and take the large τ limit in order to select the lowest ground state :

$$\langle \phi\ e^{-H\tau}\ \phi \rangle = \sum_n |\langle \phi | n \rangle|^2\ e^{-E_n t} \underset{\tau\to\infty}{=} |\langle \phi | o \rangle|^2\ e^{-E_o t} \tag{2.2}$$

from which we can deduce the lowest gluonium mass :

$$m_G = E_o \quad . \tag{2.3}$$

In practice, τ is not infinite and the separation of the lowest ground state is not clear if one has multi-degenerate gluonia masses. Bearing this remark in mind, let us go on to consider the lattice. We calculate the correlator using the Euclidian Feynman Path integral formalism :

$$\langle \phi(\tau)\ \phi(o) \rangle = Z^{-1} \int_{x,\mu} \Pi[d\ A_\mu(x)]\ \phi(\tau)\ \phi(o)\ \exp(-\beta\ S_E)\ , \qquad (2.4)$$

where $\beta \equiv 6/g^2$ and S_E is the action without g^2. Finiteness of (2.4) requires the introduction of an infrared cut-off (finite volume V) and an UV one (lattice spacing a). The gauge degrees of freedom are $SU(3)_c$ matrices :

$$U_\mu(x) \simeq \exp(i\ \ A_\mu(x)) \simeq 1 + i\ A_\mu(x)\ \ , \qquad (2.5)$$

and the lattice action goes to the correct continuum limit :

$$\beta\ S_E \underset{a\to 0}{\longrightarrow} \frac{1}{4g^2} \int d^4x\ \ G^a_{\mu\nu}\ G^{\mu\nu}_a\ . \qquad (2.6)$$

The evaluation of (2.4) is done using a Monte-Carlo algorithm for picking out typical "points" in the integration space where each point is an explicit gauge field over the lattice :

$$U^I \equiv \left\{U_\mu(x)\ ;\ \mu = 1,\ldots,4\ ;\ n = 1,\ldots,L^4\right\}^I\ . \qquad (2.7)$$

These points are generated with the distribution :

$$\prod_{n,\mu} d\ U_\mu(n)\ e^{-\beta S_E} \qquad (2.8a)$$

i.e. for $I = 1,\ldots,N$ configurations :

$$\langle \phi \, e^{-H\tau} \, \phi \rangle = Z^{-1} \int \prod_{\mu=1}^{4} \prod_{n=1}^{L^4} \left[d\, U_\mu(x) \; \phi(t) \; \phi(o) \; e^{-\beta S_E} \right]$$

$$= \frac{1}{N} \sum_{I=1}^{N} \phi(\tau) \; \phi(o) \Big|_{U=U^I} + \text{statistical error} \left(\simeq \frac{1}{\sqrt{N}} \right) . \qquad (2.8b)$$

Continuum physics is expected to be reached from (2.8) when $aL \gg \xi \gg a$, where ξ is a typical non-perturbative QCD length scale (presumably 1 Fermi like the hadron size), i.e. in this case we have small finite volume effects. In the continuum limit, we also expect to have asymptotic scaling, i.e. independence on the UV cut-off. This can be controlled from the value of the $\beta \equiv 6/g^2$ function which is constant in this asymptotic regime. This is reached for $\beta \geqslant 6$. At the moment, not all lattice results satisfy these two criteria.

2. THE LARGE N_c LIMIT AND THE $U(1)_A$ PROBLEM

Many features of QCD appear to be understood in the ideal large N limit[54] where QCD planar diagrams dominate the Green's functions while the other ones are down by powers $\frac{1}{N}$ and/or $\left(\frac{n}{N}\right)$ for $SU(n)_F \times SU(N)_c$. Of particular interest is the analytical solution of some equations[55] based on the vacuum average of the Wilson loop[56] :

$$W(C) = \mathrm{Tr}\left\{ P \left[\exp ig \oint_C dx^\mu \; A_\mu(x) \right] \right\}, \qquad (2.9)$$

which simplify considerably in the large N limit of QCD[57]. This seems to be the way to understand analytically the confinement problem but the "transfer" of the information obtained in the lattice to continuum theory still remains an open problem[58].

Let us discuss much more phenomenological uses of the large

N limit. In the $\frac{1}{N}$ counting rules[54], a gluon line can be represented by a pair of (quark anti-quark) lines having two opposite orientations. The gluon vacuum polarization is normalized to one in such a way that $g^2 \sim 1/\sqrt{N}$. Therefore, one can deduce that the meson (quark anti-quark) mass is of the order 1 and that its decay amplitude (f_π) is of the order $1/\sqrt{N}$. The meson two-point correlator goes like N and in general a k-point meson correlator behaves as $(N)^{(2-k)/2}$. In the $N \to \infty$ world, baryons are built with N quarks. Therefore, a pion-nucleon interaction which goes like f_π N behaves as $\sqrt{N}$ and baryon masses as N. A much more delicate problem is the $U(1)_A$ anomaly[59] with which the η' meson and the so-called θ vacuum of QCD are associated. For $SU(N)_c$ Yang-Mills, the θ vacuum modifies the action as :

$$Z = \int dA_\mu \exp i \int \mathrm{Tr} \left\{ -\frac{1}{4} G^2_{\mu\nu} + \theta \frac{g^2}{16\pi N} G_{\mu\nu} \tilde{G}_{\mu\nu} \right\}, \tag{2.10}$$

in a path integral formulation, where θ is normalized in such a way that the action has a smooth large N limit. θ is an angular variable and the associated generating function is unchanged by the shift θ into $\theta + 2\pi$. From (2.10), by differentiating with θ, one can derive a relation between the vacuum energy and the gluonium two-point correlator :

$$\frac{d^2E}{d\theta^2} = \frac{1}{N^2} \left(\frac{g^2}{16\pi^2} \right)^2 \lim_{q \to 0} U(q) \quad , \tag{2.11}$$

with :

$$U(q) = i \int d^4x \, e^{iqx} \left\langle 0 \mid \mathbb{T} \; G\tilde{G}(x) \; \left(G\tilde{G}(0)\right)^+ \mid 0 \right\rangle . \tag{2.12}$$

As $G\tilde{G}$ is a total divergence, its matrix elements vanish at zero momentum in perturbation theory. However, the large N counting rules discussed previously suggest that U(q) behaves as N^2, i.e. the 1/N expansion suggests a θ-dependence of the vacuum energy. Following Witten and Veneziano [60], the θ-dependence of the vacuum energy can be removed in the world with massless quarks. But large N counting suggests that their effects are down as 1/N (quark loop corrections). Writing :

$$U(q) = U_0 \text{(no quarks)} + U_1 \text{(one quark loop)} + \\ + U_2 \text{ (two quark loops)} + \ldots \tag{2.13a}$$

one can see that :

$$U_0 \sim N^2 \ , \ U_1 \sim N \ \text{ and } \ U_2 \sim 1 \ . \tag{2.13b}$$

How, then, does U(0) vanish ? The problem can be formulated in terms of a spectral function saturated by hadron poles :

$$U(q) = \sum_{\text{gluonium}} \frac{N^2 f_G^2}{q^2 - M_G^2} + \sum_{\text{mesons}} \frac{N f^2}{q^2 - m^2} \ , \tag{2.14a}$$

where M_G and m are the gluonium and meson masses. f_G and f are the decay amplitudes normalized as :

$$\langle 0 | G\tilde{G} | \text{gluonium} \rangle = N f_G$$
$$\langle 0 | G\tilde{G} | \text{mesons} \rangle = \sqrt{N} f \ , \tag{2.14b}$$

where f_G and f are of order one in the large N limit. From (2.14), the only way that U(0) vanishes in the presence of massless quarks is that the meson propagator behaves as N at q = 0. The $\eta'(958)$ is the lightest natural candidate having this property. Then, one requires :

$$\sum_{\text{gluonium - } M_G^2} \frac{N^2 f_G^2}{} \simeq \frac{N f^2}{M_{\eta'}^2} , \tag{2.15}$$

i.e. $M_{\eta'}^2 \sim 1/N$. More explicitly, one can associate the η' to the divergence of the $U(1)_A$ current which in the massless limit for $SU(n)_F$ reads :

$$\partial_\mu A^\mu = \frac{2n}{N} \frac{g^2}{16\pi^2} G\tilde{G} , \tag{2.16a}$$

with :

$$\left\langle 0 \left| \partial_\mu A^\mu \right| \eta' \right\rangle = \sqrt{n} \; . \; \sqrt{2} \, f_\pi M_{\eta'}^2 . \tag{2.16b}$$

Using a straightforward algebraic manipulation in (2.15) and (2.16), one can express f in terms of f_π. Using the relation between $V_o(0) = \frac{d^2E}{d\theta^2}|_{\theta=o}$ in the world without quarks, one can deduce the mass formula :

$$M_{\eta'}^2 \simeq \frac{2n}{f_\pi} \left(\frac{d^2E}{d\theta^2}\right)_{\theta=o}^{\text{No-quarks}} , \tag{2.17}$$

which has been generalized by Veneziano[60] for n derivatives. One can check the N-dependence of $M_{\eta'}^2$ from (2.17). $\left(\frac{d^2E}{d\theta^2}\right)$ is of the order one in the normalization (2.10) while $f_\pi \sim \sqrt{N}$ from large N counting, i.e. $M_\eta^2 \sim 1/N$ as expected. The vanishing of the η' mass at $N = \infty$ is closely related to the vanishing of the QCD anomaly (quark triangle loop) at $N = \infty$. The η'differs however from other Goldstone bosons as it gets a non-zero contribution in order 1/N. This is because the anomaly is of order 1/N and we know that Goldstone boson masses squared are a linear function of the chiral symmetry breaking parameters very similar to the fact that m_π^2 goes as $(m_u + m_d)$.

3. THE LARGE N LIMIT AND THE SKYRME MODEL

Let us now discuss another phenomenological approach in the large N limit. The Skyrme model [61] was revived after Witten's discovery[62] that the solitons of the non-linear σ model have the quantum numbers of QCD baryons provided that one includes the Wess-Zumino term[63]:

$$\mathcal{L}_{WZ} = \frac{N_c}{240\pi^2} \int d^5x \, \epsilon_{ijk\ell m} \, \mathrm{Tr}\{U^+\partial^i U \, U^+\partial^j U \, U^+\partial^k U \, U^+\partial^\ell U \, U^+\partial^m U\} \qquad (2.18)$$

where $U = \exp\left(\frac{\vec{\pi}.\vec{\lambda}}{f_\pi}\right)$ is the pion unitary matrix (f_π = 93 MeV and $\mathrm{Tr}\,\lambda_a\lambda_b = 2\,\delta_{ab}$). The Skyrme Lagrangian reads in its minimal version

$$\mathcal{L}_s = \mathcal{L}_\sigma + \mathcal{L}_{WZ} + \frac{1}{32e^2}\,\mathrm{Tr}\left[\partial_\mu U U^+ ,\ \partial_\nu U U^+\right]^2 + \ldots \qquad (2.19)$$

where ... indicates some other possible combinations and higher derivative terms.

$\mathcal{L}_\sigma$ is the familiar term of the σ model (see 1.21) whilst the last term is necessary to prevent the classical soliton from shrinking to zero size. Let us discuss this model in the simple $SU(2)_F$ case. One should note that if one writes U as $1 + \pi_a \lambda_a + \ldots$ the Wess-Zumino term is proportional to $\epsilon_{ijk\ell m} \cdot \mathrm{Tr}(\lambda_a \lambda_b \lambda_c \lambda_d \lambda_e)$. The fact that $\mathcal{L}_{WZ}$ is antisymmetric in Lorentz indices also needs its antisymmetry with isospin indices b,c,d,e. This is not possible for $SU(2)_F$. Then $\mathcal{L}_{WZ}$ vanishes. The classical solution of (2.19) can be written as the Skyrme ansatz :

$$U = A(t)\left[U_o \equiv \exp\left(i\, F(r)\, \frac{\vec{\lambda}}{2}\, \frac{\vec{x}}{|x|}\right)\right] A^{-1}(t) \quad , \qquad (2.20)$$

where F(t) satisfies the boundary conditions $F(o) = \pi$, $F(\infty) = 0$ corresponding to $U_o(r)$ having a unit topological charge and then a baryon number one. The soliton mass is obtained by substituting U_o in $\mathcal{L}_s$. Then, one obtains the static mass[64]:

$$M_s \simeq \frac{8\pi}{e} f_\pi \left[\int_0^\infty d\tilde{r}\, \tilde{r}^2 \left\{\frac{1}{8}\left[F'^2 + \frac{2\sin^2 F}{\tilde{r}^2}\right] + \frac{\sin^2 F}{2\tilde{r}^2}\left[\frac{\sin^2 F}{\tilde{r}^2} + 2F'^2\right]\right\} \equiv \mu \simeq 73\right] \qquad (2.21a)$$

and the non-linear equation :

$$\left(\frac{1}{4}\tilde{r}^2 + \sin^2 F\right) F'' + \frac{1}{2}\tilde{r}\, F' + (\sin 2F) F'^2 - \frac{1}{4}\sin 2F - \frac{\sin^2 F}{\tilde{r}^2}\sin 2F = 0 \quad , \qquad (2.21b)$$

which can be solved numerically where $\tilde{r} \equiv r.2e\, f_\pi$.
The moment of inertia λ of the soliton can be obtained by substituting $U(r,t)$ in $\mathcal{L}_s$. Expressing $A(t)$ in terms of the SU(2) matrices $a_o + i\, \vec{a}\, \frac{\vec{\lambda}}{2}$ with $\Sigma\, a_i^2 = 1$, one obtains :

$$\mathcal{L}_S = -M_S + 2\lambda \sum_{0}^{3} \left(\dot{a}_i\right)^2 \quad , \qquad (2.22a)$$

with :

$$\lambda = \frac{\pi}{3e^2 f_\pi} \left[\left[\int_0^\infty d\tilde{r}\, \tilde{r}^2 \sin^2 F \left[1+4 \left(F'^2 + \frac{\sin^2 F}{\tilde{r}^2} \right) \right] \right] \equiv L \simeq 51 \right] . \qquad (2.22b)$$

Introducing the conjugate momenta $\pi_i = \dfrac{\partial \mathcal{L}_s}{\partial \dot{a}_i} = 4\lambda \dot{a}_i$, one has the Hamiltonian :

$$H_S = \pi_i \dot{a}_i - \mathcal{L}_S = M_S + \frac{1}{8\lambda} \left[\sum_{i=0}^{3} \pi_i^2 \equiv -\nabla^2 \right] \quad , \qquad (2.23a)$$

with the eigenvalues :

$$E = M_S + \frac{1}{8\lambda}\, \ell(\ell+2) \quad \text{with} \quad \ell = 2J. \qquad (2.24b)$$

Eq (2.24) gives an expression of the N and Δ masses, which leads to :

$$M_S = \frac{1}{4} (5\, M_N - M_\Delta) \qquad \lambda = \frac{3}{2(M_\Delta - M_N)} . \qquad (2.25)$$

(2.25) was originally used in Ref. 64) to get :

$$e \simeq 5.45 \quad \text{and} \quad f_\pi \simeq 65 \text{ MeV} \quad , \qquad (2.26)$$

by giving the experimental values of M_N and M_Δ.
As one can see, f_π comes out badly $\left(f_\pi^{exp} \simeq 93 \text{ MeV}\right)$ even in some more sophisticated versions of the model. Another badly predicted quantity

is the axial coupling g_A of the nucleon ($g_A^S \simeq 0.61$, which is about half of the data). The excuse might be that one probably cannot ask too much from a model based on the $N_c \to \infty$ realizations of QCD. There are other attempts to fix the Skyrme parameter (monopole catalysis of the proton decay[65], integration of the non-Abelian QCD anomaly[66], $\pi\pi$ [67] and πN[68] scattering data, QSSR[69] to which we will return later). The results agree with the previous one in (2.26).There have also been some attempts to incorporate vector mesons into the model by a "gauging"[70], but the origin of the ρ mass is not yet understood.

Less fundamental but more popular approaches are the bag and potential models.

4. BAG MODELS

I will not give a status report of the bag models as there are many versions in the literature [71]. I will try to present the main idea behind the models. In this scheme, confinement is separated into two phases : the non-perturbative vacuum outside the bag and the perturbative one inside it. Let us illustrate our discussion by the MIT bag[71,72]. The Lagrangian is

$$\mathcal{L}_{MIT} = \int_V d^3x \, \mathcal{L}_{QCD} - BV - \frac{1}{2}\int_{\delta V} d^2\delta \, \bar{\psi}\psi \, . \qquad (2.27)$$

B is the vacuum energy density which can be expressed as the pressure balance condition :

$$B = \frac{1}{2}\vec{n}\,\nabla(\bar{\psi}\psi) - \frac{1}{4}G_{\mu\nu}G^{\mu\nu} \quad , \qquad (2.28)$$

where $\vec{n}$ is the outward normal of the bag ; the positivity of B does not allow the bag to expand. The last term in (2.27) expresses the quark equation of motion, with the boundary condition $\left(n \equiv \left(0,\vec{n}\right)\right)$

$$\left(i \vec{n} \vec{\gamma} + 1\right)\psi = 0 \quad \Longrightarrow n_\mu \ \vec{\bar\psi} \ \gamma^\mu \ \psi = 0 \quad , \tag{2.29a}$$

which means that there is no net flow of colour current through the bag surface. This does not allow the bag to collapse. The gauge invariant boundary condition is the magnetic superconductor one :

$$n_\mu \ G^{\mu\nu} = 0 \quad . \tag{2.29b}$$

a) Static cavity approximation for light hadrons

In the MIT approach, a static sphere cavity approximation is used. Hadrons can be obtained by exciting various colour singlet combinations of one-particle modes in the bag. The energy of a particular state is the number and kind of excited modes and the size of the bag :

$$E(R) = \sum_i \frac{m_i x_i}{R} + \sum_{i \leqslant j} C_{ij} \frac{\alpha_s}{R} + \frac{4\pi}{3} R^3 \ B - \frac{Z_0}{R} \ , \tag{2.30}$$

for n_i constituents i. The first term is a kinetic energy, the second one an interaction energy. The third term is a volume energy. Z_0 is an ad hoc additional parameter needed for a better data fit which is of the order one. The bag radius R is determined for each hadron in such a way that :

$$M_{HAD} \simeq \min_R E(R) . \tag{2.31}$$

As one can see from (2.30), too large R increases the volume energy whilst too small R enhances the quark kinetic energy. For hadrons, R comes out at about $(0.2 \text{ GeV})^{-1}$, let us say the inverse of Λ_{QCD} ! However, chiral symmetry is not seen from the Lagrangian (2.27), i.e. some other versions with chiral transformations are needed, the so-called chiral bags [71]. Another problem, but this time for nuclear physics, is the fact that the nucleon radius of about 1 fm seems too high compared with the one of 0.5 fm expected from nucleon-nucleon interactions where the nucleons are expected to move freely inside the nucleus. However, this difference might be due to the inability of the model to describe the energy of higher radial and orbital excitations of the nucleons. In fact, when two nucleons interact, the energy should be evaluated correctly. Among others, the originality (or the problem) of the bag models is that they lead to many exotics (cryptoexotic $u\bar{u}s\bar{s}$, dibaryons...). The introduction of the notion of constituent gluons (which is, however, gauge dependent) by analogy with the constituent quarks, leads to the existence of gluonia, hybrids or hermaphrodites [71,73]... Gluonia low-lying states are classified as transverse electric $(TE)^2$ and magnetic (TE).(TM) states in this approach :

$$(TE)^2 = J^{PC} = 0^{++},\ 2^{++}\ ,\ldots$$

$$(TE)\ (TM)\ J^{PC} = 0^{-+},\ 2^{-+}\ . \tag{2.32}$$

b) <u>Adiabatic cavity approximation for heavy quarks.</u>

The static cavity approximation is no longer good for heavy quark systems. One assumes instead that bag surfaces move faster than the quarks (adiabatic approximation) which is very analogous to the Born-Oppenheimer approximation in molecular physics (nucleus and electron clouds are replaced by quarks and coloured gluons). For two quarks of distance r much smaller than the hadron radius, one has in a dipole approximation :

$$E(R) = 2\ M_q - \frac{\alpha_s C}{r} + \alpha_s \frac{Cr^2}{4R^3} + \frac{4\pi}{3} B\ R^3\ . \tag{2.33}$$

A minimization of E(R) with respect to R leads to a Coulombic (linear) potential at short (large) distances. One can extend the same approximation to heavy baryons and to heavy-light quark systems. For the latter, one has to assume that the bag is centred around the heavy quark whilst light quarks move up to the bag boundary. However, the predictions for the heavy-light quark systems are not quite successful [71] . Other horizontal applications of the bag models (string tension, quark-gluon plasma...) can be seen in some bag-dedicated reviews[71]. There have been some attempts to derive the bag from QCD first principles[18,74,75] but the results are not yet conclusive.

5. POTENTIAL APPROACHES

The potential models applied to heavy-quark systems are less fundamental but very successful. The simplest QCD-like potential is the Cornell potential[76]:

$$V_c(r) = -\frac{4}{3}\frac{\alpha_s(r)}{r} + \sigma . r \quad , \tag{2.34}$$

where the corresponding Schrödinger equation can be easily solved. σ is the string tension which can be fixed from the Regge trajectories slope. $\alpha_s(r)$ is the QCD coupling kept constant in the Cornell version but running in the Richardson[77] potential :

$$V_R(q) = -\frac{4}{3} \times \frac{8\pi}{(-\beta_1)} \cdot \frac{1}{q^2 \log\left(\frac{q^2}{\Lambda^2} + 1\right)} \quad , \tag{2.35a}$$

where for $q^2 \leqslant \Lambda^2$:

$$V_R \sim -\frac{8\pi}{q^2}\sigma \quad . \tag{2.35b}$$

The charmonium and upsilon data fix[76]

$$\Lambda \simeq 400 \text{ MeV} \quad \text{and} \quad \sigma \simeq 0.15 \text{ GeV}^2 \quad . \tag{2.35c}$$

A much more phenomenological model is the Martin potential[78]:

$$V(r) = A + B\,r^{n} \quad , \qquad (2.36)$$

where the fit fixes n at about 0.1. Martin's potential is neither Coulombic at short distance nor linear at long distance. The reason might be that the form of the potential is strongly constrained only inside the region 0.1 to 1 Fermi but its slight modification outside this region does not affect the results as the wave functions are small. In fact, Martin's potential tends to have a logarithmic behaviour in this critical region. A lattice evaluation of the $Q\bar{Q}$ potential indicates that the potential behaves like (2.34) but becomes (2.36) at intermediate energies[53].

There are some impressive sets of inequalities which can be derived from the concavity property of the potentials[79]. For instance let us give the observed hierarchy :

$$E(1S) < E(1P) < E(2S) \ . \qquad (2.37)$$

The first inequality is satisfied for a local potential as the centrifugal barrier is repulsive in this case. The second inequality is only satisfied for a Coulombic potential if $(r\,V(r))'' > 0$. More generally :

$$E(n,\ell) < (>)\ E(n-1,\ \ell+1) \quad \text{if} \quad (r\,V)'' > 0\ (< 0) \ . \qquad (2.38)$$

The flavour independence assumption leads to the concavity relation :

$$2E\left(Q\bar{q}\right) > E\left(Q\bar{Q}\right) + E\left(\bar{q}q\right) \quad , \qquad (2.39)$$

which is satisfied by the observed masses.

However, despite the great phenomenological success of potentials models, some difficulties arise in attempting to relate them to field theory. Leutwyler and Voloshin[80] criticize the locality of the potentials whilst Bell and Bertlmann[19] do not observe their flavour independence.

6. DISCRETIZED LIGHT-CONE QUANTIZATION METHOD

This method has been advocated by Ref. 81 for obtaining bound--state spectra and wave functions. It is based on the quantization of field theory at equal light cone time $\tau = t+x/c$ rather than at equal time t. The simplest case of 1+1 dimensions where fermions interact with scalar (instead of gluon) fields has been studied. For definiteness, one considers the interaction term $\lambda \bar{\psi} \varphi \psi$. The model consists of a diagonalization of finite matrices in Fock space. The three constants of motion are the total light cone energy $P^- = E-P$, the total light cone momentum $P^+ = E+P$ and the charge Q. P^- and P^+ are components of a Lorentz vector to which the invariant $M^2 = P^+P^-$ corresponds. Since Q and $P^{\pm}$ commute naturally, they can be diagonalized simultaneously. The problem can be formulated in Fock space by introducing two parameters which are the box size L and the ultraviolet cut-off Λ. Since on the light cone one has a finite number of Fock states which can have the same light cone momentum and charge, the mass matrix has a finite dimension. The eigenvalues of the mass matrix, as well as its eigenfunctions and more generally the physical results should be independent of the box size L and the momentum cutt-off Λ at least when $L,\Lambda \to \infty$. The level of refinement of the method is controlled by the "harmonic resolution" K which corresponds to the ratio of the box size L over the Compton wavelength of physical particle $\lambda_c = 2\pi h/Mc$. In fact, for large L, one is close to the continuum limit and at the same time the spectrum of physical particles with mass close to M is more complex. The method might resemble the lattice approach but here there is no doubling of fermion fields. The derivative of the fermion field is represented by a 1/n factor in the massive part of the Hamiltonian contrary to the case of lattice gauge theory which uses the nearest-neighbour approximation.

The method seems to present some attractive features but, like lattice gauge theory, it is condemned to a formidable numerical manipulation at least in the real 3+1 space-time dimensions and for "vector" gluons.

CHAPTER 3

SVZ EXPANSION: THEORETICAL FOUNDATIONS

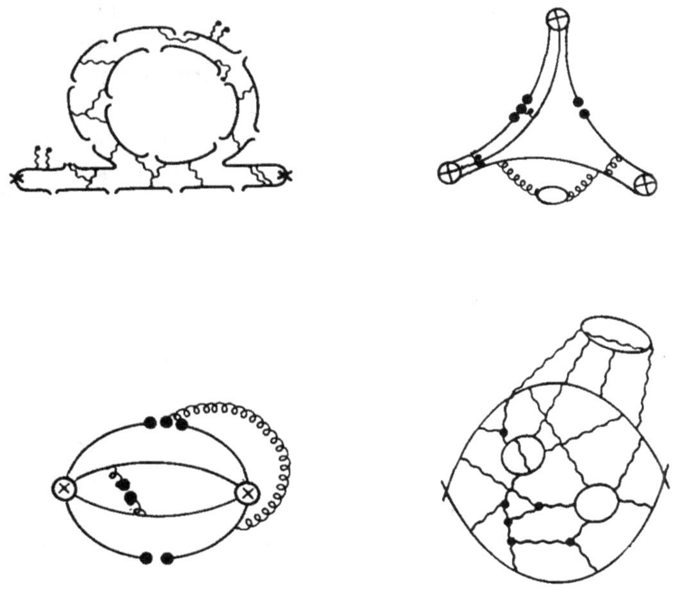

Some typical non-perturbative QCD graphs.

Contrary to previous approaches, with the exception of lattice gauge theories and less probably the discretized light cone quantization, the SVZ approach is based on the first principles of QCD as it involves in a natural way the QCD universal perturbative parameters (Λ and quark masses) and non-perturbative parameters (condensates).

1. SVZ EXPANSION AND VACUUM CONDENSATES

For definiteness, let us illustrate our discussion from the hadronic correlator :

$$\Pi(q^2) = i \int d^4x \, e^{iqx} \left\langle 0 \mid \mathbb{T} \, J_H(x) \, (J_H(o))^+ \mid 0 \right\rangle , \tag{3.1}$$

associated with the current $J_H(x)$ of quarks and/or gluons.
It should also be borne in mind that the sum rules discussed in Chapter 1 equate the spectral function measurable at low energy with its real part in principle calculable in QCD at sufficiently large Euclidian momentum or quark mass squared. In this chapter we shall concentrate on the QCD part of the sum rule. According to SVZ, the breaking of the usual perturbation theory at low q^2 is due to the manifestation of non-perturbative terms which appear as power corrections in the OPE of Green's function à la Wilson[82]. In this way, one can write :

$$\text{Re}\, \Pi(q^2) \simeq C_{\mathbb{1}} \, \mathbb{1} + \sum_{n=2,\dots} C_{2n} \frac{\langle O_{2n} \rangle}{(q^2)^n} + \dots \tag{3.2}$$

where C_{2n} are Wilson coefficients calculable perturbatively via Feynman diagram techniques (short wavelength fluctuations). O_{2n} are high-dimension operators with non-vanishing expectation values induced by the non-perturbative large fluctuations. Up to dimension six, these vacuum condensates read :

$$\left\langle O_4^{\psi} \right\rangle = m \left\langle \bar{\psi} \psi \right\rangle \qquad \text{quark condensate}$$

$$\left\langle O_4^G \right\rangle = \left\langle \alpha_s \, G_a^{\mu\nu} \, G_{\mu\nu}^a \right\rangle \qquad \text{gluon condensate}$$

$$\left\langle O_6^\psi \right\rangle = \left\langle \bar{\psi} \, \Gamma_1 \, \psi \, \bar{\psi} \, \Gamma_2 \, \psi \right\rangle \qquad \text{four-quark condensate}$$

$$\left\langle O_6^G \right\rangle = g^3 \, f_{abc} \left\langle G_a^{\mu\nu} \, G_{\nu\rho,b} \, G_{\mu,c}^{\rho} \right\rangle \quad \text{triple gluon condensate}$$

$$\left\langle O_6^D \right\rangle = g^2 \left\langle \left(D_\mu \, G_{\alpha\mu}^a \right) \left(D_\nu \, G_{\alpha\nu}^a \right) \right\rangle$$

$$\left\langle O_6^M \right\rangle = mg \left\langle \bar{\psi} \, \sigma^{\mu\nu} \frac{\lambda_a}{2} \, \psi \, G_{\mu\nu}^a \right\rangle \qquad \text{mixed condensate.} \tag{3.3}$$

The first one is the well-known quark vacuum condensate responsible for the spontaneous breaking of chiral symmetry. It is related to the light quark masses through the PCAC relation in (1.18). The other operators are new ones. Using a dilute gas instanton approximation[83) (DIGA), SVZ parametrize the condensates in terms of the dilute gas density[18)]:

$$d(\rho) \simeq (\text{cst.}) \, . \left(\frac{2\pi}{\alpha_s(\rho)} \right)^6 \exp \left(\frac{-2\pi}{\alpha_s(\rho)} \right) , \tag{3.4}$$

where ρ is the instanton radius. For instance, by writing the t'Hooft instanton solution of the Yang-Mills equation :

$$G_{\mu\nu}^a = \frac{4 \, \eta_{\mu\nu}^a \, \rho^2}{g \left[(x-x_o)^2 + \rho^2 \right]^2} \tag{3.5a}$$

where x_o is the instanton position and $\eta_{\mu\nu}^a$ is the t'Hooft symbol with the properties :

$$\eta_{\mu\nu}^a = \begin{cases} \epsilon_{a\mu\nu} & \mu,\nu \neq 4 \\ -\delta_{a\nu} & \mu = 4 \\ \delta_{a\mu} & \nu = 4 \end{cases} \qquad \bar{\eta}_{\mu\nu}^a = \begin{cases} \epsilon_{a\mu\nu} \\ \delta_{a\nu} \\ -\delta_{a\mu} \end{cases}$$

$$\eta^a_{\mu\nu} = \frac{1}{2}\epsilon_{\mu\nu\alpha\beta}\,\eta^a_{\alpha\beta} \qquad \bar{\eta}^a_{\mu\nu} = -\frac{1}{2}\epsilon_{\mu\nu\alpha\beta}\,\bar{\eta}^a_{\alpha\beta}$$

$$\eta^a_{\mu\nu} = -\eta^a_{\mu\nu} \qquad \eta^a_{\mu\nu}\,\bar{\eta}^b_{\mu\nu} = 0 \qquad \eta^a_{\mu\nu}\,\eta^b_{\mu\nu} = 4\,\delta^{ab}$$

$$\eta^a_{\mu\nu}\,\eta^a_{\gamma\lambda} = \delta_{\mu\gamma}\,\delta_{\nu\lambda} - \delta_{\mu\lambda}\,\delta_{\nu\gamma} + \epsilon_{\mu\nu\gamma\lambda} \quad , \tag{3.5b}$$

one can write the gluon condensates as :

$$\langle O_{2n}\rangle \equiv \left\langle \left(G^a_{\mu\nu}\right)_1 \cdots \left(G^a_{\mu\nu}\right)_n \right\rangle = \int_0^{\rho_c} \frac{d\rho}{\rho^{2n+1}}\, d(\rho) \quad , \tag{3.6}$$

where ρ_c is the critical size cut-off of the instanton. Introducing the approximate relation[84]

$$\frac{2\pi}{\alpha_s(\rho)} = \frac{2\pi}{\alpha_s(\rho_c)} + 11\log\left(\frac{\rho_c}{\rho}\right) \quad , \tag{3.7}$$

(3.6) becomes :

$$\langle O_{2n}\rangle = \frac{1}{\rho_c^{2n}}\,\frac{1}{(11-2n)}\left(\frac{2\pi}{\alpha_s(\rho_c)}\right)^6 \exp\left(-\frac{2\pi}{\alpha_s(\rho_c)}\right)\sum_{k=o}^{6}\left(\frac{11\,\alpha_s(\rho_c)}{2(11-2n)}\right)^k \frac{6\,!}{(6-k)\,!}, \tag{3.8}$$

which indicates a critical dimension :

$$2n = 11 \quad , \tag{3.9}$$

which separates the large-size instanton phase ($2n \leqslant 11$) controlled by the confinement radius ρ_c to the small one (controlled by $1/q^2$). There is no accurate control of the small-size instanton effects in the OPE. We shall simply assume that we can neglect them like over-high dimension condensates in the range of q^2 values ($\gtrsim 1\ GeV^2$) where the QSSR analysis is done. Large-size instantons i.e. low-dimension condensates might be roughly evaluated from the above DIGA, but as can be seen from (3.8) the numerical value depends crucially on the infrared cut-off ρ_c and on the α_s. For a $\rho_c \simeq 1/200$ MeV fixed from the phenomenological estimate of $\langle \alpha_s G^2 \rangle$ (see next chapter), one can deduce from (3.8):

$$g^3 f_{abc} \langle G^a_{\mu\nu} G^b_{\mu\rho} G^c_{\rho\mu} \rangle \simeq (1 \text{ GeV}^2) \langle \alpha_s G^2 \rangle , \qquad (3.10)$$

which should only be considered as a very rough estimate. A much more involved estimate comes from $SU(2)_c$ lattice calculations. In fact, Di Giacomo et al [53] have demonstrated from lattice Monte-Carlo simulations that indeed the gluons condense. In addition, they have shown that these gluon condensates possess a correlation length λ defined through its relation with the condensate correlation function :

$$\lim \phi(|R| \to \infty) \equiv \langle 0 \mid \alpha_s G_{\mu\nu} W(x,0) G_{\mu\nu} \mid 0 \rangle \simeq \exp\left(-\frac{|x|}{\lambda}\right) , \qquad (3.11)$$

where $W(x,o)$ is the Wilson loop in (2.9). A SU(2) lattice measurement gives :

$$\langle \alpha_s G^2 \rangle \lambda^4 \simeq 6.3 \; 10^{-3} \; ; \quad \lambda \simeq 0.2 \text{ fm}$$

and

$$\left| \langle g^3 f_{abc} G^a_{\mu\nu} G^b_{\nu\rho} G^c_{\rho\mu} \rangle \right|^{1/6} \simeq 1.3 \left(\langle \alpha_s G^2 \rangle \right)^{1/4} \qquad (3.12)$$

These values would imply :

$$\langle \alpha_s G^2 \rangle \simeq 0.06 \text{ GeV}^4$$

$$\left| \langle g^3 f_{abc} G^a_{\mu\nu} G^b_{\nu\rho} G^c_{\rho\mu} \rangle \right| \simeq (1.2 \text{ GeV}^2) \langle \alpha_s G^2 \rangle , \qquad (3.13)$$

which appears to be consistent with the phenomenological value of $\langle \alpha_s G^2 \rangle$

(see next chapter) and with the DIGA in (3.10). However, a more realistic estimate needs the effects of fermions for $SU(3)_c$. For future applications, we shall use :

$$\left|\left\langle g^3 f_{abc} G^a_{\mu\nu} G^b_{\nu\rho} G^c_{\rho\mu}\right\rangle\right| \simeq (1 \pm 0.5)\ \text{GeV}^2 \left\langle \alpha_s G^2 \right\rangle \qquad (3.14)$$

and the remaining condensates defined in (3.3) will be estimated phenomenologically.

Now let us turn to the question of the "rigorous" validity of the SVZ assumption for the OPE. In fact in (3.2), one assumes explicitly that one can separate unambiguously the perturbative Wilson coefficients (short-wavelength fluctuations) from the non-perturbative condensates (large fluctuations). This assumption has been tested in some QCD-like models by various authors[85-93].

2. SVZ EXPANSION IN THE $\lambda\varphi^4$ MODEL

For a simple pedagogical reason, we shall study this question in the example of scalar-field theory. We start from the bare Lagrangian of $\lambda\varphi^4$ theory :

$$\mathcal{L}_\varphi = \frac{1}{2}(\partial_\mu \varphi_B)^2 - \frac{1}{2} m_B^2 \varphi_B^2 - \frac{\lambda_B}{4!}\varphi_B^4 , \qquad (3.15)$$

where φ is the scalar field, m is its mass and λ its coupling. The index B corresponds to bare quantities. We know that in the case $m^2 < 0$, we have a spontaneous breaking mechanism where the field φ acquires a non-vanishing expectation value which is non-analytical in the coupling constant, i.e. the model mimics non-perturbative effects. However to simplify our discussion, let us ignore renormalization effects[85] and work with $m_B^2 > 0$, i.e. we have no condensate which breaks spontaneously the symmetry. We study the scalar propagator :

$$\mathcal{D}(q) = i \int d^4x\, e^{iqx} \langle 0 | T\, \varphi(x)\, \varphi(0) | 0 \rangle . \qquad (3.16)$$

In the first way, we use the standard perturbative expansion in λ_B

$$D(q) =$$

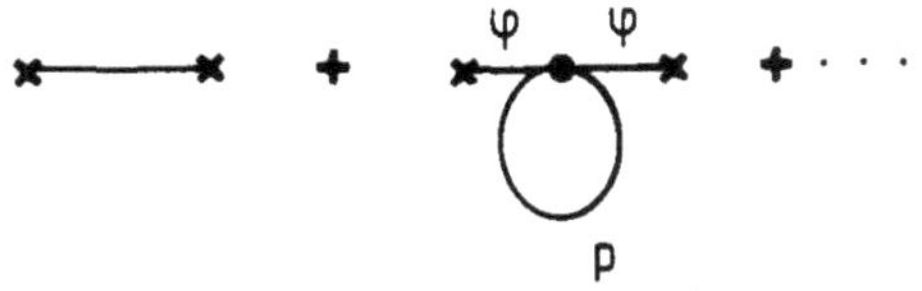

(3.17)

Using, for instance, a Pauli-Villars regularization (the following conclusion is independent of the choice of regularization), one obtains :

$$\mathbb{D}(q^2) \simeq \frac{1}{q^2 - m_B^2}\left\{1 + \frac{\lambda_B}{32\pi^2}\,\frac{M^2 - m_B^2 \log\left(M^2/m_B^2\right)}{q^2 - m_B^2}\right\}$$

$$\underset{-q^2 \gg m_B^2}{\simeq} \frac{1}{q^2} + \frac{1}{q^4}\left\{m_B^2 + \frac{\lambda_B}{32\pi^2}\left(M^2 - m_B^2 \log\frac{M^2}{m_B^2}\right)\right\} , \qquad (3.18)$$

where M^2 is the U.V. arbitrary scale.

In the second way, one evaluates the propagator using the SVZ expansion for $-q^2 \gg m_B^2$. Therefore :

$$\mathbb{D}(q^2) \simeq C_1\, \mathbb{1} + C_\varphi \langle \varphi^2 \rangle + \ldots \ . \qquad (3.19)$$

One introduces a renormalization point ν to separate the long and short wavelength fluctuations[85]. The Wilson coefficient C_1 comes from the perturbative graph and corresponds to the short fluctuations $(p > \nu)$:

$$C_1 = \frac{1}{q^2} + \frac{1}{q^4}\left\{m_B^2 + \frac{\lambda_B}{32\pi^2}\left(M^2 - \nu^2 - m_B^2 \log\frac{M^2}{\nu^2}\right)\right\} . \qquad (3.20)$$

The Wilson coefficient C_φ is obtained from the Feynman graph associated to the φ^2 "condensate".

$$C_\varphi \simeq$$  $$= \frac{\lambda_B}{2q^4} \, . \qquad (3.21)$$

The condensate $\langle \varphi^2 \rangle$ corresponds to the evaluation of the tadpole-like graph for $p < \nu$ (large fluctuations). Therefore :

$$\langle \varphi^2 \rangle \simeq \frac{1}{16\pi^2} \left(\nu^2 - m_B^2 \log \frac{\nu^2}{m_B^2} \right) \, . \qquad (3.22)$$

One can easily see that at this level of approximation the SVZ-expansion in Eqs (3.19) to (3.22) and the usual series in Eq (3.18) coincide. However, it is interesting to check whether this coincidence continues to hold in higher orders of perturbation theory.

The main point from the above analysis is the introduction of a ν scale which separates the low and high q^2-behaviour of the propagator. According to Ref. 85), previous authors[86)] have not performed this separation carefully. Now, once we understand this "pure" perturbative case, it is an easy task to analyze the case of spontaneous breaking $m^2 > 0$.

There are some other QCD-like models[87)] such as the Schwinger two-dimensional gauge theories[88)], the CP^{N-1} model[89)] which are known to have instantons and θ vacua, The Gross-Neveu model[90)] with dynamical chiral symmetry breaking and the free two-dimensional non-linear σ models[91,85)].The last two models have the nice asymptotic freedom property like QCD. We shall concentrate on this case in the following section.

3. SVZ EXPANSION IN THE TWO DIMENSIONAL O(N) σ MODEL

The σ model Lagrangian density can be written as :

$$\mathcal{L} = \frac{N}{2f}\left\{\partial_\mu \sigma_a \partial_\mu \sigma^a - \frac{\alpha(x)}{\sqrt{N}}\left(\sigma_a \sigma^a - \frac{N}{f}\right)\right\} , \qquad (3.23)$$

where α is an auxiliary field ; $a \equiv 1,\dots,N$ is the colour index ; f is the bare coupling. The σ^a field is defined on the unit sphere ($\sigma^2 = 1$) . The great advantage of the model is that it can be exactly solved in the large N-expansion. At the classical level, the O(N) symmetry is *spontaneously broken where N-1 massless Goldstone bosons* emerge. This symmetry is restored at the quantum level. Then one gets N massive bosons with mass :

$$m = (cst) \;.\; \nu \exp\left(\frac{1}{\beta_1 \; f_R}\right) , \qquad (3.24)$$

which is an analytic function of the running coupling :

$$f(\nu) = \frac{1}{-\beta_1 \;\log \dfrac{\nu^2}{m^2}} , \qquad (3.25)$$

where $\beta = \beta_1 \; f + \dots$ is the usual β function. Integrating over the σ fields in (3.23), one obtains the non-local effective action :

$$S_{eff} = \frac{N}{2}\left\{\mathrm{Tr}\;\log\left(-\partial^2 + \frac{\alpha}{\sqrt{N}}\right) - \frac{1}{f\;N}\int d^2x\;\frac{\alpha(x)}{\sqrt{N}}\right\} . \qquad (3.26)$$

The expansion around the saddle point of this action gives a 1/N expansion very similar to perturbation theory but with a more complicated propagator for the composite field α. The stationary point of the generating functional in $\alpha(x)$ should be independent of x owing to Lorentz invariance. It is :

$$\alpha(x) = \sqrt{N}\, m^2 + \alpha_d \tag{3.27}$$

where α_d describes quantum fluctuations of the α field. Ref. 91) studies the vacuum expectation value $\langle \alpha^2 \rangle$, the "spin wave condensate" which is the analogue of the QCD gluon condensate. If one introduces the scale ν in order to separate the short and large fluctuations, then[85] :

$$\langle \alpha^2(\nu) \rangle = \int_{-q^2 < \nu^2} \frac{d^2 q}{(2\pi)^2} D^\alpha(q) \quad , \tag{3.28}$$

where $D^\alpha(q)$ is the α propagator defined as :

$$\begin{aligned} D^\alpha(q) &= i \int d^4x\, e^{iqx} \langle 0 \mid T\, \alpha(x)\, \alpha(0) \mid 0 \rangle \\ &= (2\pi)^2\, \delta^2(q)\, \langle \alpha \rangle^2 - \frac{4\pi\, q^2 (1+\beta)}{\log \dfrac{\beta+1}{\beta-1}} \quad , \end{aligned} \tag{3.29}$$

where $\beta \equiv \sqrt{1 + \dfrac{4m^2}{q^2}}$. The first term is the factorized term $N\, m^4$. We shall be interested in the non-factorized one here. This term is down by $1/N$ compared to $\langle \alpha^2 \rangle$. Then, after the change of variable :

$$x(q^2) = \left(\sqrt{1 + \frac{q^2}{4m^2}} + \sqrt{\frac{q^2}{4m^2}} \right)^4 \quad , \tag{3.30a}$$

one can rewrite :

$$\langle \alpha^2(\nu) \rangle_{N.F} = - m^4 \int_1^{x(\nu^2)} \frac{dx}{x^2} (x-1)^2 \frac{1}{\log x} \quad , \tag{3.30b}$$

which gives the "condensate" :

$$\langle \alpha^2(\nu) \rangle_{NF} \underset{\nu^2 \gg m^2}{=} -2\,\nu^4\, e^{-L}\, E_1(L) - \frac{4}{L}\nu^2 m^2 + 2\,m^4 \left[\gamma_E + \log L - \frac{1}{L} + \frac{4}{L^2}\right] , \qquad (3.31)$$

where $L = 4 \log\frac{\nu}{m} \gg 1$, $e^{-L}\, E_1(L) \underset{L\to\infty}{\simeq} \sum_{0}^{\infty} n!/L^{n+1}$.

The second step is to evaluate (in principle) the Wilson coefficients of the two-point correlator for $-q^2 > \nu^2$. It has the SVZ expansion :

$$i \int d^2x\, e^{iqx} \langle 0 \mid T\, \alpha(x)\, \alpha(0) \mid 0 \rangle = C_{\mathbb{1}}(\nu^2, q^2)\, \mathbb{1} + C_2\, (\nu^2, q^2)\, \langle \alpha(\nu) \rangle + C_4(\nu^2, q^2)\, \langle \alpha(\nu)^2 \rangle , \qquad (3.32)$$

where we recall that $\alpha(x) \equiv f(\nu)\, (\partial\sigma)^2(\nu)$ is the RGI current. This RGI of the two-point correlator has been exploited by SVZ in order to give constraints on the coefficients C_i for the cancellation of the ν-dependence. The ν^4 term in (3.31) should be cancelled by the one in $C_{\mathbb{1}}\cdot\mathbb{1}$ which is present in a cut-off regularization procedure ; the $m^2\nu^2$ term of (3.31) should be compensated for by $C_2(\nu^2)\langle\alpha(\nu)\rangle$ where ν^2 comes from $C_2(\nu^2)$ and m^2 from $\langle\alpha(\nu)\rangle$. The cancellations of the ν term appearing in m^4 are less obvious. Here one needs to know the anomalous dimension of the $\alpha^2(\nu)$ operator to next-to-leading order in 1/N. Therefore, the improved RGE form of $\alpha^2(\nu)$ is :

$$\alpha^2(\nu) = \left(\exp \int_{f(\mu)}^{f(\nu)} df\, \frac{\gamma(f)}{\beta(f)}\right) \alpha^2(\mu) \qquad (3.33a)$$

where $\alpha^2(\mu)$ is ν-independent. Using the perturbative expression :

$$\gamma/\beta \simeq \frac{a_{-1}}{f} + a_0 + a_1 f + \ldots \qquad (3.33b)$$

and integrating, one has :

$$\alpha^2(\nu) \simeq \alpha^2(\mu) \cdot \left\{1 + \frac{1}{N} a_{-1} \log f(\nu) + a_0 f(\nu) + \frac{1}{2} a_1 f^2(\nu) + \ldots\right\}. \qquad (3.33c)$$

The vanishing of the ν-dependence in (3.31) fixes :

$$a_{-1} = -2 \quad ; \quad a_0 = -\frac{1}{4\pi} \quad ; \quad a_1 = \frac{1}{4\pi^2} . \qquad (3.33d)$$

Had we used dimensional regularization as in Ref. 91), we would have noticed that the ν^4 and $m^2\nu^2$ terms were absent. Divergences would appear as ϵ poles. The coefficient of these poles is minus the one of the $\log \nu^2/m^2$ term (see Chapter 1). Therefore, it is sufficient to replace the log-term appearing in the regularization-scheme independent m^4 term in (3.31) to get $a^2(\nu)$ in the $\overline{MS}$ scheme :

$$\langle \alpha^2(\nu) \rangle_{NF}^{\overline{MS}} \simeq 2m^4 \; [-\log \epsilon + \gamma_E + \mathcal{O}(\epsilon)] \; . \qquad (3.34)$$

Eq (3.34) can help in understanding David's claim[91]: the condensate develops two opposite imaginary parts which are due (from 3.34) to the $\log(\epsilon \pm io)$ when ϵ goes to zero. But this is not the end of the story! David also expresses the perturbative Wilson coefficient $C_{\mathbb{1}}$ in terms of its Borel transform B :

$$C_{\mathbb{1}} \equiv \sum_i f_R^i a_i = \frac{1}{f_R} \int_0^\infty db \; e^{-b/f_R} \left(B(b) \equiv \sum_i \frac{b^i}{i!} a_i \right) , \qquad (3.35)$$

where B presents "IR renormalon" singularities located at $b = 2n/\beta_1$, (n = 1,2,...,) which would correspond to non-perturbative IR divergences of the order m^2, m^4 ... for $C_{1\!\!1}$ according to the conjecture of Parisi [92]. Therefore $C_{1\!\!1}$ has two possible determinations with opposite imaginary parts $C^{\pm}_{1\!\!1}$ corresponding to the integration in the Borel plane of (3.35) above (+) or below (-) the singularity. Using the running coupling $f(\nu)$, David shows that the Borel transform B(b) of $C_{1\!\!1}$ is related to the discontinuity at the first cut $s_o = 0$ of the Mellin transform of the α propagator :

$$\mathcal{M}(s) = \int_0^\infty d\nu\ \nu^{-s-1}\ D(\nu x) \quad . \tag{3.36}$$

This allows him to conclude that the imaginary part of the condensate in (3.34) is cancelled by that of $C_{1\!\!1}$ or equivalently renormalon singularities of the condensate are cancelled by those of $C_{1\!\!1}$. This cancellation is expected to occur for higher dimension condensates rendering the SVZ expansion well defined ! Cancellations of IR renormalons in connection with previous discussions have also been explicitly analyzed in QCD by Mueller[93]. We conclude from these previous paragraphs that the SVZ expansion is meaningful but good definitions of the condensates can still be lacking. In the next paragraphs we shall give some "good" definitions of these condensates within perturbation theory and renormalization group approaches.

4. DEFINITIONS OF $m\langle\bar\psi\psi\rangle$ AND $\langle\alpha_s G^2\rangle$ IN QCD

In QCD, one might expect that the lowest dimension quark $m\langle\bar\psi\psi\rangle$ and gluon $\langle\alpha_s G^2\rangle$ condensates have small perturbative pieces which might come from the quark-mass corrections. This is indeed the case of $m\langle\bar\psi\psi\rangle$ if we examine the Ward identity in (1.121). The proper definition of the quark condensate might be :

$$(m_i \pm m_j)\left\langle \bar{\psi}_i \psi_i \pm \bar{\psi}_j \psi_j \right\rangle = (m_i \pm m_j)\left\{\left\langle \bar{\psi}_i \psi_i \pm \bar{\psi}_j \psi_j \right\rangle_{NP} + \right.$$

$$\left. + \frac{3}{4\pi^2} \cdot \left(m_i^3 Z_i + m_j^3 Z_j\right)\right\} \quad , \tag{3.37a}$$

where :

$$Z_i \equiv 1 + \log \frac{\nu^2}{m_i^2} + \left(\frac{2\alpha_s}{3\pi}\right)\left(5 + 5 \log \frac{\nu^2}{m_i^2} + 3 \log^2 \frac{\nu^2}{m_i^2}\right) \quad . \tag{3.37b}$$

This combination is the one useful for cancelling the m^4 mass singularities in practical QCD calculations, as can be seen, for instance, in Refs 46) and 94). However, due to the numerical smallness of the perturbative contributions, it is no matter to use (3.37) or just the non-perturbative $m\left\langle\bar{\psi}\psi\right\rangle$ in the estimate of the quark condensate. The definition in Eq.(3.37) does not contradict David's claim that the $\left\langle\bar{\psi}\psi\right\rangle$ condensate is protected by chiral symmetry which is true in the chiral limit $m_i \to 0$.

For the gluon condensate, we do not yet have in QCD any explicit examples showing the required perturbative piece for cancelling mass singularities. The renormalization group invariant combination defined in (1.114) :

$$\beta(\alpha_s)\langle G^2\rangle + 4\,\gamma_m(\alpha_s)\,m\left\langle\bar{\psi}\psi\right\rangle , \tag{3.38}$$

is already sufficient to do a consistent calculation up to order m^4 in the perturbative graph. This can easily be understood from the analysis of the vector correlator to order $\alpha_s\, m^4$ in Ref.46) as all m^4 singularities have been already eaten by the $m\left\langle\bar{\psi}\psi\right\rangle$ contributions. One should go presumably to order $\alpha_s^2\, m^4$ in order to see the exact structure of the perturbative contribution to the gluon condensate in addition to the one appearing in (3.38) from $\alpha_s\, m\langle\bar\psi\,\psi\rangle$. The definition in (3.38) fits with David's expectation that the "true" gluon condensate would have the form :

$$\langle\alpha_s G^2\rangle + \lambda^4 \log\frac{\nu^2}{\lambda^2}\ , \tag{3.39}$$

where λ is an infrared scale independent of ν. Perturbative QCD can indeed indicate that the perturbative piece of the gluon condensate behaves as :

$$\langle\alpha_s G^2\rangle\Big|_{\mathrm{Pert}} \sim \alpha_s\, m^4 \log\frac{m^2}{\nu^2}\ , \tag{3.40}$$

where the coefficient is fixed from (3.37) if only the perturbative pieces come from $m\langle\bar\psi\,\psi\rangle$. The smallness of the term in (3.39) justifies a posteriori its neglect in the phenomenological estimate of the gluon condensate.

For heavy quark systems, the condensate $\langle\bar Q\, Q\rangle$ can be shown to be related to the gluon condensate as :

$$M_Q\langle\bar Q\, Q\rangle = -\frac{1}{12\pi}\langle\alpha_s G^2\rangle - \frac{1}{1440\pi^2}\, g^3\,\frac{\langle G^3\rangle}{M_Q^2} - \frac{g^2}{120\pi^2}\,\frac{\langle(DG)^2\rangle}{M_Q^2} \tag{3.41}$$

i.e. $\langle\bar Q\, Q\rangle$ vanishes like $1/M_Q$. $g^3\langle G^3\rangle$ and $g^2\langle(DG)^2\rangle$ are short-hand notations for O_6^G and O_6^D in (3.3). The first term in (3.41) has been originally obtained by SVZ[18)] while higher dimension corrections have been evaluated in Ref 95, 96) using a $1/M_Q$ SVZ expansion of the $\langle\bar Q\, Q\rangle$

one-point function.

5. HIGHER GLUONIC CONDENSATES

A discussion of the triple gluon condensate $g^3\langle G^3\rangle$ has been anticipated in paragraph 3.1 in connection with instanton calculus and lattice Monte-Carlo simulations, which has lead to the phenomenological value in (3.14). The renormalization of the $g^3\langle G^3\rangle$ condensate has been studied in Ref.42) using the background field techniques outlined in Chapter 1. It has been shown that $\langle G^3\rangle$ does not mix with the class of dimension-six operators which survive after use of the equation of motion. Its anomalous dimension for $SU(N)_c \times SU(n)_F$ is :

$$\gamma_G = \frac{1}{6}(2 + 7N) \tag{3.42}$$

and the resulting RGI condensate is :

$$\left\langle \bar{O}_G \right\rangle = (\alpha_s)^{-(\gamma_G/\beta_1)} \cdot \langle O_G \rangle \tag{3.43}$$

where β_1 is the first coefficient of the β function. The log-dependence in (3.43) is always forgotten in the QSSR phenomenology, but fortunately its effects are harmless because of the small corrections due to the higher condensates in the sum rules analysis.

The O_6^D condensate in (3.3) reduces to the quark condensate after use of the equation of motion. One gets :

$$O_6^D \equiv g^2\left\langle D_\mu G^a_{\alpha\mu} D_\nu G^a_{\alpha\nu}\right\rangle = g^4 \left\langle \left(\sum_\psi \bar{\psi} \gamma_\mu \frac{\lambda_a}{2} \psi\right)^2 \right\rangle . \tag{3.44}$$

The dimension-eight gluon condensates can be expressed in terms of the eight quantities :

$$\Phi_1 = \left\langle \mathrm{Tr}\ (G^2)\ \mathrm{Tr}\ (G^2) \right\rangle$$

$$\Phi_2 = \left\langle \mathrm{Tr}\ \left(G_{\nu\mu}\ G^{\rho\mu}\right)\ \mathrm{Tr}\ G^{\nu}_{\tau}\ G^{\tau}_{\rho} \right\rangle$$

$$\Phi_3 = \left\langle \mathrm{Tr}\ G_{\nu\mu}\ G^{\tau\rho}\ \mathrm{Tr}\ G^{\nu\mu}\ G_{\tau\rho} \right\rangle$$

$$\Phi_4 = \left\langle \mathrm{Tr}\ G_{\nu\mu}\ G^{\tau\rho}\ \mathrm{Tr}\ G^{\nu}_{\tau}\ G^{\mu}_{\rho} \right\rangle$$

$$\Phi_5 = \left\langle \mathrm{Tr}\ \left(G_{\nu\mu}\ G^{\mu\rho}\ G_{\rho\tau}\ G^{\tau\nu}\right) \right\rangle$$

$$\Phi_6 = \left\langle \mathrm{Tr}\ \left(G_{\nu\mu}\ G^{\nu\mu}\ G_{\rho\tau}\ G^{\rho\tau}\right) \right\rangle$$

$$\Phi_7 = \left\langle \mathrm{Tr}\ \left(G_{\nu\mu}\ G^{\nu\rho}\ G^{\mu\tau}\ G_{\rho\tau}\right) \right\rangle$$

$$\Phi_8 = \left\langle \mathrm{Tr}\ G_{\nu\mu}\ G^{\rho\tau}\ G^{\nu\mu}\ G_{\rho\tau} \right\rangle \ , \qquad (3.45)$$

or their combinations.
Bagan et al[95] have studied constraints among these condensates by an explicit evaluation of the trace and using the symmetry properties of the colour indices. They derive for $N_c = 3$:

$$\Phi_5 + 2\,\Phi_7 = \Phi_2 + \frac{1}{2}\,\Phi_4 \ ,$$

$$\Phi_8 + 2\,\Phi_6 = \Phi_3 + \frac{1}{2}\,\Phi_1 \ , \qquad (3.46)$$

i.e. only six $\langle G^4 \rangle$ condensates are independent. Comparing their results with the ones derived from the factorization assumptions[97] (true in the large N limit) :

$$\Phi_1 = \langle G^2 \rangle^2\ \frac{1}{4}\left(1 + \frac{1}{3(N^2-1)}\right) \qquad \Phi_2 = \frac{\langle G^2 \rangle^2}{4}\left(\frac{1}{4} + \frac{1}{3(N^2-1)}\right)$$

$$\phi_3 = \frac{\langle G^2\rangle^2}{4}\left(\frac{1}{6} + \frac{7}{6}\frac{1}{(N^2-1)}\right) \qquad \phi_4 = \frac{\langle G^2\rangle^2}{4}\left(\frac{1}{12} + \frac{1}{2(N^2-1)}\right)$$

$$\phi_5 = \frac{\langle G^2\rangle^2}{4}\frac{1}{N}\left(\frac{1}{2} - \frac{1}{12(N^2-1)}\right) \qquad \phi_6 = \frac{\langle G^2\rangle^2}{4N}\left(\frac{7}{6} - \frac{1}{6(N^2-1)}\right)$$

$$\phi_7 = \frac{\langle G^2\rangle^2}{4N}\left(\frac{1}{3} - \frac{1}{4}\frac{2}{(N^2-1)}\right) \qquad \phi_8 = \frac{\langle G^2\rangle^2}{4N}\left(\frac{1}{3} - \frac{1}{N^2-1}\right), \qquad (3.47)$$

they realize that (3.47) satisfies (3.46) for $N = 3$. But as can be seen in the large N limit where one should consider the factorization result, the coefficients of ϕ_5 to ϕ_8 are not correct if one compares them with those deduced from (3.46). Therefore, the authors invalidate the factorization assumptions of the $\langle G^4\rangle$ condensates. Instead, the authors propose a set of relations but based on the $\frac{1}{M_Q}$ study of the one-point function of four-quark condensates, where they start from the factorization assumption of the four-quark condensates. However, the validity of such an assumption has also been questioned on phenomenological grounds [25,98,100,101]. Therefore, the accuracy of the relations of ϕ_1 to ϕ_8 with ϕ_2, which we shall quote, can become questionable :

$$\phi_1 = \frac{1}{4}\langle G^2\rangle^2$$

$$\phi_3 = -\frac{1}{16}\langle G^2\rangle^2 + 2\,\phi_2$$

$$\phi_4 = -\frac{1}{32}\langle G^2\rangle^2 + \phi_2$$

$$\phi_5 = -\frac{1}{192}\langle G^2\rangle^2 + \frac{1}{2}\,\phi_2$$

$$\phi_6 = \frac{1}{12} \langle G^2 \rangle^2$$

$$\phi_7 = -\frac{1}{192} \langle G^2 \rangle^2 + \frac{1}{2} \phi_2$$

$$\phi_8 = -\frac{5}{48} \langle G^2 \rangle^2 + 2 \phi_2 \ . \qquad (3.48)$$

In particular, if (3.48) is a good approximation, it indicates that the ϕ_5 condensate, which is the leading effect in charmonium sum rules, has been overestimated by the factorization assumption.

6. THE MIXED CONDENSATE

The renormalization of the mixed $O_4 \equiv g\left\langle \bar{\psi}\, \sigma^{\mu\nu} \frac{\lambda_a}{2} \psi\, G^a_{\mu\nu} \right\rangle$ condensate has been studied in Ref. 42), where it has been shown that O_4 mixes under renormalization with the other operators :

$$O_1 \equiv i\, m^2 \left\langle \bar{\psi} \psi \right\rangle \quad ; \quad O_2 \equiv -\frac{1}{4}\, i\, m \langle G^2 \rangle \quad ;$$

$$O_3 \equiv -m\, \bar{\psi}\, (\hat{D} + im)\, \psi \qquad (3.49)$$

and the only RGI combination which one can form is :

$$\left\langle \bar{O}_5 \right\rangle = (\alpha_s)^{\gamma_5 / -\beta_1} \cdot \langle O_4 + x\, O_1 + y\, O_2 \rangle \ , \qquad (3.50a)$$

where for $SU(3)_c$:

$$x = -\frac{1944}{315} \quad ; \quad y = -\frac{72}{63} \quad ; \quad \gamma_5 = \frac{1}{3} \ , \qquad (3.50b)$$

i.e. working only with $\langle O_4 \rangle$ is valid to leading order in the quark mass. The O_4 one-point function has also been studied, again using a heavy quark mass expansion. The expression would read :

$$m g \left\langle \bar{\psi} \sigma^{\mu\nu} \frac{\lambda_a}{2} \psi G^a_{\mu\nu} \right\rangle = -\frac{5}{24}\left(\frac{\alpha_s}{\pi}\right) \cdot g\langle G^3 \rangle - \frac{\left\langle \alpha_s G^2 \right\rangle}{\pi} m^2 \log\left(\frac{\nu^2}{m^2} - 1\right), \quad (3.51)$$

where various authors[95,102] agree with the first simple local contribution but Bagan et al [95] object to the existence of the last term. They claim that this term comes from a diagram which cannot represent a quark condensation. In the chiral limit, which is a good approximation for light-quark systems, this discrepancy is absent. However, in the case of massive quarks, one should also take into account the mixing of $\langle O_4 \rangle$ with the operators in (3.49). Eq. (3.51) might just reflect the non-RGI of the O_4 operator ! Despite this problem one can fix the sign of the triple-gluon condensate in the Euclidian region from (3.51) as :

$$g \langle G^3 \rangle > 0 , \quad (3.52)$$

which is already a non-trivial constraint. There is a much more phenomenological parametrization of the mixed condensate for light quarks. This is[103] :

$$\left\langle g \bar{\psi} G \psi \right\rangle \equiv g \left\langle \bar{\psi} \sigma^{\mu\nu} \frac{\lambda_a}{2} \psi G^a_{\mu\nu} \right\rangle = M_0^2 \left\langle \bar{\psi} \psi \right\rangle , \quad (3.53)$$

where M_0^2 is a parameter fixed from the data. Baryon sum rules[99b,103] suggest a value around 1 GeV^2 where the uncertainties are related to the choice of the interpolating baryon operators. Use of the quark equation of motion which transforms (3.53) into $\left\langle \bar{\psi} D_\alpha D^\alpha \psi \right\rangle$ where D_α is the covariant derivative and the assumption that the average off-shellness of the vacuum gluons and quarks is equal (!) lead to[97]:

$$M_0^2 \equiv \left\langle \bar{\psi} D^2 \psi \right\rangle / \left\langle \bar{\psi} \psi \right\rangle \simeq \langle DG\, DG \rangle / \langle G^2 \rangle \simeq 0.3 \text{ GeV}^2 , \quad (3.54)$$

which should be a very crude estimate.
However, a recent analysis of the heavy-light quark systems spectra which are sensitive to the mixed condensate gives a strong constraint[104]:

$$M_0^2 = (0.80 \pm 0.01)\ \mathrm{GeV}^2 \ , \tag{3.55}$$

fixed by the observed B and B^* masses. A lattice Monte-Carlo calculation gives[105] a value around 1 GeV^2 but it is always difficult to appreciate the accuracy of the lattice results involving fermion fields. The mixed condensate has also been related to the other condensates by a Cauchy-Schwarz-like inequality[106]:

$$\langle g\bar\psi G\psi\rangle^2 \leqslant 16\pi \left\langle \alpha_s G^2\right\rangle \left|\left\langle \bar\psi \gamma_5 \psi \bar\psi \gamma_5 \psi\right\rangle\right| \ , \tag{3.56}$$

but the extraction of the value of M_0^2 from (3.56) is affected by the uncertainty on the value of the gluon condensate and of the four-quark condensate.

7. THE FOUR-QUARK CONDENSATE

The renormalization of the four quark $\left\langle \bar\psi \Gamma_1 \psi \bar\psi \Gamma_2 \psi\right\rangle$ condensate has also been studied in Ref 42) (see also Ref. 107)). In the chiral limit, one can form nine independent dimension-six condensates (in addition to $g\langle G^3\rangle$ which does not mix with the others).
These operators are :

$$O_2 = g^2\ \bar\psi \psi \bar\psi \psi$$

$$O_3 = g^2 \left(4\ \bar\psi\psi\ \bar\psi\psi + 11\ \bar\psi \gamma^\mu \psi \bar\psi \gamma_\mu \psi\right)$$

$$O_4 = g^2 \left(6\ \bar\psi\psi\ \bar\psi\psi + \frac{11}{2}\ \bar\psi \sigma^{\mu\nu} \psi \bar\psi \sigma_{\mu\nu} \psi\right)$$

$$O_5 = g^2\left(4\ \bar\psi\psi\ \bar\psi\psi - 11\ \bar\psi\ \gamma^\mu\ \gamma^5\ \psi\ \bar\psi\ \gamma_\mu\ \gamma_5\ \psi\right)$$

$$O_6 = g^2\left(\bar\psi\psi\ \bar\psi\psi + 11\ \bar\psi\ \gamma_5\ \psi\ \bar\psi\ \gamma_5\ \psi\right)$$

$$O_7 = -\frac{1}{2}\left(\partial_\mu\ G_{\mu\nu,a} + g\ f_{abc}\ A^\mu_b\ G^c_{\mu\nu}\right)\cdot\left(\partial^\rho\ G^\nu_{\rho,a} + g\ f_{amn}\ A^\rho_m\ G^\nu_{\rho,n}\right)$$

$$-\frac{1}{2}\ g\left(\partial_\mu\ G^{\mu\nu}_a + g\ f_{abc}\ A^\mu_b\ G_{\mu\nu,c}\right)\bar\psi\ \gamma^\nu\ \frac{\lambda^a}{2}\ \psi$$

$$O_8 = i\ \bar\psi\ \hat{D}^3\ \psi$$

$$O_9 = ig\ \bar\psi\ \sigma^{\mu\nu}\ \frac{\lambda^a}{2}\ G_{\mu\nu,a}\ \hat{D}\ \psi$$

$$O_{10} = g\left(\partial^\mu\ G_{\mu\nu,a} + g\ f_{abc}\ A^\mu_b\ G_{\mu\nu,c}\right)\bar\psi\ \frac{\lambda_a}{2}\ \gamma^\nu\ \psi$$

$$-\frac{1}{2}\ g^2\ \left[\bar\psi\psi\ \bar\psi\psi - \frac{1}{2}\left(1 - \frac{2}{N}\right)\bar\psi\ \gamma^\mu\psi\ \bar\psi\ \gamma_\mu\psi\right.$$

$$\left. -\frac{1}{2}\ \bar\psi\ \gamma^\mu\gamma^5\ \psi\ \bar\psi\ \gamma_\mu\gamma_5\psi - \bar\psi\ \gamma_5\psi\ \bar\psi\ \gamma_5\psi\right] , \qquad (3.57)$$

where summation over colour indices is understood. O_3 to O_6 are written in such a way that their vacuum condensates vanish if one uses the vacuum saturation assumption of SVZ[18]:

$$\left\langle\bar\psi\ \Gamma_1\psi\ \bar\psi\ \Gamma_2\psi\right\rangle = \frac{1}{N^2}\ [\mathrm{Tr}\ \Gamma_1\ \mathrm{Tr}\ \Gamma_2 - \mathrm{Tr}\ (\Gamma_1\Gamma_2)]\left\langle\bar\psi\ \psi\right\rangle^2 . \qquad (3.58)$$

The operators O_7 to O_{10} vanish after use of the equations of motion. An explicit calculation using the background field method shows that O_2

mixes with all other operators. The situation becomes much better in the large N limit. The only remaining renormalization constant is the diagonal one, to which corresponds the anomalous dimension :

$$\gamma_\phi \ (N \to \infty) \simeq \frac{143N}{33} \quad , \tag{3.59a}$$

and the corresponding RGI is

$$\langle \bar{O}_2 \rangle = (\alpha_s)^{(\gamma_\phi/-\beta_1)} \cdot \langle O_2 \rangle \ . \tag{3.59b}$$

The four-quark condensates have been estimated from the vector[25,100,101], axial[98], baryons[99] and from the observed B and B^* masses[104]. All of the above analysis indicates a violation of the factorization assumption in (3.54) by a factor larger than or equal to two, but more probably not far from two if one expects that the sum rules approach is able to reproduce the properties of the known lowest ground states of hadrons, as we shall see later.

CHAPTER 4

METHODS FOR THE EVALUATION OF THE WILSON COEFFICIENTS OF CONDENSATES

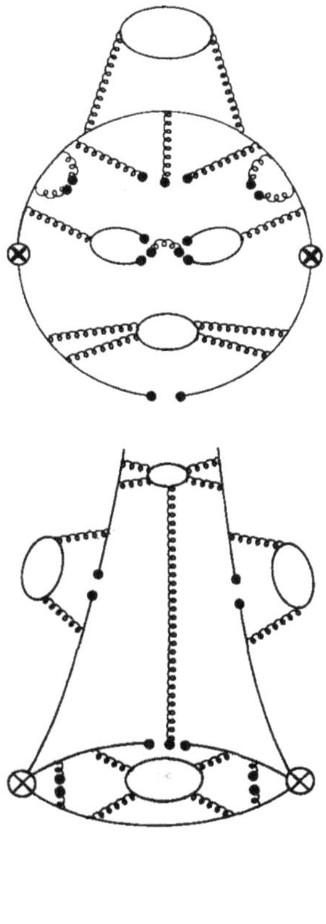

robot-type diagrams in the SVZ-expansion.

There are nice expositions of these methods in the existing literature [96,97,29,47b,108-115]. We shall not try to supplement them but, for pedagogical reasons, we might even repeat some of the discussions given in the literature. Let us recall that the SVZ expansion provides a parametrization of the vacuum structure of *QCD* in terms of the condensates where the vacuum fields play the role of external fields. Therefore, the evaluation of a Green's function of some local colourless currents is reduced to its evaluation in external gluon and/or quark fields, assuming that the field is weak. Weak field here means that its average intensity is smaller than the characteristic momentum value, i.e. in this case the expansion in a power series makes sense. The whole procedure is formalized within the framework of the Wilson's operator expansion. The Wilson coefficients C_{2n} are usually evaluated with Feynman diagram techniques. Different methods exist in the literature :

1. THE FOCK-SCHWINGER FIXED-POINT TECHNIQUE

This is the most used and probably the most convenient method. It is based on the Fock-Schwinger choice of gauge[116,117]:

$$(x - x_o)\, A^a_\mu(x) = 0 \ , \qquad (4.1)$$

mostly used in QED. $A^a_\mu(x)$ is the four-potential and x_o is an arbitrary choice of coordinate which plays the role of gauge. As (4.1) explicitly breaks translational invariance, its restoration for gauge-invariant quantities provides a double check of the validity of the calculation, i.e. one should expect that terms dependent on x_o cancel each other. For convenience, in the algebraic manipulations, one has, unfortunately, to take $x_o = 0$ from the very beginning. Therefore, one can express the gauge field $A^a_\mu(x)$ in terms of the gluon-strength tensor $G^a_{\mu\nu}$ as :

$$A^a_\mu(x) = \int_0^1 d\alpha \ . \ \alpha\, G^a_{\rho\mu}(\alpha x)\, x^\rho \ . \qquad (4.2)$$

Eq (4.2) can be derived from the identity :

$$A_\mu(z) = \frac{\partial}{\partial z_\mu}(A_\rho(z)\ z_\rho) - z_\rho \frac{\partial A_\rho(z)}{\partial z_\mu} , \qquad (4.3a)$$

where from Eq. (4.1) at $x_o = 0$:

$$- z_\rho \frac{\partial A_\rho}{\partial z_\mu} = - z_\rho\ G_{\mu\rho} - z_\rho \frac{\partial A_\mu(z)}{\partial z_\rho} , \qquad (4.3b)$$

one can deduce from (4.3a) and (4.3b) :

$$A_\mu(z) + z_\rho \frac{\partial A_\mu(z)}{\partial z_\rho} = z_\rho\ G_{\rho\mu}(z) . \qquad (4.3c)$$

The substitution of $z \equiv \alpha x$ in (4.3c) immediately shows that (4.3c) is a full derivative :

$$\frac{d}{d\alpha}[\alpha\ A_\mu\ (\alpha x)] , \qquad (4.3d)$$

which gives (4.2) after integration. A Taylor expansion of $G_{\rho\mu}$ *in* (4.2) around $x^\mu = 0$ gives :

$$A^a_\mu(x) = \sum_{x=0}^{\infty} \frac{1}{n!(n+2)} \cdot x^\rho\ x^{\nu_1}\ldots x^{\nu_n}\ \partial_{\nu_1}\ldots\partial_{\nu_n}\ G^a_{\rho\mu}(0) . \qquad (4.4a)$$

Using $A_1(0) = 0$, the gauge condition implies a relation between the ordinary and covariant derivatives :

$$x_{\nu_1}\ \partial_{\nu_1}\ G_{\rho\mu}(0) = x^{\nu_1}\left[D_{\nu_1}(0),\ G_{\rho\mu}(0)\right] . \qquad (4.4b)$$

In the same way, using $x^{\nu_1} x^{\nu_2} \partial_{\nu_1} A_{\nu_2}(0) = 0$, one has :

$$x_{\nu_1} x_{\nu_2} \partial_{\nu_1} \partial_{\nu_2} G^a_{\rho\mu}(0) = x^{\nu_1} x^{\nu_2} \left[D_{\nu_1}(0), \left[D_{\nu_2}(0), D_{\rho\mu}(0)\right]\right] \qquad (4.4c)$$

and so on. Finally, we obtain the very useful formula in terms of covariant derivatives :

$$A_\mu(x) = \sum_{n=0} \frac{1}{n!} \frac{1}{(n+2)} x^\rho x^{\nu_1} \ldots x^{\nu_n} \left[D_{\nu_1}(0), \left[D_{\nu_2}(0), \left[\ldots \left[D_{\nu_n}, G_{\rho\mu}(0)\right] \ldots\right]\right]\right]. \qquad (4.5)$$

From (4.5), one can immediately form the gluon condensate :

$$\begin{aligned} A_\mu(x)\, A_\nu(y) &= \frac{1}{4} x^\lambda x^\ell G_{\lambda\mu} G_{\rho\nu} + \ldots \\ &= \frac{1}{4d(d-1)} x^\lambda y^\ell \left[g_{\lambda\ell} g_{\mu\nu} - g_{\lambda\nu} g_{\mu\rho}\right] . G^{\alpha\beta} G_{\alpha\beta} \\ &+ \ldots \end{aligned} \qquad (4.6)$$

where $d \equiv 4-\epsilon$ is the space-time dimension. Analogous arguments give the Taylor expansion of the quark fields :

$$\psi(x) = \sum_n \frac{1}{n!} x^{\nu_1} \ldots x^{\nu_n} D_{\nu_1}(0) \ldots D_{\nu_n}(0)\, \psi , \qquad (4.7a)$$

$$\bar{\psi}(x) = \sum_n \frac{1}{n!} x^{\nu_1} \ldots x^{\nu_n} \bar{\psi}(0)\, D^+_{\nu_1} \ldots D^+_{\nu_n}(0) , \qquad (4.7b)$$

with :

$$\bar{\psi}(0)\, \partial^+_{\nu_1} = \partial_{\nu_1} \bar{\psi} . \qquad (4.7c)$$

From (4.7), one can form the quark condensates for $SU(N)_c$

$$\left\langle \bar{\psi}^{F}_{i\alpha}(x)\ \psi^{F'}_{j\beta}(0) \right\rangle = \frac{1}{4N}\ \delta_{FF'}\ \delta_{\alpha\beta} \left\{ \left[\delta_{ij} + \frac{i}{4}\ m_F\ x^{\mu}\ (\gamma_{\mu})_{ij} \right] \left\langle \bar{\psi}\ \psi \right\rangle \right.$$

$$- \frac{i}{16}\ x^2 \left(\delta_{ij} + \frac{i}{6}\ m_F\ x^{\mu}\ (\gamma_{\mu})_{ij} \right) \left\langle \bar{\psi}\ \sigma^{\mu\nu}\ \frac{\lambda_a}{2}\ G^a_{\mu\nu}\ \psi \right\rangle$$

$$\left. + \frac{i}{288}\ x^2\ x^{\mu}\ (\gamma_{\mu})_{ij}\ g^2 \left\langle \bar{\psi}\ \gamma^{\ell}\ \frac{\lambda_a}{2}\ \psi \sum_f \bar{\psi}_f\ \gamma_{\ell}\ \frac{\lambda_a}{2}\ \psi_f \right\rangle \right\} \quad . \qquad (4.8)$$

Eq. (4.8) indicates that the "propagation" of the $\left\langle \bar{\psi}\ \psi \right\rangle$ condensate induces contributions due to the mixed and four-quark condensates. Thus great care must be taken in evaluating the Wilson coefficients of high-dimension condensates.

One can also form from (4.5) and (4.7) the mixed condensate :

$$\left\langle \bar{\psi}_i(x)\ A_\rho(z)\ \psi_j(0) \right\rangle = \frac{1}{2}\ z^{\mu} \left\langle \bar{\psi}_i\ G_{\mu\rho}\ \psi_j \right\rangle + \frac{1}{2}\ x^{\nu}\ z^{\mu}\ .$$

$$. \left\langle \bar{\psi}_i\ D^{+}_{\nu}\ G_{\mu\rho}\ \psi_j \right\rangle + \ldots$$

$$= \frac{z^{\mu}}{96} \left\{ \left[\sigma_{\mu\rho} - \frac{m_F}{2}\ (x_\mu\ x_\rho - x_\rho x_\mu) + i\ \frac{m_F}{2}\ x^{\nu}\ \sigma_{\mu\rho}\ x_\nu \right]_{ij} \right. .$$

$$. \left\langle \bar{\psi}\ \sigma_{\tau k}\ G^{\tau k}\ \psi \right\rangle + \left[i \left(-\frac{2}{3}\ z_\mu\ \gamma_\rho + \frac{2}{3}\ z_\rho\ \gamma_\mu \right) + \frac{1}{2}\ x^{\nu}\ \gamma_\nu\ \sigma_{\mu\rho} \right]_{ij} .$$

$$\left. .\ g^2 \left\langle \bar{\psi}\ \gamma^{\alpha}\ \frac{\lambda_a}{2}\ \psi \sum_f \bar{\psi}_f \gamma_{\alpha} \psi_f \right\rangle \right\} \quad . \qquad (4.9)$$

Again (4.9) indicates that the "propagation" of the mixed condensate induces a quartic condensate.

For calculational purposes, one also likes to have the expression of the quark propagator in momentum space. For the non-perturbative effects, propagators describe a propagation in external gluon and/or quark fields. For external gluon fields, one can solve the Dirac equation :

$$\left(i\, \partial_\mu\, \gamma^\mu + g\, A_\mu(x) - m\right) S(x,y) = \delta^{(4)}\,(x-y)\ , \tag{4.10}$$

where $A_\mu \equiv \lambda_a/2\ A^a_\mu(x)$. In the present case, $A_\mu(x)$ is small compared to the scale $(x-y)$. Within this assumption, one can represent $S(x,y)$ in the Taylor series :

$$\begin{aligned} i\, S(x,y) &= i\, S^{(o)}(x-y) + g \int d^4z\ i\, S^{(o)}(x-z)\ i\, \hat{A}(z)\ i\, S^{(o)}(z-y) \\ &+ g^2 \int d^4z'\ d^4z\ i\, S^{(o)}(x-z')\ i\, \hat{A}(z')\ i\, S^{(o)}(z'-z)\ i\, \hat{A}(z)\ i\, S^{(o)}(z-y) \\ &+ \dots \end{aligned} \tag{4.11}$$

Taking its Fourier transform, one gets :

$$\begin{aligned} S(p) &= \int S(x,o)\ e^{ipx}\ d^4x \\ S^+(p) &= \int S(o,x)\ e^{-ipx}\ d^4x \end{aligned} \tag{4.12}$$

which graphically is :

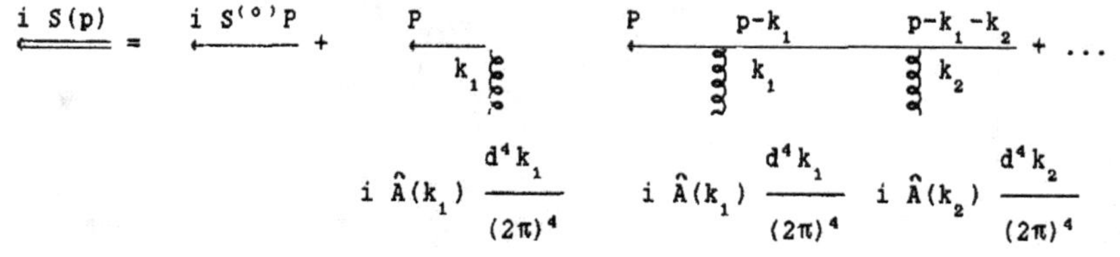

(4.13)

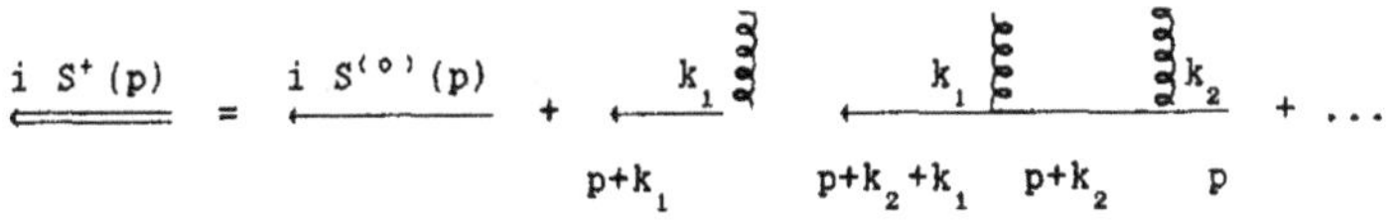
i S⁺(p)
=
i S⁽ᵒ⁾(p)
+
k₁
p+k₁
k₁
k₂
p+k₂+k₁
p+k₂
p
+ ...

where

$$A_\mu(k) = \int A_\mu(z)\, e^{ikz}\, d^4z = -\frac{i(2\pi)^4}{2} G_{\rho\mu}(o) \frac{\partial}{\partial k_\rho} \delta^{(4)}(k)$$

$$+ (-i)^2 \frac{(2\pi)^4}{3} (D_\alpha G_{\rho\mu}(o)) \frac{\partial^2}{\partial k_\rho\, \partial k_\alpha} \delta^{(4)}(k) + \ldots \qquad (4.14)$$

with a trivial integration over a δ function.

However, taking $y = 0$ in the very beginning as in (4.12) cannot be done if one has to study a correlator associated with current with derivative, such as $\bar\psi \overleftrightarrow{D} \psi$ ... In this case, one should perform first the derivative of the quark propagator in external fields :

$$\frac{\partial}{\partial y_\alpha} i\, S(o,y) = \frac{\partial}{\partial y_\alpha} i\, S^{(o)}(x-y) + g \int d^4z\, i\, S^{(o)}(x-z)\, i\, \hat{A}(z) \frac{\partial}{\partial y_\alpha} \cdot$$

$$i\, S^{(o)}(z-y) + \ldots$$

$$= -\frac{\partial}{\partial x_\alpha} i\, S^{(o)}(x-y) + g \int d^4z\, i\, S^{(o)}(x-z)\, i\, \hat{A}(z) \left(-\frac{\partial}{\partial z_\alpha}\right) \cdot$$

$$i\, S^{(o)}(z-y) + \ldots \qquad (4.15)$$

and then put $y = 0$ at the end. The corresponding Fourier transform is :

$$\int d^4x\, e^{ipx} \left[\frac{\partial}{\partial y_\alpha} i\, S(x,y)\right]_{y=o} = i\, P_\alpha\, i\, S^{(o)}(p)$$

$$+ g \int i\, S^{(0)}(p)\; i\, \hat{A}(k) \frac{d^4k}{(2\pi)^4}\; i(p-k)_\alpha\; i\, S^{(0)}(p-k) + \ldots \qquad (4.16)$$

Within the above materials and the properties of condensates given in Chapter 3, the reader is already able to evaluate the Wilson coefficients. Before applying these above rules from an explicit example, let me very briefly discuss alternative methods.

2. THE PLANE WAVE METHOD

This method exploits the fact that Wilson's expansion is an operator identity, i.e. one can single out a given operator by sandwiching it between appropriate states. For instance let the two-point correlator associated with the quark current

$$J^\Gamma(x) = \bar{\psi}\, \Gamma\, \psi \ , \qquad (4.17)$$

which possesses the Wilson expansion (omitting Lorentz indices) in Eq. (3.2) :

$$\mathrm{Re}\,\Pi(q^2) = C_{\mathbb{1}}\mathbb{1} + C_m\, \bar{\psi}\psi + C_G\, G^a_{\alpha\beta}\, G^{\alpha\beta}_a + D_G \left\{ G_{\alpha\delta} G^\alpha_\beta q^\delta q^\beta = \frac{1}{4} q^2 G^{\alpha\beta}_a\, G^a_{\alpha\beta} \right\}. \quad (4.18)$$

The first unit term corresponds to the usual perturbative calculation which one obtains by sandwiching the correlator between the vacuums. The next term is obtained by sandwiching the correlator between one-quark states and corresponds to the quark-current scattering diagram :

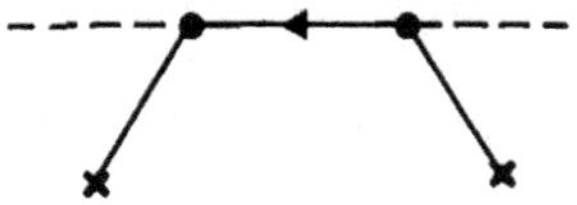

- - - - - current

———— quark (4.19)

The Wilson coefficient C_G can be obtained by sandwiching the correlator between one-gluon states. The problem can be reduced to a calculation of the forward gluon scattering amplitude on a colour-singlet current. From Lorentz invariance, this amplitude can be decomposed as:

$$T^{\mu\nu}(q,k) = i \int d^4x \, e^{iqx} \left\langle k, \mu \left| \mathbb{T} J^{\Gamma}(x) \, (J^{\Gamma}(o))^{\dagger} \right| k, \nu \right\rangle$$

$$= F_1^{\mu\nu} \, C(q,k) + F_2^{\mu\nu} \, D(q,k) \qquad (4.20)$$

where :

$$F_1^{\mu\nu} = 4 \, (k^2 g^{\mu\nu} - k^{\mu}k^{\nu}) \equiv \left\langle k, \mu \left| G^a_{\alpha\beta} \, G_a^{\alpha\beta} \right| k, \nu \right\rangle$$

$$F_2^{\mu\nu} = 2\left(k^2 q^{\mu} q^{\nu} - (k.q) \, (q^{\mu}k^{\nu} + q^{\nu}k^{\mu}) + g^{\mu\nu}(k.q)^2\right) - q^2(k^2 g^{\mu\nu} - k^{\mu}k^{\nu})$$

$$\equiv \left\langle k, \mu \left| G_{\alpha\delta} \, G^{\alpha\beta} \, q^{\delta} q^{\beta} - \frac{1}{4} q^2 \, G^a_{\alpha\beta} \, G_a^{\alpha\beta} \right| k, \nu \right\rangle \qquad (4.21)$$

and corresponds to the diagrams :

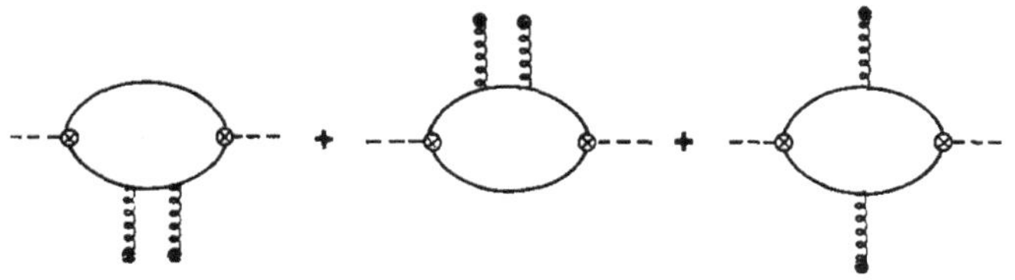

. (4.22)

A comparison of (4.18) and (4.20) gives :

$$D_G(q^2) = C_G(q, \, k) \Big|_{k_\alpha = o} \qquad (4.23)$$

From a practical point of view, the plane-wave method is convenient in the case where we have external weak quark fields as in (4.19). For many "weak" external gluon fields, the extraction of a particular operator from various possible ones with the same dimensions becomes very difficult. This is (in some sense) the major inconvenience of this method.

3. THE SCHWINGER OPERATOR "FORMAL" METHOD FOR HEAVY QUARKS

This method has been nicely discussed in Ref. 96). It mainly consists of the calculation of the Dirac operator Det $\left\| i\, D_\mu\, \gamma^\mu - x \right\|$ in an external gluon field A^μ, which enters as a pre-exponential factor in the vacuum-to-vacuum transition :

$$\langle O_1 | O_2 \rangle = \int D\, A_\mu\, D\bar{\psi}\, D\psi \exp i \left\{ \int d^4x \left[-\frac{1}{4} G^a_{\mu\nu} G^{\mu\nu}_a + \bar{\psi}\,(i\,\gamma_\mu D_\mu - m)\,\psi \right] \right\}$$

$$= \int D\, A_\mu \,\text{Det} \left\| i\,\gamma_\mu\, D^\mu - m \right\| \exp \left\{ \frac{i}{4} \int d^4x\, G^a_{\mu\nu}\, G^{\mu\nu}_a \right\}$$

$$= \int d\, A_\mu \exp \left\{ i \int d^4x \left[-\frac{1}{4} G^a_{\mu\nu}\, G^{\mu\nu}_a - i \log \text{Det} \left\| \hat{P} - m \right\| \right] \right\} , \quad (4.24)$$

where the operator P is defined as :

$$\left\langle x \left| \hat{P}^{ab}_{\alpha\beta} \right| y \right\rangle = (\gamma_\mu)_{\alpha\beta} \left[i\, \delta^{ab} \frac{\partial}{\partial x_\mu} + g(\gamma_c)_{ab}\, A^c_\mu(x) \right] \delta(x-y) . \quad (4.25)$$

In order to regularize the divergences appearing in the evaluation of the determinant in a gauge invariant way, one can introduce an auxiliary Pauli-Villars fermion field quantized as a boson. Then, the extra-term added into the classical action is :

$$\Delta S = - i \log D_{Reg} = - i \operatorname{Tr} \log \left\| \frac{(\hat{P}-m)_A \left(\hat{P}-\Lambda_{uv}\right)_{A=0}}{(\hat{P}-m)_{A=0} \left(\hat{P}-\Lambda_{uv}\right)_A} \right\| \tag{4.26}$$

The aim is now to give a series expansion of ΔS in terms of $\left(G/m^2\right)$ where $G^a_{\mu\nu} \ll m^2$. Λ_{uv} is an ultraviolet mass regulator. This can be done by working with :

$$\begin{aligned} P &\equiv - \frac{1}{m} \frac{d}{dm} \Delta S \\ &= - \frac{i}{m} \left\{ \operatorname{Tr} \left[\frac{1}{(\hat{P}-m)(\hat{P}+m)} (\hat{P}+m) \right]_A - \operatorname{Tr} [\quad]_{A=0} \right\} \end{aligned} \tag{4.27}$$

which, after eliminating the trace of odd numbers of γ matrices, becomes :

$$P = -i \operatorname{Tr} \left[\frac{1}{p^2 - m^2 + ig \dfrac{\sigma_{\mu\nu}}{Z} G^{\mu\nu}} - \frac{1}{(p^2 - m^2)_{A=0}} \right] \tag{4.28}$$

and with the help of :

$$[P_\mu, P_\nu] = ig \frac{\lambda_a}{2} G^a_{\mu\nu} \equiv ig\, G_{\mu\nu} \tag{4.29a}$$

where :

$$\sigma_{\mu\nu} = \frac{1}{2} (\gamma_\mu \gamma_\nu - \gamma_\nu \gamma_\mu) \quad . \tag{4.29b}$$

Now one can expand (4.28) in terms of the magnetic term σG. It becomes :

$$P \equiv \sum_{n=1} P_n$$

$$= - i \, Tr \left\{ \frac{1}{(P^2 - m^2)_A} - \frac{1}{(P^2 - m^2)_{A=0}} \right\}$$

$$- i \, Tr \frac{1}{(P^2 - m^2)^2} O$$

$$- i \, Tr \frac{1}{(P^2 - m^2)^2} O \frac{1}{(P^2 - m^2)} \left(- \frac{ig}{2} \sigma G \right)$$

$$- i \, Tr \frac{1}{(P^2 - m^2)^2} O \, . \, \frac{1}{(P^2 - m^2)} \, . \, O + \ldots \, \mathcal{O}(G^5) \, , \tag{4.30}$$

where :

$$O \equiv \left(- \frac{ig}{2} \sigma G \right) \frac{1}{(P^2 - m^2)} \left(- \frac{ig}{2} \sigma G \right) . \tag{4.31}$$

The first term corresponds to the effective action of a scalar particle and represents a scalar loop in external fields :

$$\frac{d}{dm^2} \Delta S_s = - i \, Tr \left\{ \frac{1}{(P^2 - m^2)_A} - \frac{1}{(P^2 - m^2)_{A=0}} \right\} . \tag{4.32}$$

By noting that this regularized expression does not change under the shift $p \to p+q$ in the same way as convergent integrals over space-time momentum are invariant under the analogous shift of momentum, one can rewrite (4.32) as :

$$\frac{d}{dm^2}\Delta S_s = -i\,\mathrm{Tr}\left\{\frac{1}{[(p+q)^2 - m^2]_A} - \frac{1}{[(p+q)^2 - m^2]_{A=0}}\right\} , \qquad (4.33)$$

which one can expand in q assuming that q is small. However, (4.32) is independent of q. This means that coefficients of q should vanish identically. Considering, for instance, the coefficient of q^2, one gets the sum rule :

$$\mathrm{Tr}\left\{\frac{1}{(p^2-m^2)_A^2} - \frac{1}{(p^2-m^2)_{A=0}^2}\right\} = \mathrm{Tr}\left\{\frac{1}{(p^2-m^2)^2}P_\mu\frac{1}{(p^2-m^2)}P_\mu\Big|_A - \frac{1}{(p^2-m^2)^2}P_\mu\frac{1}{(p^2-m^2)}P_\mu\Big|_{A=0}\right\} , \qquad (4.34)$$

which one can rewrite as :

$$\mathrm{Tr}\left\{\frac{1}{(p^2-m^2)_A^3} - \frac{1}{(p^2-m^2)_{A=0}}\right\} = -\frac{1}{m^2}\mathrm{Tr}\left\{\frac{1}{(p^2-m^2)^4}\left[p^2, p_\mu\right] \cdot \left[p^\mu - [p^2, p^\mu]\frac{1}{(p^2-m^2)} + \ldots\right]\right\} \qquad (4.35)$$

after use of the relations :

$$\frac{1}{(p^2-m^2)^2}p^\mu\frac{1}{(p^2-m^2)}p^\mu = \frac{1}{(p^2-m^2)^3}p^2 + \frac{1}{(p^2-m^2)^3}[p^2,p^\mu].\frac{1}{(p^2-m^2)}p^\mu, \qquad (4.36a)$$

$$\frac{1}{(p^2-m^2)}\,p^\mu \equiv \frac{1}{p^2-m^2}\,P^\mu\,(p^2-m^2)\,\frac{1}{p^2-m^2}$$

$$= P_\mu\,\frac{1}{p^2-m^2} - [p^2,\,p^\mu]\,\frac{1}{(p^2-m^2)^2} + \left[p^2,\,\left[p^2,p_\mu\right]\right]\frac{1}{(p^2-m^2)^3} \qquad (4.36b)$$

Now let us concentrate on the evaluation of (4.35). In this case and later on we shall use the properties :

$$[P_\alpha,G_{\alpha\mu}] = i\,D_\alpha\,G_{\alpha\mu} = 0\ , \qquad (4.37a)$$

$$G_{\mu\nu}\,D^4\,G_{\mu\nu} = D^2\,G_{\mu\nu}\,D^2\,G_{\mu\nu}\ , \qquad (4.37b)$$

from the equation of motion (no sources for external fields), and the Bianchi identity

$$D_\alpha\,G_{\mu\nu} + D_\nu\,G_{\alpha\mu} + D_\mu\,G_{\nu\alpha} = 0\ . \qquad (4.37c)$$

They imply the relations :

$$\begin{aligned}\left(D^2\,G_{\mu\nu}\right)(D_\alpha\,D_\alpha\,G_{\mu\nu}) &= 2\left(D^2\,G_{\mu\nu}\right)(D_\alpha\,D_\mu\,G_{\alpha\nu})\\ &= 2\left(D^2\,G_{\mu\nu}\right)[D_\alpha,D_\mu]\,G_{\alpha\nu}\\ &= -\,2ig\left(D^2\,G_{\mu\nu}\right)(G_{\alpha\mu}\,G_{\alpha\nu})\\ &= -\,4g^2\,(G_{\alpha\mu}\,G_{\alpha\nu})^2 \qquad (4.37d)\end{aligned}$$

and :

$$[p^2,p^\mu] = \{p_\alpha,\,[p_\alpha,p_\mu]\} = ig\,\{p_\alpha,\,G_{\alpha\mu}\} = -\,2ig\,G_{\mu\alpha}\,p_\alpha \qquad (4.37e)$$

$$\left[p^2,p_\mu\right]p_\mu - 2ig\,G_{\alpha\mu}\,P_\alpha\,p_\mu = -\,g^2\,G_{\alpha\mu}\,G_{\alpha\mu} \qquad (4.37f)$$

Therefore, the first term of (4.35) can be written as :

$$I_o = \frac{g^2}{m^2} \, Tr \left\{ \frac{1}{(p^2-m^2)^4} \, G_{\alpha\mu} \, G^{\alpha\mu} \right\} + \ldots \tag{4.38}$$

To order G^2, (4.38) reads :

$$\begin{aligned} Tr \frac{1}{(p^2-m^2)^4} G^2 &= Tr_c \int d^4x \, d^4y \left\langle x \left| \frac{1}{(p^2-m^2)^4_{A=0}} \right| y \right\rangle \left\langle y \left| G^2 \right| x \right\rangle \\ &= Tr_c \int d^4x \left\langle x \left| \frac{1}{(p^2-m^2)^4_{A=0}} \right| x \right\rangle G^2(x) \\ &= \int \frac{d^4p}{(2\pi)^4} \frac{1}{(p^2-m^2)^4} \, Tr_c \, G^2 \end{aligned} \tag{4.39}$$

where Tr_c means trace over colour indices.
Then :

$$I_o = \frac{g^2}{m^6} \frac{i}{2^5 . 3 . \pi^2} \, Tr_c \, G_{\alpha\mu} \, G_{\alpha\mu} + \ldots \tag{4.40}$$

The evaluation of the higher contribution to I_o can be done by the shift of momentum $p \to p+q$. The coefficient of the q^2 term induces the sum rule :

$$Tr \frac{1}{(p^2-m^2)^3} F(G) = \frac{1}{2^5 \pi^2 i} \frac{1}{m^2} \, Tr \, F(G) - g^2 \, Tr \left\{ \frac{1}{(p^2-m^2)^5} \, G^2 \, F(G) \right\} \tag{4.41}$$

where $F(G)$ is an arbitrary function of the field G. The first term comes from the regulator term at $A = 0$. Trace of any power of $(p^2-m^2)^{-n}$. $F(G)$ can be obtained by differentiating (4.41) with respect to m. Eq.(4.40) becomes :

$$I_o = \frac{g^2}{m^6}\,\frac{i}{2^5.3.\pi^2}\,Tr_c\left\{G_{\alpha\mu}\;G^{\alpha\mu} - \frac{g^2}{2m^4}\left(G_{\alpha\mu}\;G^{\alpha\mu}\right)^2\right\} \tag{4.42}$$

The next term in (4.35) can be written as :

$$I_1 = \frac{1}{m^2}\,Tr\left\{\frac{1}{(p^2-m^2)^5}\,[p^2,p^\mu]\;[p^2,p^\mu]\right\}$$

$$= -\frac{4g^2}{m^2}\,Tr\left\{\frac{1}{(p^2-m^2)^5}\,G_{\alpha\mu}\;p^\alpha\;p^\beta G_{\beta\mu}\right\}$$

$$= -\frac{4g^2}{m^2}\,Tr\left\{\frac{1}{(p^2-m^2)^5}\,p^\alpha\;p^\beta G_{\alpha\mu}\;G_{\beta\mu}\right\}$$

$$+ \frac{4g^2}{m^2}\,Tr\left\{\frac{1}{(p^2-m^2)^5}\,p^\alpha\left(D^\beta\;G_{\alpha\mu}\;G_{\beta\mu}\right)\right\}$$

$$= -\frac{ig^2}{2^6\,3\pi^2 m^6}\,Tr_c\;G_{\alpha\mu}\;G^{\alpha\mu} - \frac{g^3}{2^5\,3\pi^2 m^8}\,Tr_c\;G_{\alpha\mu}\;G_{\mu\nu}\;G_{\nu\alpha}$$

$$+ \frac{ig^4}{2^7\,3\pi^2 m^{10}}\,Tr\left[(G_{\mu\alpha}\;G_{\mu\alpha})^2 - \{G_{\mu\alpha},\;G_{\alpha\nu}\}^2\right]. \tag{4.43}$$

Collecting previous results, one can deduce the scalar effective action for one fermion :

$$\Delta S_s = \int d^4x\, Tr_c \left\{ - \frac{g^2}{3.2^6.\pi^2} \log \frac{\Lambda^2_{UV}}{m^2} G^2 - \frac{ig^3}{2^5 3^2 .5.\pi^2 m^2} G_{\mu\nu} G_{\nu\lambda} G_{\lambda\mu} \right.$$

$$+ \frac{g^4}{2^9 3^2 \pi^2 m^4} \left[(G_{\mu\alpha} G_{\mu\alpha})^2 + \frac{1}{5} \{G_{\mu\alpha}, G_{\alpha\nu}\}^2 + \right.$$

$$\left. \left. + \frac{1}{7} [G_{\mu\alpha}, G_{\alpha\nu}]^2 + \frac{1}{70} [G_{\mu\nu}, G_{\alpha\beta}]^2 \right] + \ldots \right\} \qquad (4.44)$$

where $G_{\mu\nu} \equiv G^{\alpha}_{\mu\nu} \lambda^a/2$, $Tr \frac{\lambda_a}{2} \frac{\lambda_b}{2} = \frac{1}{2} \delta^{ab}$, $\{\ ,\}$ and $[\ ,]$ are respectively anticommutators and commutators. Tr_c means trace over colour indices. The first coefficient in the series in (4.44) is absorbed into the charge renormalization $\left(\sim \beta^F_1 \equiv \frac{2}{3} n \right)$. The derivation of (4.44) is facilated by use of the following commutators as a generalization of the ones in (4.37) :

$$\left[P^2, \left[P^2, P_\mu\right]\right] = - 4g^2 G_{\mu\beta} G_{\beta\alpha} P_\alpha + 2ig\, D^2 G_{\mu\alpha} P_\alpha - 4g\, D_\beta G_{\alpha\mu}.P_\beta P_\alpha \qquad (4.45a)$$

$$\begin{aligned} \left[P^2, \left[P^2, \left[P^2, P_\mu\right]\right]\right] = & - 8ig\, D_\gamma D_\beta G_{\alpha\mu} P_\gamma P_\beta P_\alpha \\ & + \left(- 8ig^2 D_\beta\right) (G_{\mu\gamma} G_{\gamma\alpha}) - 4g\, D_\beta D^2 G_{\mu\alpha} + 4gD^2 D_\beta G_{\mu\alpha} \\ & + 8ig^2 D_\gamma G_{\alpha\mu} G_{\gamma\beta} + 8\, ig^2 ((D_\beta G_{\gamma\mu}) G_{\gamma\alpha}) P_\beta P_\alpha \\ & + 4\, g^2 D^2 (G_{\mu\beta} G_{\beta\alpha}) + 8\, ig^3 G_{\mu\beta} G_{\beta\gamma} G_{\gamma\alpha} \end{aligned}$$

$$- 2ig\, D^4\, G_{\mu\alpha} + 4g^2 \left(D^2\, G_{\mu\gamma}\right) G_{\gamma\alpha}$$

$$- 8\, g^2\, D_\beta\, G_{\gamma\mu}\, D_\beta\, G_{\gamma\alpha}\, P_\alpha \tag{4.45b}$$

$$\left[P^2, \left[P^2, \left[P^2, \left[P^2,\, P_\mu\right]\right]\right]\right] = 16g\, D_\delta\, D_\gamma\, D_\beta\, G_{\alpha\mu}\, P_\delta\, P_\gamma\, P_\beta\, P_\alpha$$

$$+ 8ig\, D^2\, D_\gamma\, D_\beta\, G_{\alpha\mu}\, P_\gamma\, P_\beta\, P_\alpha - 16\, g^2\, D_\gamma\, D_\beta\, G_{\alpha\mu}$$

$$.(G_{\gamma\delta}\, P_\delta\, P_\beta\, P_\alpha + P_\gamma\, G_{\beta\delta}\, P_\delta\, P_\alpha + p_\gamma\, P_\beta\, G_{\alpha\delta}\, P_\delta)$$

$$+ \left(16g^2\, D_\delta\, D_\beta\, (G_{\mu\gamma}\, G_{\gamma\alpha}) - 8ig\, D_\delta\, D_\beta\, D^2\, G_{\mu\alpha}\right)$$

$$+ 8ig\, D_\delta\, D^2\, D_\beta\, G_{\alpha\mu} - 16g^2\, D_\delta\, (D_\gamma\, G_{\alpha\mu}\, G_{\gamma\mu})$$

$$- 16g^2\, D_\delta\, ((D_\beta\, G_{\gamma\mu})\, G_{\gamma\alpha})\, P_\delta\, P_\beta\, P_\alpha + \ldots \tag{4.45c}$$

$$\left[P^2, \left[P^2, \left[P^2, \left[P^2, \left[P^2, P_\mu\right]\right]\right]\right]\right] = 32ig\, D_\rho\, D_\delta\, D_\gamma\, D_\beta\, G_{\alpha\mu}\, P_\rho\, P_\delta\, P_\beta\, P_\alpha$$

$$+ \ldots \tag{4.45d}$$

Now, one comes to the evaluation of the magnetic terms in (4.30). This can easily be done once one succeeds in interchanging σG and $(p^2-m^2)^{-1}$ in such a way that one ends up with the expression of the type :

$$\mathrm{Tr} \left\{ \frac{1}{(p^2-m^2)^n} F(G) \right\} \quad , \tag{4.46}$$

where F(G) is an arbitrary function of G. This interchange can be done with the help of the commutators :

$$[P^2,\, \sigma G] = \{P_\alpha, [P_\alpha,\, \sigma G]\} = -\, D^2 \sigma G + 2i\, (D_\alpha \sigma G)\, P_\alpha \quad ,$$

$$\left[P^2, \left[P^2, \sigma G\right]\right] = D^4 \sigma G - 2i\, D_\alpha\, D^2 \sigma G\, P_\alpha + 4g\, D_\beta \sigma G\, G_{\beta\alpha}\, P_\alpha$$

$$- 2i\, D^2\, D_\alpha \sigma\, G\, P_\alpha - 4\, D_\beta\, D_\alpha \sigma G\, P_\beta\, P_\alpha \quad ,$$

$$\left\{P^2, \left[P^2, \left[P^2, \sigma G\right]\right]\right\} = - 8i\, D_\gamma\, D_\beta\, D_\alpha \sigma G\, P_\gamma\, P_\beta\, P_\alpha + \text{Higher dimensions.} \quad (4.47)$$

One can deduce from the above formulae (4.37, 4.45 and 4.47) the useful expansions :

$$P_4 = - i\, \mathrm{Tr}(p^2 - m^2)^{-5} \left(- \frac{ig}{2}\, \sigma G\right)^4$$

$$= - \frac{g^4}{3.2^{10} \pi^2 m^6} \int d^4x \left\{ \mathrm{Tr}(\sigma G)^4 \equiv \mathrm{Tr}_c \left[48\, (G_{\mu\alpha}\, G_{\mu\alpha})^2 \right.\right.$$

$$\left.\left. + 8\, [G_{\mu\nu}, G_{\alpha\beta}]^2 - 16\{G_{\mu\nu}, G_{\nu\alpha}\}^2 - 48[G_{\mu\alpha}, G_{\alpha\nu}]^2 \right] \right\} \quad (4.48a)$$

$$P_3 = \frac{ig^3}{2^8.3.\pi^2 m^4} \int d^4x\, \mathrm{Tr}(\sigma G)^3 - \frac{g^4}{2^4.3.\pi^2.m^6} \int d^4x\, \mathrm{Tr}[G_{\mu\alpha}, G_{\alpha\nu}]^2$$

with :

$$\mathrm{Tr}(\sigma G)^3 \equiv 2^5\, \mathrm{Tr}_c\, G^3 \, . \quad (4.48b)$$

$$P_2 = \mathrm{Tr}_c \int d^4x \left\{ - \frac{g^2}{2^4 \pi^2 m^2} G^2 + \frac{g^4}{2^5.3.\pi^2.m^6} (G^2)^2 \right.$$

$$\left. + \frac{g^4}{2^3.3.5.\pi^2.m^6} [G_{\mu\alpha}, G_{\alpha\nu}]^2 + \frac{g^4}{2^6.3.5.\pi^2.m^6} [G_{\mu\nu}, G_{\alpha\beta}]^2 \right\} \quad (4.48c)$$

Assembling <u>all</u> pieces of (4.30), one obtains the Heisenberg-Euler Lagrangian :

$$\Delta S_{eff} = -\frac{1}{2}\int_{m^2}^{\Lambda^2_{UV}} dm^2 \ P(m^2)$$

$$= -\frac{1}{32\pi^2}\int d^4x \ Tr_c \left\{\frac{2}{3} g^2 \ G^2_{\mu\nu} \log \frac{\Lambda^2_{UV}}{m^2} - \frac{2}{45} ig^3 \ . \right.$$

$$. \ G_{\mu\nu} \ G_{\nu\lambda} G_{\lambda\mu} \ \frac{1}{m^2} + \frac{g^4}{18}\left[(G_{\mu\nu} \ G_{\mu\nu})^2 - \frac{7}{10}\{G_{\mu\alpha}, \ G_{\alpha\nu}\}^2\right.$$

$$\left.\left. - \frac{29}{70}[G_{\mu\alpha}, \ G_{\alpha\nu}]^2 + \frac{8}{35}[G_{\mu\nu}, \ G_{\alpha\beta}]^2 \right| \frac{1}{m^4}\right\} . \qquad (4.49)$$

For a practical calculation, the Schwinger Operator "formal" method can be used for the evaluation of the correlator built from heavy-quark currents. Here the algorithm is to write the heavy-quark propagators by including the anomalous σG terms. The second step is to perform an expansion of these propagators in terms of σG. For let instance,

$$\Psi(q^2) = i\int e^{iqx} \ d^4x \left\langle T \ \bar{\psi}(x) \ (i) \ \psi(x), \ \bar{\psi}(o)(i) \ \psi(o)\right\rangle \qquad (4.50)$$

We can write it as :

$$\Psi(q^2) = i\int d^4x \ Tr\left\{(i) \left\langle x \left| \frac{1}{\left(p^2 - m^2 + \frac{ig}{2} \ \sigma G\right)} (\hat{p}+m) \right| y\right\rangle \right. .$$

$$. (i) \left\langle y \left| \frac{1}{(p+q)^2-m^2 + \frac{ig}{2}\sigma G} (\hat{p}+\hat{q}+m) \right| x \right\rangle \Bigg\}_{y=o} \tag{4.51}$$

Now do the σG expansion, take the trace over Lorentz indices and integrate over y. You will then be dealing with scalar operators of the type :

$$Tr(p^2-m^2)^{-1} \left(G(p^2-m^2)^{-1}\right)^n P_\alpha (P^2-m^2)^{-1} \left(G(p^2-m^2)^{-1}\right)^m P_\alpha \tag{4.52}$$

which you already know how to evaluate.

4. APPLICATIONS OF THE FOCK-SCHWINGER FIXED-POINT TECHNIQUE

a) The two-point correlator of the pseudoscalar light-quark current

In order to illustrate the method, we first evaluate the non-perturbative contributions appearing in the Wilson expansion (3.2) of the two-point correlator :

$$\psi_5(q^2) = i \int d^4 x \left\langle 0 \left| \mathbb{T}\, \partial_\mu A^\mu(x)^u_d \left(\partial_\mu A^\mu(o)^u_d\right)^+ \right| 0 \right\rangle \tag{4.53a}$$

associated with the light-quark current divergences :

$$\partial_\mu A^\mu = (m_u + m_d) : \bar{\psi}_u (i \gamma_5) \psi_d \quad . \tag{4.53b}$$

We have already discussed in Chapter 1 the evaluation of its perturbative part by means of the standard covariant gauge technique within the $\overline{MS}$ scheme. For the evaluation of the Wilson coefficients induced by the non-perturbative condensates, we shall see that the Fock-Schwinger technique is convenient and even superior. The Wilson expansion of $\psi_5(q^2)$ reads up to the dimension-six condensates :

$$\psi_5(q^2) = C_{\mathbb{1}}\, \mathbb{1} + \left[C_\psi \left\langle \bar\psi \psi \right\rangle + C_G \left\langle \alpha_s\, G^2 \right\rangle \right] / q^2$$

$$+ \left[C_M\, g \left\langle \bar\psi\, \sigma^{\mu\nu}\, \lambda_a / 2\, G^a_{\mu\nu} \right\rangle + \right.$$

$$+ C_{3G}\, g^3 f_{abc} \left\langle G^a_{\mu\nu}\, G^b_{\nu\lambda}\, G^c_{\lambda\mu} \right\rangle +$$

$$\left. + C_{4\psi} \left\langle \bar\psi\, \Gamma_1 \psi\, \bar\psi\, \Gamma_2\, \psi \right\rangle \right] / (q^2)^2 \qquad (4.54)$$

where we know from section 1 that for large $-q^2$:

$$C_{\mathbb{1}}\ (-q^2 \to \infty) \simeq \frac{3}{8\pi^2} (m_i + m_j)^2\ (-q^2)\ \log - \frac{q^2}{\nu^2} + \ldots \qquad (4.55)$$

Now let us discuss successively the evaluation of each of the other Wilson coefficients :

i) <u>The $\langle \bar\psi \psi \rangle$ quark condensate</u>

One should start with Wick's theorem and leave one pair of $\langle \bar\psi \psi \rangle$ without contraction. Therefore :

$$\psi_5(q^2) = (m_u + m_d)^2\ (\gamma_5)_{ij}\ (\gamma_5)_{k\ell}(-i) \int d^4 x\ e^{iqx} \left\{ \underline{d(x)_{\alpha j}\ \bar d(o)_{\beta k}} \cdot \right.$$

$$\cdot \left\langle 0 \right|\ :\ \bar u\ (x)_{\alpha i}\ u(o)_{\beta \ell}\ : \left| 0 \right\rangle + \underline{u_{\beta\ell}(o)\ \bar u(x)_{\alpha i}} \cdot \left\langle 0\ \right|\ :\ \bar d(o)_{\beta k}\ d(x)_{\alpha j}$$

$$: \mid 0\rangle \Big\}. \tag{4.56}$$

Use the definition of the propagator.

$$\bar{\Psi}^{F}_{\alpha i}(x)\ \psi^{F}_{\beta j}(y) = i\ \delta_{\alpha\beta}\ \delta^{FF'}\ \ S_{ij}(x-y)$$

$$= \delta^{FF'}\ \delta_{\alpha\beta}(i) \int \frac{d^4R}{(2\pi)^4}\ \delta_{ij}(p)\ e^{-ip(x-y)} \tag{4.57a}$$

with :

$$S^{F}_{ij}(p) = \frac{1}{\hat{p}-m_F + i\epsilon'}\ . \tag{4.57b}$$

Then :

$$\psi_5(q^2) = (m_u+m_d)^2 \int d^4x \int \frac{d^4p}{(2\pi)^4}\ e^{-i(p-q)x} \Big\{\Big\langle 0 \Big|: \bar{u}(x)_{\alpha i} u(0)_{\beta\ell} :\Big| 0\Big\rangle.$$

$$.\Big(\gamma_5\ S^d(p)\ \gamma_5\Big)_{i\ell} + \Big\langle 0 \Big|: \bar{d}(0)_{\beta k}\ d(x)_{\alpha j}\ :\Big| 0\Big\rangle \Big(\gamma_5 S^u(p)\ \gamma_5\Big)_{kj}\Big\}\ . \tag{4.58}$$

In terms of Feynman diagrams, (4.58) reads :

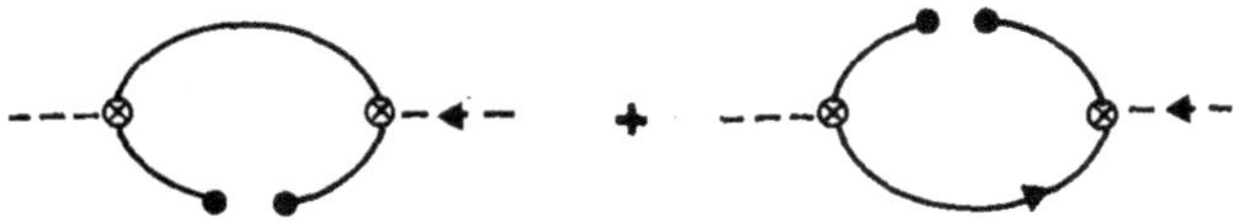

(4.59)

where —• •— means that the two-quark fields "condense" at the same point, so that a Taylor expansion in x_μ of $\Big\langle 0 \Big|: \bar{\Psi}(x)\ \psi(0)\ :\Big| 0\Big\rangle$ makes sense. One can use (4.8) where for the moment we shall limit ourselves to the first two terms of the expansion. After straightforward algebra, $\psi_5(q^2)$ then becomes :

$$\psi_5(q^2) = (m_u+m_d)^2 \cdot \frac{3}{12}\left\{\left\langle \bar{u}\, u\right\rangle \left[\mathrm{Tr}\ \gamma_5\ S^d(q)\ \gamma_5 - \right.\right.$$

$$\left. - \frac{1}{4} m_u \left[- \frac{\partial}{\partial p_\lambda} \mathrm{Tr}\left(\gamma_5\ S^d(p)\ \gamma_5\ \gamma^\lambda\right)\right]_{p=q}\right] + \left\langle \bar{d}\, d\right\rangle \cdot$$

$$\left. \cdot \left[\mathrm{Tr}\ \gamma_5\ S^u(q)\ \gamma_5 - \frac{1}{4} m_d\left(- \frac{\partial}{\partial p_\lambda} \mathrm{Tr}\ \gamma_5\ S^u(q)\ \gamma_5\ \gamma^\lambda\right)_{p=q}\right]\right\} . \tag{4.60}$$

Using the property :

$$- \frac{\partial}{\partial p_\lambda} S(p) = S(p)\ \gamma_\lambda\ S(p) . \tag{4.61}$$

One then gets the final result :

$$\psi_5(q^2)\Big|_{\bar{\psi}\psi} = (m_u+m_d)^2 \frac{1}{q^2} \cdot \left[\left(m_d - \frac{m_u}{2}\right)\left\langle \bar{u}\, u\right\rangle + \left(m_u - \frac{m_d}{2}\right)\left\langle \bar{d}\, d\right\rangle\right] . \tag{4.62}$$

Notice that the minus sign is due to the γ_5 chirality flip which acts on the term $\frac{\partial}{\partial p^\lambda}$ in (4.60), i.e. for the scalar current this minus sign becomes a plus sign.

ii) <u>The $\left\langle \alpha_s\ G^2\right\rangle$ gluon condensate</u>

Let us evaluate the effects of the gluon condensate. We shall be concerned with the quark propagators in external fields. Diagramatically, our work is to evaluate :

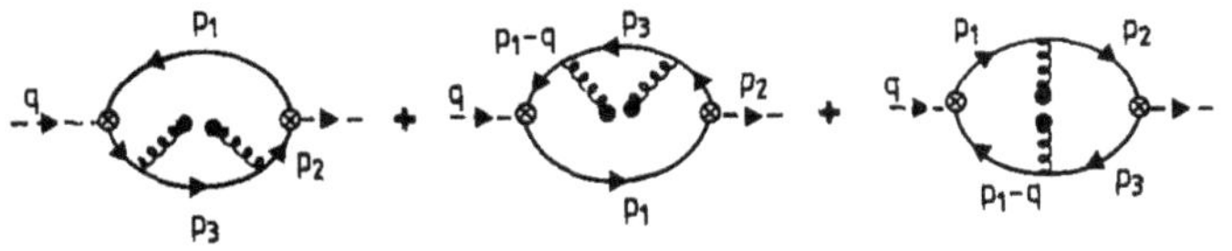

(4.63)

As usual, we apply Wick's theorem where all quark fields should be contracted but not the gluon ones. The notation • means that the gluon fields are placed at the same point so that one can use the Taylor expansion in (4.6). Using the usual Feynman rules, one can deduce from (4.63) :

$$\psi_5(q^2)\Big|_G = (m_u+m_d)^2(-i)\,\frac{g^2}{2}\int d^4y\, d^4z \int \prod_{i=1}^{3} \frac{d^4p_i}{(2\pi)^4}\,.\,\langle 0 \,|: A^a_\lambda(y)\, A^a_\rho(z) :|\, 0\rangle$$

$$\Big\{\mathrm{Tr}\Big[\gamma_5\, S(p_1+q)\,\gamma^\lambda\, S(p_3)\,\gamma^\rho\, S(p_2)\,\gamma_5\, S(p_1)\Big]\, e^{i(q+p_1-p_3)y}\, e^{i(p_3-p_2)z}$$

$$+\,\mathrm{Tr}\Big[\gamma_5 S(p_1)\,\gamma_5 S(p_2)\,\gamma^\rho S(p_3)\gamma^\lambda S(p_1-q)\Big]\, e^{i(-p_1+p_3+q)y}\, e^{i(p_2-p_3)z}$$

$$+\,\mathrm{Tr}\Big[\gamma_5 S(p_1)\,\gamma^\rho S(p_2)\,\gamma_5 S(p_3)\,\gamma^\lambda S(p_1-q)\Big]\, e^{i(q-p_1+p_3)y}\, e^{i(p_1-p_2)z}, \quad (4.64)$$

where we have forgotten the flavour indices u,d as we shall work in the chiral limit $m_u = m_d = 0$. Now one takes advantage of (4.6) which is valid in the Schwinger gauge. Substituting it in (4.64), one gets :

$$\psi_5(q^2) = -\frac{i}{16n(n-1)} \left\langle g^2 G_a^{\mu\nu} G^a_{\mu\nu} \right\rangle [g_{\nu\tau} g_{\lambda\rho} - g_{\nu\rho} g_{\lambda\tau}] \, .$$

$$\cdot \int \frac{d^4 p_1}{(2\pi)^4} \left\{ 2 \frac{\partial}{\partial p_{1\nu}} \frac{\partial}{\partial p_{2\tau}} \mathrm{Tr}\left[\gamma_5 \, S(p_1+q)\, \gamma^\lambda \, S(p_3)\, \gamma^\rho \, S(p_2)\, \gamma_5 \, S(p_1)\right]_{\substack{p_2=p_3\\=p_1+q}} \right.$$

$$\left. + \frac{\partial}{\partial p_{3\nu}} \frac{\partial}{\partial p_{2\tau}} \mathrm{Tr}\left[\gamma_5 S(p_1)\, \gamma^\rho S(p_2)\, \gamma_5 S(p_3)\, \gamma^\lambda S(p_1-q)\right]_{\substack{p_2=p_1\\=p_3=p_1-q}} \right\} , \quad (4.65)$$

where we have used the fact that the two self-energy-like diagrams give the same contribution. Using (4.61) and the properties of Dirac matrices and integrals in the appendix, one can deduce :

$$\psi_5(q^2)\Big|_G = -\left(\frac{m_u + m_d}{q^2}\right)^2 \left(\frac{1}{8\pi}\right) \left\langle \alpha_s \, G^2 \right\rangle \, . \quad (4.66)$$

iii) <u>The</u> $\left\langle \bar\psi \, \sigma^{\mu\nu} \frac{\lambda_a}{2} \psi \, G^a_{\mu\nu} \right\rangle$ <u>mixed condensate</u>

This contribution corresponds to the diagrams :

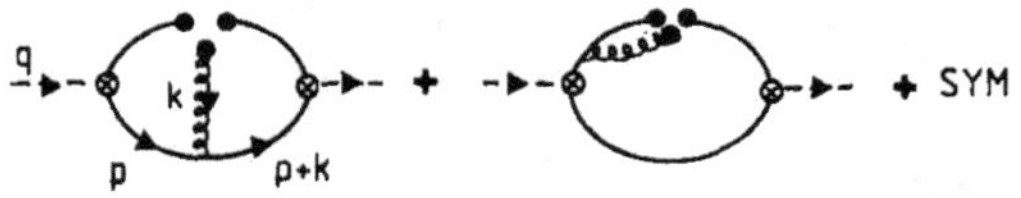

(4.67)

As before, one writes the Wick product where two quark fields should be contracted. Then, the first diagram gives :

$$\psi_5(q^2)\Big|_M^a = -(m_u+m_d)^2 \int d^4x\, d^4y \,.\, e^{iqx} \,.\, \int \frac{d^4k}{(2\pi)^4}\frac{d^4p}{(2\pi)^4} \,.$$

$$\left\langle 0 \left| :\ \bar{u}(x)_{\alpha i}\ A^a_\mu\ (y)\ u(o)_{\beta\ell}\ : \right| 0 \right\rangle g(\gamma^\mu)_{mn}\ (\gamma_5)_{ij}\ (\gamma_5)_{k\ell}\ \cdot$$

$$.\ e^{-ip(x-y)}\ e^{-i(p+k)y}\ .\ S^d_{jm}(p)\ S^d_{nk}(p+k)\ +\ (u \leftrightarrow d)\ . \qquad (4.68)$$

Now, we use (4.9) and the property in (4.61) and do the Dirac algebra. The self-energy-like diagram in (4.67) can be obtained by considering the "propagation" of the $\langle\bar{\psi}\,\psi\rangle$ condensate put in a weak external field (4.8). Using iteratively the property in (4.61), and doing the Dirac algebra, one obtains the desired result. The mixed condensate contributes as :

$$\psi_5(q^2)_M = -\frac{(m_u+m_d)^2}{2(q^2)^2}\ .\ \left[m_d\left\langle\bar{\psi}_u\ G\ \psi_u\right\rangle - m_u\left\langle\bar{\psi}_d\ G\ \psi_d\right\rangle\right] \qquad (4.69a)$$

with the shorthand notation :

$$\left\langle\bar{\psi}_i\ G\ \psi_i\right\rangle \equiv g\left\langle\bar{\psi}_i\ \sigma^{\mu\nu}\ \frac{\lambda_a}{2}\ G^a_{\mu\nu}\ \psi_i\right\rangle\ . \qquad (4.69b)$$

In the case of the scalar current, the contribution of the mixed condensate can be deduced from (4.69) by substituting m_d with $-m_d$.

iv) <u>Four-quark condensates</u>

We have here two classes of diagrams which contribute to the four-quark condensates.

<u>Class 1</u> are those where the gluon fields once contracted give a hard momentum gluon propagator :

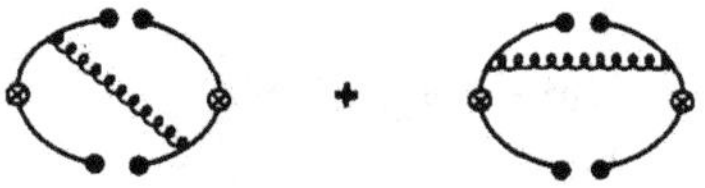

(4.70)

The computation of these diagrams can be done using standard perturbation theory, i.e. by writing the Wick's product, contracting the gluon fields and two pairs of quark fields and by taking the vacuum expectation values (v.e.v.) of the four-quark operators. Then, one obtains :

$$\psi_5(q^2)\Big|^1_{4\psi} = \frac{(m_u+m_d)^2}{(q^2)^2}\,\pi\alpha_s \left\langle \bar{u}\,\sigma^{\mu\nu}\,\gamma_5\,\frac{\lambda_a}{2}\,u - \bar{d}\,\sigma^{\mu\nu}\,\gamma_5\,\frac{\lambda_a}{2}\,d\right\rangle^2 \qquad (4.71)$$

Class 2 are those where the momentum of the gluon propagator is zero. This contribution is represented by diagrams :

(4.72)

The first two diagrams are generated by the propagation of the $\langle\bar{\psi}\,\psi\rangle$ condensate in Eq. (4.8). The third diagram is generated by the mixed condensate in Eq. (4.9). Evaluation of these diagrams leads to :

$$\psi_5(q^2)\Big|^2_{4\psi} = \frac{(m_u+m_d)^2}{(q^2)^2}\,\frac{\pi\alpha_s}{6}\left\langle\left(\bar{u}\,\gamma_\mu\,\frac{\lambda_a}{2}\,u + \bar{d}\,\gamma_\mu\,\frac{\lambda_a}{2}\,d\right)\sum_{u,d,s}\bar{\psi}\,\gamma_\mu\,\frac{\lambda_a}{2}\,\psi\right\rangle \qquad (4.73)$$

If one uses vacuum saturation and $SU(2)_F$ symmetry for the condensates, the sum of the contribution in (4.71) and (4.73) reads :

$$\psi_5(q^2)\Big|_{4\psi} \simeq \frac{(m_u+m_d)^2}{(q^2)^2} \cdot \frac{112}{27}\pi\alpha_s \cdot \langle\bar\psi\psi\rangle^2 \qquad (4.74)$$

v) Triple gluon condensate $\langle g f_{abc} G^a_{\mu\nu} G^b_{\nu\ell} G^c_{\ell\mu}\rangle$

This contribution, which has been analyzed in Refs 111,115), comes from the diagrams :

(4.75)

Here, one uses the quark propagator in external fields (4.13) and writes the gluon fields in terms of the field strengths (4.5, 4.15) in order to form the triple gluon condensates. The calculation can be done using standard perturbation theory. In the chiral limit (m = 0), this effect is zero for any currents formed by quark bilinears.

By adding each non-perturbative contribution to the perturbative one in Chapter 1, we obtain the complete expression of the pseudoscalar two-point correlator up to dimension-six condensates. Radiative corrections to some of these condensate contributions have been computed in Refs 118) and 119), where the coefficients are small compared to the leading-order approximation. This gives an indication of the quite good convergence of the QCD series.

b) The two-point correlator of heavy quarks

i) General procedure

We shall be concerned with the generic correlator :

$$\psi_\Gamma(q^2) = i \int d^4x \, e^{iqx} \left\langle 0 \left| T \, J^\Gamma(x) \, (J^\Gamma(0))^\dagger \right| 0 \right\rangle \, , \qquad (4.76)$$

where Γ is any Dirac matrix.
Here, the techniques are not similar to the case of light-quark systems. The reason is that we can no longer neglect the quark masses which are the most important scale in this channel. In addition, the Wigner-Weyl realization of chiral symmetry for heavy-quark systems implies that the $\langle \bar\psi \psi \rangle$ condensate vanishes as the inverse quark mass and it is correlated to the gluon condensates as in Eq. (3.41). Therefore the gluon condensates are the most relevant non-perturbative terms in the analysis.

Here the Fock-Schwinger gauge remains the most convenient working gauge. In the light-quark sector we had the option of working in the x or p space. Here, the p space appears to be the most convenient one, as in x space one encounters transcendental functions in the analysis. Then for the evaluation of the Wilson coefficients of the gluon condensates, one can repeat the algorithm used to obtain $\langle \alpha_s G^2 \rangle$ for light quarks discussed in the momentum space. One repeatedly uses the property in (4.61). Then, the correlator is typically of the form :

$$\psi_\Gamma(q^2, G\ldots G) \sim \langle gG\ldots gG \rangle \int \frac{dk}{(2\pi)^4} \, \mathrm{Tr} \, \frac{\left(\Gamma\ldots(\hat k, \hat q, m) \ldots \Gamma\ldots\right)}{(k^2-M^2)((k+q)^2-M^2)^m} \, . \qquad (4.77)$$

The trace can be done using either the Schoonship[120] or Reduce[121] algebraic program. Therefore, it is convenient to expand the numerator in powers of the denominator factor k^2-M^2 and $(k+q)^2-M^2$, where the integrand becomes nicely simplified. After a Feynman parametrization, one encounters integrals of the type :

$$I_n^{\alpha\beta}(-q^2, M^2) = \int_0^1 dx \; . \; \frac{x^\alpha (1-x)^\beta}{[-q^2 x(1-x) + M^2]^n} \; . \qquad (4.78)$$

By noting the symmetry $x \to (1-x)$, one can re-expand (4.78) in $x(1-x)$ and deduce the recursive relation :

$$I_n^{\alpha\alpha} \equiv I_n^\alpha = \frac{1}{Q^2}\left(I_{n-1}^{\alpha-1} - M^2 \; I_n^{\alpha-1}\right) \; . \qquad (4.79)$$

This leads to the basic integral :

$$J_n = \int_0^1 \frac{dx}{\left(1 - x(1-x) \; q^2/M^2\right)^n} \; , \qquad (4.80a)$$

which reads :

$$J_n = \frac{(2n-3)!!}{(n-1)!}\left[\left(\frac{v^2-1}{2v^2}\right)^n \sqrt{v} \; \log \frac{v+1}{v-1} + \sum_{k=1}^{n-1} \frac{(k-1)!}{(2k-1)!!}\left(\frac{v^2-1}{2v^2}\right)^{n-k}\right] \; , \qquad (4.80b)$$

where $v \equiv \left(1 - \frac{4m^2}{q^2}\right)^{1/2}$.

Now let us apply this general result for the evaluation of the $\left\langle \alpha_s G^2 \right\rangle$ and $g f_{abc}\left\langle G^a_{\mu\nu} \; G^b_{\nu\rho} \; G^c_{\rho\mu}\right\rangle$ condensates contribution to the vector current correlator which we write as :

$$\Pi^{\mu\nu} = -(g^{\mu\nu}q^2 - q^\mu q^\nu)\left\{C_{\mathbb{1}}\mathbb{1} + C_G\left\langle \alpha_s G^2 \right\rangle + C_{3G}\left\langle g^3 f_{abc} G^a_{\mu\nu} \; G^b_{\nu\rho} \; G^c_{\rho\mu}\right\rangle + \right.$$

$$C_{jj}\left\langle g^4\ J^a_\mu\ J^a_\mu\right\rangle + \ldots\Big\}\ , \tag{4.81}$$

where :

$$g\ J^\mu_a \equiv D_\mu\ G^{\mu\nu}_a = -\frac{1}{2}g \sum_{u,d,s} \bar{\psi}\,\gamma^\nu\,\frac{\lambda_a}{2}\,\psi \quad , \tag{4.82}$$

by virtue of the equation of motion of gluon fields.

ii) $\left\langle \alpha_s G^2\right\rangle$ <u>of the vector correlator</u>

We use the method outlined earlier and the property :

$$G^a_{\alpha\beta}\ G^{a'}_{\gamma\rho} = \frac{1}{96}\,\delta^{aa'}\,(g_{\alpha\gamma}g_{\beta\rho} - g_{\alpha\rho}g_{\beta\gamma})\left\langle G_{\mu\nu}G^{\mu\nu}\right\rangle \ . \tag{4.83}$$

The algorithm is very similar here to the light-quark case which we have discussed in p space. The first two self-energy-like figures in (4.63) give [95b]:

$$C^a_4 = -\frac{1}{96\pi}\,\frac{\left\langle \alpha_s G^2\right\rangle}{q^4}\left[2\,\frac{(5v^4+3)}{v^4} + \frac{(v^2-1)^2(5v^2+3)}{v^5}\log\frac{v-1}{v+1}\right] \ . \tag{4.84}$$

The vertex-like diagram contributes as :

$$C^b_4 = \frac{\left\langle \alpha_s G^2\right\rangle}{48\pi q^4}\left[\frac{2(1+v^2)}{v^2} + \frac{(v^2-1)^2}{v^3}\log\frac{v-1}{v+1}\right] \ . \tag{4.85}$$

However, in the case $v \neq 1$, each set of diagrams develops a non-transverse part :

$$q^{\mu}q^{\nu}\,\Pi^{b}_{\mu\nu} = -\frac{1}{16\pi}\,\frac{\langle\alpha_s G^2\rangle}{q^4}\,\frac{1-v^2}{v^2}\left[1+\frac{1+v^2}{2v}\,\log\frac{v-1}{v+1}\right] \qquad (4.86)$$

which vanishes in the sum. The sum of the transverse contributions can be expressed in terms of the basic integral in (4.80). It is :

$$C_4 \equiv C_4^a + C_4^b = \frac{\langle\alpha_s G^2\rangle}{24\pi q^4}\,(-1 + 3J_2 - 2J_3) \qquad (4.87)$$

which is more useful for further phenomenological uses.

iii) Four-quark and triple-gluon condensates

The four-quark condensates contribute through the diagrams :

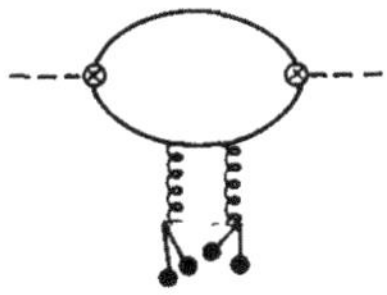

(4.88)

via the equation of motion in (4.82). The triple-gluon condensates contribute from the diagrams in (4.75). We want to obtain explicit expressions of the v.e.v :

$$\langle D_\alpha D_\beta G_{\mu\nu} G_{\rho\sigma}\rangle\,,\ \langle D_\alpha G_{\mu\nu} D_\beta G_{\rho\sigma}\rangle \ \text{and}\ \langle g^3 f_{abc}\, G^a_{\mu\nu} G^b_{\alpha\beta} G^c_{\rho\sigma}\rangle\,. \qquad (4.89)$$

We use the colour trace due to two and three λ matrices, the equation of motion in (4.82) and the Bianchi identity in (4.37). Using the commutation relation :

$$D_\mu D_\nu G^a_{\alpha\beta} - D_\nu D_\mu G^a_{\alpha\beta} = g f^{abc} G^b_{\alpha\beta} G^c_{\mu\nu}, \qquad (4.90a)$$

one can write :

$$D^2 G^a_{\mu\nu} = g\left(2f_{abc} G^b_{\mu\alpha} G^c_{\nu\alpha} + D_\nu j^a_\mu - D_\mu j^a_\nu\right) , \qquad (4.90b)$$

which becomes :

$$\left\langle D_\nu j^a_\mu G^a_{\mu\nu}\right\rangle = - g \left\langle j^a_\mu j^a_\mu\right\rangle . \qquad (4.90c)$$

Using the above relations, one obtains[114]:

$$\left\langle 0 \left| f_{abc} G^a_{\mu\nu} G^a_{\alpha\beta} G^c_{\rho\sigma} \right| 0\right\rangle = \frac{1}{24}\left\langle 0 \left| f_{abc} G^a_{\gamma\delta} G^b_{\delta\epsilon} G^c_{\epsilon\gamma}\right| 0\right\rangle$$

$$\times \left(g_{\mu\sigma}g_{\alpha\nu}g_{\beta\rho} + g_{\mu\beta}g_{\alpha\rho}g_{\sigma\nu} + g_{\alpha\sigma}g_{\mu\rho}g_{\nu\beta} + g_{\rho\nu}g_{\mu\alpha}g_{\beta\sigma}\right.$$

$$\left. - g_{\mu\beta}g_{\alpha\sigma}g_{\rho\nu} - g_{\mu\sigma}g_{\alpha\rho}g_{\nu\beta} - g_{\alpha\nu}g_{\mu\rho}g_{\beta\sigma} - g_{\beta\rho}g_{\mu\alpha}g_{\nu\sigma}\right) , \qquad (4.91a)$$

$$\left\langle 0 \left| G^a_{\mu\nu} G^a_{\alpha\beta;\rho\sigma}\right| 0\right\rangle = 20^- g_{\rho\sigma}(g_{\mu\beta}g_{\alpha\nu} - g_{\mu\alpha}g_{\nu\beta})$$

$$+ 0^-(g_{\mu\beta}g_{\alpha\sigma}g_{\rho\nu} + g_{\alpha\nu}g_{\mu\rho}g_{\beta\sigma} - g_{\alpha\sigma}g_{\mu\rho}g_{\nu\beta} - g_{\rho\nu}g_{\mu\alpha}g_{\beta\sigma})$$

$$+ 0^+(g_{\mu\sigma}g_{\alpha\nu}g_{\beta\rho} + g_{\mu\beta}g_{\alpha\rho}g_{\sigma\nu} - g_{\mu\sigma}g_{\alpha\rho}g_{\nu\beta} - g_{\rho\beta}g_{\mu\alpha}g_{\nu\sigma}), \qquad (4.91b)$$

where

$$0^{\pm} = \frac{1}{72}\left\langle 0 \left| g^2 J^a_\mu J^a_\mu \right| 0\right\rangle \pm \frac{1}{48}\left\langle 0 \left| g f_{abc} G^a_{\mu\nu} G^b_{\mu\lambda} G^c_{\lambda\mu}\right| 0\right\rangle. \qquad (4.91c)$$

This result can be used in (4.77) in order to get the desired contribution of dimension-six operators. One obtains[18,114]:

$$C_{3G} = -\frac{1}{72\pi^2(q^2)^3}\left(\frac{2}{15} + 4J_2 - \frac{31}{3}J_3 + \frac{43}{5}J_4 - \frac{12}{5}J_5 + \frac{q^2}{10M^2}\right)$$

$$G_{JJ} = -\frac{1}{36\pi^2(q^2)^3}\left(\frac{41}{45} + \frac{2}{3}J_1 - J_2 - \frac{4}{9}J_3 - \frac{26}{15}J_4 + \frac{8}{5}J_5 - \frac{q^2}{3M^2}J_1 + \frac{3q^2}{5M^2}\right), \quad (4.92)$$

which are useful for further applications of the sum rules. The contributions of dimension-eight operators have also been obtained in Ref. 122) using the same technique.

c) Heavy versus light quark expansion

Here we wish to compare (see e.g. Ref. 95b) coefficient functions of the gluon condensates $\langle\alpha_s G^2\rangle$ obtained in the limit $v = 1$ of the heavy-quark correlator with the one which one obtains directly starting from the $m = 0$ case of light quarks. Let us recall that in the light-quark case :

$$G^a_G\ (m = 0) = 0$$

$$G^b_G\ (m = 0) = \frac{1}{12\pi}\frac{\langle\alpha_s G^2\rangle}{q^4} \quad (4.93)$$

where, as before a and b refer respectively to the self-energy and vertex-like diagrams in (4.63). Let us now (naively) take the limit $v = 1$ of the heavy-quark results in (4.84) and (4.85). One obtains :

$$G^a_G(v \to 1) = -\frac{1}{6\pi}\frac{\langle\alpha_s G^2\rangle}{q^4}$$

$$G^b_G(v \to 1) = \frac{1}{12\pi} \frac{\left\langle \alpha_s G^2 \right\rangle}{q^4} . \qquad (4.94)$$

The two limits do not coincide for the self-energy-like diagrams ! So, there is some missing contribution in this latter diagram. In fact, it must be remembered that the quark-condensate $\left\langle \bar\psi \psi \right\rangle$ contribution which is correlated to the $\left\langle \alpha_s G^2 \right\rangle$ condensate through Eq. (3.40) has not yet been computed here. A straightforward calculation gives :

$$C_{\bar\psi\psi}(m = 0) = \frac{2}{q^4} m \left\langle \bar\psi \psi \right\rangle$$

$$C_{\bar\psi\psi}(v) = \frac{8}{3} \frac{2+v}{(1+v)^2} \frac{m\left\langle \bar\psi\psi \right\rangle}{q^4} , \qquad (4.95)$$

where the two results coincide for v = 1. Using (3.41), we obtain the part of the gluon condensate which cancels C^a_G $(v \to 1)$ given in Eq. (4.94).

The lesson is that one cannot take directly the v = 1 limit of the heavy-quark correlator in order to get the light-quark mass result without paying attention to the "masked" contribution of the quark condensate. In fact, the gluon condensate contribution as presented in the heavy-quark result already includes in it the $\left\langle \bar\psi\psi \right\rangle$ contribution. Some other relations between the quark and gluon condensates of higher dimensions have been presented in Chapter 3.

d) Treatment of mass singularities in the $\overline{MS}$ scheme

The above remark also holds for checking explicitly the cancellation of mass singularities in the evaluation of the Wilson coefficients. This technical point has been discussed in Ref.123), where one wishes to understand the commutativity of the operation : taking the limit $m \to 0$ before or after the momentum integration calculation. More explicitly, let us discuss the Wilson coefficient of the gluon condensate for the vector correlator built from the heavy-light currents.

$$J_\mu(x)^i_j = \bar{\psi}_i \gamma_\mu \psi_j . \tag{4.96}$$

The evaluation of the coefficient is done by putting the quark propagator in an external field and using the Fock-Schwinger gauge. Keeping the quark-mass terms and taking the limit $q^2 \to -\infty$ after integration, one would obtain for the transverse part of the correlator :

$$C^T_G = \left(1 - \frac{m_i}{m_j} - \frac{m_j}{m_i}\right) \frac{\langle \alpha_s G^2 \rangle}{12\pi\, q^4} . \tag{4.97}$$

The mass term can be cancelled by subtracting the quark condensate $\langle \bar{\psi} \psi \rangle$ contribution which is related to $\langle \alpha_s G^2 \rangle$ via Eq. (3.41), in much the same way as C^a_G. Eq. (4.94) coincides with the one in Eq. (4.93). The quark-condensate contribution is :

$$C_{\bar{\psi}\psi} = m_i \langle \bar{\psi}_j \psi_j \rangle + m_j \langle \bar{\psi}_i \psi_i \rangle . \tag{4.98}$$

But do we get the same result if we take $m_i = m_j = 0$ before loop integration ? In this case, one encounters integrals of the types :

$$\int \frac{d^n k}{(2\pi)^n} \left(\frac{q^2}{(k^2 - i\epsilon')}\right)^a \left(\frac{q^2}{(k+q)^2 - i\epsilon'}\right)^b =$$

$$i\left(\frac{q^2}{4\pi}\right)^{n/2}\frac{\Gamma\left(a+b-\frac{n}{2}\right)\Gamma\left(\frac{n}{2}-a\right)\cdot\Gamma\left(\frac{n}{2}-b\right)}{\Gamma(a)\,\Gamma(b)\,\Gamma(n-a-b)}\,, \qquad (4.99)$$

where mass singularities can appear as poles in $\frac{1}{\epsilon}$. $(n \equiv 4-\epsilon)$. In the case considered here the limit $n \to 4$ which is taken at the very end of the calculation gives the infrared finite result :

$$C_G^T = \frac{\left\langle\alpha_s\, G^2\right\rangle}{12\pi\, q^4}\,, \qquad (4.100)$$

in agreement with previous result. In general, the extension of this method to the Wilson coefficients of higher dimension condensates can be done provided one takes care to mix the operators under renormalizations in accordance with Ref. 42). This mixing provides the constraints among different condensates presented in Chapter 3, which are derived using a heavy-quark expansion of a one-point function[85,123a,b].

CHAPTER 5

RHO MESON SPECTRAL SUM RULES AND THE DETERMINATION OF THE VACUUM CONDENSATES

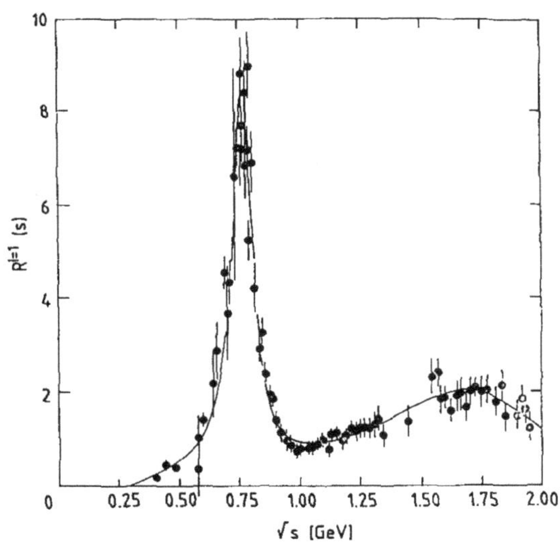

I = 1 Hadrons total cross-section from electron positron machine.

Phenomenological uses of QCD spectral sum rules for the ρ-meson systems are discussed in this chapter. Among other channels, the ρ-two-point correlator is accurately known in QCD while its spectral part is measured in the $e^+e^- \to$ Hadrons (I = 1) reaction. Therefore, one might expect a priori that this channel could be a good laboratory for testing the sum rules.

1. THE TWO-POINT CORRELATOR

We shall be concerned with the transverse two-point correlator :

$$\Pi^{\mu\nu} = i \int d^4x \; e^{iqx} \left\langle 0 \left| \mathbb{T} \, J^{\mu}_{\rho}(x) \left(J^{\nu}_{\rho}(0)\right)^{+} \right| 0 \right\rangle \equiv -(g^{\mu\nu}q^2 - q^{\mu}q^{\nu})\, \Pi_{\rho}(q^2) \quad (5.1a)$$

associated with the current :

$$J^{\mu}_{\rho} = \frac{1}{2}\left(\bar{u}\,\gamma^{\mu}u - \bar{d}\,\gamma^{\mu}d\right) \quad . \quad (5.1b)$$

$\Pi_{\rho}(q^2)$ is the most accurate QCD expression of two-point correlators known at present. It is known to three[124] and four[38c] loops in the chiral limit. Mass corrections to $\Pi_{\rho}(q^2)$ are known completely to lowest[34] and[47] α_s order. Perturbative m_s^2 corrections are available to order α_s^2 [38b] while m_s^4 contributions are known up to order α_s [47]. These effects have already been discussed in Chapter 1.

Non-perturbative corrections due to $\langle \bar{\psi}\psi \rangle$ [118,119,47] and $\langle \alpha_s G^2 \rangle$ [119] condensates are known to order α_s. Dimension-six[18,100] and some of the dimension-eight [124] condensates contributions are also known. The QCD expression of the R.G.I. first derivative of $\Pi_{\rho}(q^2)$ reads in the $\overline{MS}$ scheme :

$$8\pi^2 \, \frac{q^2 \, \overline{d\Pi_{\rho}(q^2)}}{dq^2} = C_{\mathbb{1}}\mathbb{1} + 2\,\frac{C_4\langle O_4\rangle}{q^4} +$$

$$+ 3 \frac{C_6 \langle O_6 \rangle}{q^6} + 4 \frac{C_8 \langle O_8 \rangle}{q^8} + \ldots \tag{5.2}$$

where :

$$C_{\mathbb{1}}\mathbb{1} = 1 - \bar{a}_s + (1.986 - 0.115n)\, \bar{a}_s^2 + (70.985 - 1.200\, n$$

$$- 0.05\, n^2)\, \bar{a}_s^3 - \left(\sum_i Q_i\right)^2 1.679\, \bar{a}_s^3$$

$$+ \frac{\bar{m}^2}{q^2} \left[6 + 28\, \bar{a}_s + (296.15 - 12.25n)\, \bar{a}_s^2\right] - \sum_{u,d,s} \frac{\bar{m}_j^2}{q^2} 1.05\, \bar{a}_s^2$$

$$+ 2 \frac{\bar{m}^4}{q^4} \cdot \left(\frac{1}{15+2n}\right) \cdot \left(- \frac{72}{a_s} + (357 - 23n)/4\right) \tag{5.3}$$

is the perturbative contribution up to m^4 terms where $\bar{a}_s \equiv \bar{\alpha}_s/\pi$ is the running coupling to three loops in Eq. (1.93). Here, one should notice that $m^4 \log \frac{m^2}{\nu^2}$ which appeared in the m^4 term has been eaten by the $m\langle\bar{\psi}\, \psi\rangle$ condensate according to the relation in Eq. (3.37). The non-perturbative contributions are :

$$C_4 \langle O_4 \rangle = 8\pi^2\, m_u \langle \bar{u}\, u \rangle \left(1 + \frac{\bar{a}_s}{3}\right) + 3\pi \left\langle \alpha_s\, G^2 \right\rangle \left(1 + \frac{7}{6}\, \bar{a}_s\right) \tag{5.4}$$

$$C_6\langle O_6\rangle = 4\pi^3\left[\alpha_s\left\langle\bar{u}\,\gamma_\mu\,\gamma_5\,\frac{\lambda_a}{2}\,u - \bar{d}\,\gamma_\mu\,\gamma_5\,\frac{\lambda_a}{2}\,d\right\rangle^2 - \frac{2}{9}\,\alpha_s\left\langle\left(\bar{u}\,\gamma_\mu\,\frac{\lambda_a}{2} + \bar{d}\,\gamma_\mu\,\frac{\lambda_a}{2}\right)\right.\right.$$

$$\left.\left. \cdot \sum_{u,d,s} \bar{\psi}\,\gamma_\mu\,\frac{\lambda_a}{2}\,\psi\right\rangle\right] \tag{5.5a}$$

which can be conveniently written as

$$C_6\langle O_6\rangle = \frac{896}{81}\,\pi^3 k\,\alpha_s\,\left\langle\bar{\psi}\,\psi\right\rangle^2 \quad . \tag{5.5b}$$

where $k = 1$ if one uses the vacuum saturation assumption (Eq. 3.58) and $SU(2)_F$ symmetric condensates.

The leading dimension-eight condensate effects can be expressed in terms of Φ_i given in (3.45). One has[125]:

$$C_8\langle O_8\rangle \simeq (4\,\Phi_6 - 196\,\Phi_8 - 292\,\Phi_5 - 534\,\Phi_7) \quad , \tag{5.6a}$$

where we have added to the constant term the ones induced by the log term after a q^2 derivative. Eq. (5.6) can be reduced to $\langle G^2\rangle^2$ by using the modified factorization in Eq. (3.48). This leads for ($N_c = \infty$) to :

$$C_8\langle O_8\rangle \simeq \frac{\pi^2}{648}\left\langle\alpha_s\,G^2\right\rangle^2 \cdot \frac{75}{3} \tag{5.6b}$$

2. VACUUM CONDENSATES FROM MOMENT RATIOS $R_\rho(\tau)$

As a first phenomenological application of our previous discussions, let us analyze the ρ-meson sum rule moment ratio (Eq. 1.40)

$$R_\rho(\tau) \equiv -\frac{d}{d\tau} \log \int_0^\infty dt e^{-t\tau} \frac{1}{\pi} \text{Im}\, \Pi_\rho(t) \ , \tag{5.7}$$

One should notice that contrary to the Laplace $F(\tau)$ sum rule in Eq. (1.39), $R(\tau)$ is not sensitive to the leading α_s contribution to the unit-perturbative operator. The *QCD expression of the moment in the chiral limit* is :

$$R_\rho(\tau) = \tau^{-1}\left\{1 + 2\, C_4\langle O_4\rangle \tau^2 - \frac{3}{2} C_6\langle O_6\rangle \tau^3\right\} . \tag{5.8}$$

The spectral function $\text{Im}\,\Pi_\rho(t)$ is related to the e^+e^- into $I = 1$ Hadrons total cross-section via the optical theorem :

$$\frac{1}{\pi} \text{Im}\, \Pi_\rho(t) \equiv \frac{t}{16\pi^3\alpha^2} \sigma_{e^+e^- \to I=1}(t) \ . \tag{5.9}$$

The data of $\sigma_{e^+e^- \to I=1}$ normalized to $\sigma_{e^+e^- \to \mu^+\mu^-}$ is shown in Fig. 5.1 up to $\sqrt{t} \simeq 2$ GeV.

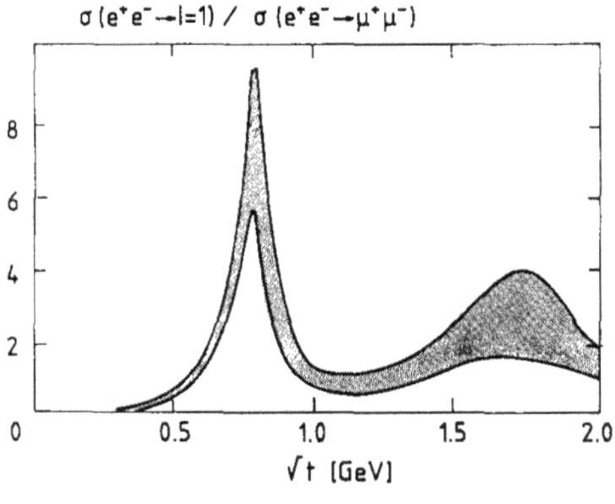

Fig. 5.1 : $R = \dfrac{\sigma(e^+e^- \to I=1)}{\sigma\left(e^+e^- \to \mu^+\mu^-\right)}$ *versus the centre-of-mass energy* $\sqrt{t}$.

Beyond this value of t, we parametrize the continuum using the discontinuity of the QCD Feynman diagrams. This phenomenological choice of the continuum threshold will be justified from the stability criterion of the output parameters and for the asymptotic consistency of the QCD and experimental sides of the sum rules. (We compare the phenomenological and QCD sides of the moment $R_\rho(t)$ in the region $[0, \tau]_{MAX}$ by using the MINUIT fitting program with the two parameters $C_4\langle O_4\rangle$ and $C_6\langle O_6\rangle$. We seem to find an optimal value of $\tau \simeq 1\ \mathrm{GeV}^{-2}$ where information from the lowest resonance is optimal and at the same time the OPE is still reliable.

This corresponds to the set of values[100]:

$$C_4\langle O_4\rangle = (3.5 \pm 0.9)\ 10^{-2}\ \mathrm{GeV}^4$$

$$C_6\langle O_6\rangle \simeq \frac{3}{2}\ (0.18 \pm 0.06)\ \mathrm{GeV}^6 \quad . \qquad (5.10)$$

Using the pion PCAC relation in Eq.(1.18) for the estimate of the $m\langle\bar\psi\psi\rangle$ quark condensate, we also deduce at this optimal value of τ :

$$\langle\alpha_s\, G^2\rangle = (0.04 \pm 0.01)\ \mathrm{GeV}^4 \quad . \qquad (5.11)$$

We give the behaviour of these results versus the changes in τ_{MAX}. The results are quite stable and indicate that larger values of $C_4\langle O_4\rangle$ require larger values of $C_6\langle O_6\rangle$. This can be understood as their contributions tend to compensate in the moment $R(\tau)$.

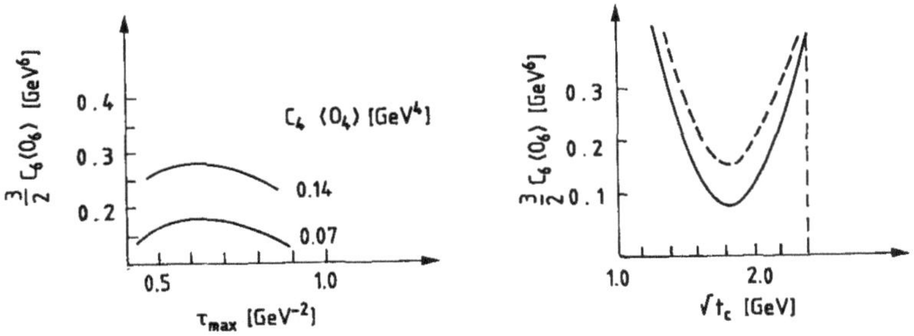

Fig. 5.2 and 5.3 : τ- and t_c-behaviours respectively of the QCD condensates from moment ratios.

However, the results in Eq. (5.10) are sensitive to the changes of the continuum threshold $\sqrt{t_c}$. The additional requirement that these values are *minimally sensitive* to the choice of the unphysical parameter $\sqrt{t_c}$ requires that their derivative with respect to $\sqrt{t_c}$ vanishes. This corresponds to $\sqrt{t_c} \simeq 2$ GeV which corresponds to the phenomenological threshold. This value is also in the range required for a minimal sensitivity of R_ρ^{EXP} in the choice of $\sqrt{t_c}$ $\left(\frac{d\ R_\rho^{EXP}}{dt_c} = 0\right)$. The error bars in Eq. (5.10) take into account the uncertainties in the choice of τ_{MAX} and those from the e^+e^- data. However, one should, notice that in spite of the large uncertainties of the data, the experimental ratio R_ρ is quite accurate because the errors tend to compensate in the ratio of the sum rules. The resulting fit of $R_\rho(t)$ is shown in Fig. 5.4 :

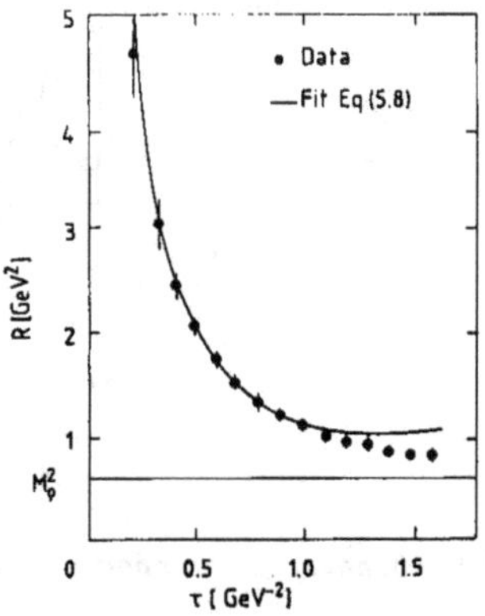

Fig. 5.4 : Fitted value of $R_\rho(\tau)$ *versus* τ *and comparison with the data.*

3. THE $R_\rho(\tau)$ RESULT VERSUS OTHER ESTIMATES

Let us now compare the estimate in (5.10) with some other results. The value of $\langle \alpha_s G^2 \rangle$ obtained in (5.10) is consistent with the so-called "canonical" value obtained from charmonium[18,29] which we shall see later on. The value of $\langle \alpha_s G^2 \rangle$ also agrees with the one obtained from the Laplace sum rule in Eq. (1.39) but the value of $C_6\langle O_6 \rangle$ obtained in Ref. 126) seems to be underestimated. One possible source of this discrepancy is that the analysis of Ref. 126) is also sensitive to the α_s perturbative term contrary to R_ρ. Therefore, from their fit, one might lose the accuracy for the determination of $C_6\langle O_6 \rangle$. When using the value of $\langle \alpha_s G^2 \rangle$ in the estimate of $C_8\langle O_8 \rangle$, it is reassuring that the neglected term is small. This can be a good indication of the convergence of the QCD series. The ratio of the $C_6\langle O_6 \rangle$ over the $C_4\langle O_4 \rangle$ condensates have also been determined directly from FESR-like exponential moments. The result is[101]:

$$C_6\langle O_6\rangle \Big/ C_4\langle O_4\rangle \simeq (2.86 \pm 0.04)\ \mathrm{GeV}^2 \qquad (5.12)$$

and agrees within the errors with the one in Eq. (5.10) and Fig. 5.3. FESR of the type in Eq. (1.43) have also been used for the estimate of the above condensates. The advantage of FESR is that one can express each condensate in terms of the spectral function and the perturbative part of Π_ρ. One gets the constraints[23].

$$C_2\langle O_2\rangle = 8\pi^2 \int_0^{t_c} dt\ \frac{1}{\pi}\ \mathrm{Im}\ \Pi_\rho(t) - t_c\ F_2(t_c) \qquad (5.13a)$$

$$C_4\langle O_4\rangle = 8\pi^2 \int_0^{t_c} dt\ t\ \frac{1}{\pi}\ \mathrm{Im}\ \Pi_\rho(t) - \frac{t_c^2}{2}\ F_4(\tau_c)$$

$$- C_6\langle O_6\rangle = 8\pi^2 \int_0^{t_c} dt\ t^2\ \frac{1}{\pi}\ \mathrm{Im}\ \Pi_\rho(t) - \frac{t_c^3}{3}\ F_6(t_c) \qquad (5.13b)$$

where :

$$F_{2n}(t_c) = 1+ \frac{1}{-\beta_1 \log t_c/\Lambda^2} + \left(\frac{1}{-\beta_1 \log t_c/\Lambda^2}\right)^2 \left[F_3 - \frac{\beta_2}{\beta_1} \log \frac{t_c}{\Lambda^2} - \frac{\beta_1}{2n}\right] + .. \qquad (5.13c)$$

The drawback of the method is its sensitivity on the high t-behaviour of the spectral function where the data are inaccurate. The resulting estimate of the condensates fitted from the data is[28]:

$$C_4\langle O_4\rangle \simeq 0.07 - 0.19\ \mathrm{GeV}^4$$

$$C_6\langle O_6\rangle \simeq 0.3 - 0.5\ \mathrm{GeV}^6 \quad . \qquad (5.14)$$

The absolute values of the condensates in Eq. (5.14) are at least a factor two higher than the ones in Eq. (5.10) although the relative strengths of $C_4\langle O_4\rangle / C_6\langle O_6\rangle$ coming from the three different analyses agree with each other. In trying to test the accuracy of the results in Eq. (5.14), we show the different behaviour of the heat evolution test for each set of condensates at different time τ. This is shown in Fig. 5.5 from Ref. 25) for σ = 0.1, 0.5 and 1 GeV^4. The discrepancy between the curves is pronounced for small σ where the QCD series are unreliable. For reasonable σ values, the two sets of condensates in Eqs (5.10) and (5.14) give (taking into account the accuracy for the theoretical derivation of $G(\hat{t},\sigma)$) almost (within 10%) the same curves. Then we are tempted to conclude that the "heat evolution test" might not be sufficient for discriminating the two values in (5.10) and (5.14) and we may still need some other tests.

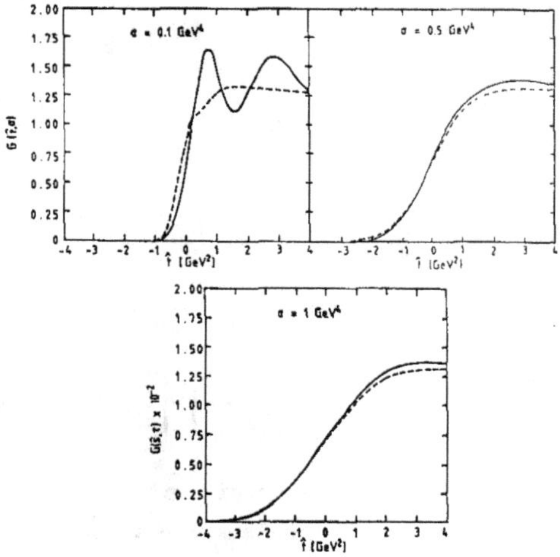

Fig. 5.5 : Heat evolution test for the sets of QCD condensates obtained from FESR at different values of σ *:*

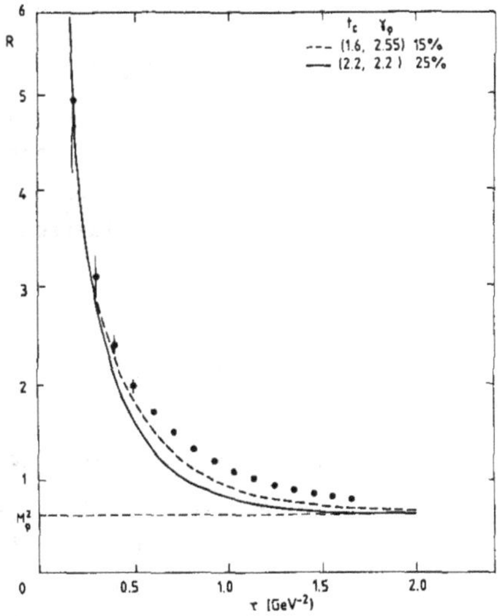

Fig. 5.6 : $R_\rho(\tau)$ *for different parametrization of the spectral functions :* ϕ *data ; --- and — duality ansatz.*

4. MODELLING THE DATA USING A QCD-DUALITY ANSATZ

Due to the complexity and to the absence of the data in some channels, it appears necessary to introduce a simple model to replace the data. The model often discussed in the literature is the duality ansatz parametrization of the spectral function. In this context, one has :

$$\frac{1}{\pi} \text{Im } \Pi_\rho(t) = \frac{M^2_\rho}{4\gamma^2_\rho} \delta\left(t-M^2_\rho\right) + \theta(t-t_c) \text{ "QCD continuum"} \quad , \qquad (5.15)$$

where the first term is the lowest resonance contribution whilst the second term takes into account <u>all</u> discontinuities coming from QCD diagrams. γ_ρ is the coupling constant normalized as in Eq. (1.29b).

One might expect that this simple ansatz would give a good description of the spectral integral for Laplace sum rules-like moments, owing to the exponential weight which enhances the lowest ground-state effects. This is not so in the case of FESR. We test the accuracy of (5.15) using the moment ratio $R_\rho(\tau)$. The asymptotic coincidence of the two sides of $R_\rho(\tau \to 0)$ (QCD and within the parametrization in (5.15) leads to the FESR lowest moment in (5.13a). This gives in the chiral limit :

$$t_c \simeq \frac{2\pi^2}{\gamma_\rho} M_\rho^2 \,/\, F_2(t_c) \tag{5.16}$$

Given $\gamma_\rho^{EXP} \simeq 2.55$, we deduce $t_c \simeq 1.7$ GeV^2. The behaviour of $R_\rho(\tau)$ using (5.15) is given in Fig. 5.6. We compare this curve with the full data (crossed points) obtained from Fig. 5.1 and adding a QCD continuum for $\sqrt{t} \geqslant 2$ GeV. One can realize that in the region $\tau \geqslant 0.6$ GeV^{-2}, the model tends to underestimate the moments by about 15%. This can be understood because the QCD continuum gives only the average of the complex structures (ρ', ρ''...) in the energy region between 1 and 2 GeV. However, the agreement is quite good in the region where the OPE of QCD applies, i.e. $\tau \leqslant 3$ GeV^{-2} and might explain the success of this simple model in various uses of the Laplace sum rules[18,29,83,125,126]. It is quite clear that trying to push the agreements of this simple model with the data in the region where the OPE of QCD does not apply is rather demanding, since in that region one should take into account the low-energy behaviour of the spectral function which has a great effect there. However, in the interesting domain the neglect of this threshold effect is a good approximation.

5. THE ACCURACY OF THE ρ-MASS PREDICTION FROM THE MOMENT RATIO

The meson mass can be extracted from $R_\rho(\tau)$ by comparing the "phenomenological" expression deduced from (5.15) with its QCD side. This can be done using a least square fit method like MINUIT or FUMILI which minimizes the distance

$$\left| R_\rho^{QCD}(\tau) \quad - \quad R_\rho^{EXP} \ (\tau) \right| \tag{5.17}$$

Another method which is numerically "simpler" is the modified moment[18] (Eq. 1.41). Contrary to $R(\tau)$ in Eq. (1.40) which has a positivity property from which one can deduce the optimal upper bound on the lowest ground-state mass (minimum or inflexion point of R), the modified $R^c_\rho(\tau)$ moment in Eq. (1.41) does not generally have such a property. Within the duality ansatz in Eq. (5.15), the LHS of $R^c_\rho(\tau)$ represents the lowest ground-state mass squared.

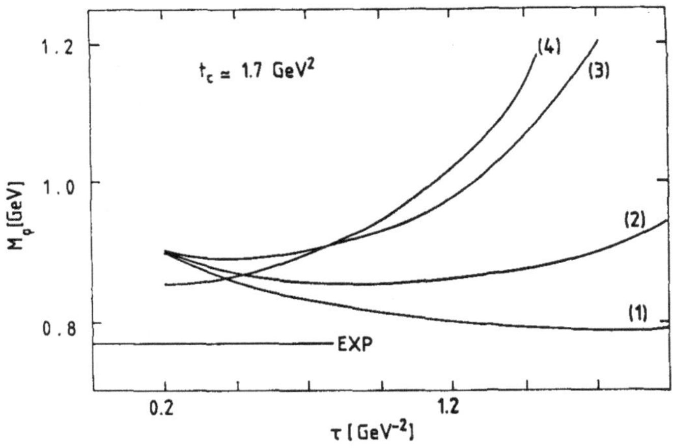

Fig. 5.7 : Predicted value of M_ρ from $R^c(\tau)$ versus the LSR scale τ for different values of the condensates in Table 5.1.

Fixing $t_c \simeq 1.7$ GeV^2 from Eq. (5.16), we give the prediction on M_ρ in Fig. 5.7 versus the sum rule variable τ for each set of condensates predicted previously from the data (Table 5.1) :

set	$C_4\langle O_4\rangle$ GeV^4	$C_6\langle O_6\rangle$ [GeV^6]	Remarks
1	0.035	0.06	Factorization
2	0.035	0.12	(Ref.100)
3	0.07	0.3	
4	0.15	0.5	(Ref.25)

Table 5.1 sets of condensates

It will be realized that at the given canonical value of $C_4\langle O_4\rangle$, the increases of $C_6\langle O_6\rangle$ shift the position of the minimum. Stability is obtained for (1) at too low values of $\tau \simeq 1.4 - 1.6$ GeV^{-2} where neglected non-perturbative contributions can show up ($C_8\langle O_8\rangle$ effects are still negligible!) ,although the ρ-mass prediction is good. Set 2 gives stable results for wider and smaller values of $\tau(\simeq 0.4 \sim 1$ GeV$^{-2})$. One can cross-check that in these regions, one still has the "fiducial" region or "sum rule window" (small continuum <u>and</u> condensate contributions). For condensate sets (3) and (4), stability in τ is obtained at too low values of $\tau \simeq 0.2 - 0.4$ GeV^{-2} , owing to the large values of these condensates. However, in this region the continuum contribution to the sum rule is important. So, one cannot satisfy the "fiducial" constraint by using Sets 3 and 4. In these cases, the result should be sensitive to the form of the continuum which is an unlikely situation. We give the behaviour of the M_ρ minima in τ versus the continuum threshold in Fig. 5.8. We notice that for t_c smaller than 1.7 GeV^2, the curves (3) and (4) loose the τ-stability.

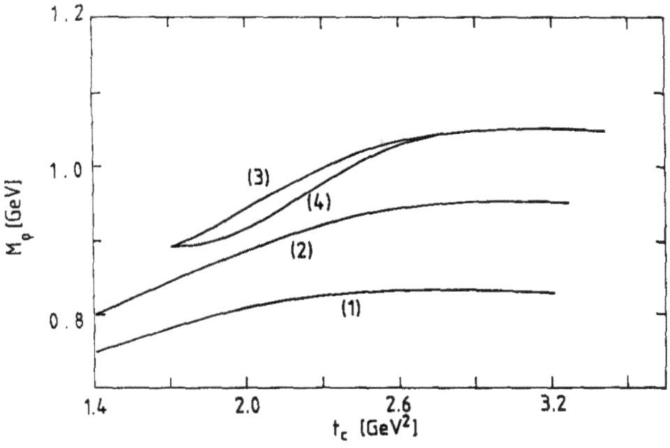

Fig. 5.8 : Behaviour of the τ minima for M_ρ versus the continuum threshold t_c.

The stability in t_c is obtained for $t_c \geqslant 2.6$ GeV^2 which is too high compared to the FESR constraint in Eq. (5.16). "Plateau" in t_c means that in this region, we have a complete dominance of the lowest ground state to the sum rule. However, owing to the nearby resonances in the 1-2 GeV energy region, it is impossible to satisfy this complete ρ-dominance unless one overestimates the resonance mass. Let us, note, however that the t_c-stability can be satisfied for real values of the resonance mass in the heavy-quark sector when working with the moments in (1.55). For light-quark systems, we might conclude that optimal estimate of the resonance mass is obtained within :

i) The FESR lowest dimension constraint on the value of the continuum threshold t_c ;

ii) The τ-stability of the moments.

We shall test these two requirements in some other channels as well.

6. **PREDICTION OF THE ρ-COUPLING CONSTANT γ_ρ**

We again use the duality ansatz in Eq. (5.15) and work for instance with the Laplace sum rule :

$$\int_0^\infty dt\ e^{-t\tau}\ \frac{1}{\pi}\ \mathrm{Im}\ \Pi_\rho(t). \tag{5.18}$$

One gets the sum rule for γ_ρ :

$$\frac{1}{\gamma_\rho^2} = \frac{e^{M_\rho^2\tau}}{\left(M_\rho^2\ \tau\right)} \cdot \frac{1}{2\pi^2}\left\{\left(1 + \frac{\bar{\alpha}_s}{\pi}(\tau)\right)\left(1 - e^{-t_c\tau}\right) + C_4\langle O_4\rangle\tau^2 - \frac{1}{2}C_6\langle O_6\rangle\tau^3 + \ldots\right\}, \tag{5.19}$$

where we have retained the most relevant QCD contributions. We give the predicted value of γ_ρ in Fig. 5.9 versus τ for the sets of condensates given in the Table 5.1.

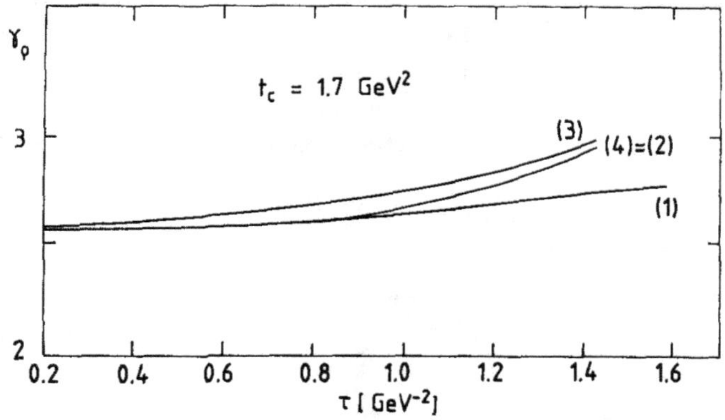

Fig. 5.9 : Value of the ρ-couplig γ_ρ versus the sum rule scale τ.

One can notice that for values of $\tau \leqslant 1$ GeV^{-2}, the result is almost insensitive to the condensate values. We have also studied the t_c-dependence of the minimum. We found that for t_c larger than 1.7 GeV^2, we loose the stability in τ. Then, by requiring that the optimal prediction of γ_ρ, which is obtained for the values $t_c \approx 1.4 - 1.7$ GeV^2, is τ-stable, in agreement with the FESR constraints in Eq (5.16), we have :

$$\gamma_\rho \simeq 2.5 - 2.8 \qquad (5.20)$$

for <u>all</u> sets of condensates given in the Table 5.1.

7. CONCLUSIONS : OPTIMAL RESULTS FROM THE ρ-MESON SUM RULES

i) The ratio of the condensates in Eq (5.12) seems to be a common prediction of different e^+e^- data fits.

ii) The "fiducial" region or "sum rule window" condition for the ρ-meson mass prediction excludes the sets 3 and 4 of condensate values in Table 1, although these values seem to be favoured by the heat evolution test in Fig. 5.5. However, the accuracy of this test is not known. Thus it appears that only set 2 satisfies the two conditions i) and ii). It gives :

$$\left\langle \alpha_s\, G^2 \right\rangle \simeq 0.04 \text{ GeV}^4$$

$$\alpha_s \left\langle \bar\psi\, \psi \right\rangle^2 \simeq 3.5\ 10^{-4} \text{ GeV}^6 \quad , \qquad (5.21)$$

which correspond to a "canonical" value of the gluon condensate and to a violation by a factor two of the vacuum saturation assumption. However, owing to the spread of the QCD predictions on these quantities, it is conservative to consider the values of these condensates within a factor two higher but, as we have shown previously, too large values lead to inconsistencies.

iii) The last condition needed to get the optimal values of the resonance parameters within the duality ansatz in Eq (5.15) is the value of the continuum threshold which also influences the τ-stability of the predictions. The value of γ_ρ is τ stable for the t_c values in the range given by FESR in (5.17). At this FESR value of t_c, one predicts :

$$\gamma_\rho \simeq 2.5 - 2.8 \quad \text{and} \quad M_\rho \simeq (0.8 - 0.85)\ \text{GeV} \ , \qquad (5.22)$$

which is quite good compared to the data. The 10% error can be considered as a typical uncertainty of the moment prediction given the t_c value fixed by FESR. As we have already noticed, contrary to the heavy-quark sector (see following chapters), the optimal prediction of M_ρ cannot be obtained inside the t_c plateau where a single dominance of the sum rule by the lowest ground state is a good approximation. The reason is that the exponential factor entering into the sum rule is not operative enough to kill the contributions of the complex structure in the 1-2 GeV region.

CHAPTER 6

WEINBERG AND AXIAL-VECTOR SUM RULES IN QCD

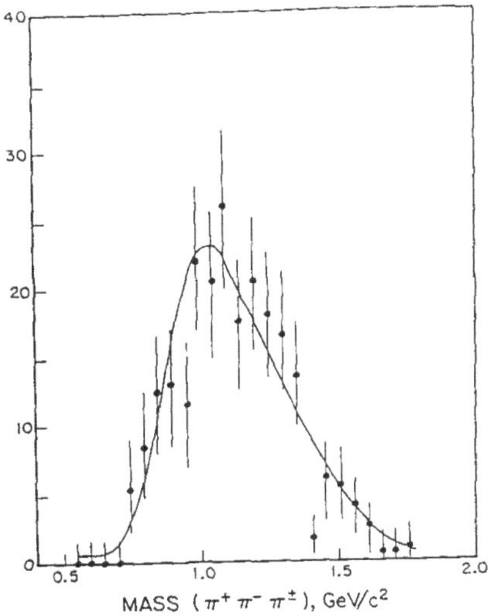

Prof. Steven Weinberg looking at the A1 resonance from tau decay

1. WEINBERG SUM RULES IN QCD

a) QCD corrections to the Weinberg sum rules

We have discussed in Chapter 1.2 the Weinberg sum rules based on the assumed asymptotic realizations of $SU(2)_L \times SU(2)_R$ chiral symmetry (Eqs (1.25, 1.26)). With the advent of QCD it is now possible to give a quantitative control of these assumptions. Quark-mass corrections to (1.25) and (1.26) have been studied in Ref. 128). Radiative corrections to the $\langle \bar{\psi} \psi \rangle$ condensate have been discussed in Ref. 118), while the effect of the four-quark condensates has been included in Ref. 129). It should be noted that radiative corrections have been shown to conserve the $SU(N)_L \times SU(N)_R$ symmetry in the chiral limit[130]. The two-point correlator entering into the first Weinberg sum rule reads[128]:

$$Q_{ij} \equiv \Pi^{(1)}_{LR} + \Pi^{(0)}_{LR} \simeq \left(\frac{\alpha_s}{\pi}\right) \frac{1}{\pi^2} \left\{\frac{m_i m_j}{q^2} + \mathcal{O}\left(\frac{m^2}{q^2}\right)^2 \log - \frac{q^2}{m^2}\right\} , \quad (6.1)$$

where $\Pi^{(k)}_{LR} \equiv \Pi^{(k)}_V - \Pi^{(k)}_A$ is the difference of correlators associated respectively with the vector and axial currents. Eq. (6.1) indicates that the combination of spectral functions having the quantum numbers of the spin 1 + 0 hadrons, is only broken by the mass terms to order α_s/π. As a consequence, the first Weinberg sum rule is superconvergent in QCD.

The perturbative contribution to the two-point correlator entering into the second Weinberg sum rule, reads[128]:

$$\Pi^{(1)}_{LR} = - \Pi^{(0)}_{LR} = - \frac{3}{2\pi^2} \frac{m_i m_j}{-q^2} \left\{\log - \frac{q^2}{\nu^2} + \mathcal{O}(1) + \right.$$

$$\left. \frac{m_i^2}{-q^2} \log \frac{-q^2}{m_i^2} + \frac{m_j^2}{-q^2} \log \frac{-q^2}{m_j^2} \right\} . \qquad (6.2)$$

This sounds bad as the second Weinberg sum rule is quadratically divergent for finite quark mass. Some possible improvements of the second sum rule can be done in many ways. For instance Ref. 128) has proposed the simplest superconvergent combination :

$$\Delta^{(1,0)}_{ijk} \equiv \Delta^{(1,0)}_{ij} - \frac{m_j}{m_k} \Delta^{(1,0)}_{ik} \qquad (6.3)$$

with $\Delta^{(n)}_{ij} \equiv \Pi^{(n)}_{LR}$.

We shall also see that the sum-rule approach can improve the second sum rules.

b) The Laplace transform and FESR versions of the WSR

Using the results of Refs 128, 118 and 129), one can deduce the Laplace transform of the original Weinberg sum rules in Eqs (1.25) and (1.26) :

$$\int_0^\infty dt\, e^{-t\tau} \frac{1}{\pi} \left(\mathrm{Im}\, \Delta^{(1)}_{ud} + \mathrm{Im}\, \Delta^{(0)}_{ud} \right) \simeq \left(\frac{\alpha_s}{\pi} \right) \frac{1}{\pi^2} \left\{ - \bar{m}_u \bar{m}_d + \right.$$

$$\left. \frac{8}{3} \pi^2 m_d \langle \bar{u} u + \bar{d} d \rangle \tau - \frac{32}{9} \pi^3 \rho \langle \bar{u} u \rangle^2 \tau^2 \right\} \qquad (6.4)$$

and

$$\int_0^{\infty} dt\ t\ e^{-t\tau}\ \frac{1}{\pi}\ \mathrm{Im}\ \Delta^{(1)}_{ud} = \frac{3}{2\pi^2}\ \tau^{-1}\ \bar{m}_u\ \bar{m}_d\ -\ 2\left\langle m_u\ \bar{d}\ d + m_d\ \bar{u}\ u\right\rangle , \qquad (6.5)$$

where $\rho = 1$ if one uses the vacuum saturation assumption of the four-quark condensates. The leading chiral breaking term in Eq.(6.4) is the linear mass term which is related to $f_\pi^2\ m_\pi^2$ via pion PCAC. As the sum-rule scale τ is different from zero $\left(\mathcal{O}(1\ \mathrm{GeV}^{-2})\right)$, one can expect improved predictions of the resonance parameters from the WSR in Eqs (6.4) and (6.5). The FESR version of the WSR reads

$$\int_0^{t_c} dt\ \frac{1}{\pi}\ \mathrm{Im}\ \left(\Delta^{(1)}_{ud} + \Delta^{(0)}_{ud}\right) \simeq \left(\frac{\alpha_s}{\pi}\right) \frac{1}{\pi^2} \left(-\bar{m}_u\ \bar{m}_d\right) ,$$

$$\int_0^{t_c} dt\ t\ \frac{1}{\pi}\ \mathrm{Im}\ \Delta^{(1)}_{ud} \simeq \frac{3}{2\pi^2}\left(\bar{m}_u\ \bar{m}_d\right) .\ t_c + \ldots\ , \qquad (6.6)$$

which would correspond to $\tau \to 0$. The accuracy of the Laplace and FESR versions of the WSR has been tested by using the spectral functions of the $\tau \to \nu_\tau$ + "Hadrons" data[131] (odd minus even numbers of pions), which are known in a limited kinematical range ($t \leqslant 2.5\ \mathrm{GeV}^2$). This test is shown in Figs 6.1 and 6.2 respectively for the first (after subtracting the pion pole contribution) and second WSR versus the cut--off t_c of the integral and for two values of the Laplace sum-rule scale τ. One can see that in the case of the first sum rule, the Laplace transform approaches nicely the chiral limit prediction. In the case of the second WSR, there is less agreement with the chiral limit prediction. This is mainly due to the quadratic divergence of the integral. However, it indicates that for t_c about 2 GeV^2 the prediction is optimal for the Laplace and FESR, the latter being more dependent on t_c than the former. At this value of t_c, one could deduce for $\tau \simeq 0.7 - 1\ \mathrm{GeV}^{-2}$ a prediction for f_π from Fig. 6.1 :

$$f_\pi \simeq 85 - 102 \text{ MeV} \quad , \tag{6.7}$$

where the range of values in Eq. (6.7) covers that from FESR.

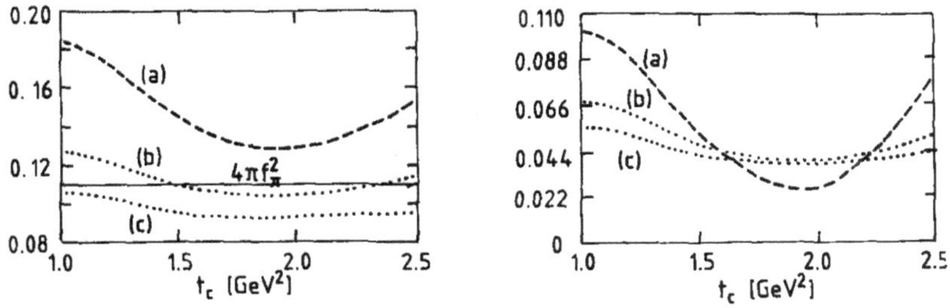

Figs 6.1 and 6.2 : 1st and 2nd WSR --- compared to its Laplace version ... for τ = 0.7 GeV^{-2} (b) and τ = 1 GeV^{-2} (c).

From this test, we find it appropriate to discuss the Laplace version of the WSR within the duality ansatz for the spectral function.

c) The A_1-coupling and f_π from the WSR

Let us reconsider the predictions in Eq.(1.28) by using the WSR in Eqs (6.4) and (6.5) where we shall use the duality ansatz parametrization of the spectral function in Eq. (5.15). QCD continuum is due to the discontinuity of the QCD diagrams. Then, this contribution cancels in the first sum rule while in the second it is related to the $\bar{m}_u \bar{m}_d$ term. This cancellation of continuum might explain the success of the lowest mass resonance saturation of the WSR. The lowest mass resonance contribution to the spectral function is due to the

$\rho(1^{--})$, $A_1(1^{++})$, $\pi(0^{-+})$ and $a_0(0^{++})$according to the low-energy realization of the $SU(2)_L \times SU(2)_R$ symmetry. Their couplings to the currents are :

$$\left\langle 0 \left| \bar{u}\ \gamma_\mu(\gamma_5)\ d \right| \rho(A_1) \right\rangle = \epsilon_\mu \sqrt{2}\ \frac{M^2_{\rho(A_1)}}{2\ \gamma_{\rho(A_1)}}$$

$$\left\langle 0 \left| \bar{u}\ \gamma_5\ (\gamma_5)\ d \right| a_0\ (\pi) \right\rangle = \sqrt{2}\ f_{a(\pi)}\ P_\mu \quad . \tag{6.8}$$

Then, we obtain from Eqs (6.4) and (6.5)[129]:

$$e^{-M^2_\rho \tau}\ \frac{M^2_\rho}{2\gamma^2_\rho} - e^{-M^2_{A_1} \tau}\ \frac{M^2_{A_1}}{2\gamma^2_{A_1}} - 2\ f^2_\pi\ e^{-M^2_\pi\ \tau} \simeq$$

$$(-)\ \frac{\bar{\alpha}_s}{\pi} \cdot \tau \left(\frac{8}{3}\ m^2_\pi\ f^2_\pi + \frac{32}{9}\ \pi\ \rho \left\langle \bar{u}\ u \right\rangle^2 \tau \right) \quad ; \tag{6.9a}$$

$$e^{-M^2_\rho\ \tau}\ \frac{M^4_\rho}{2\gamma^2_\rho} - e^{-M^2_{A_1}\ \tau}\ \frac{M^4_{A_1}}{2\gamma^2_{A_1}} \simeq 2\ f^2_\pi\ m^2_\pi \quad . \tag{6.9b}$$

We have neglected the small contribution of the 0^{++} resonance as it is proportional to the light-quark mass difference $(m_d - m_u)$. Notice that the four-quark condensate $\left\langle \bar{u}\ u \right\rangle^2$ gives the leading QCD correction in the chiral limit $m^2_\pi \to 0$. Following Ref. 129), we shall test the sensitivity of the sum rules to the resonance contributions for different τ values, as we cannot hope to have τ-stability of the results within our approximations. We have, in principle, two unknowns γ_{A_1} and M_{A_1} for given values of τ and for ρ and π parameters fixed from the data. We can also use the data on the A_1 mass and deduce the relevant chiral symmetry order parameter f_π. This strategy has been used in Refs 129) and 132). Using $M_{A_1} \simeq 1.27$ GeV, we deduce from Eq. (6.9b) for $\tau \simeq (0.7 \sim 1.6)\ \text{GeV}^2$:

$$\gamma_{A_1} \simeq (1.2 \sim 1.9)\ \gamma_\rho \ . \tag{6.10}$$

Using this result in Eq.(6.8) and $\gamma_\rho \simeq 2.5$ from the data, we deduce:

$$f_\pi \simeq (74 - 96)\ \text{MeV} \quad , \tag{6.11}$$

in good agreement with the data $f_\pi = 93.3$ MeV. If, instead, we had used the FESR version of the WSR, we would have obtained a higher value of 119 MeV for f_π. Eq. (6.11) indicates that the WSR really provides a measurement of the chiral symmetry order parameter f_π.

d) $\pi^+ - \pi^0$ mass difference from the WSR

Hadronic contributions to the $\pi^+ - \pi^0$ mass difference have been derived by Das et al[11] using current algebra techniques within the assumed $SU(2)_L \times SU(2)_R$ realization of chiral symmetry at short distance. They obtained the well-known sum rule :

$$m_{\pi^+} - m_{\pi^0} \simeq \frac{6\pi\alpha}{f_\pi^2 i} \int \frac{d^4q}{(2\pi)^4} \frac{1}{q^2} \int_0^\infty \frac{dt\ t}{q^2+t-i\epsilon} \frac{1}{\pi} \left(\text{Im}\,\Pi_V^{(1)} - \text{Im}\,\Pi_A^{(1)}\right) , \tag{6.12}$$

in terms of the spectral functions entering into the second WSR. Using a lowest resonance saturation, plus the approximate KSFR relation[16], one obtains

$$m_{\pi^+} - m_{\pi^0} \simeq \frac{3\ \alpha\ M_\rho^2\ \log 2}{4\pi\ m_\pi} \tag{6.13}$$

in good agreement with the data. This agreement is, however, quite surprising as Eq. (6.12) diverges at large t. The study of the short-distance behaviour of Eq. (6.12) within QCD has been done in the literature[133]. A more phenomenological approach is to test the cancellation mechanism in Eq. (6.12) by using the available data from τ decay. This has been done by writing Eq. (6.12) as a FESR-like expression[131]:

$$m_{\pi^+} - m_{\pi^0} \simeq \frac{3\,\alpha}{65\,\pi^2\,f_\pi^2}\int_0^{t_c} dt\; t\,\frac{1}{\pi}\left(\mathrm{Im}\,\Pi_A^{(1)} - \mathrm{Im}\,\Pi_V^{(1)}\right)\log t \qquad (6.14)$$

and study the influence of the cut-off t_c on the prediction. The result in Fig. 6.3 shows the cancellation of the high-energy parts of the spectral function, though they are individually important, and explains why the naive relation in Eq. (6.13) works nicely. The "optimal" prediction from Eq. (6.14) is reached at $t_c \simeq 1.8\ \mathrm{GeV}^2$ where:

$$m_{\pi^+} - m_{\pi^0} \simeq 4.6\ \mathrm{MeV} \qquad (6.15)$$

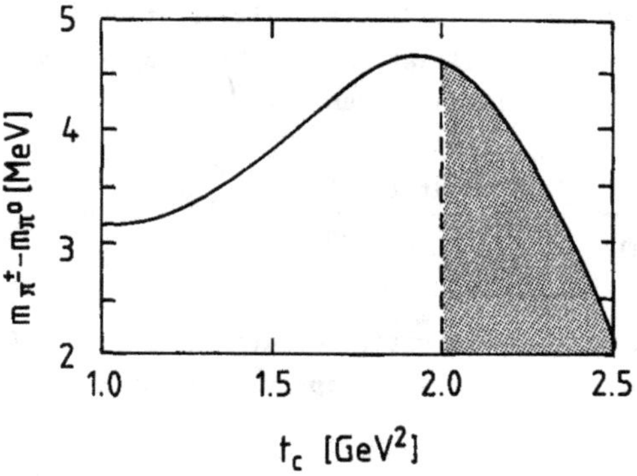

Fig. 6.3 : The π^+-π^0 mass difference versus the continuum threshold t_c.

2. THE AXIAL-VECTOR SUM RULE

a) A_1 coupling and upper bound for f_π

Alternatively, one can also determine f_π from the sum of the 1^{++} and 0^{-+} spectral functions[18] in the chiral limit $m_\pi^2 = 0$. The strategy can be understood as follows :

Write the axial correlator :

$$\Pi^{\mu\nu}(q^2) = i \int d^4x \, e^{iqx} \left\langle 0 \left| A^\mu(x)_j^i \left(A^\nu(o)_j^i\right)^+ \right| 0 \right\rangle \qquad (6.16a)$$

in terms of the invariants $\Pi^{(1)}$ and Π :

$$\Pi^{\mu\nu} = - g^{\mu\nu} q^2 \Pi_A^{(1)} + \Pi_A q^\mu q^\nu , \qquad (6.16b)$$

i.e. $\Pi_A^{(0)} \equiv \Pi_A^{(1)} - \Pi_A$ is the longitudinal part of the correlator. In the chiral limit the axial current is conserved. Therefore Im $\Pi^{(0)}$ vanishes. Keeping the leading quark-mass term, one has :

$$\Pi_A^{(1)} - \Pi_A = - \frac{(m_u+m_d)\left\langle \bar{u} u + \bar{d} d \right\rangle}{q^2} , \qquad (6.17)$$

which is the part of $\Pi^{(0)}$ coming from pion PCAC. Using Eqs (6.16) and (6.17), one can deduce a sum rule for Π_A :

$$\int_0^\infty \frac{dt}{t} e^{-t\tau} \frac{1}{\pi} \mathrm{Im}\, \Pi_A(t) \simeq \frac{\tau^{-1}}{4\pi} \left\{ 1 + \frac{\bar{\alpha}_s}{\pi}(\tau) + \frac{\pi}{3} \left\langle \alpha_s G^2 \right\rangle \tau^2 + \left(\frac{11}{7}\right) \frac{1}{2} C_6 \langle O_6 \rangle \tau^3 \right\} \qquad (6.18a)$$

where $C_6 \langle O_6 \rangle$ has been defined in Eq. (5.5b). As stated previously, this sum rule involves the π and A_1 mesons :

$$\frac{1}{\pi} \mathrm{Im}\, \Pi_A(t) = 2 f_\pi^2 \, \delta\left(t-m_\pi^2\right) + \frac{M_{A_1}^2}{2\gamma_{A_1}^2} \delta\left(t-M_{A_1}^2\right) + \Theta(t-t_c) \left(1 + \frac{\bar{\alpha}_s}{\pi}(t)\right) \quad (6.18b)$$

and can be used in order to determine f_π. One can use either the prediction of γ_{A_1} in Eq. (6.10) or use the data from $\tau \to \nu_\tau\, 3\pi$ decay or try to determine A_1 using another sum rule. We might try the last alternative by working with $q^2\, \Pi_A^{(1)}$ as SVZ[18]. The corresponding sum rule is :

$$\int_0^\infty dt\, e^{-t\tau} \frac{t}{\pi} \mathrm{Im}\, \Pi_A^{(1)}(t) = \frac{\tau^{-2}}{4\pi^2} \left\{1 + \frac{\bar{\alpha}_s}{\pi}(\tau) - \frac{\pi}{3}\left\langle \alpha_s\, G^2 \right\rangle \tau^2 - \left(\frac{11}{7}\right) C_6 \langle O_6 \rangle \tau^3 \right\}. \quad (6.19)$$

Using the duality ansatz, one can write a sum rule for γ_{A_1}. We use Set 2 of condensate values in Eq. (5.18). A numerical study of the constraint shows that $\gamma_{A_1}(\tau)$ has τ stability (Fig. 6.4), a condition which is necessary for an optimal prediction. The position of the τ minima decreases for increasing t_c. Like the case of the ρ, we use the lowest constraint from FESR :

$$4\pi^2 \frac{M_{A_1}^4}{2\gamma_{A_1}^2} \simeq \left(\frac{1}{2}\right) t_c^2\, F_4(t_c) - \frac{\pi}{3}\left\langle \alpha_s\, G^2 \right\rangle \quad (6.20)$$

which we show in Fig. 6.4 , F_4 being defined in Eq.(5.13c). The common solution from the two constraints is :

$$\gamma_{A_1} \simeq 4.2 \quad \text{and} \quad t_c \simeq 2.3\ \text{GeV}^2 \tag{6.21}$$

in remarkable agreement with the WSR result in Eq. (6.10).

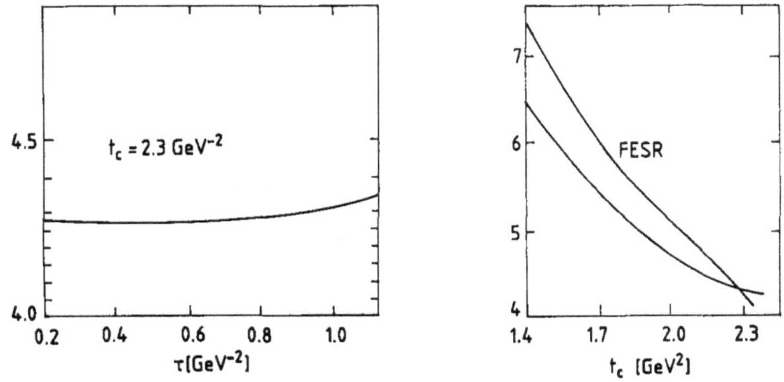

Fig. 6.4 : a) γ_{A_1} *versus the LSR scale* τ. *b) LSR and FESR results versus* t_c.

Let us now use the previous value of γ_{A_1} in the Π_A sum rule in Eq. (6.18) and try to determine f_π. As usual, we study the τ-stability of the prediction. The result does not satisfy this test, though at a particular value of $\tau \simeq M_\rho^{-2}$, we obtain the experimental value of f_π. Thus, we may conclude that the sum rules in Eq. (6.18) cannot give a reliable prediction of f_π within the approximation used. Then we try to derive a bound of f_π from Eq. (6.18) by using the positivity of the spectral function[129]. The result is much better. τ- and t_c-stabilities are reached (Fig. 6.5a,b and c) from which one deduces :

$$f_\pi \leqslant 111\ \text{MeV} \tag{6.22}$$

which is a stringent and rigorous bound.

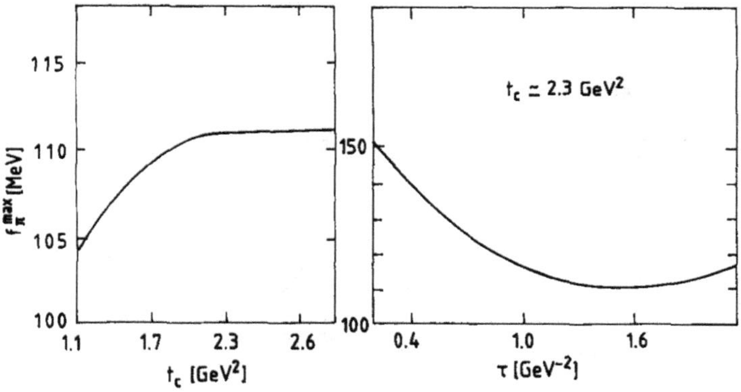

Fig. 6.5 : Upper bound on f_π versus t_c and τ from LSR.

b) The A_1 mass from the ratio of the moments

The analysis has been done in Ref. 29) by working with moments associated with $t\,\Pi_A^{(1)}$ and Π_A. We shall limit ourselves to the moment associated with $\Pi_A^{(1)}$ which involves only the 1^{++} mesons. We use the duality ansatz and work with the FESR-like moment $R_c(\tau)$ defined in Eq. (1.41) which is equal in this parametrization to $M^2_{A_1}$:

$$M^2_{A_1} \simeq 2\tau^{-1}\left\{\frac{\left(1+\frac{\bar{\alpha}_s}{\pi}\right)\left[1-e^{-t_c\tau}\left(1+t_c\tau+\frac{(t_c\tau)^2}{2!}\right)\right]+\frac{1}{2}\left(\frac{11}{7}\right)C_6\langle O_6\rangle\tau^3}{\left(1+\frac{\bar{\alpha}_s}{\pi}\right)\left[1-e^{-t_c\tau}(1+t_c\tau)\right]-\frac{\pi}{3}\left\langle\alpha_s G^2\right\rangle\tau^2-\frac{11}{7}C_6\langle O_6\rangle\tau^3}\right\} \tag{6.23}$$

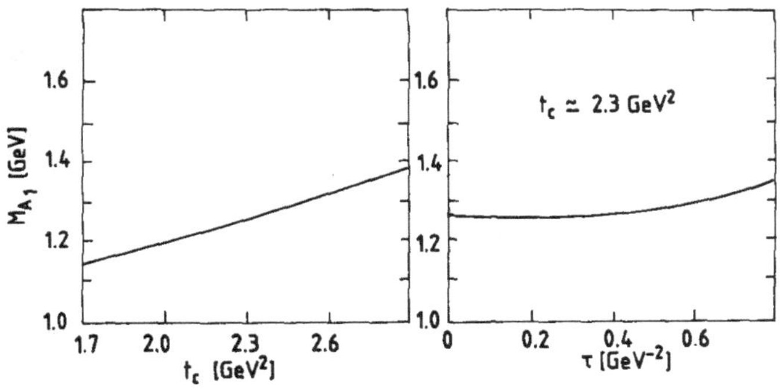

Fig. 6.6 : A_1 mass from LSR versus t_c and τ.

We show the t_c-behaviour of the τ minima and the τ-behaviour of the prediction in Fig. 6.6. The mass prediction increases with t_c. Using the t_c value from FESR in Eq. (6.21), we obtain :

$$M_{A_1} \simeq 1.28 \text{ GeV} \tag{6.24}$$

which is again in remarkable agreement with the data. We test the effects of the condensates on the prediction. For a given value of t_c, we use Set 3 given in Table 5.1. The A_1-mass prediction increases by about 5% and the stability in τ happens at very small τ values $\leqslant 0.1$ GeV^{-2}. These features related to high values of the condensates have already been encountered for the ρ case. The value of γ_{A_1} and M_{A_1} obtained previously can be used for predicting the τ decay into 3π. The branching ratio normalized by $\tau \to \nu_\tau e\bar{\nu}_e$ reads :

$$R_{A_1} = \cos^2\theta_c \frac{12\pi}{M_\tau^2} \int_0^{M_\tau^2} dt \left(1 - \frac{t}{M_\tau^2}\right)^2 \left(1 + \frac{2t}{M_\tau^2}\right) . \text{ Im } \Pi_A^{(1)}(t) \tag{6.25a}$$

where in a narrow width approximation :

$$\frac{1}{\pi} \text{ Im } \Pi_A^{(1)} (t) = \frac{M^2_{A_1}}{2\gamma^2_{A_1}} \delta\left(t - M^2_{A_1}\right) . \tag{6.25b}$$

Thus we deduce :

$$R_{A_1} \simeq 0.8 . \tag{6.25c}$$

This prediction is in agreement with the $\tau \to \nu_\tau + 3\pi$ (res) data.

3. ON THE DETERMINATION OF THE GLUON CONDENSATE FROM THE HYBRID AXIAL SUM RULES

We have seen in the previous chapter that the determination of the gluon condensate $\langle \alpha_s G^2 \rangle$ from $e^+e^- \to I = 1$ hadrons is not accurate. The reason is that the $\langle \alpha_s G^2 \rangle$ contribution is a small correction to the unit operator. In order to enhance the $\langle \alpha_s G^2 \rangle$ effect, Ref.134) suggests the study of the off-diagonal correlator :

$$\Pi^{\mu\nu} = i \int d^4x \, e^{iqx} \langle 0 | \mathbb{T} A_\mu(x) \left(\tilde{A}_\nu(0)\right)^+ | 0 \rangle \tag{6.26}$$

built from the axial $A_\mu \equiv \bar{u} \gamma_\mu \gamma_5 d$ and hybrid : $\tilde{A}_\nu \equiv g \bar{u} \gamma_\rho . \frac{\lambda a}{2} \tilde{G}^a_{\nu\rho} d$ currents. In the chiral limit, only the transverse piece of $\Pi^{\mu\nu}$ remains :

$$\Pi^{\mu\nu} \equiv (g^{\mu\nu} q^2 - q^\mu q^\nu) \Pi(q^2) . \tag{6.27}$$

Its QCD expression reads[134] :

$$\Pi(q^2 \equiv -Q^2) \simeq \frac{\alpha_s}{18\pi^3} Q^2 \log \frac{Q^2}{\nu^2} - \frac{1}{6\pi Q^2} \left\langle \alpha_s Q^2 \right\rangle - \frac{64\pi}{27 Q^4} \alpha_s \left\langle \bar{u} u \right\rangle^2 . \qquad (6.28)$$

As we work to leading order in α_s, we shall not include the anomalous dimension induced by the $\tilde{A}$-current. The lowest mass resonance contributions to the sum rules can be introduced as :

$$\left\langle 0 \left| \tilde{A}_\nu \right| \pi \right\rangle = -i \sqrt{2}\, f_\pi\, \delta^2_\pi\, p_\nu \quad ,$$

$$\left\langle 0 \left| \tilde{A}_\nu \right| A_1 \right\rangle = \sqrt{2}\, \frac{M^2_{A_1}}{2\,\gamma_{A_1}} \cdot \delta^2_A\, \epsilon_\nu \quad , \qquad (6.29)$$

where δ_i are constants to be fixed from the sum rules. We include (to a first approximation) in the QCD continuum the contribution of an axial hybrid meson $\tilde{A}_1$ which is expected to be in the range 1.6 - 2.1 GeV (see next chapter). From (6.28), one can derive the Laplace transforms in the chiral limit :

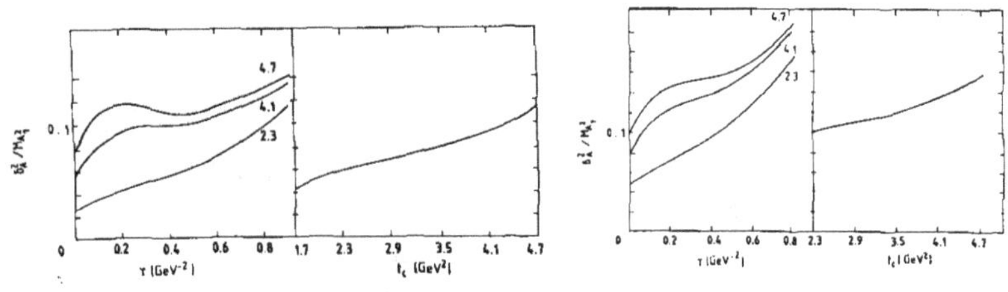

Fig. 6.7 : δ_A from LSR versus τ and t_c for sets 2 and 3 of the condensates given in Table 5.1.

$$- 2 f_\pi^2 \delta_\pi^2 + \frac{M_{A_1}^2}{2\gamma_{A_1}^2} \cdot \delta_A^2 e^{-\tau M_{A_1}^2} \simeq \frac{\bar{\alpha}_s}{18\pi^3} \cdot \tau^{-2} (1 - \rho_1)$$

$$- \frac{1}{6\pi} \left\langle \alpha_s G^2 \right\rangle - \frac{64}{27} \pi \tau \alpha_s \left\langle \bar{u} u \right\rangle^2 , \tag{6.30}$$

$$\frac{M_{A_1}^4}{2\gamma_{A_1}^2} \delta_A^2 e^{-\tau M_{A_1}^2} \simeq \left(\frac{\bar{\alpha}_s}{9\pi^3} \right) \tau^{-3} (1-\rho_2) + \frac{64\pi}{27} \alpha_s \left\langle \bar{u} u \right\rangle^2 , \tag{6.31}$$

where $\rho_n \equiv e^{-t_c \tau} \left(1 + t_c \tau + \ldots \frac{(t_c \tau)^n}{n!} \right)$.

We also have the FESR constraints :

$$- 2 f_\pi^2 \delta_\pi^2 + \frac{M_{A_1}^2}{2\gamma_{A_1}^2} \delta_A^2 \simeq \left(\frac{\bar{\alpha}_s}{18\pi^3} \right) \left(\frac{t_c^2}{2} \right) - \frac{1}{6\pi} \left\langle \alpha_s G^2 \right\rangle , \tag{6.32}$$

$$\frac{M_{A_1}^4}{2\gamma_{A_1}^2} \delta_A^2 \simeq \left(\frac{\bar{\alpha}_s}{18\pi^3} \right) \cdot \frac{t_c^3}{3} - \frac{64\pi}{27} \alpha_s \left\langle \bar{u} u \right\rangle^2 . \tag{6.33}$$

We use Eq. (6.31) for fixing δ_A. The prediction is given in Fig. 6.7 versus τ and t_c values using the values of the four-quark condensates from Sets 2 and 3 of Table 5.1. There is an inflexion point for $\tau \simeq 0.4$ GeV^{-2} which becomes a minimum for higher values of t_c. The behaviour of this minimum versus t_c in Fig. 6.7 shows that there is a stability for $t_c \simeq 3.5$ GeV^2. In this range one deduces :

$$\begin{aligned} \delta_A^2 &\simeq (0.13 - 0.15) \text{ GeV}^2 \quad \text{(set 2)} \\ &\simeq (0.20 - 0.23) \text{ GeV}^2 \quad \text{(set 3)} . \end{aligned} \tag{6.34a}$$

One can double-check the previous results. We use Eq. (6.34) in Eq. (6.33), and deduce the value of t_c :

$$t_c \simeq (3.7 - 4.2)\ \mathrm{GeV}^2 \qquad (6.34b)$$

which is consistent with the t_c-stability in Fig. 6.7. However, the t_c used in Ref.134) fails this test. Using these values in the FESR in Eq. (6.32), we obtain for the values of the gluon condensate in Table (5.1) :

$$\delta_\pi^2 \simeq (0.17 - 0.24)\ \mathrm{GeV}^2 \quad (\text{Set } 2)$$

$$\simeq (0.48 - 0.55)\ \mathrm{GeV}^2 \quad (\text{Set } 3) \qquad (6.35)$$

The result for Set 2 again agrees with Ref.134) but with larger errors despite the inconsistent values of t_c used in Ref.134). As in Ref 134), we also test the consistency of these results by analyzing the longitudinal part of the "pure" hybrid correlator :

$$\tilde{\Pi}_{\mu\nu} = i \int d^4x\ e^{iqx} \left\langle 0 \left| \mathbb{T}\ \tilde{A}_\mu(x) \left(\tilde{A}_\nu(0)\right)^+ \right| 0 \right\rangle \qquad (6.36)$$

The associated sum rule is :

$$2\, f_\pi^2\, \delta_\pi^4 + \frac{M_{A_1}^2}{2\gamma_{A_1}^2}\, \delta_A^4\, e^{-M_{A_1}^2 \tau} \simeq \frac{\alpha_s}{80\pi^3}\, \tau^{-3}\, (1-\rho_2) + \frac{\tau^{-1}}{72\pi} \left\langle \alpha_s\, G^2 \right\rangle$$

$$+ \frac{8\pi}{9}\, \alpha_s \left\langle \bar{u}\, u \right\rangle^2 \qquad (6.37)$$

As before, we absorb into the QCD continuum the contributions of hybrid states 1^{++} and 0^{--}, the latter having a mass of about 3 GeV. We use the value of δ_A in Eq. (6.34) and the corresponding values of the condensates, and we fit δ_π^2 . We show the result in Fig. 6.8 where we see nice τ- and t_c-stabilities. We obtain with very good precision :

$$\delta^2_\pi \simeq 0.24\ \text{GeV}^2 \quad \text{(set 2)}$$

$$\simeq 0.33\ \text{GeV}^2 \quad \text{(set 3)} \quad . \tag{6.38}$$

A comparison of the results in Eqs (6.35) and (6.38) does not favour higher values of the condensates. Our results confirm Ref.134)'s claim that high values of the condensates lead to inconsistencies. However, while confirming the "canonical" value of the gluon condensate, our detailed numerical fit does not exclude underestimation by a factor of about two of the four-quark operator from the vacuum saturation assumption. The arguments raised by Ref.134) for the value of the four-quark condensate remain obscure to us. Needless to say, the choice of the sum-rule scale of about 1.3 GeV^{-2} used in Ref. 134) to relate the π-pole to $\alpha_s \langle \bar{u}\, u \rangle^2$ is too large. This leads to great uncertainty for the corresponding prediction.

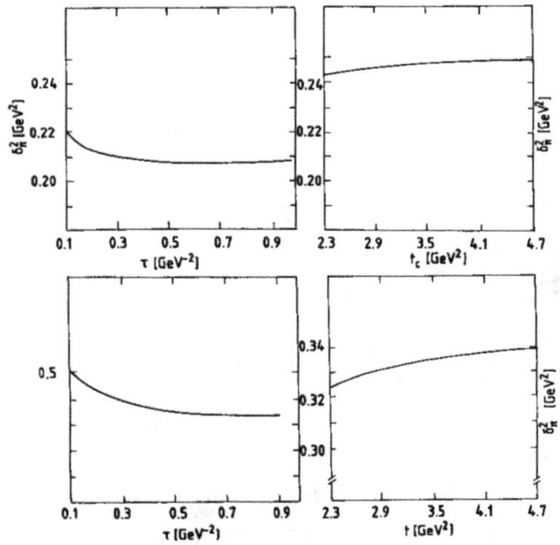

Fig. 6.8 : δ^2_π from LSR versus τ and t_c for sets 2 and 3 of condensate values in Table 5.1.

CHAPTER 7

SU(2) CHIRAL AND FLAVOUR SYMMETRY BREAKINGS : (PSEUDO)SCALAR AND OMEGA-RHO MIXING SUM RULES

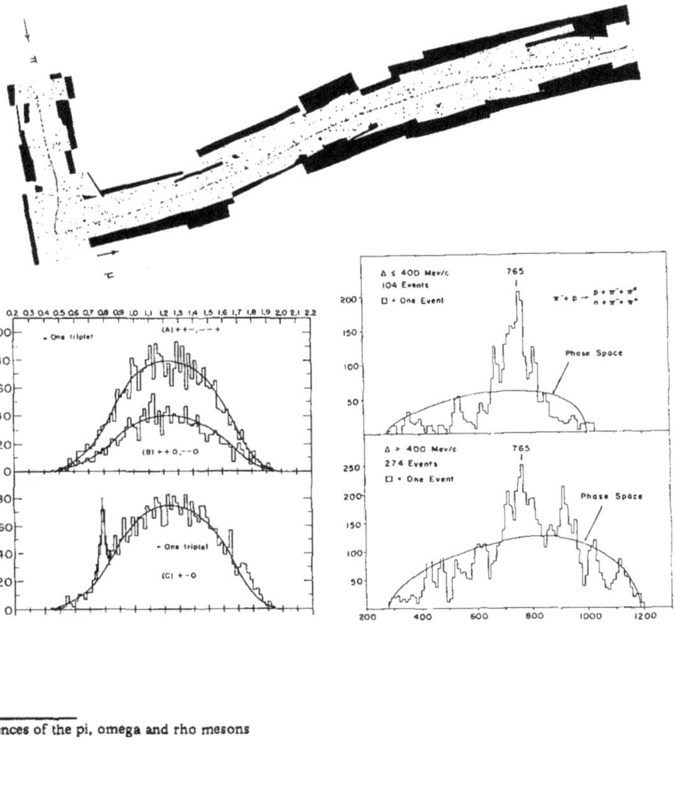

First evidences of the pi, omega and rho mesons

1. LIGHT-QUARK MASSES FROM THE PSEUDOSCALAR SUM RULES

a) Two-point correlators in QCD

We shall be concerned with the two-point correlator :

$$\psi_5(q^2)^i_j = i\int d^4x\, e^{iqx} \left\langle 0 \left| \mathbb{T}\, \partial_\mu A^\mu(x)^i_j \left(\partial_\mu A^\mu(o)^i_j\right)^+ \right| 0\right\rangle \qquad (7.1a)$$

built from the divergence of the axial current :

$$\partial_\mu A^\mu(x)^i_j = (m_i+m_j)\,\bar\psi_i(i\gamma_5)\,\psi_j \quad . \qquad (7.1b)$$

We have discussed in detail the QCD evaluation of these two-point functions in Chapters 1 and 4. We recall these expressions :

$$\psi_5(q^2)^u_d = \frac{3}{8\pi^2}(m_u+m_d)^2\left\{-q^2\log\frac{-q^2}{\nu^2}\left[1+2\,\frac{m_u^2+m_d^2-m_um_d}{-q^2}+\left(\frac{\alpha_s}{\pi}\right)\left(\frac{17}{3}-\log\frac{-q^2}{\nu^2}\right)\right]+\frac{8\pi^2}{3q^2}\left[\left(m_d-\frac{m_u}{2}\right)\left\langle\bar u\,u\right\rangle+\left(m_u-\frac{m_d}{2}\right)\left\langle\bar d\,d\right\rangle\right]-\frac{\pi}{3q^2}\left\langle\alpha_s\,G^2\right\rangle-\frac{1}{2q^4}\left[m_d\left\langle\bar u\,G\,u\right\rangle-m_u\left\langle\bar d\,G\,d\right\rangle\right]+\frac{896}{81}\,\frac{\pi^3}{q^4}\,\rho\,\alpha_s\left\langle\bar u\,u\right\rangle^2\right\} \quad . \qquad (7.2)$$

For the light u and d quarks, the contributions proportional to the quark masses can be neglected.

It is clear from (7.2) that the second derivative of $\psi_5(q^2)^u_d$ obeys an unsubtracted dispersion relation :

$$\psi_5''(q^2) = 2\int_0^\infty \frac{dt}{(t-q^2)^3}\ \frac{1}{\pi}\ \mathrm{Im}\ \psi_5(t) \tag{7.3}$$

and a homogeneous RGE :

$$\left\{-\frac{\partial}{\partial t} + \beta(\alpha_s)\frac{\partial}{\partial\alpha_s} - (1+\gamma_m)\ x_i\ \frac{\partial}{\partial x_i}\right\}\psi_5''(q^2) = 0 \quad . \tag{7.4}$$

We also know the low-energy behaviour of the two-point correlator from current algebra Ward identity (Eq. 1.17) :

$$\psi_5(o)_d^u = -(m_u+m_d)\left\langle \bar{u}\ u + \bar{d}\ d\right\rangle . \tag{7.5}$$

We shall also consider in connection with (7.3) the subtracted quantity :

$$\frac{\psi(q^2) - \psi(0)}{q^2} = \int_0^\infty \frac{dt}{t}\ \frac{1}{t-q^2}\ \frac{1}{\pi}\ \mathrm{Im}\ \psi(t). \tag{7.6}$$

b) **The instanton contribution to the pseudoscalar correlator**

There is as yet no reliable estimate of the instanton effect. ***According to the dilute gaz parametrization of this effect, one might*** **expect an instanton contribution[83]:**

$$\psi_5(q^2)_{inst} \simeq \left(\bar{m}_u + \bar{m}_d\right)^2 \frac{n_c}{m_{eff}^2}\left|\int d^4x\ e^{iqx}\ \bar{\psi}_o(x)\ (i\gamma_5)\ \psi_o(x)\right|^2 , \tag{7.7}$$

which is obtained by the substitution of zero modes $\psi_o(x)$ in the pseudoscalar currents and by integration over x and the instanton position. m_{eff} is the quark effective mass. Using an integral representation for

the MacDonald functions appearing in Eq. (7.7) after x-integration, one may obtain the Laplace transform :

$$\mathcal{L}\,\psi_5(q^2) = \left(\bar{m}_u+\bar{m}_d\right)^2 \cdot \frac{n_c\sqrt{2}}{m^2_{eff}}\,\tau^{-1}\int_0^\infty\int_0^\infty d\alpha d\beta\,\cosh\alpha\,\cosh\beta\left(\frac{\rho_c^5 t^3}{\tau^{5/2}}-\frac{3\rho_c^3}{2\tau^{3/2}}\,t\right)$$

$$\cdot\,\exp\left(-\frac{\rho_c^2\,t^2}{\tau}\right) \qquad (7.8)$$

with $t \equiv \frac{1}{2}(\cosh\alpha + \cosh\beta)$ and ρ_c is the confinement radius and $d(\rho_c)$ is the instanton density (Eq. 3.4). Using reasonable values of these parameters, a quantitative estimate of Eq. (7.8) indicates that the instanton contribution is of the same order as the power corrections for $\tau \simeq 1$ GeV^{-2}, while it dominates the sum rule for $\tau \gtrsim 2$ GeV^{-2}.

c) **Parametrization of the spectral function in a narrow width approximation**

The lowest ground state (the π) is a Goldstone boson but not its first excitation (the π'). Therefore, one cannot expect that the simple duality ansatz "one resonance" + "QCD continuum" gives a good approximation of the spectral function in this particular channel, unless one works at large $\tau \simeq 2$ GeV^{-2} in the case of the Laplace sum rule in order to enhance the π effect in the sum rule. However, according to previous discussion, one might expect that the effects of small-size instantons break the OPE at this large τ value. Therefore, it would be difficult to find a compromise "fiducial region" within the single-pion saturation of the spectral function. As it is not a straightforward matter to improve the accuracy of the dilute gas instanton estimate, which we consider at this level as a "qualitative one", we choose to improve the spectral part of the sum rule by current algebra control of the π' contribution. By the inclusion of the π', the spectral function in a narrow width approximation reads:

$$\frac{1}{\pi} \operatorname{Im} \psi_5(t) = 2 f_\pi^2 m_\pi^4 \left[\delta\left(t-m_\pi^2\right) + \frac{F_{\pi'}^2 M_{\pi'}^4}{f_\pi^2 m_\pi^4} \delta\left(t-M_{\pi'}^2\right) \right]$$

$$+ \text{"QCD continuum"} . \tag{7.9}$$

In the chiral limit $m_\pi^2 \to 0$, one knows from current algebra that $F_{\pi'} \sim m_\pi^2$ so that formally the π and π' contribution to the spectral function is of the same strength, i.e. one cannot include the π'-effect in the QCD continuum. This anomalous feature is a consequence of the Goldstone nature of the pion and indicates that <u>one cannot have a naive copy of the ρ-meson case here.</u> However, there are no data on $F_{\pi'}$ owing to the technical difficulty of measuring $\pi' \to \mu\nu$ or of selecting the 0^{-+} channel in $\tau \to \nu_\tau 3\pi$ decay which is dominated by the A_1 meson. Some theoretical estimates of $F_{\pi'}$ and $M_{\pi'}$ give for the combination [14,135]

$$r_\pi \equiv \frac{F_{\pi'}^2 M_{\pi'}^4}{f_\pi^2 m_\pi^4} \simeq 6 - 8 , \tag{7.10}$$

which indicates the relevance of the π' for the estimate of the quark mass. Eq. (7.9) can be sufficient for the sum-rule analysis but one still might not be satisfied with the narrow width approximation and might like to include the finite width corrections[136] and the final-state interactions as in Ref. 94).

d) **Finite-width corrections to the sum rule**

One can start from the Lagrangian in Eq. (1.21) plus mass term :

$$\mathcal{L}_{eff} = -\frac{1}{4} f_\pi^2 \operatorname{Tr}\left(\partial_\mu V \partial^\mu U^+\right) + V \operatorname{Tr}\{M U + (MU)^+\} \quad (7.11)$$

where

$$M \equiv \begin{pmatrix} m_u & & 0 \\ & m_d & \\ 0 & & m_s \end{pmatrix} \quad \text{and} \quad V \equiv \frac{f_\pi^2 m_\pi^2}{2(m_u+m_d)} \simeq \frac{f_\pi^2 M_K^2}{2(m_u+m_s)} \quad (7.12)$$

are respectively the quark-mass matrix and a phenomenological constant. Using the low-energy realizations of the axial currents in (1.27), and keeping the one and three-pion final states[137], one obtains :

$$\frac{1}{\pi} \operatorname{Im} \psi_5(t) = 2 m_\pi^4 f_\pi^2 \left\{\delta\left(t-m_\pi^2\right) + \Theta\left(t-9m_\pi^2\right) \frac{1}{18} \frac{1}{\left(16\pi^2 f_\pi^2\right)^2} \cdot \right.$$

$$\cdot \int_{4m_\pi^2}^{\left(\sqrt{t}-m_\pi\right)^2} ds \sqrt{1-\frac{4m_\pi^2}{s}} \sqrt{1-\frac{\left(\sqrt{s}-m_\pi\right)^2}{t}} \sqrt{1-\frac{\left(\sqrt{s}+m_\pi\right)^2}{t}} \cdot$$

$$\left. \cdot \left\{\frac{5}{2} + \frac{1}{2} \frac{1}{\left(t-m_\pi^2\right)^2} \left[(t-3s)^2 + 3(t-s)^2\right] + \frac{1}{t-m_\pi^2} 3\left(s-m_\pi^2\right) - t+s\right\}\right\}, \quad (7.13)$$

which tends to the current algebra result :

$$\lim_{m_\pi^2 \to 0} \frac{1}{\pi} \operatorname{Im} \psi_5(t) \simeq 2 m_\pi^4 f_\pi^2 \left\{\delta\left(t-m_\pi^2\right) + \frac{t}{6\left(16\pi^2 f_\pi^2\right)^2}\right\}. \quad (7.14)$$

The inclusion of the finite width of the π' modifies this result as :

$$\frac{1}{\pi}\,\mathrm{Im}\,\psi_5(t) = 2\,f_\pi^2 m_\pi^4 \left\{ \delta\left(t-m_\pi^2\right) + \frac{M_{\pi'}^4}{6\left(16\pi^2 f_\pi^2\right)^2} \quad (1+\gamma^2)\frac{t}{\left(M_{\pi'}^2-t\right)^2+\gamma^2 M_{\pi'}^4} \right\} \tag{7.15}$$

A test of this parametrization within a Veneziano-like model leads to the constraints obtained from low-energy processes[136]:

$$\frac{\Gamma_{\pi'}}{M_{\pi'}} \equiv \gamma \simeq 0.1 \qquad M_{\pi'} \simeq 1.1 \text{ GeV} \ , \tag{7.16}$$

which one could compare with the data :

$$\gamma \simeq 0.15 - 0.30 \quad ; \quad M_{\pi'} \simeq 1.3 \text{ GeV} \ . \tag{7.17}$$

e) **F.E.S.R. predictions for (m_u+m_d)**

We can also derive a set of constraints from FESR. The two lowest moments which are independent of the subtraction constant $\psi_5(0)$ read[24,137]

$$\mathcal{M}^{(0)} \equiv \int_0^{t_c} dt\,\frac{1}{\pi}\,\mathrm{Im}\,\psi_5(t) = \frac{3}{16\pi^2}\left(\bar{m}_u+\bar{m}_d\right)^2 t_c^2 \left[1+R_2(t_c) + \frac{2\pi}{3}\frac{\left\langle\alpha_s G^2\right\rangle}{t_c^2}\right] \tag{7.18}$$

$$\mathcal{M}^{(1)} \equiv \int_0^{t_c} dt\,\frac{t}{\pi}\,\mathrm{Im}\,\psi_5(t)$$

$$= \frac{3}{24\pi^2}\left(\bar{m}_u+\bar{m}_d\right)^2 .\ t_c^3 \left[1 + R_3(t_c) - \frac{896}{27}\pi^3\rho\,\frac{\alpha_s\left\langle\bar{u}\,u\right\rangle^2}{t_c^3}\right] \tag{7.19}$$

where

$$R_n(t_c) = \frac{\bar{\alpha}_s(t_c)}{\pi}\left[\frac{17}{3} + \frac{2}{n} - \frac{2}{\beta_1}\left(\gamma_2 - \frac{\gamma_1\beta_2}{\beta_1}\right) + \frac{2\gamma_1\beta_2}{\beta_1^2}\log\log\frac{t_c}{\Lambda^2}\right] \tag{7.20}$$

Following Refs 84) and 132), the authors of Ref. 137) study the ratio of the moments :

$$R_{FESR} \equiv \frac{\mathcal{M}^{(1)}}{\mathcal{M}^{(0)}} , \tag{7.21}$$

for the values of γ and $M_{\pi'}$ from the data (Eq. 7.17) and by comparing the QCD and phenomenological sides of R_{FESR} . They obtain for $M_{\pi'} \simeq 1.3$ GeV (Fig. 7.1) :

$$\gamma \simeq 0.15 \quad \text{and} \quad t_c \simeq (1.7 - 2.4)\ \text{GeV}^2 , \tag{7.22}$$

where one should note that larger values of γ need higher values of the condensates. The value of t_c obtained in (7.22) agrees with the position of the π'' pole, while the value of γ approaches the one expected from a Veneziano-like dual model in Eq.(7.16). One obtains the prediction in Fig. (7.2) from which one can deduce

$$\left(\hat{m}_u + \hat{m}_d\right) \simeq (24 - 25)\ \text{MeV} , \tag{7.23}$$

where the error is due to the uncertainties of t_c for given values of γ.

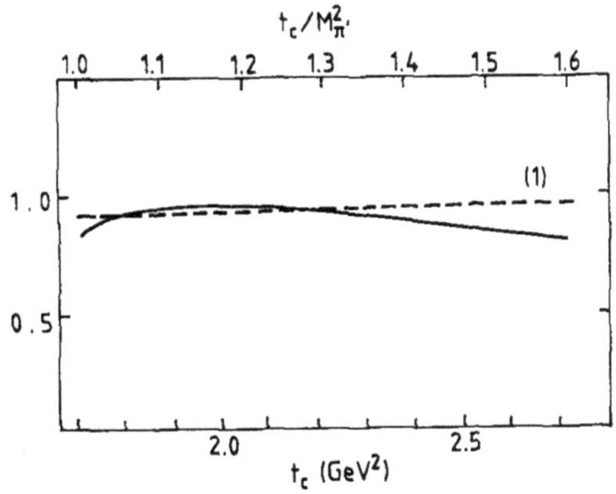

Fig. 7.1 : --- QCD behaviour of the FESR versus t_c for Set 2 of condensates — Hadronic behaviour of the LHS for γ = 0.15.

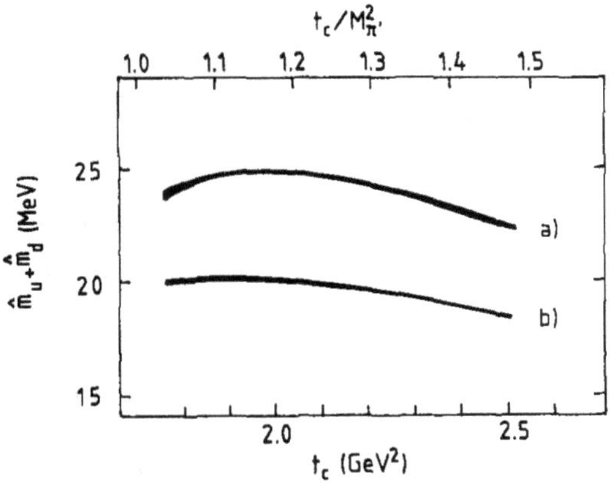

Fig. 7.2 : The invariant mass $\hat{m}_u + \hat{m}_d$ with t_c found in Fig. 7.1 and for Set 2 and Set 3 of QCD condensates (curves a) and b) respectively).

By allowing τ to increase up to 0.23, one would obtain the prediction in curve (b). Assembling these previous results, we might conclude the conservative estimate from the FESR[137]:

$$\left(\hat{m}_u + \hat{m}_d\right) \simeq (22.5 \pm 2.9)\ \text{MeV} \quad . \tag{7.24}$$

f) $(m_u + m_d)$ from the Laplace transform

From the Laplace transform of Eqs (7.2) and (7.9), one can deduce the quark-mass sum rule to two[20] and three[139] loops for $SU(3)_F$:

$$\left(\hat{m}_u + \hat{m}_d\right)^2 \simeq \frac{16\pi^2}{3} m_\pi^4 f_\pi^2 \exp\left[- m_\pi^2 \tau\right] \left(1 + r_\pi . \exp\left[\left(m_\pi^2 - M_{\pi'}^2\right) \tau\right]\right) \tau^2 \left(\frac{L}{2}\right)^{8/9} .$$

$$. \left\{\left[1 - (1+t_c\tau) \exp\left[- t_c\tau\right]\right] \left[1 + \frac{2.94}{L} - 0.7 \frac{\log L}{L} + \right.\right.$$

$$\left. + \frac{3.21}{L^2} - 4.93 \frac{\log L}{L^2} + 0.52 \left(\frac{\log L}{L}\right)^2\right] - 2\left(\bar{m}_u^2 + \bar{m}_d^2 - \bar{m}_u \bar{m}_d\right)\tau +$$

$$\left. + \tau^2 \left[\frac{\pi}{3}\left\langle \alpha_s G^2 \right\rangle - \frac{8\pi^2}{3}\left[\left(\bar{m}_d - \frac{\bar{m}_u}{2}\right)\left\langle \bar{u}u \right\rangle + (u \leftrightarrow d)\right] + \frac{896}{81} \pi^3 \rho\, \alpha_s \left\langle \bar{u}u \right\rangle^2 \tau^3\right]\right\}^{-1} , \tag{7.25}$$

where $L \equiv - \log \tau \Lambda^2$; $\hat{m}_i$ is the renormalization group invariant mass in Eq. (1.91) which is related to the three-loop running mass $\bar{m}_i$ as in Eq. (1.93).

We have not included the instanton effects which are not well controlled. However, according to the estimate in Ref.83), this effect can be safely ignored for $\tau \leqslant 1\ \text{GeV}^{-2}$. We show the analysis in Figs 7.3 and 7.4 for the narrow-width approximation ($r_\pi \simeq 6$) and for finite width π' with the parameters in Eqs (7.16) and (7.17). The analysis of the τ-stability (Fig. 7.3) at given values of t_c shows that in the finite width parametrization the prediction

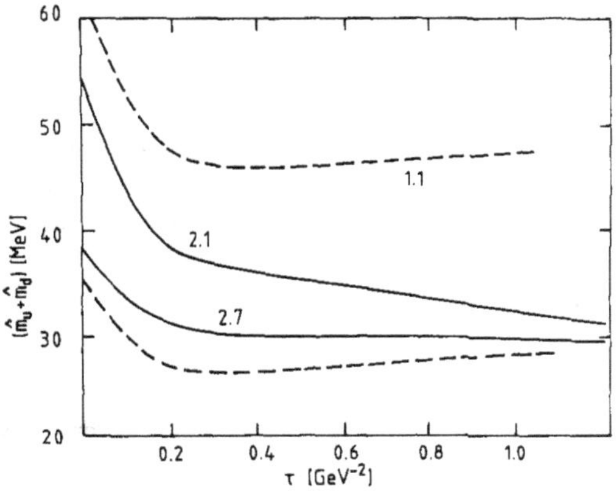

Fig. 7.3 : $\left(\hat{m}_u+\hat{m}_d\right)$ from LSR versus τ — narrow-width approximation for two values of t_c --- finite-width parametrization for two values of the π' parameters (7.16, 7.17) and for t_c = 2.1 GeV^2.

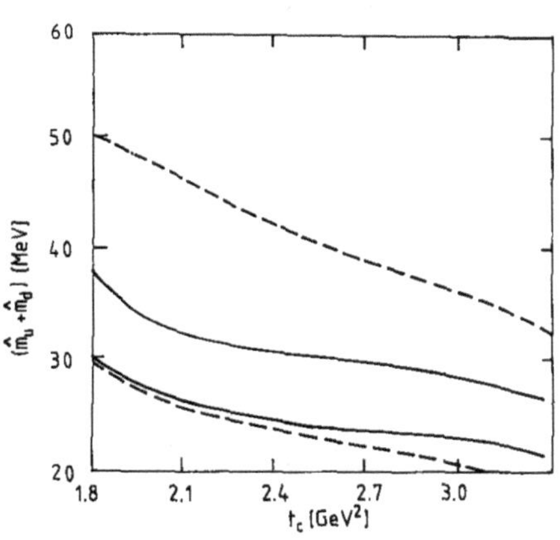

Fig. 7.4 : Behaviour of the τ minima versus t_c for two values of the π' mass.

has a large plateau in τ while in the narrow-width approximation the stability is indicated by an inflexion point for $t_c \leqslant 2.5$ GeV^2 and becomes a plateau for large t_c. The stability in t_c is studied in Fig. 7.4. It will be noticed that with the parametrization in Eq. (7.15), the result has a stronger t_c-dependence in than within the narrow width parametrization. For this latter case, one even has a stability in t_c which is inside the range of values fixed by duality in Eq. (7.22). Another weak point of the finite-width parametrization for the Laplace sum rule is the strong dependence of the result on the value of the π' mass. Therefore, we might conclude from the above comparison, that the finite-width parametrization as discussed earlier does not provide a real improvement of the quark-mass estimate obtained from the Laplace sum rule. In the τ- and t_c-stability regions satisfied by the narrow-width approximation, we obtain the estimate :

$$\left(\hat{m}_u + \hat{m}_d\right) \simeq (27.5 \pm 5.8)\ \mathrm{MeV} \quad , \qquad (7.26)$$

where we have included in the errors the 8% uncertainties induced by $\Lambda \simeq (150 \pm 50)$ MeV. This result is in good agreement with the FESR result in Eq. (7.15). We can also compare Eq. (7.26) with the one which one would obtain by using the parametrization in Eq. (7.15) :

$$\left(\hat{m}_u + \hat{m}_d\right) \simeq (33 \pm 11.6)\ \mathrm{MeV} \quad . \qquad (7.27)$$

Here, one should notice that the error is mainly due to the π' mass value. Contrary to the FESR case, the error induced by γ is negligible for LSR.

One can combine these different estimates and deduce the best value of the quark mass from the LSR and FESR methods :

$$\left(\hat{m}_u + \hat{m}_d\right) \simeq (24.0 \pm 2.5)\ \mathrm{MeV} \ . \qquad (7.28)$$

This result agrees with the optimal bound given in Ref. 20) and might indicate a posteriori that it may make sense to neglect instanton effects at $\tau \simeq 1.6\ \mathrm{GeV}^{-2}$ where this bound has been obtained !

Using the three-loop expression of the running mass in Eq. (1.93), one can deduce from Eq. (7.28) :

$$\left(\bar{m}_u + \bar{m}_d\right) (1\ \mathrm{GeV}) \simeq (14.0 \pm 1.1)\ \mathrm{MeV}\ . \qquad (7.29)$$

We can combine the result in Eqs (7.28) and (7.29) with the one from the chiral effective Lagrangian result[14]:

$$\frac{m_d - m_u}{m_d + m_u} \simeq (0.28 \pm 0.03)\ . \qquad (7.30)$$

Thus, we obtain the quark-mass difference for $\Lambda = (150 \pm 50)$ MeV :

$$\hat{m}_d - \hat{m}_u \simeq (6.6 \pm 0.9)\ \mathrm{MeV}$$

$$\left(\bar{m}_d - \bar{m}_u\right) (1\ \mathrm{GeV}) \simeq (3.9 \pm 0.5)\ \mathrm{MeV}\ . \qquad (7.31)$$

We therefore deduce the improved values of the up and down quark masses to three loops :

$$\hat{m}_d \simeq (15.4 \pm 0.8)\ \mathrm{MeV} \quad \text{and} \quad \hat{m}_u = (8.7 \pm 0.8)\ \mathrm{MeV}\ , \qquad (7.32a)$$

$$\bar{m}_d(1\ \mathrm{GeV}) \simeq (9.2 \pm 0.5)\ \mathrm{MeV} \quad \text{and} \quad \bar{m}_u(1\ \mathrm{GeV}) \simeq (5.2 \pm 0.5)\ \mathrm{MeV}\ . \qquad (7.32b)$$

2. $\psi_5(0)$ AND DEVIATION FROM PION PCAC

In the previous paragraphs, we have used a sum rule independent of the subtraction constant $\psi_5(0)$ in order to estimate values of the quark masses in Eq. (7.32).

Instead, we can work with the once subtracted quantity in Eq. (7.6) in order to estimate $\psi_5(0)$. As first done in Ref. 140), we work with the Laplace transform of Eq. (7.6) :

$$\int_0^\infty \frac{dt}{t} e^{-t\tau} \frac{1}{\pi} \mathrm{Im}\, \psi_5(t) = \psi_5(0) + \frac{3}{8\pi^2} \frac{\left(\hat{m}_u+\hat{m}_d\right)^2}{L^{8/9}} \left\{\left(1 - e^{(t_c\tau)}\right) \cdot \right.$$

$$\tau^{-1} \left[1 + \frac{2}{-\beta_1 L}\left(\frac{17}{3} + 2\gamma_E + \frac{2}{-\beta_1}\left(\gamma_2 - \gamma_1 \frac{\beta_2}{\beta_1}\right) + \right.\right.$$

$$\left.\left. \frac{2\gamma_1\beta_2}{\beta_1^2} \log 2L\right]\right. + 2\, m_d^2 \log \tau\, \bar{m}_d^2$$

$$\left. + \frac{8}{3}\pi^2\tau\, m_d \left(\langle u\, \bar{u}\rangle - \frac{1}{2}\langle \bar{d}\, d\rangle\right) - \frac{\pi}{3}\tau\langle\alpha_s\, G^2\rangle - \frac{1}{2}\frac{896}{81}\pi^3\tau^2\rho\, \alpha_s\langle \bar{u}\, u\rangle^2\right\} \qquad (7.33)$$

where $L \equiv -\log \tau \Lambda^2$ and we use standard notations for the other parameters. We can also work with a modified sum rule[135] by formally taking the difference between Eq. (7.33) and the unsubracted sum rule $\mathcal{L}\, \psi_5(q^2)$used previously to get the quark masses. The modified sum rule reads :

$$\int_0^\infty \frac{dt}{t} e^{-t\tau} (1-t\tau) \frac{1}{\pi} \mathrm{Im}\, \psi_5(t) \simeq \psi_5(0) + \frac{3}{8\pi^2} \frac{\left(\hat{m}_u+\hat{m}_d\right)^2}{L^{8/9}} \tau^{-1} \quad .$$

$$\cdot \left\{(t_c\tau)\, e^{-t_c\tau} + 2\, \bar{m}_d^2\, \tau \log \tau\, m_d^2 - \right.$$

$$\frac{16\pi^2}{3}\tau^2 m_d \left\{\langle \bar{u}\, u\rangle - \frac{1}{2}\langle \bar{d}\, d\rangle\right\} - \frac{2\pi^2}{3}\tau^2 \langle \alpha_s G^2\rangle \tau^2$$

$$- \frac{3}{2}\left(\frac{896}{81}\right)\pi^3 \rho\, \alpha_s \langle \bar{u}\, u\rangle^2 \tau^3\Bigg\} \ . \qquad (7.34)$$

We show the predictions of $\psi_5(0)$ from Eqs (7.33) and (7.34) in Fig. 7.5 where the curves almost coincide.

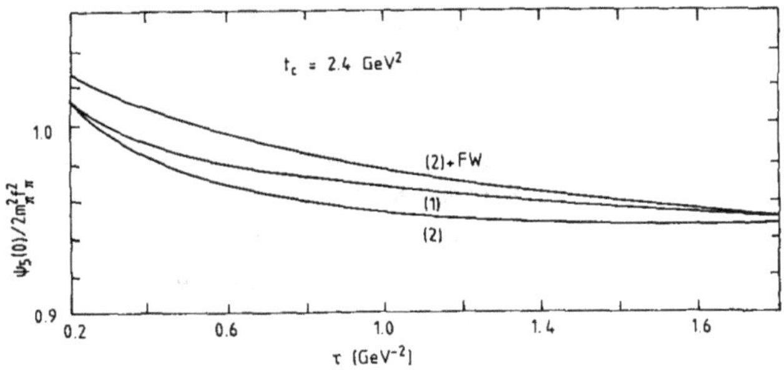

Fig. 7.5 : $\psi_5(o)$ versus τ at given t_c for two sets of the QCD condensates in Table 5.1.

We have used $t_c \simeq 2.4$ GeV^2 inside the duality region in Eq. (7.22) where the prediction of the sum of the quark masses stabilizes (Fig. 7.4). (However, the result is not sensitive to the value of t_c.) We use a NWA and the one in Eq. (7.15). The sum rule in Eq. (7.34) has a much better τ-stability using the NWA. The τ-stability of the same sum rule is reached for slightly larger values of τ if one uses the chiral Lagrangian parametrization. In this case, the effect of the π' mass is still important. In the τ-stability region where the different sum rules yield a unique result, we obtain :

$$\psi_5(0) \equiv 2\, m_\pi^2\, f_\pi^2\, (1 - \rho_\pi) \tag{7.35a}$$

where :

$$\rho_\pi \simeq (5 \pm 0.5)\ 10^{-2} \tag{7.35b}$$

which is to be compared with the FESR result $(4\pm1)\ 10^{-2}$ obtained from the moments :

$$\int_0^{t_c} \frac{dt}{t} \frac{1}{\pi} \operatorname{Im} \psi_5(t) = \psi_5(0) + \frac{3}{8\pi^2} \left(\bar{m}_d + \bar{m}_u\right)^2 t_c\ [1 + R_1(t_c)] \ . \tag{7.36}$$

The analytical continuation method[141] gives a value $(7.7 \pm 0.7)10^{-2}$ which is higher than previous results. (An overestimate of the quark masses has also been obtained from this method). These facts already indicate the degree of accuracy of this method. The result in Eq. (7.35) can be used in chiral perturbation theory[142] in order to fix the arbitrariness of the analytical chiral corrections to the PCAC relation, these terms being related to the arbitrary high-energy couplings from the different counterterms of the chiral Lagrangian.

3. SCALAR SUM RULES : THE $a_o(980)$ AND $\psi(0)$

We shall study the two-point correlator :

$$\psi(q^2)^i_j = i \int d^4x\, e^{iqx} \left\langle 0 \left| \mathbb{T}\, \partial_\mu V^\mu(o)^i_j \left(\partial_\mu V^\mu(o)^i_j\right)^+ \right| 0 \right\rangle \tag{7.37a}$$

built from the divergence of the vector current :

$$\partial_\mu V^\mu(x)^i_j = (m_i - m_j)\, \bar{\psi}_i(i)\, \psi_j \ . \tag{7.37b}$$

The expression of the two-point correlator can be deduced from (7.2) by replacing m_d with $-m_d$ and the coefficient 896 in front of the four-quark condensate with - 1408.

The associated spectral function can be saturated by the $a_o(980)$ plus a QCD continuum. However, the nature of the a_o needs a much better clarification as it is also assumed to be a four-quark state. In this chapter, we shall limit ourselves to the $\bar{q}q$ nature of the a_o and a posteriori we shall test the validity of this assumption by comparing the sum-rule result with the data. In a narrow-width approximation, we have :

$$\left\langle 0 \left| \partial_\mu V^\mu(x)^u_d \right| a_o \right\rangle = \sqrt{2}\, f_a\, M_a^2 \quad . \tag{7.38}$$

where f_{a_o} is the a_o coupling normalized as $f_\pi = 93.3$ MeV. A parametrization based on the chiral Lagrangian expression of the vector current :

$$\partial_\mu V^\mu(x)^u_d = -i \left(\frac{m_d - m_u}{m_d + m_u}\right) m_\pi^2 \left[\sqrt{\frac{2}{3}}\, \pi^- \eta + K^\circ K^-\right] \tag{7.39a}$$

leads to :

$$\frac{1}{\pi} \operatorname{Im} \psi(t)^u_d = \frac{1}{32\pi^2} \left(\frac{m_d - m_u}{m_d + m_u}\right)^2 \left\{ \frac{2}{3} \lambda^{1/2} \left(1, \frac{M_\eta^2}{t}, \frac{m_\pi^2}{t}\right) \Theta\left(t - (M_\eta + m_\pi)^2\right) \right.$$

$$\left. + \Theta\left(t - 4M_K^2\right) \sqrt{1 - \frac{4M_K^2}{t}} \right\} \frac{M_a^2(1 + \gamma_a)}{\left(t - M_a^2\right)^2 + M_a^4 \gamma_a^2} \quad , \tag{7.39b}$$

where $\gamma_a \equiv \Gamma_a / M_a$, i.e. with this parametrization one cannot obtain useful information from the sum rule unless one introduces new input. This can be done by relating the normalization factor in Eq. (7.39) to the kaon hadronic tadpole[143]:

$$\left(\Delta M_K^2\right)_{TAD} \equiv M^2_{K^+} - M^2_{K^o} \simeq 2\, m_\pi^2 \left(\frac{m_d - m_u}{m_d + m_u}\right) \tag{7.40}$$

where [144] $\left(\Delta M_K^2\right)_{TAD} \simeq (5.3 \pm 0.8)\ 10^{-3}\ \text{GeV}^4$ from approaches based on the chiral Lagrangian. Therefore, using Eq.(7.40) in (7.39), one might deduce from the sum rule an estimate of the light-quark mass difference. The result, though higher than the one from Eqs (7.30 + 7.32), is not inconsistent with the latter, bearing in mind the uncertainties appearing in the parametrization of the spectral function.

Much more suited to the phenomenology of the a_o is the duality ansatz parametrization. The sum rule can be used in this case for the determination of f_a and M_a once the value of $\hat{m}_d - \hat{m}_u$ obtained previously is introduced. An estimate of f_a from the Laplace transform has been done in Refs 145) and 139) while M_a has been determined from the moment ratio[29,146].

The decay constant f_a can be obtained from the sum rule

$$\int_0^\infty dt\ e^{-t\tau}\ \frac{1}{\pi}\ \text{Im}\ \psi(t) \tag{7.41}$$

which gives in the chiral limit :

$$f_a \simeq \frac{e^{M_a^2\tau/2}}{M_a^2} \cdot \frac{\sqrt{3}}{4\pi}\left(\bar{m}_d - \bar{m}_u\right)\tau^{-1}\ \{[1 - (1 + t_c\tau)\exp(-t_c\tau)]\ .$$

$$\cdot \left[1 + \frac{2.94}{L} - 0.7\,\frac{\log L}{L}\right] + \tau^2\,\frac{\pi}{3}\left\langle\alpha_s G^2\right\rangle - C_6\langle O_6\rangle\,\tau^3\Big\}^{1/2} \tag{7.42b}$$

with :

$$C_6\langle O_6\rangle = \frac{1048}{81}\,\pi^3\,\rho\,\alpha_s\left\langle\bar{u}\,u\right\rangle^2 \ . \tag{7.42c}$$

The a_o mass is obtained from the ratio of (7.42a) with its first derivative in τ. This gives

$$M_a^2 \simeq \frac{\int_0^{t_c} t\,dt\; e^{-t\tau}\,\frac{1}{\pi}\,\mathrm{Im}\,\psi_5(t)}{\int_0^{t_c} dt\; e^{-t\tau}\,\frac{1}{\pi}\,\mathrm{Im}\,\psi_5(t)} \simeq \qquad (7.43a)$$

$$2\tau^{-1}\left\{\frac{(1-\rho_2)\left(1+\frac{2.49}{L}-\frac{0.7\log L}{L}\right)+\frac{1}{2}C_6\langle O_6\rangle\tau^3}{(1-\rho_1)\left(1+\frac{2.94}{L}-\frac{0.7\log L}{L}\right)+\tau^2\frac{\pi}{3}\left\langle\alpha_s G^2\right\rangle - C_6\langle O_6\rangle\tau^3}\right\}, \qquad (7.43b)$$

with :

$$\rho_n = \left(1+(t_c\tau)+\frac{(t_c\tau)^2}{2!}+\ldots+\frac{(t_c\tau)^n}{n!}\right)e^{-t_c\tau}\,. \qquad (7.43c)$$

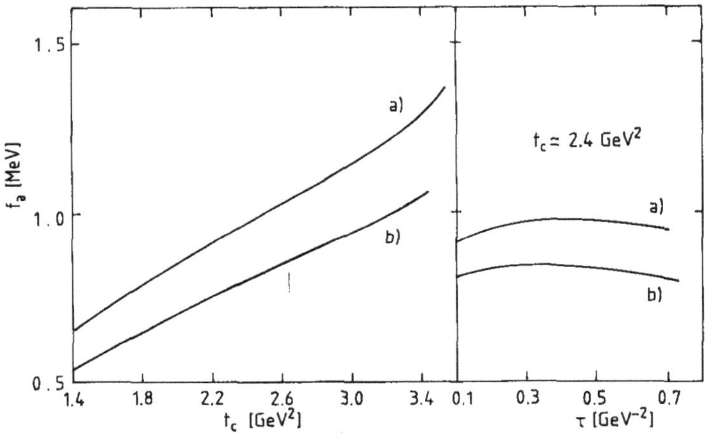

Fig. 7.6 : f_a versus t_c and τ : a) quark mass in Eq. 7.26 from the pseudoscalar : b) weighted average in Eq. 7.32.

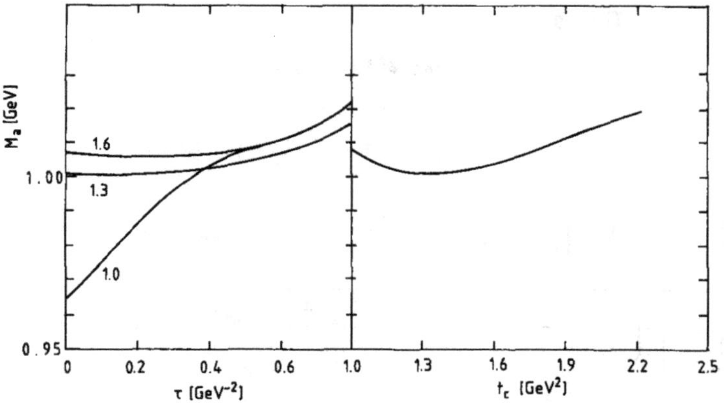

Fig. 7.7 : M_a versus τ and t_c.

The two sum rules are shown respectively in Figs 7.6 and 7.7 versus the τ variable and the continuum threshold t_c. The mass and the coupling stabilize for τ at between 0.4 GeV^{-2} and 0.8 GeV^{-2}. M_a has a clear t_c-stability around 1.3 - 1.6 GeV^2 while f_a has an inflexion point for $t_c \simeq 2.4 - 3\ GeV^2$. The t_c values for M_a are in the range expected from FESR[137] at which we conclude the optimal estimate:

$$M_a \simeq (1 - 1.05)\text{GeV} \tag{7.44}$$

in good agreement with the data. Uncertainties in Λ and condensates do not affect this prediction. The value of f_a is less certain as the t_c values corresponding to the inflexion point seem too high, though these values correspond to the ones obtained from a two-parameter fit[145].For a much more conservative result, we consider the values of

f_a in the range $t_c \simeq 1.4 - 3\ GeV^2$ as the ones obtained from QSSR :

$$f_a \simeq (0.5 - 1.2)\ MeV \tag{7.45a}$$

which agrees within the errors with the fitted value in Ref. 145) corresponding to higher values of the quark mass.

It will, however, be noticed that this value of f_a is lower than the one derived from a chiral Lagrangian[147] using a duality ansatz parametrization of the spectral integral ($f_a \simeq 1.6$ MeV) and the one ($f_a \simeq 1.8$ MeV) using the hadronic kaon tadpole mass difference plus an a_0 dominance of the $K\bar{K}$ form factor[143].

$$f_a \simeq \left(M^2_{K^0} - M^2_{K^+}\right)_{Tad} \frac{1}{\sqrt{2}} \cdot \frac{1}{\sqrt{3/2}.g_{a_0\eta\pi}} \simeq 1.8\ MeV \quad . \tag{7.45b}$$

In view of these uncertainties, we consider different predictions for f_a in the range :

$$f_a \simeq (0.5 - 1.8)\ MeV \tag{7.45c}$$

We can also estimate the subtraction constant $\psi(0)^u_d$:

$$\psi(0)^u_d \equiv -(m_d - m_u) \left\langle \bar{d}d - \bar{u}u \right\rangle \tag{7.46}$$

using sum rules similar to those used in Eqs (7.33) and (7.34) but this time replacing m_d with $-m_d$ and changing the coefficient of the $\left\langle \bar{u}\, u\right\rangle^2$ condensate. The analysis is shown in Fig. 7.8 for $t_c \simeq 1.6\ GeV^2$ where we use the correlated values of t_c, f_a and $(m_d - m_u)$ in Fig. 7.7. The analogue of Eq. (7.34) has no τ-stability while the scalar analogue of Eq. (7.33) presents an inflexion point for $\tau \simeq 1\ GeV^{-2}$.

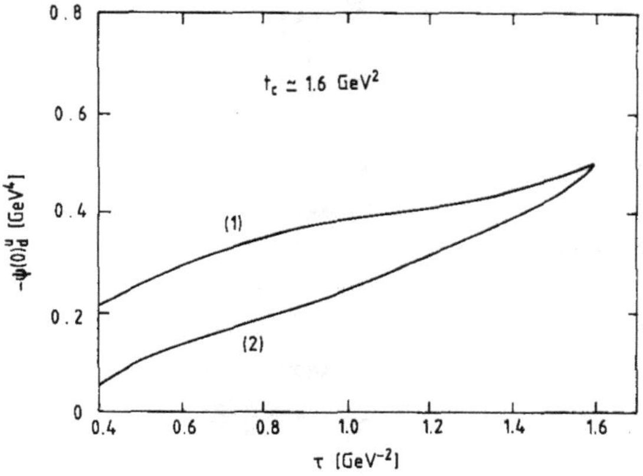

Fig. 7.8 : $\psi(0)^u_d$ versus τ for t_c corresponding to the optimal value of M_a. (1) ≡ Eq (7.33) ; (2) ≡ Eq (7.34).

This change of τ-behaviour compared to the pseudoscalar case is due to the sign of the dimension-six condensate effects. Taking as the optimal result the value of the inflexion point and the one where the two sum rules meet, we obtain :

$$\psi(0)^u_d \simeq -(0.4 - 0.5)\ 10^{-6}\ \text{GeV}^4 \quad . \tag{7.47}$$

The result in Eq.(7.47) is insensitive to the change of $\Lambda \simeq 100$-200 MeV, of t_c inside the duality region, of an increase of the condensate values by a factor two and of f_a correlated to t_c via Fig. 7.7. This result agrees with previous results[148,139,149,137].

4. SU(2) BREAKING OF THE CONDENSATE RATIO $\langle \bar{d}d \rangle / \langle \bar{u}u \rangle$

Now we can use our estimate of $\psi_5(0)$ in Eq. (6.60) and of $\psi(0)$

in Eq. (7.47) in order to deduce the ratio of condensate $\langle \bar{d}d\rangle / \langle \bar{u}u\rangle$. We have :

$$\frac{\psi(0)^u_d}{\psi_5(0)^u_d} \equiv \left(\frac{m_d - m_u}{m_d + m_u}\right)\left(\frac{r-1}{r+1}\right) \tag{7.48a}$$

with $r \equiv \langle \bar{d}\, d\rangle / \langle \bar{u}\, u\rangle$. For the quark-mass ratio in Eq. (7.30), we then deduce :

$$r \equiv \frac{\langle \bar{d}\, d\rangle}{\langle \bar{u}\, u\rangle} \simeq 1 - 9\ 10^{-3} \quad , \tag{7.48b}$$

which is in agreement with previous estimates[148,139,149,137]. Eq. (7.48) shows a small SU(2) flavour-breaking for the condensate.
Using the value in Eq. (7.48), the pion PCAC relation and the value of the quark mass in Eq. (7.32), we deduce the value of the light-quark condensate to three loops in units of MeV :

$$\hat{\mu}_u = (188.9 \pm 6.6) \qquad \langle -\bar{u}\, u\rangle^{1/3}\ (1\ \text{GeV}) \simeq (223.9 \pm 7.8)$$

$$\hat{\mu}_d = (188.3 \pm 6.6) \qquad \langle -\bar{d}\, d\rangle^{1/3}\ (1\ \text{GeV}) \simeq (233.3 \pm 7.8). \tag{7.49}$$

5. THE ω-ρ MIXING FROM QSSR

The ω-ρ mixing has been analyzed by SVZ[18]. We follow their discussion by working with the off-diagonal correlator

$$\Pi^{\rho\omega}_{\mu\nu} = i \int d^4x\ e^{iqx} \left\langle 0 \left| J^\rho_\mu(x) \left(J^\omega_\nu(0)\right)^\dagger \right| 0 \right\rangle$$

$$\equiv -\left(g_{\mu\nu}\, q^2 - q_\mu\, q_\nu\right) \Pi^{\rho\omega}(q^2)\ , \tag{7.50a}$$

where :

$$J^\rho_\mu = \frac{1}{2}\left(\bar{u}\,\gamma_\mu u - \bar{d}\,\gamma_\mu d\right)$$

$$J^\omega_\mu = \frac{1}{6}\left(\bar{u}\,\gamma_\mu u + \bar{d}\,\gamma_\mu d\right)\ . \tag{7.50b}$$

The *QCD* contributions to the *OPE* of the correlator come essentially from a one-photon exchange (disconnected diagram), two mass insertions, and the quark condensates $\left(\langle\bar{\psi}\,\psi\rangle,\ \langle\bar{\psi}\,\Gamma_1\psi\,\bar{\psi}\,\Gamma_2\psi\rangle\right)$. To leading order, one obtains[18]:

$$\Pi^{\rho\omega}(q^2) = \frac{1}{12}\left\{-\frac{\alpha}{16\pi^3}\log\frac{Q^2}{\nu^2} + \frac{3}{2\pi^2}\,\frac{m_d^2 - m_u^2}{Q^2} + \right.$$

$$\left. 2\,\frac{m_u\langle\bar{u}\,u\rangle - m_d\langle\bar{d}\,d\rangle}{Q^4} - \frac{C_6\langle O_6\rangle}{Q^6}\right\} \tag{7.51a}$$

where :

$$C_6\langle O_6\rangle = 2\pi\,\alpha_s\left[\left\langle\left(\bar{u}\,\gamma_\alpha\gamma_5\,\frac{\lambda_a}{2}\,u\right)^2 - \left(\bar{d}\,\gamma_\alpha\gamma_5\,\frac{\lambda_a}{2}\,d\right)^2\right\rangle\right.$$

$$\left. + \frac{2}{9}\left\langle\left(\bar{u}\,\gamma_\alpha\,\frac{\lambda_a}{2}\,u\right)^2 - \left(\bar{d}\,\gamma_\alpha\,\frac{\lambda_a}{2}\,d\right)^2\right\rangle\right]$$

$$\simeq \frac{112\pi}{81} k \alpha_s \left[\left\langle \bar{u} u \right\rangle^2 - \left\langle \bar{d} d \right\rangle^2 \right] \tag{7.51b}$$

with k = 1 if one uses the vacuum saturation assumption of the four-quark condensate. Therefore, one can deduce the Laplace transform :

$$\int_0^\infty dt\ e^{-t\tau} \frac{1}{\pi} \operatorname{Im} \Pi^{\rho\omega}(t) \simeq \frac{\tau^{-1}}{12} \cdot \left\{ \frac{\alpha}{16\pi^3} + \frac{3}{2\pi^2} \left(m_d^2 - m_u^2 \right) \tau - 2\left(m_d \left\langle \bar{d} d \right\rangle - m_u \left\langle \bar{u} u \right\rangle \right) \tau^2 - C_6 \langle O_6' \rangle \tau^3 \right\} \tag{7.52a}$$

$$\int_0^\infty dt\ t\ e^{-t\tau} \frac{1}{\pi} \operatorname{Im} \Pi^{\rho\omega}(t) \simeq \frac{\tau^{-2}}{12} \left\{ \frac{\alpha}{16\pi^3} + 2 \left(m_d \left\langle \bar{d} d \right\rangle - m_u \left\langle \bar{u} u \right\rangle \right) \tau^2 + 2\ C_6 \langle O_6 \rangle \tau^3 \right\} . \tag{7.52b}$$

Following SVZ, the spectral function can be parametrized using the standard two-component mixing formalism :

$$|\omega\rangle = |\omega_B\rangle + \epsilon |\rho_B\rangle$$

$$|\rho\rangle = |\rho_B\rangle - \epsilon |\omega_B\rangle \ , \tag{7.53}$$

where ω_B and ρ_B are pure isospin (0,1) eigenstates. ϵ is the mixing parameter :

$$\epsilon = \frac{\delta_{\omega\rho}}{\left(M_\omega - \frac{1}{2} i\ \Gamma_\omega \right)^2 - \left(M_\rho - \frac{1}{2} i\ \Gamma_\rho \right)^2} \tag{7.54}$$

which can be large due to the almost degenerate ω and ρ masses . $\delta_{\omega\rho}$ is an off-diagonal mass which also includes in it the $\rho \to \gamma \to \omega$ transition. One can estimate this contribution as :

$$\delta_{\omega\rho} \quad (\rho \to \gamma \to \omega) \simeq \frac{\pi\alpha \; M_\rho^2}{3 \; \gamma_\rho^2} \tag{7.55}$$

where the SU(3) relation for the mass and coupling has been used ($M_\omega = M_\rho$, $\gamma_\omega \simeq 3\gamma_\rho$). This measured quantity should be subtracted when one compares the theoretical prediction with the data. The spectral function reads :

$$\frac{12}{\pi} \text{ Im } \Pi^{\omega\rho}(t) = f^{\rho\omega}\left(M_\rho^2\right) \delta\left(t - M_\rho^2\right) - f^{\rho\omega}\left(M_\omega^2\right) \delta\left(t - M_\omega^2\right) \tag{7.56a}$$

where :

$$f^{\rho\omega}\left(M_\rho^2\right) \equiv \epsilon\left(M_\rho^2\right) \cdot \frac{M_\omega^2}{4\gamma_\rho \; \gamma_\omega} \; . \tag{7.56b}$$

By introducing :

$$M^2 \equiv \frac{1}{2}\left(M_\rho^2 + M_\omega^2\right) \qquad \delta \, M^2 = M_\omega^2 - M_\rho^2$$

$$\xi = \frac{3}{\gamma_\rho \gamma_\omega} \; \frac{\delta_{\omega\rho} \quad (\rho \to \gamma \to \omega) \; - \; \delta_{\omega\rho}}{M^2}$$

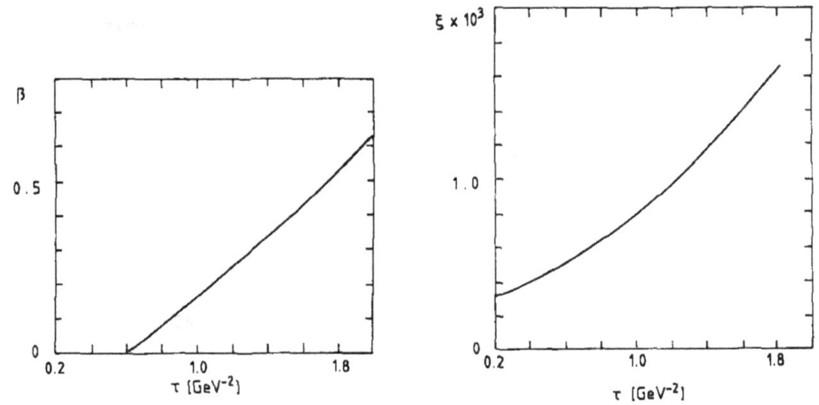

Fig. 7.9 : τ behaviour of β and ξ from Eq. (7.58) for $t_c \simeq 1.7$ GeV² (see chapter 5).

$$f^{\rho\omega} \equiv \frac{1}{2}\left[f^{\rho\omega}\left(M_\rho^2\right) + f^{\rho\omega}\left(M_\omega^2\right)\right] \simeq \frac{M^4 \xi}{\delta M^2} \quad ,$$

$$\beta = \frac{f^{\rho\omega}\left(M_\rho^2\right) - f^{\rho\omega}\left(M_\omega^2\right)}{M_\rho^2 - M_\omega^2} \; \frac{M^2}{f^{\rho\omega}} \quad , \tag{7.57}$$

one obtains to leading order in δM^2 :

$$\int_0^\infty dt \; e^{-t\tau} \, \frac{1}{\pi} \, \mathrm{Im}\, \Pi^{\rho\omega}(t) = \frac{1}{12} \, \xi \, M^2 (M^2 \tau - \beta) \, e^{-\tau M^2} + \text{"continuum"}$$

$$\int_0^\infty dt \; e^{-t\tau} \, \frac{t}{\pi} \, \mathrm{Im}\, \Pi^{\rho\omega}(t) = \frac{1}{12} \, \xi \, M^4 (M^2 \tau - 1 - \beta) \, e^{-\tau M^2} + \text{"continuum"} \tag{7.58}$$

We solve the two sum rules in Eq. (7.58) and deduce a sum rule for β and for ξ. The result is given in Fig. 7.9 for $t_c \simeq 1.7$ GeV² as suggested by the duality constraint from the ρ-meson systems in Chapter 5.

However, the prediction is not quite sensitive to t_c around this value. As can be noticed in Fig. 7.9, there is no τ-stability at which one might extract the optimal predictions.

If one uses the value of β about 0.5 as deduced from a $SU(3)_F$ symmetry of the coupling constant, we might fix $\tau \simeq 0.8\ \mathrm{GeV}^{-2}$. At this value, we can predict :

$$\xi \simeq 1.6\ 10^{-3} \quad , \tag{7.59}$$

in good agreement with the data from diffraction[150] $(0.82 \pm 0.07)10^{-3}$ and $(1.4 \pm 0.4)10^{-3}$ from e^+e^- experiments[151], these data being deduced from those of $\delta_{\omega\rho}$.

Alternatively, we can try to use the FESR constraints, which would formally be derived from Eq. (7.58) in the limit $\tau \to 0$. In this case, one obtains the lowest dimension constraint :

$$-\beta\xi M^2 \simeq t_c \left(\frac{\alpha}{16\pi^3}\right) + \frac{3}{2\pi^2}\left(\bar{m}_d^2 - \bar{m}_u^2\right) \quad , \tag{7.60}$$

which gives a β value opposite in sign to the one obtained at small τ values. This discrepancy can signal in this case that the role of the continuum is important and cannot be approximated by the naive QCD-duality ansatz. Therefore, we improve Eq. (7.58) by using a two resonance parametrization by including the ρ'-ω' region. This leads to the change β into $\beta+\beta'$ where β' characterizes the strength of the ω'-ρ' mixing. Using a higher value of t_c of about 2-4 GeV^2, one obtains for $\xi \simeq 0.8\ 10^{-3}$ (the most accurate data) :

$$\beta' \simeq -0.6 \tag{7.61}$$

in remarkable agreement with the first set of values suggested by SVZ in their analysis ($\beta'_{SVZ} \equiv \beta'/\gamma$).

We conclude that the sum-rule approach provides a reasonable though not accurate estimate of the ρ-ω mixing.

CHAPTER 8

THE "STRANGE" STRANGE QUARK AND SU(3) BREAKINGS FROM THE LIGHT MESONS SYSTEMS

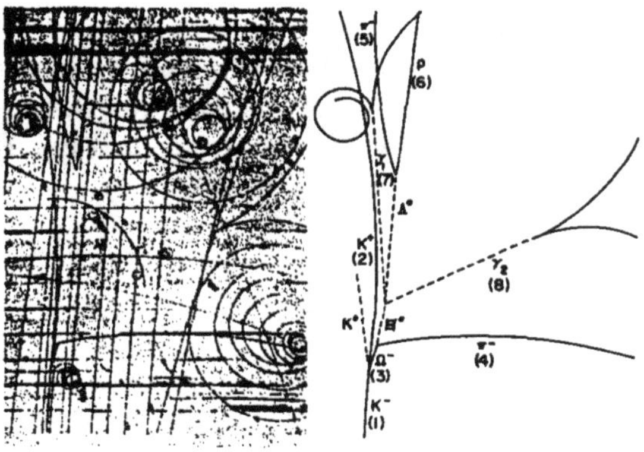

The famous 1964 BNL bubble chamber event showing the omega and its decay

The strange-quark channel is a delicate one owing to the value of its mass which is about the value of the QCD scale Λ. It is not as light as the u and d quarks, which makes disregarding the $m^2_{u,d} \ll \Lambda^2$ corrections a good approximation. It is not as heavy as the c,b and t quarks $m^2_c \gg \Lambda^2$ in order to justify the heavy-quark expansion often used in the literature. For this reason, the term strange (!) quark may be well understood. As quoted by Veneziano[60]), Gell-Mann always emphasizes that owing to the fact that $m_s \simeq \Lambda$, unexpected results (large mass mixing, large m^2_s corrections) can happen in the strange-quark sector. Gell-Mann's remarks are already supported by the effective Lagrangian approach where one needs a modified chiral expansion in this channel owing to the importance of the m^2_s correction. In this chapter, we shall explicitly study Gell-Mann's arguments within the framework of the QCD spectral sum rules.

1. THE "STRANGE" MASS

a) The pseudoscalar sum rule

We repeat the strategy used for the u and d quarks. We take into account the corrections due to the strange-quark mass which cannot be neglected. We use the value of $F_{K'}$ obtained in Ref. 135). Taking $M_{K'} \simeq 1.5$ GeV, its value indicates that :

$$r_K \equiv \frac{M^4_{K'}\, F^2_{K'}}{M^4_K\, f^2_K} \simeq r_\pi \simeq 6\text{-}8 \tag{8.1a}$$

in accordance with the value which one would expect from current algebra :

$$F_{K'} \Big/ F_{\pi'} \simeq \left(\frac{M_K}{m_\pi}\right)^2 . \tag{8.1b}$$

With the QCD inputs used for the u and d quarks and using $f_K \simeq 1.2\, f_\pi$, we show the behaviour of the invariant mass $\left(\hat{m}_s + \hat{m}_u\right)$ versus τ in Fig. 8.1. Stability in τ is reached for t_c above 3.7 GeV^2. In Fig 8.2, we show the t_c-behaviour of the τ minima. The t_c stability of the result is reached for t_c larger than 4 GeV^2. These ranges of t_c values are consistent with the rough estimate of about 3 GeV^2 deduced from FESR [24] and with the one obtained from the Regge trajectories. Then, we obtain the optimal estimate :

$$\hat{m}_s + \hat{m}_u \simeq (288 \pm 25)\ MeV\ , \tag{8.2}$$

where we have used $\Lambda = (0.15 \pm 0.05)$ GeV ; $r_K = 7 \pm 1$; $t_c \geqslant 4$ GeV^2, $\left\langle \bar{s}\, s\right\rangle / \left\langle \bar{u}\, u\right\rangle = (0.5 - 1)$. These parameters induce respectively 7.6%, 4.2%, 1.8% and 1% errors which we have added quadratically. This result can be considered as an improvement of that in Ref. 139) where

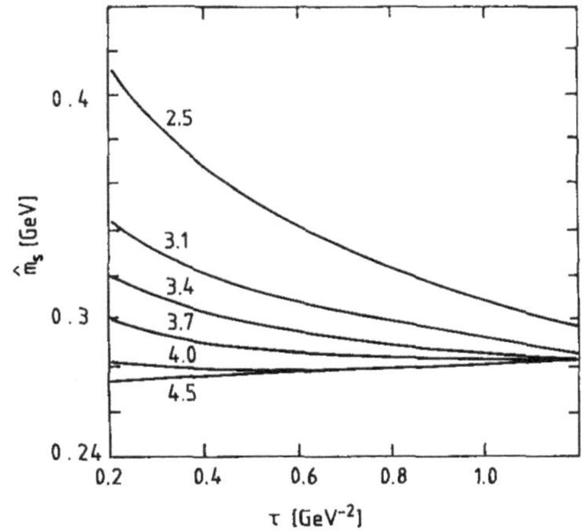

Fig. 8.1 : The invariant mass $\hat{m}_s$ versus the LSR scale τ for different values of t_c in GeV^2.

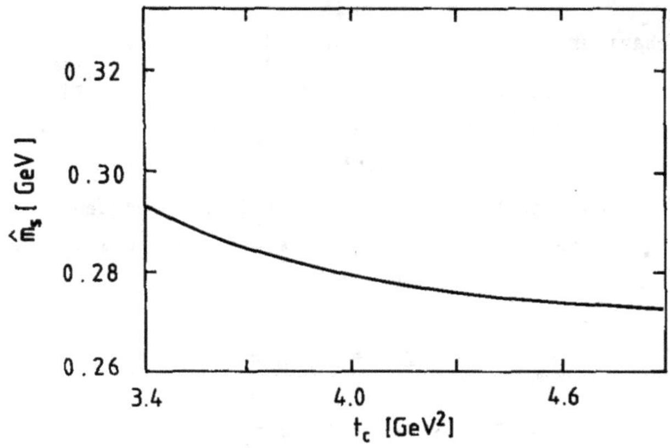

Fig. 8.2 : Optimal values of $\hat{m}_s$ *versus* t_c.

the large errors in Ref.139) are due to the non-study of the t_c-stability of the result. It is also informative to compare this result with the one obtained from FESR[24] to two-loops :

$$\hat{m}_s + \hat{m}_u \simeq (290 \pm 26) \text{ MeV} \tag{8.3}$$

or with the one obtained by combining the previous estimate of the u and d quark masses with the mass ratio from the effective chiral Lagrangian approach to the pseudoscalar, baryon mass splittings and ρ-ω mixing[142]:

$$\frac{\hat{m}_s}{\frac{1}{2}\left(\hat{m}_u + \hat{m}_d\right)} = 25.7 \pm 2.6 \quad \text{and} \quad \frac{\hat{m}_s - \left(\hat{m}_u + m_d\right)/2}{\hat{m}_d - \hat{m}_u} \simeq 43.5 \pm 2.2 \ . \qquad (8.4a)$$

The latter approach gives :

$$\hat{m}_s = (298 \pm 47) \text{ MeV} , \qquad (8.4b)$$

which is in very good agreement with previous estimates. However, the result is less accurate than the one in Eqs (8.2) and (8.3). Then, the best estimate of the strange-quark mass from the pseudoscalar sum rule to three loops is :

$$\hat{m}_s \simeq (280 \pm 25) \text{ MeV} \qquad (8.5a)$$

and :

$$\bar{m}_s \text{ (1 GeV)} = (162 \pm 15) \text{ MeV} \ . \qquad (8.5b)$$

This result rules out the possibility of having a low strange-quark running mass[152].

b) The scalar sum rule

The corresponding Laplace transform has been analyzed in Ref. 143). The spectral function has been parametrized using the data from K-π s-wave data up to $\sqrt{t} = 2$ GeV. Beyond this value, Ref. 143) uses the positivity of the spectral function from which one deduces the optimal bound

$$\hat{m}_s - \hat{m}_u \geqslant (210 \pm 40) \text{ MeV} \qquad (8.6)$$

obtained at $\tau \simeq 0.6 - 0.7 \text{ GeV}^{-2}$ where instanton effects[83] are negligible. The result is in agreement with the pseudoscalar one. The error in Eq. (8.6) is due to Λ and to the uncertainty in the input s-wave $K\pi$ amplitude where one has taken into account the interference between the resonance and the background. We shall see later on that this sum rule can also be used for deducing the decay constant K_0^* of the K_0^* resonance.

c) Can we accurately determine $\hat{m}_s$ from the φ meson ?

The Laplace transform of the φ sum rule has been discussed by SVZ [18] as a natural extension of the ρ-meson sum rule. They have used this sum rule for predicting the φ coupling and mass by using the standard strange-quark mass value of 150 MeV from current algebra. However, from a re-analysis of this sum rule, Ref. 152) has derived the accurate result :

$$\bar{m}_s \,(1\text{ GeV}) = (110 \pm 10)\text{ MeV} \tag{8.7}$$

in conflict with the (pseudo)scalar sum rules and current algebra[14] results. Motivated by this discrepancy, we have checked in Ref. 139) the stability and the accuracy of the result quoted in Eq. (8.7) when one includes radiative corrections and a $SU(3)_F$ breaking condensate and when one takes properly the value of t_c which is unambiguously fixed by FESR. Though consistent with the one in Eq. (8.7), our result does not support the error quoted.

Let us explicitly discuss this determination of m_s from the Laplace transform and FESR sum rules by including the relevant QCD corrections. These corrections are due to the two-loop gluonic corrections in the chiral limit, to the m_s^2 terms to three loops[38] and to the m_s^4 term to two loops[47] . We omit the known radiative corrections [118,119] to the vacuum condensates owing to their negligible contributions. Therefore after the leading-log resummation we get the sum rule for $SU(3)_F$ in the $\overline{MS}$ scheme[139]:

$$\int_{t_o}^{\infty} dt\, e^{-t\tau} \frac{1}{\pi} \mathrm{Im}\, \Pi_\varphi(t) \simeq \frac{\tau^{-1}}{9} \left(\frac{1}{4\pi^2}\right) \left\{1 + \frac{4}{9L} - 0.35 \frac{\log L}{L}\right.$$

$$6\, \hat{m}_s^2 \left(\frac{2}{L}\right)^{8/9} \tau \left[1 + \frac{1}{L}\left(\frac{32}{27} + \frac{8}{9}\gamma_E - 0.7 \log L\right) + \left(\frac{1}{L}\right)^2 .\right.$$

$$\left.\left(4.2 + 4.6\,\gamma_E + 1.6 \log^2 L - 3.6 \log L\right)\right]$$

$$+ \frac{6}{7} \hat{m}_s^4 \left(\frac{2}{L}\right)^{16/9} \tau^2 \left(11 - 7\,\gamma_E - 9L\right)$$

$$+ \tau^2 \left[\frac{\pi}{3}\left\langle \alpha_s\, G^2\right\rangle + 8\pi^2\, m_s \left\langle \bar{s}\, s\right\rangle\right] - \frac{1}{2}\left(\frac{896}{81}\pi^3\, \alpha_s \left\langle \bar{s}\, s\right\rangle^2\right)\tau^3 . \qquad (8.8)$$

where the two-point correlator is built from the current :

$$J_\varphi^\mu = \frac{1}{3}\, \bar{s}\, \gamma^\mu\, s \qquad (8.9)$$

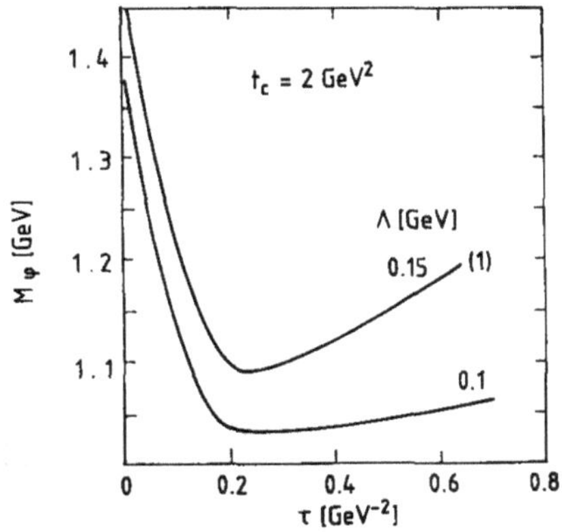

Fig. 8.3 : τ-behaviour of M_φ from the moment in Eq. (8.10).

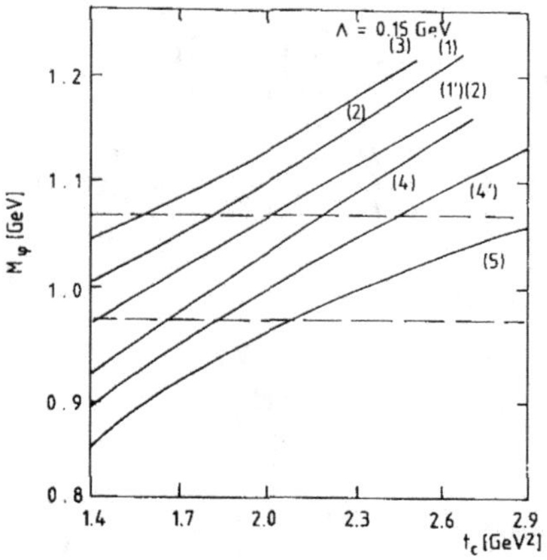

Fig. 8.4 : t_c-behaviour of the M_φ minima for the set of values of $SU(3)_F$-breaking parameters in Table 8.1.

Taking the derivative of Eq. (8.8) with respect to τ, one can easily deduce the expression of the moment :

$$R^c_\varphi = \frac{\int_{t_o}^{t_c} dt\ t\ e^{-t\tau}\ \frac{1}{\pi}\ \mathrm{Im}\ \Pi_\varphi(t)}{\int_{t_o}^{t_c} dt\ e^{-t\tau}\ \frac{1}{\pi}\ \mathrm{Im}\ \Pi_\varphi(t)} \tag{8.10}$$

and the lowest dimension FESR constraint :

$$\frac{M^2_\varphi}{\gamma^2_\varphi} \simeq \frac{1}{9\pi^2}\left\{t_c\left(1+\frac{4}{9L}-0.35\ \frac{\log L}{L}\right) - 6\ \hat{m}^2_s\left(\frac{2}{L}\right)^{8/9}\right. .$$

$$\left[1+\frac{1}{L}\ (1.98-0.7\ \log L)+\frac{1}{L^2}\ (6.4\right.$$

$$+ 1.6 \log^2 L - 3.6 \log L)] + \ldots \qquad (8.11)$$

Using $\gamma_\varphi \simeq (6.6 \pm 0.14)$, $M_\varphi = 1.02$ GeV and the value of $\hat{m}_s$ in Eq. (8.5), Eq. (8.11) allows us to get the value of t_c :

$$t_c \simeq (2.2 \pm 0.20)\ \mathrm{GeV}^2 . \qquad (8.12)$$

This value will be used for the study of the moment R^c_φ. We provide the *prediction on* M_φ *from* R_φ *in Figs 8.3 and 8.4. Fig. 8.3 gives the* τ-behaviour for the value of $\hat{m}_s$ in Eq. (8.5) and by using $\langle \bar{s} s\rangle / \langle \bar{u} u\rangle \simeq 0.5$. The stability starts at low value of τ which one can understand from the large coefficients of the SU(3)-breaking terms which tend to push to lower τ values. In Fig. 8.4, we show the effects *of the continuum threshold on the minimum for various values of the* SU(3)-breaking parameters in Table 8.1.

curves	$\hat{m}_s$	$\langle\bar{s}s\rangle/\langle\bar{u}u\rangle$	$\langle\bar{s}\Gamma s\ \bar{s}\Gamma s\rangle/\langle\bar{u}\Gamma u\ \bar{u}\Gamma u\rangle$
(1)	280	0.5	0.25 - 1
(1')	"	"	" but Λ = 100 MeV
(2)	255	"	"
(3)	280	1	1
(4)	159	1	1
(4')	"	"	" but Λ = 100 MeV
(5)	100	1	1 (Λ = 100 MeV)

TABLE 8.1

It may be unreasonable to ask 20 MeV accuracy for predicting the φ mass bearing in mind that the moment predicts the ρ-meson mass only within a 10% accuracy. *If one assumes, say,* 5% *accuracy for predicting* the φ-meson mass, one obtains the dashed horizontal lines in Fig. 8.4.

It will be noticed that situation (3) is unlikely. The value of $\hat{m}_s$ in Eq. (8.5) suggests a breaking of the $SU(3)_F$ symmetry for the condensates. A decrease of the strange mass gives a better agreement with the data. Various scenarios in Table 8.1 are not excluded by the 5% reasonable accuracy of the sum rules for the t_c values given by the FESR constraints in Eq. (8.12). Scenario (4) preferred by Ref. 152) is inside this range but it is clear that the accuracy argued by Ref. 152) is far from being realized when one includes the effects of various parameters in the uncertainties. In particular the effects of t_c are important. These large uncertainties can explain the spread of results found in the literature[152,153,154] which should be consistent with each other if the authors take care to include all sources of uncertainties in their results[139]. Now, we are in a good position to answer the question raised at the beginning of this section :

"We cannot determine accurately the strange mass from the φ-meson sum rule owing to the large uncertainties induced by the input parameters".

From Fig. 8.4, we deduce from the sum rule :

$$\hat{m}_s \simeq (80 - 280) \text{ MeV} \quad . \tag{8.13}$$

For a further check, we have also worked with the complete (non-expanded in $\frac{m_s^2}{t}$) expression of the two-point correlator to lowest order for testing the validity of the expansion in Eq. (8.8) where the m_s^2 term has a large coefficient. We found that the prediction for M_φ tends to decrease, but by a negligible amount.

It will be noticed that we have used here a simpler numerical procedure than in Ref. 139) in order to test the method independence of the result. In Ref. 139) a two-parameter fit $\left(\hat{m}_s \text{ and } \langle \bar{s} s \rangle\right)$ has been performed and where the four-quark condensate has been (at the first step) implicitly related to $\langle \bar{s}\, s \rangle$ via the factorization hypothesis. We have required as in Ref. 152) an (unrealistic) accuracy of 1% for predicting the φ-meson mass in order to have the strongest constraints on the parameter fit. The fitting procedure is done inside the range $[0, \tau_{max}]$ where τ_{MAX} is about 1 GeV^{-2}. The result is shown in Fig. 8.5 versus the value of t_c. The t_c-behaviour of the result can be compared to the one in Fig. 8.4. Using the value of t_c fixed by FESR in Eq. (8.12), we obtain for an (unrealistic) 1% accuracy for the φ mass :

$$\hat{m}_s \simeq (80 - 225)\ MeV \longrightarrow \bar{m}_s(1\ GeV) \simeq (46 - 130)\ MeV$$
$$m_s \langle \bar{s}\, s \rangle = -\ (1. - 15)\ 10^{-3}\ GeV^4 \qquad (8.14)$$

where the strong dependence on t_c is the main source of errors for $\hat{m}_s$.

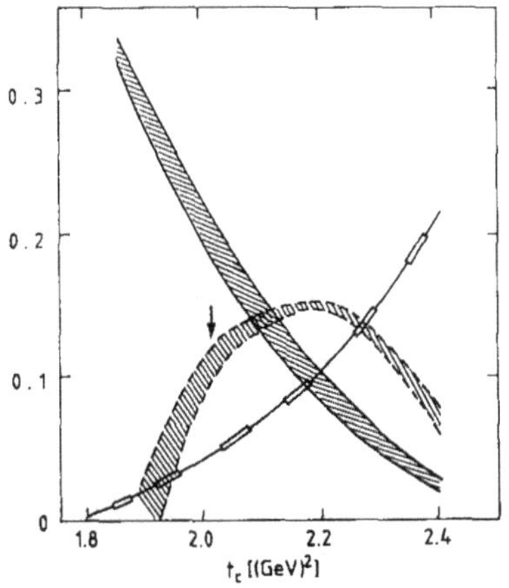

Fig. 8.5 : t_c-behaviour of the correlated set of — $\hat{m}_s$ and --- $m_s\langle \bar{s}s \rangle$ from a two-parameter fit of the φ-meson mass squared for Λ = 100 MeV ; —▭— : χ^2/NDF ; ⟶ pseudoscalar result.

However, the prediction for $m_s\langle\bar{s}\,s\rangle$ is much more precise. Its absolute value is lower than the naive kaon PCAC estimate based on the $SU(3)_F$ symmetry for the condensate :

$$m_s\langle\bar{s}\,s\rangle \simeq - M_K^2\, f_K^2 \simeq - 3\; 10^{-3}\ \mathrm{GeV}^4 \quad . \tag{8.15}$$

The value of $m_s\langle\bar{s}\,s\rangle$ in Eq. (8.14) is in agreement with the one in Ref. 152). We shall come back to this point later.

d) **An improved Gell-Mann-Okubo-like mass formula**

An alternative derivation of the strange-quark mass can be obtained from a Gell-Mann-Okubo "improved" mass formula[153], improved in the sense that in addition to the usual linear m_s term which is responsible for the ρ-K^*-φ mass spacing we include in the analysis the m_s^2 term and some other QCD corrections. We can either work with the difference[153]

$$\delta_{\varphi\rho} \equiv R_\varphi^c - R_\rho^c \simeq M_\varphi^2 - M_\rho^2 \tag{8.16}$$

or the ratio :

$$r_{\varphi\rho} \equiv \frac{R_\varphi^c}{R_\varphi^c} = \frac{M_\varphi^2}{M_\rho^2} \quad , \tag{8.17}$$

in order to see the splitting in QCD. By working with Eq. (8.16), we obtain after expanding the moments :

$$\delta_{\varphi\rho} \simeq \delta_{\varphi\rho}^c + f\left(\bar{m}_s,\ \bar{\alpha}_s,\ \langle\bar{s}s\rangle\right) \quad , \tag{8.18}$$

where :

$$\delta^c_{\varphi\rho} \equiv \left(t^\rho_c\, e^{-t^\rho_c \tau} - t^\varphi_c\, e^{-t^\varphi_c \tau}\right)\left(1 + \frac{4}{9L} - 0.35\,\frac{\log L}{L}\right) , \quad (8.19a)$$

$$f\left(\bar{m}_s, \bar{\alpha}_s, \langle \bar{s}\, s\rangle\right) \simeq 6\, \hat{m}^2_s \left(\frac{2}{L}\right)^{8/9}\left[1 + \frac{1}{L}\left(\frac{56}{27} + \frac{8}{9}\gamma_E - 0.7 \log L\right)\right.$$

$$\left. + \frac{1}{L^2}\,(2.3 + 1.7\,\gamma_E - 1/9 \log L)\right]$$

$$- \frac{6}{7}\,\hat{m}^4_s \left(\frac{2}{L}\right)^{16/9} \tau(15 - 14\,\gamma_E - 18\,L)$$

$$- 16\,\pi^2\, m_s \langle \bar{s}\, s\rangle \tau + \left(\frac{3}{2}\right)\left(\frac{896}{81}\,\pi^3\right)\alpha_s \left(\langle\bar{s}s\rangle^2 - \langle\bar{u}u\rangle^2\right)\tau^2 . \quad (8.19b)$$

Only the $SU(3)_F$-breaking terms remain, so we are not aware of the uncertainties due to $\langle \alpha_s\, G^2\rangle$. $\delta^c_{\varphi\rho}$ takes account of the difference of continuum thresholds for the φ and ρ channels which we know from previous FESR studies to be :

$$t^\rho_c \simeq 1.7\ \text{GeV}^2 \quad ; \quad t^\varphi_c \simeq 2.2\ \text{GeV}^2 . \quad (8.20a)$$

We have studied the effects of t_c on the predictions. The effects are negligible for :

$$t^\rho_c \simeq (1.4 - 2.5)\ \text{GeV}^2 ,$$

$$t^\varphi_c \simeq (0.5 - 0.7)\ \text{GeV}^2 + t^\rho_c . \quad (8.20b)$$

This is already a great advantage compared to the analysis based on the individual moment R^c_φ. We study the τ-behaviour of the predictions for different scenarios given in Table 8.2 where we study the deviation from the central value (dashed line) in Fig. 8.6.

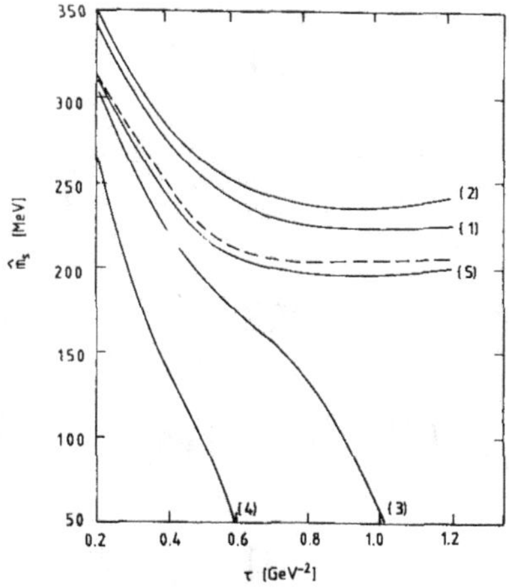

Fig. 8.6 : τ-behaviour of $\hat{m}_s$ from the GMO-like mass formula.

curve	$\delta_{\varphi\rho}$	Λ	$\langle\bar{s}s\rangle/\langle\bar{u}u\rangle$	$\langle\bar{s}\Gamma s\ \bar{s}\Gamma s\rangle/\langle\bar{u}\Gamma u\ \bar{u}\Gamma u\rangle$
...	0.4	0.15	0.5	$(0.5)^2$
(1)	0.44	"	"	"
(2)	0.4	0.1	"	"
(3)	"	0.15	"	1
(4)	"	"	1	1
(5)	"	"	"	" but $\delta t_c = 0.7\ GeV^2$

One can see from Fig. 8.6 that for a SU(3) breaking of the $\langle\bar{s}s\rangle$ and four-quark condensates, one has a better τ-stability and reasonable values of $\hat{m}_s$. In the case of $SU(3)_F$ symmetric condensates, the τ-stability becomes an inflexion point and the value of $\hat{m}_s$ is lower. Such behaviours of the SU(3)-breaking terms are consistent with the intuitive picture of SU(3) breaking. The conclusion from the above analysis can be rendered sharper if one introduces new information on the size of the $\langle\bar{s}s\rangle$ and $\langle\bar{s}\Gamma s\ \bar{s}\Gamma s\rangle$ condensates. The former is expected to deviate from $SU(3)_F$ symmetry as we know from the study of R^c_φ. The latter is not under good control. Assuming that its value is in the range :

$$\langle\bar{s}\Gamma s\ \bar{s}\Gamma s\rangle / \langle\bar{u}\Gamma u\ \bar{u}\Gamma u\rangle \simeq 0.25 - 1 \tag{8.21}$$

we deduce from Fig. 8.6 by adding the errors quadratically :

$$\hat{m}_s \simeq (205 \pm 50)\ \text{MeV}\ . \tag{8.22}$$

We conclude that the best information from the R^φ_c and Gell-Mann-Okubo-like sum rules is :

$$m_s \langle\bar{s}s\rangle = -\ (1 - 1.5)\ 10^{-3}\ \text{GeV}^4\quad , \tag{8.23a}$$

$$\hat{m}_s \simeq (201 \pm 40)\ \text{MeV}\quad , \tag{8.23b}$$

which come respectively from Eqs (8.14) and (8.22). Eq. (8.23a) suggests a large deviation from the kaon PCAC estimate of $-\ 3.10^{-3}$ GeV^4 for $m_s\langle\bar{s}s\rangle$ while Eq. (8.23b) leads to the running mass at 1 GeV to three loops.

$$\bar{m}_s\,(1\ \text{GeV}) \simeq (116 \pm 24)\ \text{MeV}\ . \tag{8.24}$$

e) The "value" of the strange mass and comparison with other approaches

At this point, it is tempting to compare these results with the one obtained from the (pseudo)scalar sum rule in Eq. (8.5). It is interesting to notice that owing to the large coefficient of the m_s^2 term and the importance of the linear term in the vector meson channels, the right splitting prefers a low value of m_s (Eq. (8.24)). On the contrary, owing to the leading quark mass dependence of the (pseudo) scalar correlators, it is the perturbative m_s^2 term which is mainly responsible for the kaon mass and controls the size of the radial excitations. As a by-product, the right properties of the (pseudo)scalar channels require a large value of m_s^2. The "unified" value of the strange-quark mass from these two channels lies in a very narrow margin. We then deduce from the pseudoscalar and vector meson sum rules and from the "hybrid determination" in Eq. (9.16) the very accurate (better than 2% !) weighted average estimate for $\Lambda = (150 \pm 50)$ MeV :

$$\hat{m}_s \simeq (266.7 \pm 14.7)\ \text{MeV} \tag{8.25a}$$

and to three loops :

$$\bar{m}_s(1\ \text{GeV}) \simeq (159.5 \pm 8.8)\ \text{MeV}. \tag{8.25b}$$

This value is in agreement with the standard current algebra value[14] but it is much more accurate. It also agrees with the range allowed in the baryon sum rules[155,99], but it is a little higher than the one from an effective potential approach[156]. The result in Eq. (25) might not include the systematic errors induced by the QSSR approaches. This source of error is expected to be less than 5-10% at the scale where the optimal value of the mass is extracted.

There are various uses of this result in weak interaction phenomenologies (Grand Unification, mixing matrices and weak kaon-decays). It is also interesting to compare the ratio of quark masses obtained from the sum rules with the one from the current algebra approach. Using our previous values of u and d quark masses, we deduce :

$$r_{32} \equiv \frac{\hat{m}_s - \hat{m}}{\hat{m}_d - \hat{m}_u} \simeq 39 \pm 6 \qquad (8.26a)$$

where $\hat{m} \equiv \left(\hat{m}_u + \hat{m}_d\right)/2$. This ratio can be compared with the current algebra weighted average :

$$r_{32} \equiv \frac{\hat{m}_s - \hat{m}}{\hat{m}_d - \hat{m}_u} \simeq 43.5 \pm 2.2 \quad . \qquad (8.26b)$$

Taking the weighted average of these two independent measurements leads to :

$$r_{32} = 43 \pm 2 \qquad (8.27)$$

which we consider as a final estimate.

2. DEVIATION FROM KAON PCAC

As in the case of $SU(2)_F$, we shall estimate the subtraction constant $\psi_5(0)^u_s$ using the same method. The K' parameters which we shall use are given in Eq. (8.1). We shall use the value of the strange--quark mass obtained in Eqs (8.5) and (8.25). We show the ratio :

$$\frac{\psi_5(0)^u_s}{2\ M_K^2\ f_K^2} \equiv 1 - \rho_K \qquad (8.28a)$$

in Fig. 8.7 versus the τ values and for possible values of $\hat{m}_s$ and $\Upsilon \equiv \langle \bar{s}\, s\rangle / \langle \bar{u}\, u\rangle$ ratio, the latter ranging from 0.5-1. The τ-stability of the results is obtained for the two sum rules Eqs (7.33 and 34) for the values of t_c in the range expected by FESR and where the prediction for $\left(\hat{m}_u + \hat{m}_s\right)$ is t_c stable. The effects of t_c in this range are also small as the τ-stability is reached for τ larger than 0.8 GeV^{-2} where the effect of the K' to the sum rule is also small, i.e. improvement of the K' effect by including finite-width correction is inessential. Taking the largest range of predictions, we obtain the optimal result :

$$\rho_K \simeq (0.29 - 0.44) \quad , \qquad (8.28b)$$

in agreement with the first findings[140] that kaon PCAC is largely violated, this result being also supported by previous authors[35,141,149,155]. We can also consider the result in Eq.(8.28) to be improvements of previous results where the τ- ant t_c-stabilities have not been carefully studied and the input on the strange-quark mass values have

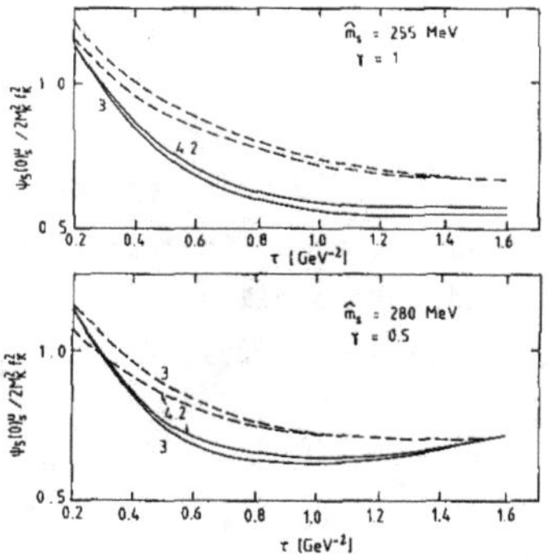

Fig. 8.7 : $\psi_5(0)^u_s/2\,M_K^2 f_K^2$ *versus* τ *and for two values of the set* $\left(\hat{m}_s,\ \gamma \equiv \langle\bar{s}s\rangle/\langle\bar{u}u\rangle\right)$.

been inaccurate. This value of $m_s\langle\bar{s}\,s\rangle$ is in agreement with the one from the φ-meson sum rule (Eq. (8.14)).

One might also deduce $\psi_5(0)^u_s$ from the FESR in Eq. (6.61) which gives to leading order :

$$\psi_5(0)^u_s \simeq 2\,M_K^2\,f_K^2\left(1 + r_K\left(\frac{M_K}{M_{K'}}\right)^2\right) - \frac{3}{8\pi^2}\left(\bar{m}_s + \bar{m}_u\right)^2 t_c . \qquad (8.29)$$

As can be seen, the estimate of $\psi_5(0)^u_s$ here depends strongly on r_K owing to the absence of the exponential factor reduction for the K' effect. But, Eq. (8.29) can tell us at least qualitatively that the m_s^2 term can be responsible for violation of kaon PCAC.

Another possible way of extracting $\psi_5(0)^u_s$ is the analysis of the correlator[157]:

$$\psi_5^\mu(q^2) = i\int d^4x\, e^{iqx}\left\langle 0\left|\mathbb{T}\,\partial_\mu A^\mu(x)^u_s\left(A^\mu(0)^u_s\right)^+\right|0\right\rangle \qquad (8.30a)$$

which, however, is related to $\psi_5(q^2)$ by the current algebra Ward identity :

$$\psi_5^\mu(q^2) = \psi_5(q^2)\,.\left(\frac{q^\mu}{q^2}\right) . \qquad (8.30b)$$

Therefore, the analyses of $\psi_5^\mu(q^2)$ and $\psi_5(q^2)$ should be equivalent.

3. SCALAR SUM RULES AND THE $K_0^*(1.35)$

We study as in the case of the $\bar{u}d$ channel (Chapter 7) the scalar sum rule in order to fix[139,145] the decay constant and the mass of the K_0^* . We use as input the value of the strange-quark mass and condensates. The prediction of $f_{K_0^*}$ versus τ and t_c is given in Fig. 8.8 :

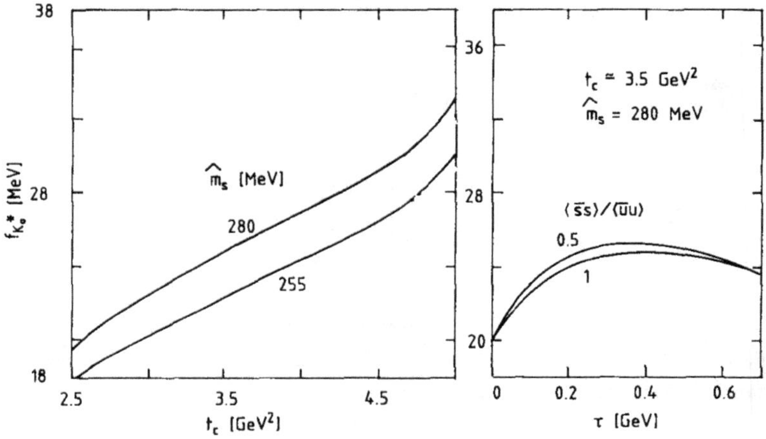

Fig. 8.8 : t_c- and τ-behaviour of $f_{K_0^*}$ for different values of $\hat{m}_s$ and $\langle\bar{s}s\rangle/\langle\bar{u}u\rangle$.

We have a nice τ minimum in the range 0.2-0.6 GeV^{-2} while there is an inflexion point in the t_c variable for $t_c \simeq 3.5 - 4.5$ GeV^2. $f_{K_0^*}$ increases obviously with $\hat{m}_s$. The dependence on $\langle\bar{s}\, s\rangle/\langle\bar{u}\, u\rangle$ is less trivial. Given $\hat{m}_s$, $f_{K_0^*}$ increases by about 1.6% for the ratio moving from 1 to 0.5. We show in Fig. 8.9 the prediction for the mass $M_{K_0^*}$ where we have τ- but not t_c-stabilities. Using the previous value of $t_c \simeq 3.5 - 4.5$ GeV^2, which is also consistent with the one from a two-parameter fit $\left(f_{K_0^*} \text{ and } t_c\right)$[139,145] analysis of the same sum rule, one obtains :

$$M_{K_0^*} \simeq 1.46 - 1.51 \text{ GeV} \tag{8.31}$$

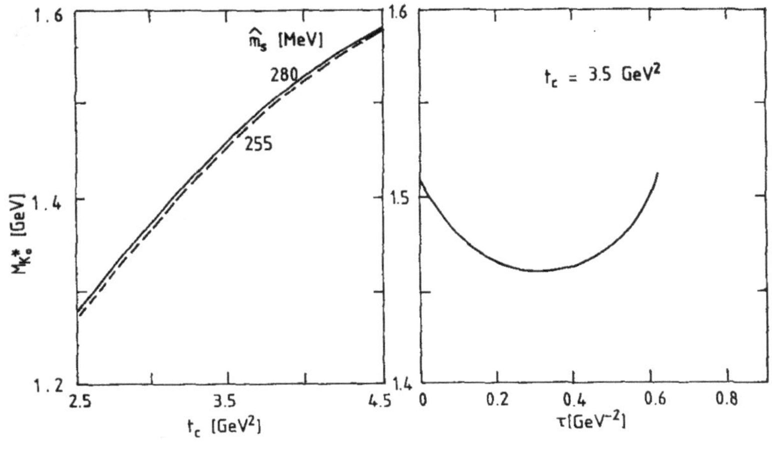

Fig. 8.9 : $M_{K_0^*}$ versus t_c and τ for two values of $\hat{m}_s$.

which is quite good compared with the data (1.35 GeV). In order to see the effect of $M_{K_0^*}$ in the prediction of $f_{K_0^*}$, we also use the largest mass in Eq. (8.31). The prediction tends to decrease by 4.5%. Taking into account all sources of uncertainties, we deduce :

$$f_{K_0^*} \simeq (25.6 \pm 3.3) \text{ MeV} \quad . \tag{8.32}$$

We can compare this result with the one from a two-parameter fit $\left(f_{K_0^*} \text{ and } t_c\right)$ of the same sum rule, which for $\hat{m}_s \simeq 250$ MeV is[139,145]:

$$f_{K_0^*} \simeq (31 \pm 3) \text{ MeV} \quad . \tag{8.33}$$

The difference between Eqs (8.32) and (8.33) can provide a gauge of we obtain

$$f_{K_0^*} \simeq (28.3 \pm 4.5) \text{ MeV} . \tag{8.34}$$

As noticed in Refs 145) and 158), $f_{K_0^*}$ is, to leading order, a linear function of the strange-quark mass.

4. $\psi(0)^u_s$ AND THE RATIO $\langle \bar{s} s \rangle / \langle \bar{u} u \rangle$

Using the result in Eq. (8.34), one can deduce the value of the subtraction constant from the sum rules analogous to Eq. (7.33) and (7.34). The prediction is shown in Fig 8.10 versus τ and t_c given the sets $\left(f_{K_0^*}, \hat{m}_s\right)$ which are (24, 255) and (33, 280) in units of MeV. One can observe that the analogue of the modified sum rule in Eq. (7.34) does not present τ- and t_c-stabilities. The inflexion point for the sum rule in Eq. (7.33) is obtained for $\tau \simeq 0.9 - 1.1 \text{ GeV}^{-2}$. It is more pronounced for the set (24, 255). We also have a t_c-stability for this sum rule. Then, we deduce the optimal prediction :

$$\psi(0)^u_s = - (9.5 \pm 1.2)\ 10^{-4} \text{ GeV}^4 \quad . \tag{8.35}$$

where the error comes from the difference between the two sets of values of $\left(f_{K_0^*}, \hat{m}_s\right)$ used.

This result is consistent with the one in Refs 139) and 145) but rules out the possibility of having $\psi(0)^u_s = 0$. This zero value allowed by Ref. 149) is related to the non-careful study of the t_c-stability and to the inaccuracy of the model used for parametrizing the spectral function for this channel. These criticisms are supported by the revised estimate of $\psi(0)^u_d$ done in Ref. 137).

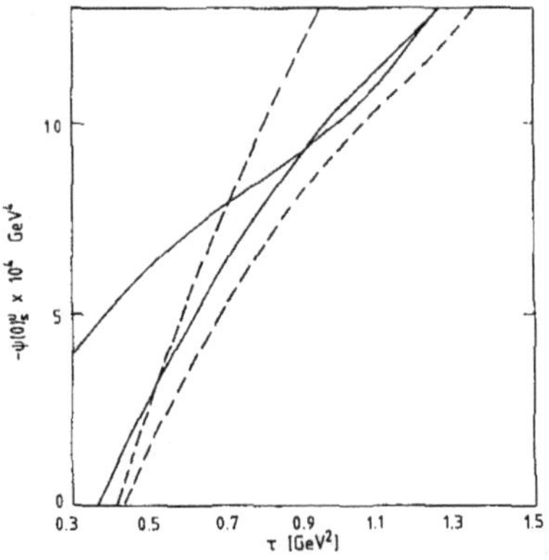

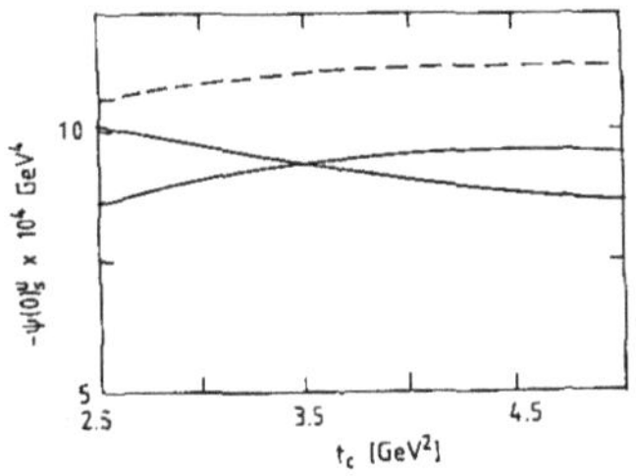

Fig. 8.10 : $\psi(0)^u_s$ versus τ and t_c : — Eq. (7.33) --- Eq. (7.34).

We can now combine this result with the value of $\psi_5(0)^u_s$ obtained in Eq. (8.28). Using their respective relation with the quark masses and condensates, we obtain :

$$\frac{\psi(0)^u_s}{\psi_5(0)^u_s}\psi \simeq \left(1 - \frac{2m_u}{m_s}\right)\left(1 - \frac{2\langle\bar{s}s\rangle}{\langle\bar{u}u + \bar{s}s\rangle}\right) \simeq -(0.19 - 0.31) \quad , \tag{8.36a}$$

from which we deduce :

$$\frac{\langle\bar{s}s\rangle}{\langle\bar{u}u\rangle} \simeq 0.51 - 0.67 \quad . \tag{8.36b}$$

Eq. (8.36) suggests a large violation of the $SU(3)_F$ symmetry for the condensates. It is also consistent with previous results quoted in Ref. 139) but it is much more accurate, the accuracy of the result being improved by the study of the t_c- and τ-stabilities which have not been carefully done in previous works. FESR analysis also supports the previous result[137,159] though one expects the numbers there to be less accurate owing to the large weight of the high-mass continuum in the FESR. Eq. (8.36) is inconsistent with the baryon sum rules estimate of about 0.8[29,155] where the accuracy and the stability of the predictions have, however, been reconsidered and questioned by the authors in Ref. 160).

Let us now leave the sum rules world and go to the effective Lagrangian framework. To order p^2, the effective Lagrangian is the $SU(3)_F$ generalization of Eq. (7.11). Using a generating functional approach, one can derive[142] :

$$\frac{\langle \bar{s}\, s\rangle}{\langle \bar{u}\, u\rangle} \simeq 1 + 3\,\mu_\pi - 2\,\mu_k - \mu_\eta + \left(m_s - \tilde{m}\right)K_1 \quad , \tag{8.37a}$$

where :

$$\tilde{m} \equiv \frac{1}{2}\,(m_u + m_d) \quad ,$$

$$\mu_p \equiv \frac{1}{32\pi^2}\,\frac{M_p^2}{F_0^2}\log\frac{M_p^2}{\nu^2} \quad , \tag{8.37b}$$

$$F_o \simeq f_\pi \simeq 93.3 \text{ MeV} \quad .$$

ν is an arbitrary subtraction associated with the pseudoscalar chiral loop and K_1 is an undetermined constant related to the high-energy constants. Eliminating these two uncontrolled numbers leads to[142]:

$$\frac{\langle \bar{d}d \rangle}{\langle \bar{u}u \rangle} \simeq 1 - \frac{m_d - m_u}{m_s - \tilde{m}} \left\{ 1 - \frac{\langle \bar{s}s \rangle}{\langle \bar{d}d \rangle} + \frac{1}{16\pi^2} f_\pi^2 \, . \right.$$

$$\left. \cdot \left[m_K^2 - m_\pi^2 \left(1 + \log \frac{M_K^2}{m_\pi^2} \right) \right] \right\} \tag{8.38a}$$

which numerically gives :

$$\left(1 - \frac{\langle \bar{d}d \rangle}{\langle \bar{u}u \rangle} \right) \cdot \left(\frac{m_s - \tilde{m}}{m_d - m_u} \right) \simeq \left(1 - \frac{\langle \bar{s}s \rangle}{\langle \bar{u}u \rangle} \right) + 0.13. \tag{8.38b}$$

We can use Eqs (7.49) and (8.27) in order to deduce the prediction

$$\frac{\langle \bar{s}s \rangle}{\langle \bar{u}u \rangle} \simeq 0.72 - 0.75 \ . \tag{8.39}$$

This result is a little bit higher than Eq. (8.36). The discrepancy between the two results might be understood once one is able to introduce properly the m_s^2 corrections into the chiral Lagrangian approach. This improvement has not yet been done[161].

5. MORE ON THE $SU(3)_F$ BREAKING FOR LIGHT MESONS

a) Tensor mesons and f'_2-f_2 splittings

The sum-rule analysis of the f_2-f'_2 splitting has been suggested in Refs 152) and 162) as a good source for measuring the value of the strange-quark mass. In fact, from previous works one can check that in this channel the behaviour of the $SU(3)_F$ -breaking term is "anomalous" in the sense that the decrease of the m_s value increases the f'_2-f_2 mass difference. The authors conclude that the experimental value can be reproduced at the cost of a low value of $\bar{m}_s$ of about 100 MeV. But, the authors do not explicitly show the sensitivity of the result on the choice of the continuum threshold t_c. Motivated by such an "anomaly", we have re-evaluated[48] the two-point correlator associated with the 2^{++} current (see Chapter 1). After correcting the QCD expression of Ref. 162), we have re-estimated[48b] the f'_2 over f_2 mass ratio using the moments :

$$r(\tau) \equiv \frac{R^c_{(3)}}{R^c_{(2)}} , \tag{8.40a}$$

where :

$$R^c_{(i)} \equiv \frac{\int_0^{t_c^{(3)}} dt\, e^{-t\tau} \frac{1}{\pi} \text{Im}\, \psi_2^{(3)}(t)}{\int_0^{t_c^{(2)}} dt\, e^{-t\tau} \frac{1}{\pi} \text{Im}\, \psi_2^{(2)}(t)} . \tag{8.40b}$$

The indices (2) and (3) refer to $SU(3)_F$ and $SU(2)_F$. Within a duality ansatz parametrization of the spectral function, it is clear that $r(\tau)$ represents the ratio of the f'_2 over the f_2 mass squared. The QCD expression of the moments can be derived from the Laplace transform :

$$\mathcal{F}(\tau,\nu) \equiv \int_{4m^2}^{\infty} dt\, e^{-t\tau} \frac{1}{\pi} \text{Im}\, \psi_2(t)$$

$$= \int_{4m^2}^{\infty} dt\, e^{-t\tau} \frac{1}{\pi} \text{Im}\, \psi_2(t)|_{Pert}$$

$$- \frac{3}{5\pi^2} \tau^{-3} \frac{\alpha_s}{9\pi} \left[\frac{113}{15} + 16 \left(\gamma_E + \log \tau\nu^2\right)\right]$$

$$- \frac{8}{9\pi} \left\langle \alpha_s\, G^2 \right\rangle \tau^{-1} + k \frac{64\pi}{9} \alpha_s \left\langle \bar{\psi} \psi \right\rangle^2$$

$$- \frac{16m}{3} g \left\langle \bar{\psi} \sigma^{\mu\nu} \frac{\lambda_a}{2} \psi\, G^a_{\mu\nu} \right\rangle + 8\, m^3 \left\langle \bar{\psi} \psi \right\rangle$$

$$- \frac{1}{3} \left(7 + 8\, \gamma_E + 8 \log \tau\, \nu^2\right) m^2 \frac{\left\langle \alpha_s\, G^2 \right\rangle}{\pi} . \qquad (8.41a)$$

which after RGI becomes

$$\mathcal{F}(\tau, \nu, \alpha_s, m) = \left(\frac{\bar{\alpha}_s(\tau)}{\bar{\alpha}_s(\nu^2)}\right)^{2\gamma^1_{11}/\beta_1} \mathcal{F}\left(\tau, \frac{1}{\sqrt{\tau}}, \alpha_s, \bar{m}\right) \qquad (8.41b)$$

$\gamma^1_{11} \equiv \frac{16}{9}$ is the anomalous dimension of the current. The spectral function $\text{Im}\, \psi_2^{Pert}(t)$ reads to lowest order :

$$\frac{1}{\pi} \text{Im}\, \psi_2^{Pert}(t) = \frac{t^2}{10\pi^2} (5 - 2v^2)v^3\, \theta(t - 4m^2) \qquad (8.42)$$

where : $v \equiv (1 - 4m^2/t)^{1/2}$ is the quark velocity. The normalization of $\psi_2(q^2)$ is the same as the one in Eqs (1.146) to (1.149).

First let us study $\mathcal{F}(\tau, \nu)$ and $R^c_{(2)}(\tau)$ corresponding to the f_2 meson. We introduce the f_2 meson coupling as :

$$\frac{1}{\pi} \text{ Im } \psi_2(t) = 2\, g_T^2\, M_T^6\, \delta\left(t-M_T^2\right) + \text{QCD continuum} \quad , \qquad (8.43)$$

which we express in terms of the renormalization group invariant $\hat{g}_T$:

$$g_T(\nu) = \frac{\hat{g}_T}{(\log \nu/V)} \gamma_{11}^1 /- \beta_1 \ . \qquad (8.44)$$

We use $\mathcal{F}(\tau,\nu)$ to extract $\hat{g}_T$. Using the usual τ- and t_c-stability criterions, we obtain[48b] (see Fig. 8.11) :

$$\hat{g}_T|_{SU(2)} \simeq (0.13 - 0.18) \quad ,$$

$$\hat{g}_T|_{SU(3)} \simeq (0.11 - 0.15) \quad , \qquad (8.45)$$

which indicates a good $SU(3)_F$ symmetry for the couplings. We use the FESR constraint :

$$2\, M_T^6 g_T^2 \simeq \frac{1}{\left(\log \sqrt{t_c/\Lambda}\right)} 2\, \gamma_{11}^1 \Big/ \beta_1\, \frac{t_c^3}{10\pi^2} \left\{1 - \frac{481}{135} \frac{\bar{\alpha}_s(t_c)}{\pi} - \frac{80\pi}{9t_c^2} \left\langle \alpha_s G^2 \right\rangle \right\} \quad (8.46a)$$

in order to obtain the value :

$$t_c \simeq 1.54\, M_T^2 \qquad (8.46b)$$

which is inside the t_c-stability of $\hat{g}_T$.

Next, we study the mass prediction $M_{f_2}^2$ using the moment $R_{(2)}^c(\tau)$. The moment presents a τ-stability for $\tau \simeq 0.2 - 0.8$ GeV^{-2}. Using the value of $t_c \simeq 2.5 - 3.5$ GeV^2 in the range suggested by Eq. (8.46), one deduces :

$$M^2_{f_2} \simeq (2 - 2.5)\ \mathrm{GeV}^2 \tag{8.47}$$

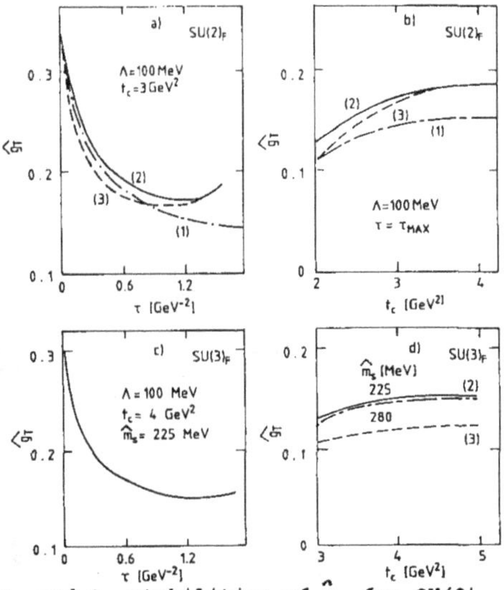

Fig. 8.11 : τ*- and* t_c*-stabilities of* $\hat{g}_T$ *for* $SU(2)_F$ *and* $SU(3)_F$.

a little bit higher than the previous result[163, 162] where the authors have not included in the QCD continuum the discontinuity due to the condensate. We now study the ratio $r(\tau)$ of moments which gives $\left(M^2_{f'_2} \big/ M^2_{f_2}\right)$. We relate the $SU(3)_F$ continuum threshold to that of $SU(2)_F$ as :

$$t_c \Big|_{SU(3)_F} \simeq t_c + \delta \tag{8.48}$$

where δ is a free parameter which we expect to be in the range 0.5-1.5 GeV^2 from Regge trajectory arguments. We show the behaviour of $r(\tau)$ in Fig. 8.12 where one can see a τ- but not a t_c-stability. The absence of this latter decreases the accuracy of the prediction. One can also see that the splitting increases with increasing value of $\hat{m}_s$ (good news!) but the effect is weaker than of t_c.

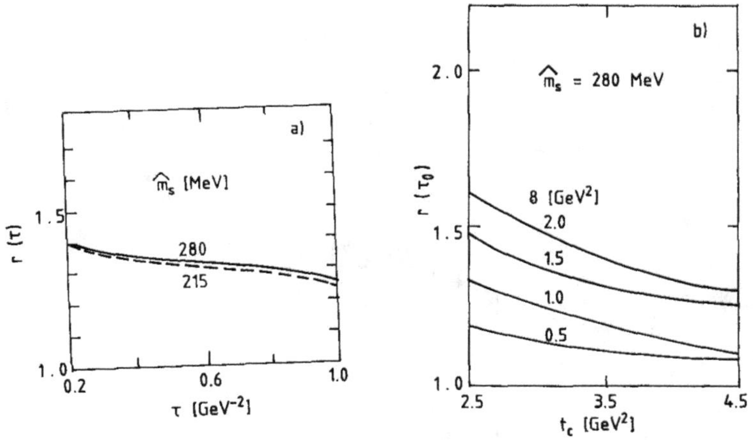

Fig. 8.12 : τ- and t_c- stabilities of the mass squared ratio $r \equiv \left(M_{f'_2}/M_{f_2}\right)^2$.

We therefore we conclude that :

One cannot deduce an accurate value of the strange-quark mass from the f_2 - f'_2 splitting.

This conclusion is in contrast to the RRY result[152] based on the incorrect QCD expression[162] of the correlator where the accuracy of the f'_2-mass prediction is indeed dual to the one of m_s.

An interesting phenomenological use of the previous coupling is the test of the low-energy theorem[163] which relates the coupling of the quark component of the energy-momentum tensor to Goldstone boson pairs. This relation is :

$$g_T(\nu)\ g_{T\pi\pi} \simeq 2\rho(\nu) \quad , \tag{8.49a}$$

where $\rho(\nu)$ is the fraction of pion momentum carried by the u and d quarks. The value of $g_{T\pi\pi}$ is 11.5 from the data :

$$\Gamma\left(f_2 \longrightarrow \pi^+\pi^- + \pi^0\ \bar{\pi}^0\right) = g^2_{T\pi\pi}\ \frac{M_T}{320\pi}\left(1 - \frac{4m_\pi^2}{M_T^2}\right)^{5/2} \quad . \tag{8.49b}$$

Using our previous estimate, we obtain :

$$\rho(\nu = 1\ \text{GeV}) \simeq 0.54 - 0.74 \tag{8.50}$$

which might be tested in lepton-pion deep inelastic scattering at momentum transfer of the order ν. However, this low-energy theorem can be put to alternative uses in the presence of a tensor gluonium as we shall see later on.

b) K^* sum rule and the $\tau \longrightarrow \nu_\tau\ K^*$decay

Let us study the sum rule associated with the $\bar{u}\ \gamma^\mu s$ current having the quantum number of the K^* meson. We shall use the complete perturbative expression of the two-point correlator to two loops given in Eq. (1.140) $\left(m_s^2 \gg m_u^2 \simeq 0\right)$. The non-perturbative contributions are the same as those of the φ except that the strength of the $m_s\left\langle\bar{\psi}\ \psi\right\rangle$ contribution is half the latter. We normalize the K^* coupling as :

$$\left\langle 0\left|\bar{u}\ \gamma^\mu s\right|K^*\right\rangle = \frac{\sqrt{2}\ M^2_{K^*}}{2\ \gamma_{K^*}}\ \epsilon\ \mu\ . \tag{8.51}$$

and we work with the one-resonance + "QCD continuum" parametrization of the spectral function. In Fig. 8.13 we show the prediction of γ_{K^*} from the Laplace sum rule analogue of Eq. (8.9). The result presents a τ inflexion point around $\tau \simeq 0.6 - 1\ \text{GeV}^{-2}$ and stabilizes for $t_c \geqslant 2\ \text{GeV}^2$. Within these conditions, we obtain :

$$\gamma_{K^*} \simeq 2 - 3 \tag{8.52}$$

for $\hat{m}_s \simeq 250$ MeV and $\Lambda \simeq (150 \pm 50)$ MeV. The large error is due to the innacurate location of the inflexion point. Using this value of γ_{K^*}, we can predict :

$$\frac{\Gamma\left(\tau \rightarrow \nu_\tau K^*\right)}{\Gamma\left(\tau \rightarrow \nu_\tau e\, \bar{\nu}_e\right)} \simeq \frac{12\pi}{M_\tau^2} \sin^2 \theta_c \int_0^{M_\tau^2} dt \left(1 - \frac{3t^2}{M_\tau^4} + \frac{2t^3}{M_\tau^6}\right) \text{Im}\, \Pi_{K^*}(t)$$

$$\simeq 0.07 - 0.15 \tag{8.53}$$

in good agreement with the data (0.10 ± 0.04). We can also compare

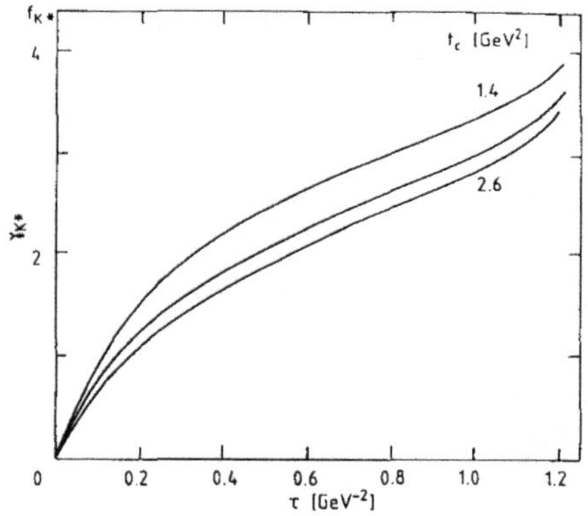

Fig. 8.13 : γ_{K^*} versus τ for two values of t_c.

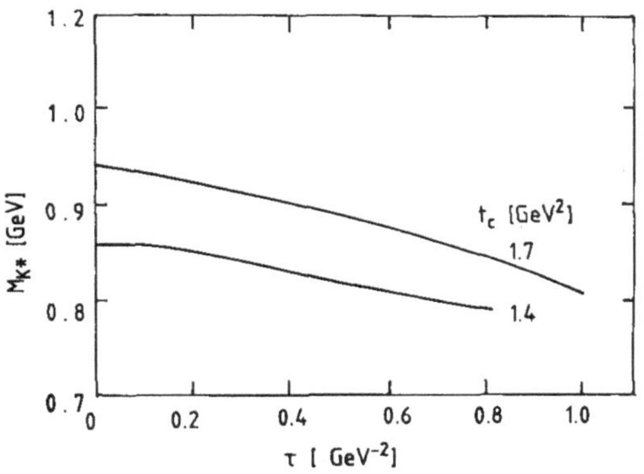

Fig. 8.14 : M_{K^*} *versus* τ *for two values of* t_c.

the prediction in Eq. (8.52) with the $SU(3)_F$ symmetry result :

$$\gamma_{K^*}\Big|_{SU(3)} \simeq \frac{\sqrt{2}}{3}\,\gamma_{\varphi} \simeq 3.1 \quad . \tag{8.54}$$

This comparison does not indicate a sharp deviation from the $SU(3)_F$ symmetry prediction in contrast to the claim of Ref. 152). This result is in agreement with the SVZ value in Ref. 18). We show in Fig. 8.14 the prediction for M_{K^*} from the moment ratio. We do not have τ-stability. Thus we cannot extract optimal information for M_{K^*} from this sum rule with our level of approximation.

c) Mass and coupling of the $f_0(1.3)$

Before concluding this chapter on SU(3) breaking for the light-meson systems, let us study the isoscalar scalar meson. Phenomenologically, the interest for the I = 0 scalar mesons is their possible mixings with scalar gluonia. In this section we shall not yet consider such mixings which should affect our predictions slightly. This particular point will be studied carefully in the next chapter dedicated to gluonia. We shall be concerned here with the $f_o(0.98)$ and $f_o(1.3)$, assumed to be associated to the $SU(2)_F$ and $SU(3)_F$ ideally mixed currents :

$$J_2 = : \frac{\tilde{\bar{m}}}{\sqrt{2}} \left(\bar{u}u + \bar{d}d\right) : \tag{8.55a}$$

$$J_3 = : m_s \, \bar{s}s : \tag{8.55b}$$

The associated two-point correlators can easily be obtained from previous chapters.

Owing to the good realizations of the SU(2) flavour and chiral symmetries, one expects that the a_o and f_o parameters are the same (up to an overall normalization). In particular, one can expect that their masses are degenerate in the same way as the ω-ρ degeneracy. For the $f_o(1.3)$, one could use a Gell-Mann-Okubo-like mass formula for predicting the splitting. If one assumes a cancellation of the continuum contribution in the relevant combination of moments, one obtains :

$$M_3^2 \simeq M_2^2 + 6\,\bar{m}_s^2 - \frac{48\pi^2}{3} m_s \left\langle \bar{s}\, s \right\rangle \tau_o \tag{8.56a}$$

or the much more involved combination :

$$M_3^2 \simeq 3\, M^2_{K_0^*} - 2\, M_2^2 - \frac{24\pi^2}{3} m_s \left\langle \bar{s}\, s \right\rangle \left(1 - 2\,\frac{\left\langle \bar{u}u \right\rangle}{\left\langle \bar{s}s \right\rangle} \right) \tag{8.56b}$$

which predicts the value :

$$M_3 \approx 1.45 \text{ GeV} \tag{8.56c}$$

(M_2 and M_3 are the masses of the $f_0(0.98)$ and $f_0(1.3)$).

This prediction is in agreement with the result obtained from an analysis of the individual moment associated with the current in Eq. (8.55b). In order to see the accuracy of the latter result, we reconsider the analysis by extracting the decay constant f_3 normalized as :

$$\langle 0 \,|J_3|\, f_0 \rangle = \sqrt{2}\, M_3^2\, f_3 \tag{8.57}$$

and the mass M_3. We give the predictions in Figs 8.15 and 8.16 . They show nice τ- but not t_c-stabilities. So, a priori, we cannot obtain an optimal prediction. If one assumes that the t_c for the K_0^* and for the

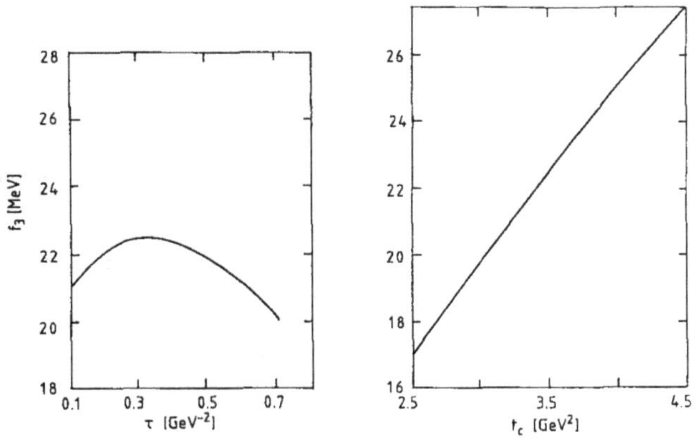

Fig. 8.15 : τ- and t_c-behaviour of the coupling f_3.

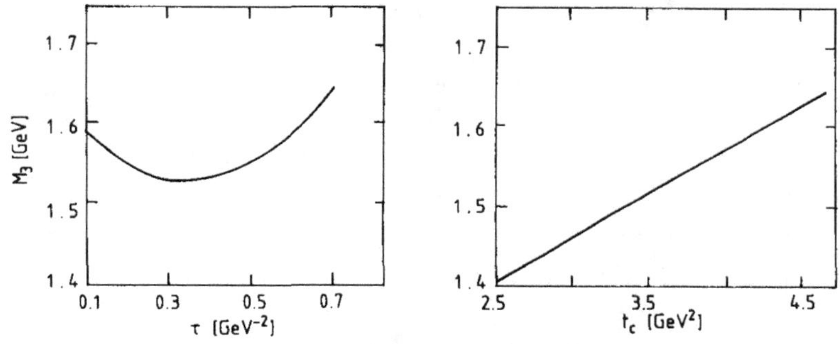

Fig. 8.16 : τ- and t_c-behaviour of M_3.

$f_o(1.3)$ is the same and if one uses $t_c \simeq (3.5 - 4.5)$ GeV², $\hat{m}_s = 255$ MeV and the value of $\langle \bar{s}s \rangle$ in Eq. (8.36), we obtain :

$$f_3 \simeq (22 - 28) \text{ MeV} \quad \text{and} \quad M_3 \simeq (1.52 - 1.65) \text{ GeV} \quad . \qquad (8.58)$$

The value of f_3 is almost the same as that of $f_{K_o^*}$, a result which one would naively expect from $SU(3)_F$ symmetry. The value of M_3 is as inaccurate as that of $M_{K_o^*}$ (Eq. (8.31)). However, despite such inaccuracy, one can conclude that the $f_o(1.3)$ mass is higher than the K_o^* mass, this splitting being mainly due to the m_s^2 term.

CHAPTER 9

HEAVY QUARK SYSTEMS

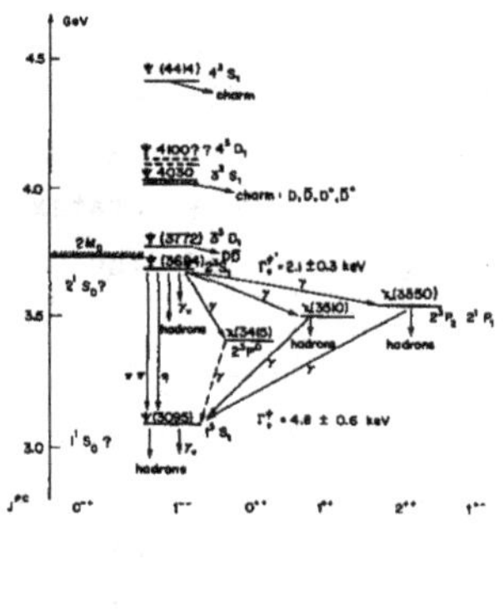

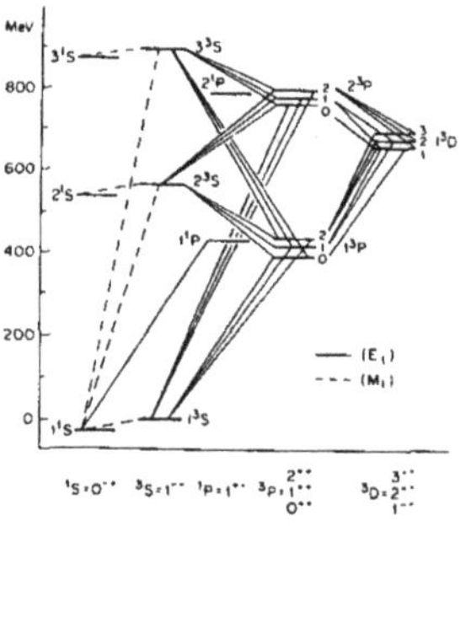

The psi and upsilon families

Heavy-quark systems are another "phase" of chiral symmetry because here we have a Wigner-Weyl type realization of it. The $\langle \bar{Q} Q \rangle$ quark condensate vanishes as the inverse of the heavy-quark mass. Eqs (3.40) and (3.51) tell us that, to leading order, this dependence is :

$$\langle \bar{Q} Q \rangle \simeq \frac{(-1)}{(12\pi M_Q)} \langle \alpha_s G^2 \rangle + \ldots + O\left(\frac{1}{M_Q^3}\right)$$

$$g\langle \bar{\Psi} G\psi \rangle \simeq -\frac{1}{\pi} \langle \alpha_s G^2 \rangle M_Q \left(\log \frac{\nu^2}{M_Q^2} - 1\right) - \frac{5}{24} \frac{\alpha_s}{\pi} \frac{g\langle G^3 \rangle}{M_Q} . \qquad (9.1)$$

We then expect that in the infinite quark mass limit we will have a degeneracy for the states belonging to the same multiplet. In the real world where the quark mass is finite, we expect to have a small splitting between the $1^{--}$and 1^{++} ($0^{-+}$and 0^{++}respectively) states owing to the still (though small) non-vanishing value of the $\langle \bar{Q} Q \rangle$ quark condensate.

Concerning the OPE of the hadronic correlator, in this chapter we can exploit the heavy-quark mass expansion techniques owing to the large value of M_Q compared to the QCD scale Λ. In this case, sum rules of the type in Eq. (1.55) become a very convenient tool. SVZ[18)] have used this sum rule at $Q_0^2 = 0$ for the charmonium systems while RRY in Ref. 29) have extended this SVZ sum rule to the case $Q_0^2 \neq 0$. On the contrary Bell and Bertlmann[19)] have preferred to work with the Laplace exponential moments which are the extreme limit $\left(Q_0^2 \to \infty\ ,\ N \to \infty\right)$ of the moments.

We shall successively discuss these different sum rules in this chapter. We shall also discuss the definition of the quark mass due to the important role of the heavy-quark mass in the sum rules analysis as we have already anticipated in subparagraph 4j of Chapter 1.

1. FEATURES OF THE $Q_0^2 = 0$ SVZ MOMENTS

a) Mass definition and the OPE

If we work with the SVZ-like $Q_0^2 = 0$ moment, it is quite clear that we have to choose the on-shell pole mass, which is regularization and renormalization-scheme invariant, as we demonstrated in Chapter 1.4 j and in Ref. 52). However, SVZ have observed that in this case the radiative correction to the unit operator is too large for the two-point correlator associated with the $\bar{c}\,\gamma^\mu c$ vector current. Therefore, they find it convenient to work with the Euclidian mass $m(-p^2 = M^2)$ in the Landau gauge. Its relation to the pole mass can be obtained from Eq (1.153) and reads in that gauge :

$$m = m(-p^2=M^2) = M(p^2=M^2)\ .\ \left\{1 - \left(\frac{\alpha_s}{\pi}\right) 4 \log 2\right\} . \tag{9.2}$$

Therefore, SVZ use an OPE in terms of m. The SVZ sum rule reads for the neutral vector current[18,97]:

$$\mathcal{M}_n^V \equiv \frac{1}{n!}\left(-\frac{d}{dQ^2}\right)^n \Pi_c \Bigg|_{Q^2=0} = \int_{4m^2}^{\infty} \frac{dt}{t^{n+1}}\ \frac{1}{\pi}\ \mathrm{Im}\, \Pi_c\ (t)$$

$$\simeq \frac{3}{4\pi^2}\ \frac{(n+1)(n-1)!}{(2n+3)!!\left(2\ m_c^2\right)^n}\ \left(1 + a_n\ \alpha_s - \frac{(n+3)!}{(n-1)!\ (2n+5)}\ \frac{\left(g^2\ G^a_{\mu\nu}\ G^a_{\mu\nu}\right)}{9\left(4\ m_c^2\right)^2}\right.$$

$$+ \frac{2}{45}\ \frac{(n+4)!\ (3n^2+8n-5)}{(n-1)!\ (2n+5)(2n+7)}\ \frac{\left(g^3\ f_{abc}\ G^a_{\mu\nu}G^b_{\nu\lambda}G^c_{\lambda\mu}\right)}{9\left(4m_c^2\right)^3}$$

$$- \frac{8(n+2)!\ (n+4)\ (3n^3 + 47n^2 + 244n + 405)}{135(n-1)!\ (2n+5)\ (2n+7)}\ \frac{\left(g^4\ j^a_\mu j^a_\mu\right)}{9\left(4m_c^2\right)^3}$$

$$+ \frac{4n(n+1)\,(n+2)\,(n+3)}{81(2n+5)(2n+7)(2n+9)\left(4\,m_c^2\right)^4}$$

$$\times \Bigg[\left(-\frac{1}{56}\,n^6 + \frac{1511}{840}\,n^5 + \frac{4639}{168}\,n^4 + \frac{4657}{40}\,n^3 - \frac{4297}{210}\,n^2 - \frac{15307}{14}\,n - 1752\right)$$

$$\left\langle g^4\ Sp\left(\hat{G}_{\mu\nu}\ \hat{G}_{\mu\nu}\ \hat{G}_{\alpha\beta}\ \hat{G}_{\alpha\beta}\right)\right\rangle$$

$$+ \left(\frac{1}{56}\,n^6 + \frac{169}{840}\,n^5 + \frac{233}{168}\,n^4 - \frac{1811}{120}\,n^3 - \frac{17491}{70}\,n^2 - \frac{14793}{14}\,n - 1416\right)$$

$$\left\langle g^4\ Sp\left(\hat{G}_{\mu\nu}\ \hat{G}_{\alpha\beta}\ \hat{G}_{\mu\nu}\ \hat{G}_{\alpha\beta}\right)\right\rangle$$

$$+ \left(\frac{23}{140}\,n^6 + \frac{4283}{420}\,n^5 + \frac{64747}{420}\,n^4 + \frac{12595}{12}\,n^3 + \frac{77234}{21}\,n^2 + \frac{226383}{35}\,n + 4512\right)$$

$$\left\langle g^4\ Sp\left(\hat{G}_{\mu\nu}\ \hat{G}_{\nu\beta}\ \hat{G}_{\alpha\beta}\ \hat{G}_{\beta\mu}\right)\right\rangle$$

$$+ \left(-\frac{23}{140}\,n^6 - \frac{2603}{420}\,n^5 - \frac{28291}{420}\,n^4 - \frac{10399}{60}\,n^3 + \frac{106756}{105}\,n^2 + \frac{213301}{35}\,n + 8592\right)$$

$$\left\langle g^4\ Sp\left(\hat{G}_{\mu\nu}\ \hat{G}_{\alpha\beta}\ \hat{G}_{\nu\alpha}\ \hat{G}_{\beta\mu}\right)\right\rangle$$

$$+ \left(\frac{3}{280}\,n^6 + \frac{1367}{840}\,n^5 + \frac{8157}{280}\,n^4 + \frac{27883}{120}\,n^3 + \frac{35121}{35}\,n^2 + \frac{161939}{70}\,n + 2232\right)$$

$$\left\langle g^5\ f_{abc}\ G^a_{\mu\nu} j^b_\nu j^c_\mu\right\rangle$$

$$- n(n+4)\left(\frac{3}{70}\,n^4 + \frac{253}{210}\,n^3 + \frac{1093}{105}\,n^2 + \frac{255}{7}\,n + \frac{1586}{35}\right)\left\langle g^3 f_{abc}\, G^a_{\mu\nu} G^b_{\nu\lambda} G^c_{\lambda\mu,\,\alpha\alpha}\right\rangle$$

$$+ (n+5)\left(\frac{9}{280} n^5 + \frac{101}{140} n^4 + \frac{1331}{168} n^3 + \frac{10583}{210} n^2 + \frac{2337}{14} n + 216\right)$$

$$\left\langle g^4 \; j^a_\mu \; j^a_{\mu,\alpha\alpha}\right\rangle \Bigg| \Bigg) \qquad (9.3)$$

where $\hat{G} = G_a \ \lambda^a/2$, λ^a are the Gell-Mann matrices, $G^a_{\mu\nu} = \partial_\mu A^a_\nu - \partial_\nu A^a_\mu + gf_{abc} \ A^b_\mu \ A^c_\nu$, and the coloured vector current of light (u,d,s) quarks j^a_μ appears owing to the equation of motion $G^a_{\mu\nu,\nu} = gj^a_\mu$. The notation $\mathcal{O}^a_{\eta\mu}$ is used for the covariant derivative $D^{ab}_\mu \ \mathcal{O}^b$.
The exact definition of the quark mass appearing in the Wilson coefficients of the quark condensate remains ambiguous until we get the radiative corrections to these Wilson coefficients. This definition might influence the value of the gluon condensate estimated from the sum rules. Another source of uncertainties is the method of estimating the strength of the high-dimension condensates. Ref. 97) claims that if one uses the vacuum saturation assumption for the dimension-eight condensates, the OPE is broken by these terms for the charmonium systems. Novikov et al[134] reply that these condensates do not factorize in the large N_c limit, which Bagan et al[95] confirm, and therefore they try to improve the factorization. The results in Eqs (3.47) and (3.48) may indicate that the strength of such condensates is lower than that obtained within the naive factorization assumption. Therefore the breaking of the OPE might not happen and the SVZ result obtained by ignoring these terms might still make sense.

Let us now discuss the sum rule in Eq. (9.3) in a much more phenomenological way.

b) Test of the duality ansatz

We study the spectral part of the sum rule in Eq. (9.3). As usual, we saturate the data by known ψ resonances plus a QCD-continuum, i.e :

$$\left\langle 0 \left| \bar{c}\, \gamma^\mu c \right| \psi \right\rangle = \sqrt{2}\, \frac{M_\psi^2}{2\gamma_\psi}\, \epsilon^\mu \quad , \tag{9.4a}$$

where γ_ψ is related to the $\psi \to e^+e^-$ width as in Eq. (1.29b). The QCD continuum is approximated by a step function :

$$\frac{1}{\pi}\, \mathrm{Im}\, \Pi_c(t)_{cont} \simeq \frac{1}{4\pi^2} \left(1 + \frac{\alpha_s}{\pi}(t) \right) \Theta(t - t_c) \; . \tag{9.4b}$$

but one can improve it by using the known complete expression given in Eq. (1.151).

The ratio of moments :

$$r_n \equiv \frac{u_n}{u_{n-1}} \tag{9.5}$$

is shown in Fig. 9.1, where it approaches the inverse mass squared of the ψ for n larger than 5-6.

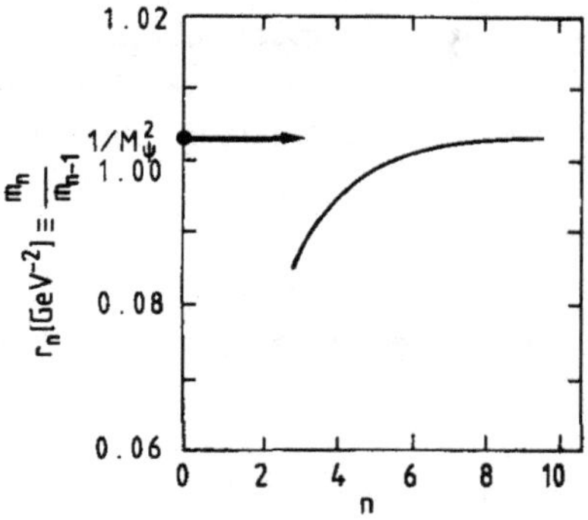

Fig. 9.1 : Data for the ratio of moments r_n versus n.

We have used the following electronic width in units of keV :

$$\Gamma(\psi) \equiv \Gamma(\psi \to e^+e^-) = 4.70$$
$$\Gamma(3886) \simeq 2.10$$
$$\Gamma(3770) \simeq 0.26$$
$$\Gamma(4030) \simeq 0.75$$
$$\Gamma(4160) \simeq 0.77$$
$$\Gamma(4415) \simeq 0.47 \quad . \qquad (9.5)$$

Our previous discussion has shown that for a value of $n \geqslant 5,6$, one expects that a "single-meson dominance" will already provide a very good description of the spectral function. That is a nice feature of the heavy quarks sum rule as the role of the continuum is much smaller here than in light-quark systems. Equivalently, the predictions for the couplings and masses of the lowest ground states present a t_c-stability here. It needs emphazing that t_c-stability, though achieved for the coupling, is not obtained in many cases for the mass predictions of the light-quark systems. Optimal predictions for the latter are obtained for the t_c values fixed by the lowest FESR constraint.

2. DETERMINATION OF THE HEAVY-QUARK MASS

It may already have been noticed that the sum rule in Eq. (9.3) is very sensitive to the value of the charm quark mass because a small deviation in m_c induces a large error in the sum rule which behaves as $(m_c)^{-2n}$ ($n \simeq$ 5-6). SVZ obtained the value of the Euclidian mass :

$$m_c\left(p^2 = -M_c^2\right) \simeq 1.26 \text{ GeV} \quad , \qquad (9.7a)$$

which has been subsequently confirmed by many authors using different methods where the convergence of the OPE is much better than that of $Q_0^2=0$ moments (ratio of sum rules [164], $Q_0^2 \neq 0$ moments[29] and exponential moments[19]). Owing to the use of the Euclidian mass for minimizing the radiative corrections entering the sum rules, the fitted value :

$$m_c\left(p^2 = -M_c^2\right) \simeq (1.26 \pm 0.02)\ \text{GeV} \quad , \qquad (9.7b)$$

should be quite insensitive to the value of the QCD scale Λ which is now expected to be about 150 MeV[165]. Using Eq. (9.2) in Eq. (9.7) one can deduce the value of the pole mass :

$$M_c\left(p^2 = M_c^2\right) = (1.45 \pm 0.05)\ \text{GeV} \quad . \qquad (9.8a)$$

This value gives the invariant mass to three loops (see Eqs (1.93) and (1.155)) :

$$\hat{m}_c \simeq (1.92 \pm 0.18)\ \text{GeV} \quad , \qquad (9.8b)$$

which can be compared with that deduced from a FESR based on a m_c^2/q^2 expansion[22c]:

$$\hat{m}_c \simeq (2.08 \pm 0.35)\ \text{GeV} \ . \qquad (9.8c)$$

A weighted average of the two results gives the best estimate :

$$\hat{m}_c \simeq (1.95 \pm 0.16)\ \text{GeV} \qquad M_c\left(p^2 = M_c^2\right) \simeq (1.47 \pm 0.05)\ \text{GeV} \quad . \qquad (9.8d)$$

A simple extension of the method to the b-quark channel gives the value of the Euclidian mass[18,19,29,164]:

$$m_b\left(p^2 = -M_b^2\right) \simeq (4.23 \pm 0.05)\ \text{GeV} \qquad (9.9)$$

from which one deduces the value of the pole mass :

$$M_b\left(p^2 = M_b^2\right) \simeq (4.67 \pm 0.10)\ \text{GeV} \quad , \qquad (9.10)$$

for Λ = (100 - 200)MeV. Owing to the heavier value of the b-quark mass even the $Q^2 = 0$ sum rule is rapidly convergent but at the same time the Coulomb-like interaction effects can become appreciable, the latter being indicated by the important role of the perturbative terms in the Υ sum rule. To improve Eq. (9.10), particular attention must be paid to the size of these Coulomb-like effects of the $\left(\frac{\alpha_s}{v}\sqrt{n}\right)^k$ type in the moments.

The value of the pole mass in Eqs(9.8) and (9.10) deduced from the Euclidian mass agree with those obtained from a direct fit of the exponential moments. The value of the b-quark pole mass in Eq. (9.10) can also be compared with that obtained from the beautiful meson B and B^* systems. This value is[104] (see also Chapter 10) :

$$M_b\left(p^2=M_b^2\right) \simeq (4.56 \pm 0.05)\text{GeV} \quad . \tag{9.11}$$

Taking the weighted average of Eqs (9.10) and (9.11) with a minimum variance as the best estimate of M_b from the sum rule, we obtain the accurate result[166]:

$$M_b\left(p^2 = M_b^2\right) \simeq (4.59 \pm 0.05)\text{GeV} \quad . \tag{9.12}$$

The pole masses are often used for the phenomenology of the heavy-quark processes (weak decays, charm production, radiative corrections in the standard model). For instance, the precision measurement of $\sin^2\theta_W$ from neutrino-scattering experiments needs a precise value of M_c which induces the main theoretical uncertainties in the analysis.

Previous values of the pole masses can also be used to get the value of the running masses which are useful in the Grand-Unification schemes as there one has to deal with the ratio of the heavy over the light quark current masses. With the help of Eqs(9.2), (1.93)and (1.155), one can deduce the values of the invariant mass for $\Lambda \simeq (150 \pm 50)$MeV :

$$\hat{m}_b \simeq (7.9 \pm 0.1)\mathrm{GeV} \quad , \tag{9.13a}$$

which agrees with the bound obtained in Ref. 22c) :

$$\hat{m}_b \leqslant 8.2 \ \mathrm{GeV} \quad . \tag{9.13b}$$

Using the relations among these different mass definitions, we then deduce from the Table in Ref. 52) :

$$\bar{m}_c(1 \ \mathrm{GeV}) \simeq (1.42 \pm 0.06)\mathrm{GeV} \quad ,$$

$$\bar{m}_b(1 \ \mathrm{GeV}) \simeq (5.87 \pm 0.06)\mathrm{GeV} \quad . \tag{9.14}$$

These values of the running masses are higher than those in Ref. 14) owing to the incorrect relation between the various mass definitions used there.

With the help of the $SU(4)_F$ symmetry assumption for the ψ and φ-wave functions, one can deduce the ratio of "current" masses[14]:

$$\frac{m_c - m_u}{m_s - m_u} \simeq (9 \pm 2) \quad . \tag{9.15}$$

Using this value in Eq. (9.14), we obtain the "hybrid determination" :

$$\hat{m}_s \simeq (224 \pm 50)\mathrm{MeV} \quad ,$$

$$\bar{m}_s \ (1 \ \mathrm{GeV}) \simeq (135 \pm 30)\mathrm{MeV} \quad . \tag{9.16}$$

This value is in good agreement with previous findings in Eq. (8.25) and indicates the consistency of the whole framework.

3. $\langle \alpha_s G^2 \rangle$ FROM POWER MOMENTS

It is certainly a fundamental question to know whether QCD is spontaneously broken by the gluon condensate $\langle \alpha_s G^2 \rangle$ where its role is important in heavy-quark processes and in gluodynamics (low-energy theorems, gluonia...) as we shall see later. Indeed, using the charmonium sum rules which we have discussed before, SVZ have shown that $\langle \alpha_s G^2 \rangle$ develops a non-vanishing expectation value, a result which is supported by lattice calculations[53] and by the light-quark systems as we discussed in Chapter 5.

In order to determine $\langle \alpha_s G^2 \rangle$, SVZ work with a ratio of moments whilst Ref. 164) uses a double ratio. In this way, the role of the quark mass in the sum rule is minimized whilst that of the gluon condensate is enhanced, a minimization of the α_s correction for the ratio of sum rules also being obtained. Fig.9.1 shows the behaviour of the theoretical ratio r_n with and without the gluon condensate $\langle \alpha_s G^2 \rangle$. At large n, $\langle \alpha_s G^2 \rangle$ stabilizes the moments ratio, where the value needed to fit the data is :

$$\Phi \equiv \frac{4\pi}{9} \langle \alpha_s G^2 \rangle \frac{1}{\left(4\, m_c^2\right)^2} \simeq 1.35\ 10^{-3} \quad . \tag{9.17}$$

Ref. 97) studied the stability of the ratio r_n when one includes the high-dimension condensates. Stability is still maintained by the $g\langle G^3 \rangle$ term but not by the $\langle G^4 \rangle$ term if one estimates these condensates via

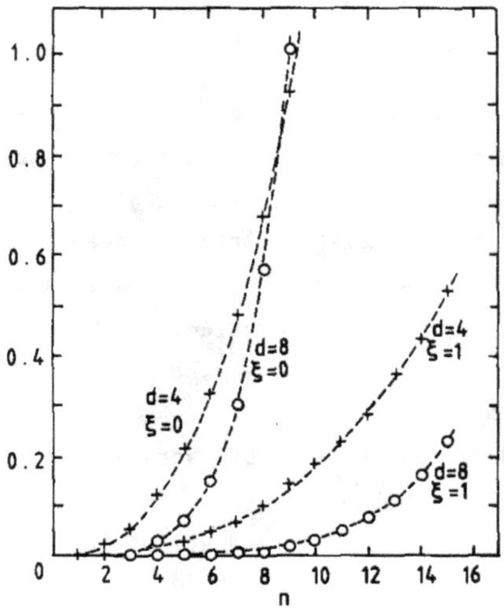

Fig. 9.2 : Relative contributions of the modulus of the d = 4, 8 condensates to the moment $\mathcal{M}_n^V(\xi)$ normalized to the unit operator .

the naive factorization assumption. As we mentioned at the beginning of this chapter, the estimate of the $\langle G^4 \rangle$ is uncertain and the strength can become much smaller than that obtained from the factorization assumption. RRY[29] have avoided this problem by working with the $Q_0^2 = 4\ m_c^2$ moments which converge rapidly. We show in Fig. 9.2 the strengths of the different operators relative to the perturbative contributions. RRY confirm the value given by SVZ in Eq. (9.17). Considering as a source of uncertainties the ambiguous definition of the quark mass appearing in the Wilson coefficients of the condensates, one can deduce from Eq. (9.17) :

$$\left\langle \alpha_s G^2 \right\rangle \simeq (0.04 - 0.07)\ \text{GeV}^4 \quad . \tag{9.18}$$

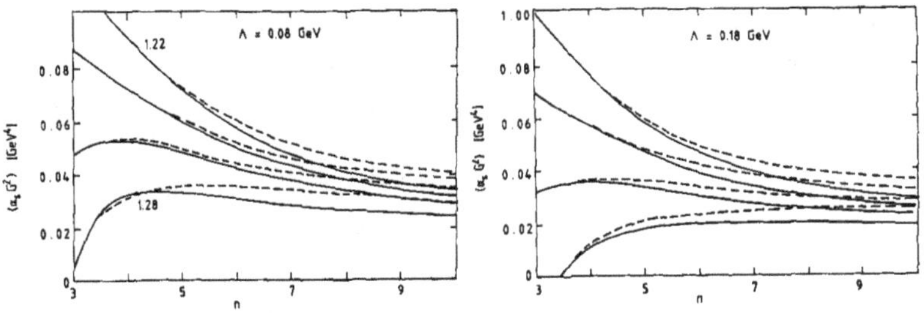

Fig. 9.3 : Predicted value of $\left\langle \alpha_s G^2 \right\rangle$ *versus n from the FESR-like moment for two different values of* Λ *and* $m_c\left(-p^2 = m_c^2\right)$.

--- *inclusion of the* $\langle g\, G^3 \rangle$ *condensate.*

An understanding of the success of the SVZ estimate based on a low-dimension condensate can be achieved if we write a FESR-like constraint for $\left\langle \alpha_s G^2 \right\rangle$. This is obtained by extracting $\left\langle \alpha_s G^2 \right\rangle$ for the theoretical moment r_n and by expressing it in terms of the other QCD parameters and r_n^{exp}. Formally, we have :

$$\left\langle \alpha_s G^2 \right\rangle \simeq \neq \left\{ \neq'(1 + a'_n \alpha_s + \ldots) - r_n^{exp} \right\} \qquad (9.19)$$

where the numbers $\neq$ can easily be deduced from r_n^{theor} . We show in Fig. 9.3 the predicted value of $\left\langle \alpha_s G^2 \right\rangle$ for two different values of Λ and of the Euclidian mass. The continuous line corresponds to the case where we do not include the effects of dimensions larger than, or equal to, six condensates in the OPE. The dashed line is the one where the $g\langle G^3\rangle$ condensate is included. It may be noticed that one has a nice *n*-stability of the estimate. If we use the Euclidian mass we deduce,

$$\left\langle \alpha_s G^2 \right\rangle \simeq (0.02 - 0.04)\ \mathrm{GeV}^4 \tag{9.20}$$

which is analogous to the SVZ result using the same inputs, the inaccuracy of our result being induced by the value of the Euclidian quark mass. We include in Fig. 9.4 the effects of dimension-eight condensates by using the naive[97)] and modified[95b)] factorization. Then we obtain :

$$\left\langle \alpha_s G^2 \right\rangle \simeq (0.035 - 0.075)\ \mathrm{GeV}^4 \quad , \tag{9.21}$$

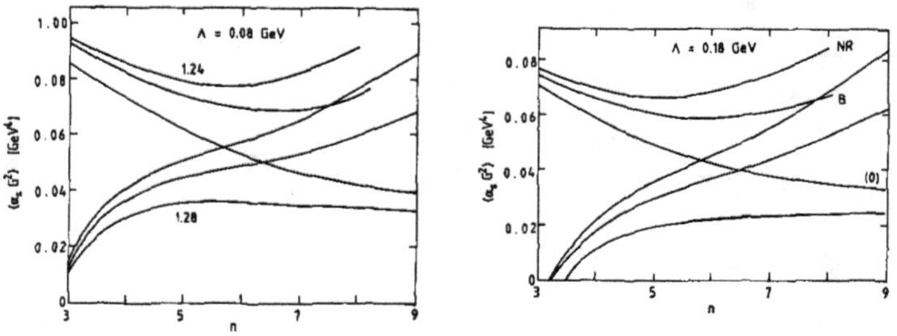

Fig. 9.4 : Inclusion of the d=8 condensates in the estimate of $\left\langle \alpha_s G^2 \right\rangle$ *:* *(O) no d=8, NR : factorization of Ref. 97) B : modified factorization of Ref. 95b) where we use the Euclidian mass.*

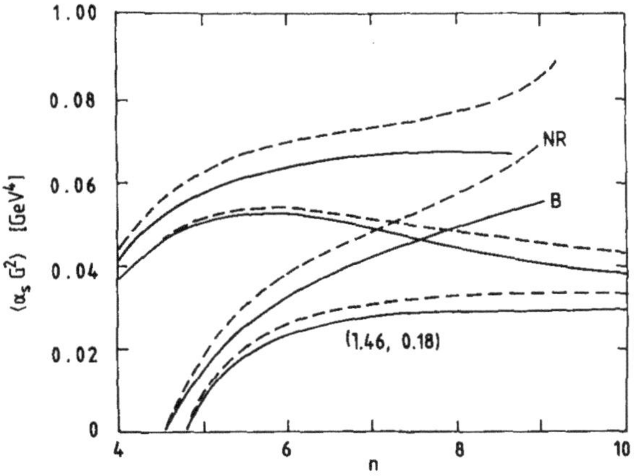

Fig. 9.5 : Analogous to Fig. 9.4 but with the pole mass.

where we notice that at fixed m_c, the dimension-eight condensates tend to increase the value of the $\langle\alpha_s G^2\rangle$ condensate. Fig. 9.5 shows the effects of the pole mass and of the correlated value of Λ. The common range of solutions from Figs. 9.4 and 9.5 is :

$$\langle\alpha_s G^2\rangle \simeq (0.035 - 0.065)\ \mathrm{GeV}^4 \qquad (9.22)$$

Eq. (9.22) agrees nicely with the one in Eq. (9.18). It also indicates that the estimate of $\langle\alpha_s G^2\rangle$ can be extracted even in the worst case where the strength of the $\langle G^4\rangle$ condensates is comparable to it. That can be achieved using appropriate FESR-like sum rules which disentangle the $\langle\alpha_s G^2\rangle$ term from other QCD parameters.

4. $\langle\alpha_s G^2\rangle$ FROM EXPONENTIAL MOMENTS

As mentioned before, Bell and Bertlmann[19)] and, subsequently,

Refs 67) and 168) have determined $\left\langle\alpha_s G^2\right\rangle$ from the exponential moment ratio in Eq. (1.40) or its variant $R_c(\tau)$ in Eq. (1.41). As in the case of the RRY moments, the strength of the $\langle G^4\rangle$ term is also negligible at the minimum of $R_c(\tau)$ (see Fig. 9.6) and validates previous analyses obtained by ignoring such terms. Ref. 19) adjusts the minimum of $R(\tau)$ to the value of the ψ mass squared and obtains :

$$\left\langle\alpha_s G^2\right\rangle \simeq (0.04 - 0.14)\ \mathrm{GeV}^4 \quad , \tag{9.23}$$

where the uncertainties are due to the value of α_s and m_c but are insensitive to the pole or Euclidian mass definition used in the analysis. At fixed m_c, the value of $\left\langle\alpha_s G^2\right\rangle$ increases when α_s decreases while at fixed α_s, $\left\langle\alpha_s G^2\right\rangle$ increases when the quark mass decreases. In our attempt to do much more careful work, we use the value of α_s evaluated at the minimum of the moments for Λ = (100-200) MeV. The minimum is obtained at $\tau_{min} \simeq 0.9\ \mathrm{GeV}^{-2}$ to which corresponds the value :

$$\alpha_s(\tau_{min}) \simeq 0.3 - 0.4 \tag{9.24}$$

and the correlated value of the pole mass in Eq. (9.8). Within this constraint, we improve the value in Eq. (9.20) to :

$$\left\langle\alpha_s G^2\right\rangle \simeq 0.04 - 0.06\ \mathrm{GeV}^4 \tag{9.25}$$

as deduced from Table 1 of the Ref. 167 a). Higher values of $\left\langle\alpha_s G^2\right\rangle$ correspond to lower values of the pole mass. The high values of $\left\langle\alpha_s G^2\right\rangle$ obtained in Ref. 168) from the vector channel might correspond to too low values of $\alpha_s \simeq 0.22$ and of the pole mass $M_c \simeq 1.37$ GeV which are inconsistent with either the one in Eq. (9.24) or the one in Eq. (9.8). However, Ref. 168) points out that the value of $\left\langle\alpha_s G^2\right\rangle$ may be influenced by high-dimension condensates and by the value of the QCD continuum threshold if one works with the FESR-like exponential moments. The former effects are similar to the case of the $Q_0^2 = 0$ moments analyzed previously. The latter effects are irrelevant for the $Q_0^2 = 0$ but they decrease the power of the prediction of the $Q_0^2 \neq 0$ and exponential moments. Therefore, it may be difficult to extract accurately from these two methods a small number like the mass-splitting between the S-P, $^3S_1$-$^1S_0$ states... Ref. 47b) has tested the reliability of the $Q_0^2 \neq 0$ moments from the ψ-η_c mass splitting. The results appear to depend on the choice of Q_0^2, of mass renormalization and continuum subtraction... but indicate in all cases that the mass splitting from the sum rules has been underestimated by at least 20% if the standard values of the QCD parameters are used. However, within the accuracy of the sum rules approach for the mass splitting, it is difficult to argue that one needs higher values of the gluon condensate for a much better fit. A careful re-examination of the mass splitting should be worked out. At least, it will be fascinating to understand why the $Q_0^2 = 0$ moment including the lowest dimension gluon condensate has successfully[169)] predicted the η_c mass well before the data. We intend to come back to this point in a later publication[170)].

5. NON-RELATIVISTIC "MAGIC MOMENTS" AND "EQUIVALENT" POTENTIALS

Non-relativistic sum rules have been discussed by Bell and Bertlmann[19)] (hereafter denoted B^2) and nicely reviewed by Bertlmann[19)]. B^2 have studied the non-relativistic limit of the "magic moment" in Eqs (1.39) and (1.40) within potential theory, since there one can have both the approximate perturbative and the exact forms of the moments. The non-relativistic moments can be deduced from the relativistic one through the changes of variables :

$$t \rightarrow E$$

$$\tau \rightarrow 4m\,\tau \;. \qquad (9.26)$$

Therefore, the non-relativistic version of Eq. (1.39) is :

$$\frac{1}{4m\tau} e^{4m^2\tau} \hat{L}\, \Pi \rightarrow \mathcal{M}(\tau) \equiv \int dE\, e^{-E\tau} \frac{1}{\pi} \mathrm{Im}\, \Pi(E) \quad . \qquad (9.27a)$$

For the neutral vector current of heavy quarks, the moment reads :

$$\mathcal{M}(\tau) = \frac{3}{2m^2} \left(\frac{m}{4\pi\tau}\right)^{3/2} \left\{1 + \frac{4}{3} \alpha_s \sqrt{4\pi m}\, \tau^2 - \frac{4\pi}{288m} \left\langle \alpha_s G^2 \right\rangle \tau^3 + \ldots\right\} \qquad (9.27b)$$

from which one can deduce the approximate expression of $R(\tau)$ in its expanded form :

$$R(\tau) = \frac{3}{2\tau} - \frac{2}{3} \alpha_s \sqrt{\pi m}\, \tau^{-1/2} + \frac{4\pi}{96m} \left\langle \alpha_s G^2 \right\rangle \tau^2 + \ldots \quad . \qquad (9.28)$$

The spectral function appearing in the sum rule can be parametrized within a potential theory as :

$$\mathrm{Im}\, \Pi(E) = \frac{3}{8m^2} \sum_n 4\pi\, |\psi_n(o)|^2\, \delta(E - E_n) \qquad (9.29)$$

with the usual notation. Therefore, the moment becomes the time-dependent Green's function :

$$\mathcal{M}(\tau) = \frac{3}{8m^2} 4\pi \left\langle \vec{x}\, | e^{-H\tau} |\, \vec{x} \right\rangle \Big|_{\vec{x}\, =\, 0} \qquad (9.30a)$$

where the Hamiltonian H of the system is :

$$H \equiv \frac{p^2}{m} + V \quad . \qquad (9.30b)$$

A perturbative evaluation of the above Green's function leads to the "equivalent" potential of Bell and Bertlmann[19].

$$V_{B^2}(r) = -\frac{4}{3}\frac{\alpha_s}{r} + \frac{\pi}{144}\left\langle\alpha_s G^2\right\rangle m\, r^4 \quad , \tag{9.31}$$

which contrasts with the phenomenological potentials used to describe successfully the quarkonia systems (see Chapter 2). In fact, effective potentials have a soft increase in r (logarithmic or at most linear) and are flavour-independent.

At large time, the moments ratio approaches the energy of the ground state from above :

$$R(\tau \to \infty) = E_o \quad . \tag{9.32}$$

B^2 have compared the exact relativistic expression of $R(\tau)$ with its approximate form in Eq. (9.28) and found a remarkable agreement. B^2 have identified the minimum of $R_{approx}(\tau)$ with the ground state mass :

$$M_o = 2m + \min_{\tau} R(\tau) \tag{9.33}$$

in order to fix the value of the gluon condensate from the charmonium system. This procedure leads them to the estimate of $\left\langle\alpha_s G^2\right\rangle$ with the improved value in Eq. (9.25).

Some remarks are in order here :

• The value of τ at which the minimum of $R(\tau)$ is obtained in the charmonium case is small enough to ensure the validity of the OPE as checked later on but large enough for the moments to be sensitive to the ground-state energy.

• In some cases like the rho meson including threshold effects, the minimum of $R(\tau)$ is replaced by a stationary inflexion point where the B^2 conclusion is still valid.

• Using V_{B^2} non-perturbatively in a Schrödinger equation, B^2 have remarked that the ground-state energy emerges very differently from the result when V_{B^2} is used perturbatively in the moments, unless one increases the gluon-condensate value by a factor 3. Therefore B^2 conjectures that :

$$\langle \alpha_s G^2 \rangle < \langle \alpha_s G^2 \rangle_{Exact.} , \tag{9.34}$$

which might be acceptable as $\langle \alpha_s G^2 \rangle$ in a totally soluble theory, might be an effective gluon condensate which includes most non-perturbative pieces not accounted for in $\langle \alpha_s G^2 \rangle$ within approximate series. However, the conjecture of B^2 can be better tested from lattice measurements.

There are various variants of the equivalent potential in Eq. (9.31). Leutwyler and Voloshin (LV)[80] have discussed a potential for very heavy quarkonia which are predominantly in a Coulombic state. Coulombic *quarkonium is like an external object placed inside the QCD vacuum* consisting of soft-gluon fluctuations which shift the energy levels and the wave functions. The Hamiltonian consists of a small perturbative piece to the Coulomb system :

$$H^1_{Coul} = \frac{\vec{P}^2}{m} + V_1 \quad \begin{cases} V_1 = -\dfrac{4}{3}\dfrac{\alpha_s}{r} \\ V_8 = \dfrac{1}{6}\dfrac{\alpha_s}{r} \end{cases} . \tag{9.35}$$

In a dipole approximation this perturbative piece reads :

$$H_{Pert} \simeq -\frac{g}{2}\left(\lambda^a_q - \lambda^a_{\bar{q}}\right) \vec{x}\, \vec{E}_a$$

where $\vec{E}^a$ is the colour electric field related to the gluon condensate as :

$$\langle 0 | g^2 \vec{E}_a \vec{E}_a | 0 \rangle = -\pi \langle \alpha_s G^2 \rangle . \tag{9.36}$$

The energy levels are determined by the quadratic Stark effect of the chromoelectric field as :

$$\delta E_{n\ell} = \langle 1|\, \langle 0|\, H_{Pert} \frac{1}{E^1 - H^8_{Coul}} H_{Pert}\, |0\rangle\, |1\rangle$$

$$= \frac{\pi \left\langle \alpha_s G^2 \right\rangle}{(m\beta)^4}\, mn^6\, \epsilon^{LV}_{n\ell} \qquad (9.37)$$

where $\beta \equiv \frac{4}{3}\alpha_s$ and $\epsilon^{LV}_{n\ell} \simeq 1$ from Leutwyler[80]. Therefore the mass of levels reads :

$$M_{n\ell} = 2m - \frac{m\beta^2}{4n^2} + \delta\, E^{LV}_{n\ell} \quad . \qquad (9.38)$$

It is interesting to note that $\delta\, E_{n\ell}$ behaves like n^6. Imposing that the perturbative shift is smaller than the Coulombic one leads to the limit of validity range for the approach :

$$m > 5 \text{ GeV} \qquad n = 1 \quad . \qquad (9.39)$$

B^2 have also investigated the LV spectrum by another equivalent potential. Using the fact that the distance in Coulombic potential should behave as :

$$\langle r \rangle^{Coul.}_n \sim n^2 \qquad (9.40)$$

and respecting the kinetic term $\vec{p}^{\,2}/m$, they obtain

$$\delta V_{BB}(r) = \frac{4\pi}{81\beta} \left\langle \alpha_s G^2 \right\rangle \left[r^3 - \frac{304}{81}\frac{r^2}{m\beta} + \frac{53}{10}\frac{r}{(m\beta)^2} - \frac{113}{100}\frac{1}{(m\beta)^3} \right] \qquad (9.41)$$

which still contrasts with effective potential models. Dosch and Marquard (DM)[80] have tried to clarify the connection between the "equivalent" potential and the potential models by assuming that quarks and gluonic backgrounds fluctuate stochastically according to a Markov process. By introducing a correlation time : T_Q for quarks and T_G for gluons as :

$$\langle x_i(\tau_1)\, x_j(\tau_2)\rangle_{n\ell} = \frac{\delta_{ij}}{3}\left\langle \vec{x}^{\,2}\right\rangle_{n\ell} \exp\left(-\frac{\tau_1-\tau_2}{T_Q}\right) ,$$

$$\left\langle E_i^a(\tau_1)\, E_j^a(\tau_2)\right\rangle_E = \frac{\delta_{ij}}{3}\frac{\delta_{ab}}{8}\left\langle \vec{E}^{\,2}\right\rangle \exp\left(-\frac{\tau_1-\tau_2}{T_G}\right) , \quad (9.42)$$

the authors conclude that :

- For $T_G \gg T_Q$, no local potential absorbing the gluon fluctuations exists. Therefore, the sum-rule approach with the B^2 equivalent potential can be applied.
- For $T_G \ll T_Q$, one can have a local potential for a rapidly fluctuating gluon field and one cannot apply the sum-rule method.

Relativistic corrections to the energy spectrum have also been investigated by the inclusion of the Breit-Fermi term :

$$H_{BF} = H_{SS} + H_{LS} + H_T + H_{Zeeman} + H_{Momentum}$$

and of the mean energy of the gluonic background :

$$H_{gluon} \to E_G = \frac{1}{T_G}$$

in the Hamiltonian. The relativistic corrected mass is[80d,e] :

$$M_{n\ell sJ}(T_G) = 2m - \frac{m\beta^2}{4n^2} + \frac{m\beta^4}{4n^3} B_{\ell sJ} + \frac{\pi\left\langle\alpha_s G^2\right\rangle}{(m\beta)^4} n^6 \left[\epsilon_{n\ell}(T_G) + \beta^2 A_{n\ell sJ}(T_G)\right] , \quad (9.43)$$

where $B_{\ell sJ}$, $\epsilon_{n\ell}(T_G)$ and $A_{n\ell sJ}(T_G)$ are some calculable functions. The LV-case is reached for $T_G \to \infty$ but in this case perturbative contributions are too large for the bottomium system. The DM case would correspond to $T_G \to \infty$, where perturbative contributions are smaller but T_G is too low ($\simeq$ 0.1 Fermi) making the dipole approximation injustified.

Like a poet, Mike Chanowitz says :

We have got a "real situation here".

CHAPTER 10

HEAVY-LIGHT QUARK SYSTEMS

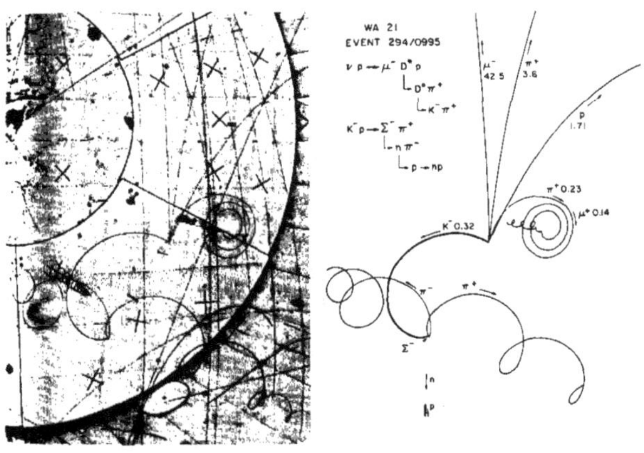

D-meson production in neutrino-hadron collision from the CERN-BEBC bubble chamber

In this section we shall study the bound states composed of one heavy h(c,b) and one light ℓ(u,d,s) quark. The study of this case needs a certain care as here the light quarks are relativistic while the heavy ones are non-relativistic. Therefore a naive extension of the methods and some models for the heavy-quark systems cannot be applied here.

Qualitatively, the main difference with the heavy-quark systems is due to the non-vanishing value of the light-quark condensate which plays a crucial role here in the splittings of the states. Heavy--light quark sum rules (HLSR) were used for the first time in Ref. 28) and improved later on by various authors in the study of the meson decay constant[171] and mass spectrum[104].

1. NON-PERTURBATIVE EFFECTS ON THE CORRELATORS

From a technical point of view, the analysis of the two-point correlator needs a certain care due to the possible presence of the light-quark mass singularities. This point has been clarified by Broadhurst and Generalis[46] once one uses the "improved" definition of the quark condensate given in Eq. (3.37). The perturbative piece in Eq. (3.37) is useful for absorbing such mass singularities. The perturbative expressions of the correlators $\psi_5(q^2)$ and $\Pi_V^{(1)}(q^2)$ associated with the currents :

$$\partial^\mu A_\mu(x)_h^\ell = (m_\ell + m_h)\,\bar\psi_\ell(i\gamma_5)\,\psi_h$$

$$V_\mu(x)_h^\ell = \bar\psi_\ell\,\gamma_\mu\,\psi_h \quad , \qquad (10.1)$$

have already been given in Section 1 to two loops. We give below the expressions of the non-perturbative contributions :

$$\psi_5(q^2)\Bigg|_{N.P} = \left(\frac{m_h^2}{m_h^2 - q^2}\right)\left[- m_h \left\langle \bar{\psi}_i \; \psi_i \right\rangle + \frac{\left\langle \alpha_s G^2 \right\rangle}{12\pi}\right]$$

$$- \frac{m_h^3 \; q^2}{\left(m_h^2 - q^2\right)^3} \; . \; \frac{1}{4} \left\langle g \; \bar{\psi}_i \; \sigma^{\mu\nu} \; \lambda_a \; \psi_i \; G^a_{\mu\nu} \right\rangle$$

$$+ \frac{m_h^2}{\left(m_h^2 - q^2\right)^2}\left[2 - \frac{m_h^2}{m_h^2 - q^2} - \left(\frac{m_h^2}{m_h^2 - q^2}\right)^2\right] \frac{\pi}{6} \alpha_s \left\langle \bar{\psi} \gamma_\mu \lambda_a \psi \sum_{i \equiv u,d,s} \left(\psi_i \gamma^\mu \lambda_a \psi_i\right)\right\rangle ;$$

$$q^2 \; \Pi_V^{(1)} \; (q^2)\Bigg|_{NP} = \left(\frac{m_h^2}{m_h^2 - q^2}\right)\left[- m_h \left\langle \bar{\psi}_i \psi_i \right\rangle - \frac{\left\langle \alpha_s G^2 \right\rangle}{12\pi}\right]$$

$$- \left(\frac{m_h^2}{m_h^2 - q^2}\right)^3 \frac{1}{m_h} \; . \; \frac{1}{4} \left\langle g \; \bar{\psi}_i \; \sigma^{\mu\nu} \; \lambda_a \; \psi_i \; G^a_{\mu\nu} \right\rangle$$

$$- \frac{m_h^2}{\left(m_h^2 - q^2\right)^2}\left[4 + 8 \; \frac{m_h^2}{m_h^2 - q^2} - 3 \left(\frac{m_h^2}{m_h^2 - q^2}\right)^2\right] \; .$$

$$. \; \frac{\pi \; \alpha_s}{18} \left\langle \bar{\psi} \; \gamma_\mu \; \lambda_a \; \psi \sum_{i \equiv u,d,s} \left(\bar{\psi}_i \; \gamma^\mu \; \lambda_a \; \psi_i\right)\right\rangle . \qquad (10.2)$$

The expressions of the two-point correlator associated with the axial partners of Eq. (10.1) can be deduced as usual after replacing m_h with $- m_h$ in the previous expressions. Here and in the following $m_h \equiv M_h$.

2. THE DECAY CONSTANTS f_D AND f_B

As a first application of the HLSR sum rules, we study the decay constants $f_{D,B}$ associated with the $D(\bar{d}c)$ and $B(\bar{d}b)$ mesons. They are normalized as $f_\pi \simeq 93.3$ MeV :

$$\left\langle 0 \left| \partial_\mu A^\mu(x)_h^d \right| H \right\rangle = \sqrt{2}\, f_H\, M_H^2 \quad , \tag{10.3}$$

where $h \equiv c,b$ and $H \equiv D,B$.

a) $Q_0^2 = 0$ moment

The lowest convergent moment is :

$$\mathcal{M}_2 = \frac{1}{2!} \frac{\partial^2 \psi_5(q^2)}{(\partial q^2)^2}\bigg|_{q^2=0} = \int_{(m_\ell+M_h)^2}^{\infty} dt \;.\; \frac{1}{(t-q^2)^3} \frac{1}{\pi} \mathrm{Im}\, \psi_5(t) \quad . \tag{10.4}$$

Using the positivity of the continuum contribution to the spectral function, this moment gives the upper bound :

$$f_D \leqslant \left(\frac{M_D}{4\pi}\right) \;.\; \left\{1 + \frac{3m_d}{M_c} + 0.751\, \bar{\alpha}_s + \ldots\right\} \quad , \tag{10.5}$$

which corresponds to :

$$f_D \leqslant 2.14\, f_\pi \quad , \tag{10.6}$$

once one includes the QCD perturbative and non-perturbative corrections. One should not worry too much here about the mass dependence of the bound as it is clear that the lowest moment cannot provide optimal information. However, the value in Eq. (10.6) is already stronger than the new charmed data upper limit of 2.6 f_π [172]. Improvement of Eq. (10.6) for the D meson cannot, however, be done, as the OPE does not converge for higher values of n. A subtraction of the continuum from Eq. (10.4) can help to reduce Eq. (10.6) but this is not yet enough to provide the optimal prediction on f_D. This explains why f_D obtained in Ref. 173) is quite high. The $Q^2 = 0$ moment is more appropriate for the B meson as the heavy-quark mass is large enough to allow one to work with higher moments $\mathcal{M}_n$. We then, obtain the constraint :

$$\mathcal{M}_n^P \equiv \frac{1}{(n+1)!}\left(\frac{d}{dq^2}\right)^{n+1} \psi_5(q^2)\Bigg|_{q^2=0} = \int_{M_b^2}^{\infty} \frac{dt}{t^{n+2}} \frac{1}{\pi} \operatorname{Im} \psi_{Pert}(t)$$
$$+ \frac{1}{\left(M_b^2\right)^{n+1}} \left\{ - M_b \left\langle \bar\psi_1 \psi_1 \right\rangle + \frac{\left\langle \alpha_s G^2 \right\rangle}{12\pi} + \frac{(n+1)(n+2)}{4M_b} M_0^2 \left\langle \bar\psi_1 \psi_1 \right\rangle \right.$$
$$\left. + \frac{(n+2)(n^2+10n+9)}{M_b^2} k \frac{\pi}{4} \alpha_s \left\langle \bar\psi \psi \right\rangle^2 \right\}, \qquad (10.7)$$

where we have used :

$$\left\langle \bar\psi_1 \psi_1 \right\rangle = - \mu_1^3 \left(\log M_b/\Lambda\right)^{\gamma_1/-\beta_1} \quad : \mu_1 = 0.192 \text{ GeV}$$
$$g\left\langle \bar\psi_1 \sigma^{\mu\nu} \lambda_a G^a_{\mu\nu} \psi_1 \right\rangle \equiv 2 M_0^2 \left\langle \bar\psi_1 \psi_1 \right\rangle$$
$$\alpha_s \left\langle \bar\psi \lambda^a \gamma_\mu \psi \sum_{i\equiv u,d,s} \bar\psi_i \lambda_a \psi_i \right\rangle \equiv - 9k \, \alpha_s \left\langle \bar\psi \psi \right\rangle^2 . \qquad (10.8)$$

$k = 1$ if one uses the factorization hypothesis $\left(\alpha_s \left\langle \bar\psi\psi \right\rangle^2 = 1.75 \; 10^{-4} \text{ GeV}^6\right)$.

We have neglected the known $g^3 f_{abc} \left\langle G^a_{\mu\nu} G^b_{\nu\rho} G^c_{\rho\mu} \right\rangle$ condensate owing to the huge numerical loop suppression $\frac{3}{720\pi^2}$ normalized to that of $\alpha_s \left\langle \bar\psi\psi \right\rangle^2$.

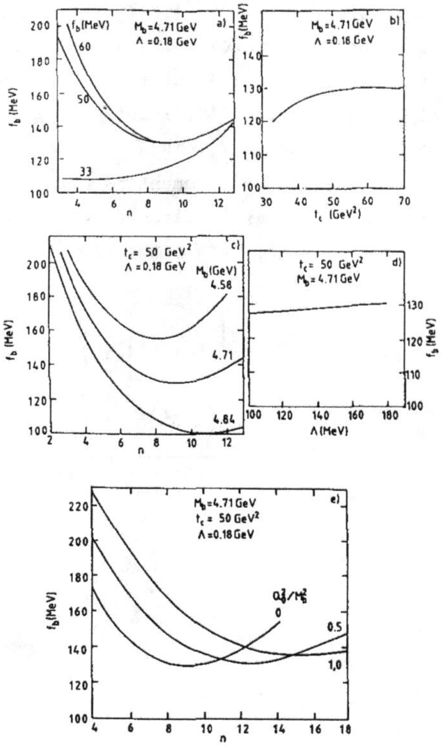

Fig. 10.1 : f_B from moments sum rules : a) versus n for different values of t_c. b) minima in a) versus t_c. c) versus M_b. d) versus Λ. e) versus Q_0^2 and n.

It should be noted that if one works with $\frac{\psi_5(q^2)}{q^2}$, the effects of $\psi_5(0)$, which are important here, must be taken into account.

The spectral part of the sum rule is parametrized using the usual duality ansatz : "one resonance" (the B) plus "QCD continuum". Transferring as usual the QCD continuum to the RHS of the sum rule, one obtains a FESR-like constraint :

$$f_B^2 \simeq \frac{1}{2}\left(M_B^2\right)^n \left\{\int_{M_b^2}^{t_c} \frac{dt}{t^{n+2}} \frac{1}{\pi} \operatorname{Im} \psi_5^{Pert}(t) + \text{Non-Perturb}\right\} . \qquad (10.9)$$

Noting that the role of the chiral $M_b\left\langle\bar{\psi}_1 \psi_1\right\rangle$ condensate is the most important one due to the heavy-quark mass in front of it, we can ignore here the effects of other condensates. We give our result in Fig. 10.1 for different values of the pole-quark mass, where one can see that the value of f_B is strongly reduced if one compares it with the n = 3 to the n = 8-10 moments where it is optimized. It will be noticed that the effects of the quark pole mass value is important while that of Λ is negligible. The t_c-stability is reached for $t_c \geqslant 40 - 50$ GeV^2 which is not too high and might be compared with the mass of the radial excitations expected from Regge trajectory arguments. The effects of the choice of moments $Q_0^2 \neq 0$ are also shown in Fig. 10.1e. These indicate that the strength of the minimum is left unchanged though it occurs at higher values of n for increasing Q_0^2. Using our optimal value of M_b in Eq. (9.12), we deduce within an accuracy better than 5% (one could add to this error the QSSR systematic uncertainty of about 10%) :

$$f_B \simeq 1.6\, f_\pi \simeq 150 \text{ MeV} \quad , \qquad (10.10)$$

which corresponds to the upper edge of the range given in Ref. 171). This is due to the lighter value of the b quark mass used here and can again be an improvement on the previous estimates as the errors in f_B become much smaller.

b) Exponential moments

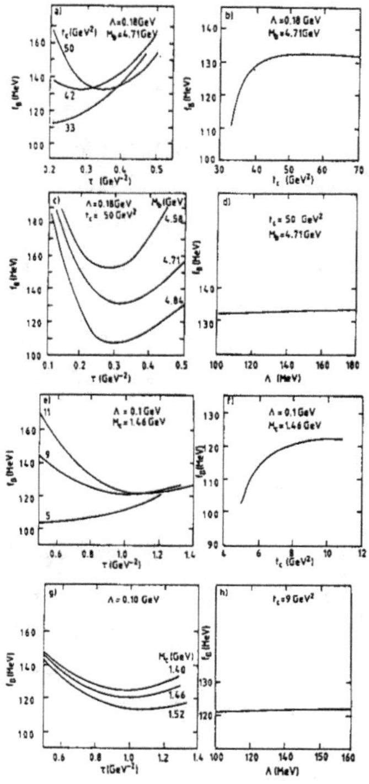

Fig. 10.2 : Laplace (Borel) sum rules. (a) Behaviour of f_B versus the sum-rule scale τ for different values of the continuum threshold t_c at given values of (Λ, M_b). (b) Behaviour of f_B obtained at the minimum of (a) versus t_c. (c) Behaviour of f_B versus τ for various values of M_b. (d) Λ effects on f_B. (e) The same as (a) but for f_D. (f) The same as (b) but for f_D. (g) The same as (c) but for f_D. (h) The same as (d) but for f_D.

The constraint from the exponential moments reads :

$$2\, f_B^2\, M_B^4\, \exp\left(-\, M_B^2\, \tau\right) \simeq \int_{M_b^2}^{t_c} dt\, \exp\, (-t\tau)\, \frac{1}{\pi}\, \mathrm{Im}\, \psi_{Pert}(t)$$

$$+\left[C_4\, \langle O_4 \rangle + C_6\, \langle O_6 \rangle\, \tau\right]\, \exp\left(-\, \tau\, M_b^2\right)\, M_b^2\ , \qquad (10.11a)$$

where

$$C_4\langle O_4 \rangle \simeq M_b\, \mu_q^3\, (-\, \log \sqrt{\tau}\, \Lambda)^{\gamma_1/\beta_1} - \left\langle \alpha_s\, G^2 \right\rangle / 12\pi,$$

$$C_6\langle O_6 \rangle \simeq -\, \frac{1}{2}\, M_b\, g\left\langle \bar{q}\sigma^{\mu\nu}\, \frac{1}{2}\, \lambda_a\, G^a_{\mu\nu}\, q\right\rangle \left(1 - \frac{1}{2}\, M_b^2\, \tau\right)$$

$$-\, \frac{8}{27}\, \pi\, \alpha_s \left\langle \bar{q}q \right\rangle^2 \left(2 - \frac{1}{2}\, M_b^2\, \tau - \frac{1}{6}\, M_b^4\, \tau^2\right)\ . \qquad (10.11b)$$

We show in Fig. 10.2 the predictions for f_B versus the sum-rule scale τ for different values of the continuum threshold t_c. Stability in t_c is also reached for $t_c \geqslant 40$ GeV^2. We also deduce with an accuracy as good as 5% (modulo the systematic uncertainties) :

$$f_B \simeq 1.6\, f_\pi \simeq 150 \text{ MeV} \quad . \qquad (10.12)$$

Here, one can also estimate f_D as the sum rule converges. Then we obtain from the exponential moments :

$$f_D \simeq (111 - 122) \text{ MeV} \quad , \qquad (10.13)$$

where the uncertainties come from the value of the c quark mass.

c) Non-relativistic exponential moment

This sum rule was used in Ref. 174) and corresponds to the case where the heavy-quark mass would be infinite. In this case, one gets the sum rule

$$2\, f_B^2\, M_B\, \exp(-\, E_r/m) \simeq \left\{\left(6m^3/\pi^2\right)\, \left[1 + 0.75\, \bar{\alpha}_s\, (M_b\ )\right]\right.$$

$$\times \left\{1 - \exp(-E_c/m)\left[1 + E_c/m + (E_c/m)^2 \frac{1}{2}\right]\right\}$$
$$+ \hat{\mu}_q^3 \left(1 - M_0^2/32m^2\right) \Big\} [\log (M_b/\Lambda)]^{\nu_1/\beta_1}$$
$$+(\pi/62m^3)\, \alpha_s \langle \bar{q}\, q\rangle^2 , \qquad (10.14)$$

where we have taken into account the log-dependence due to the anomalous dimension of the $\bar{q}\gamma_5 Q$ current and to $\langle\bar{q}q\rangle$; m is the non-relativistic Laplace sum rule variable. E_r is the energy of the lowest ground state which is expected to be around 400 MeV[174]. The continuum threshold E_c can be estimated from the FESR deduced from Eq. (10.12) for large m.

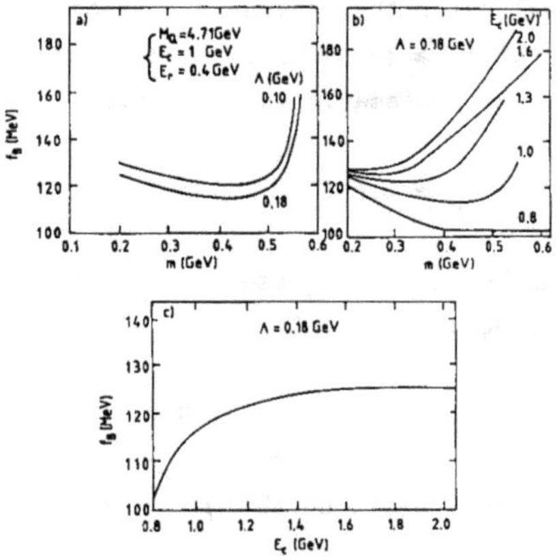

Fig. 10.3 : Non-relativistic Laplace sum rules. (a) Behaviour of f_B versus the sum-rule scale m for two values of Λ. (b) Continuum threshold E_c effects on f_B. (c) Plot of the minimum from (b) versus E_c.

$$2\, f_B^2\, M_B \simeq E_c^3 \big/ \pi^2 - \langle\bar{q}q\rangle , \qquad (10.15)$$

which would correspond to $E_c \simeq 1.2$ GeV if one uses our previous estimate of f_B. We give the results from Eq (10.11) in Fig.10.3. Fig. 10.3a shows the effect of Λ and the sum rule scale m at given (M_Q, E_c, E_r). In Fig.10.3b, we show the effects of E_c given the other set of parameters. Stability of the prediction is obtained at very low m of about 0.25-0.4 GeV. At these values, the perturbative term dominates the $\langle\bar{q}q\rangle$ ones while the effects of the $\langle\bar{q}q\rangle^2$ contributions are negligible. However, it is not clear whether high-dimension condensates will be important at such a scale. In Fig. 10.3 we show the effects of E_c on f_B. The result is insensitive to the choice of E_c for $E_c \geqslant 1.3$ GeV. Previous results obtained at very low E_c should be very sensitive to the form of the continuum. This explains why the result of Ref. 174) is too low. However, we fail to understand the normalization of f_B in Ref. 174b) when they convert their result in order to compare it with that of Ref. 174a). We have checked that Refs 174 a,b) have the same definition of f_B. The effect of E_r on f_B is also obvious. Taking $E_r \simeq (0.4 \pm 0.1)$GeV will introduce an additional 18% uncertainty on f_B. Combining various effects, we have from the non-relativistic sum rule :

$$f_B \simeq (125 \pm 28)\text{MeV}. \tag{10.16}$$

Taking the weighted average, we deduce :

$$f_B \simeq (149.1 \pm 5.2)\text{MeV} \quad . \tag{10.17}$$

to which we could add the typical 10% systematic uncertainty of the QS SR approach.

3. QCD PARAMETERS FROM THE B AND B* MASSES

We shall work either with the ratios of exponential or power moments in order to deduce the B and B^* mass squared. However, it can be easy to check that the ratio of exponential moments, even in its expanded form, does not converge here at the $\tau \simeq 1\ \mathrm{GeV}^{-2}$ values where the prediction is optimal. This non-convergence is induced by the τ-derivative of the exponential moment. Therefore, we work only with the ratio of power moments. The ratio for the B mass squared can be deduced from the moment in Eq. (10.2). That for the B^* can be obtained from the moment :

Table 10.1 :

QCD scales giving B^2 and predictions for $(B^*)^2$, with :

$$g\left\langle \bar{\psi}\sigma^{\mu\nu} \frac{1}{2}\lambda_a \psi\, G^a_{\mu\nu}\right\rangle = M_0^2 \left\langle \bar{\psi}\psi\right\rangle \quad , \quad \alpha_s \left\langle \bar{\psi}\psi\right\rangle^2 \simeq 1.75 \times 10^{-4}\ \mathrm{GeV}^6$$

Set of scales giving B^2			*Predicted value of B^* [GeV^2] (data : (28.3-28.5 GeV^2))*
$M_b\left(p^2 = M_b^2\right)$ [GeV]	k	M_0^2 [GeV^2]	
4.56	1	0.62	28.0
	2	0.75	28.8
	3	0.9	30.2
4.7	1	0.47	27.8
	2	0.51	27.3
	3	0.7	29.2
4.82	1	0.4	28.0
	2	0.5	29.6
	3	0.55	29.6

Table 10.2 :
Predictions from $q^2=0$ moments

1. QCD scales from B and B^*	$M_b\left(p^2=M_b^2\right)=4.56\pm0.05\,\mathrm{GeV} \to \Lambda = 100\pm50\,\mathrm{MeV}$ $g\left\langle\bar{\psi}\sigma^{\mu\nu}\frac{1}{2}\lambda_a\psi G^a_{\mu\nu}\right\rangle \simeq (0.80\pm0.01)\,\mathrm{GeV}^2\left\langle\bar{\psi}\psi\right\rangle$ $\left\langle\alpha_s\bar{\psi}\lambda_a\gamma_\mu\psi\sum_{i\equiv u,d,s}\bar{\psi}_i\gamma^\mu\lambda^a\psi_i\right\rangle \simeq 2\ .$ $.\left(-9\alpha_s\left\langle\bar{\psi}\psi\right\rangle^2 \equiv \text{vacuum saturation}\right)$
2. Mass splittings	$\left.\begin{array}{l} B_\delta-B \simeq 413\pm210\ \mathrm{MeB} \\ B_A^*-B^* \simeq 417\pm212\,\mathrm{MeV}\end{array}\right\} \Longrightarrow B_\delta-B \approx B_A^*-B^*$ $B_s^*-B \leq 403$ MeV for $\chi \geq 0.4$ $B_s^*-B^* \leq 452$ MeV for $\chi \geq 0.4$ $\left.\begin{array}{l} B_c-B \simeq 1.69\pm0.18\ \mathrm{GeV} \\ B_c^*-B^* \simeq 1.53\pm0.18\ \mathrm{GeV}\end{array}\right\} \Longrightarrow B_c-B\simeq B_c^*-B^*\simeq m_c$
3. Couplings and decay constants	$\gamma_{B^*}\simeq\gamma_{B_A^*} \simeq 16\pm2\ :\ \gamma_\rho\simeq2.5$ $f_{B_\delta}\simeq186\pm36\,\mathrm{MeV}\ :\ f_\pi=93.3\,\mathrm{MeV}$ $f_{B_A}\leq1.16f_B$ or $f_{B_A}\simeq1.1f_B$ if $B_s-B\simeq D_s-D$ $f_{B_c}\simeq400\pm20$ MeV $\gamma_{B_c^*} \simeq 14.0\pm1.0$

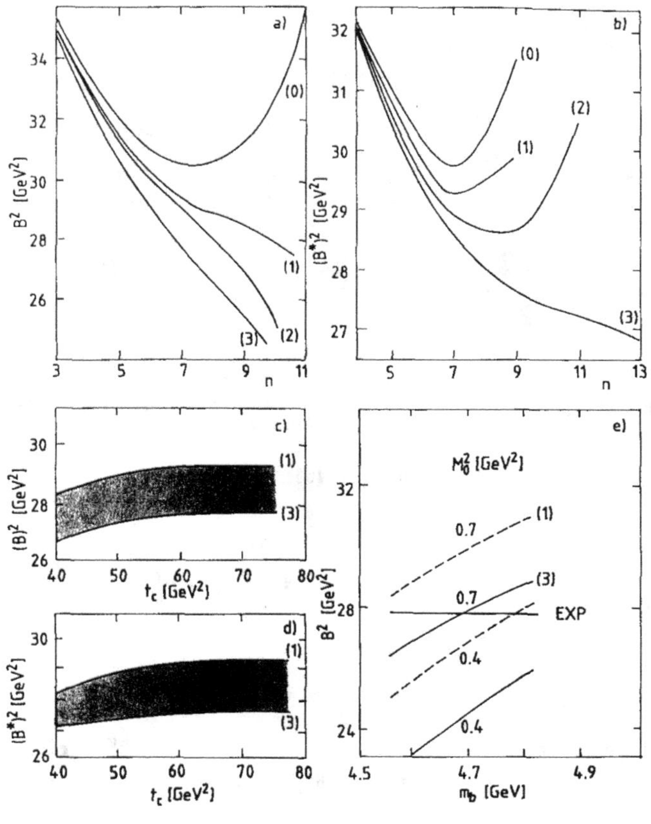

Fig. 10.4 : a-d) Mass squared of B and B^* from moments sum rules for M_b = 4.7 GeV, Λ = 0.18 GeV, t_c = 60 GeV^2 and M_0^2 = 0.6 GeV^2 for different values of the four-quark condensates (k = 0-3).
e) effects of M_b and M_0^2.

$$\mathcal{M}_n^V \equiv \frac{1}{n!}\left(\frac{d}{dq^2}\right)^n \left(q^2\, \Pi_V^{(1)}\, q^2\right)\Bigg|_{q^2=0} = \int_{M_b^2}^{\infty} \frac{dt}{t^{n+1}}\, \frac{1}{\pi}\, \mathrm{Im}\, \Pi_{Pert}^{(1)}\,(t)$$

$$+ \frac{1}{\left(M_b^2\right)^{n+2}} \left\{- M_b \left\langle \bar{\psi}_1 \psi_1 \right\rangle - \frac{\alpha_s}{12\pi} \langle G^2 \rangle + \frac{(n+1)(n+2)}{4M_b} M_0^2 \left\langle \bar{\psi}_1 \psi_1 \right\rangle \right.$$

$$\left. + \frac{(n+2)(n+4)(n-5)}{M_b^2}\, k\, \frac{\pi}{4}\, \alpha_s \left\langle \bar{\psi}\psi \right\rangle^2 \right\} \qquad (10.18)$$

The behaviour of these moments versus n is shown in Fig.10.4 for different values of the four-quark condensates. We have a nice minimum or inflexion point which makes it possible to obtain optimal results from the sum rule. However, a too high value of this condensate tends to destroy such points. We show the effects of the continuum threshold on these minima where we find that t_c-stability is reached for $t_c \geqslant 50$ GeV^2 , this value being the same as that leading to f_B. Next, we study the correlated QCD sets of parameters which are useful for obtaining the observed B and B^* masses. These are shown in Table 10.1 where one can see that a simultaneous fit of the B and B^*masses constrains the QCD parameters to remain in very narrow ranges. We have the typical choices of sets :

set 1 : M_b = (4.56±0.05) GeV ; k = 2 M_0^2 = (0.75-0.80) GeV^2

set 2 : M_b = (4.56-4.82) GeV ; k = 1 M_0^2 = (0.4-0.6)GeV^2

set 3 : $M_b\left(p^2=-M_b^2\right)$=(4.23±0.05) GeV ; k = 2 M_0^2 = (0.80-0.85)GeV^2 (10.19)

The additional requirement on the independency of the prediction for the choice of the quark-mass definition to be used in the sum rule leads to the value given in Table 10.2. These are the values which we shall use in the following analysis. However, in order to see the effects due to the change of the quark masses, for a pedagogical reasons, we also show the prediction when we use the value 4.7 GeV of the pole mass.

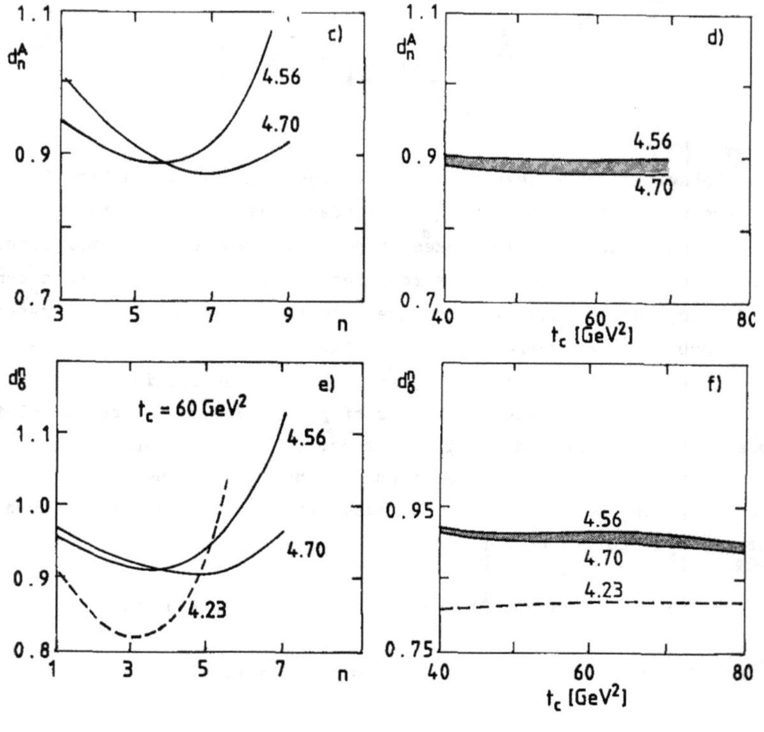

Fig. 10.5 : Mass squared ratios of the $1^{--}/1^{++}$ and $0^{-+}/0^{++}$ versus n and t_c and for two values of the b-pole mass.
--- prediction corresponding to the Euclidian mass.

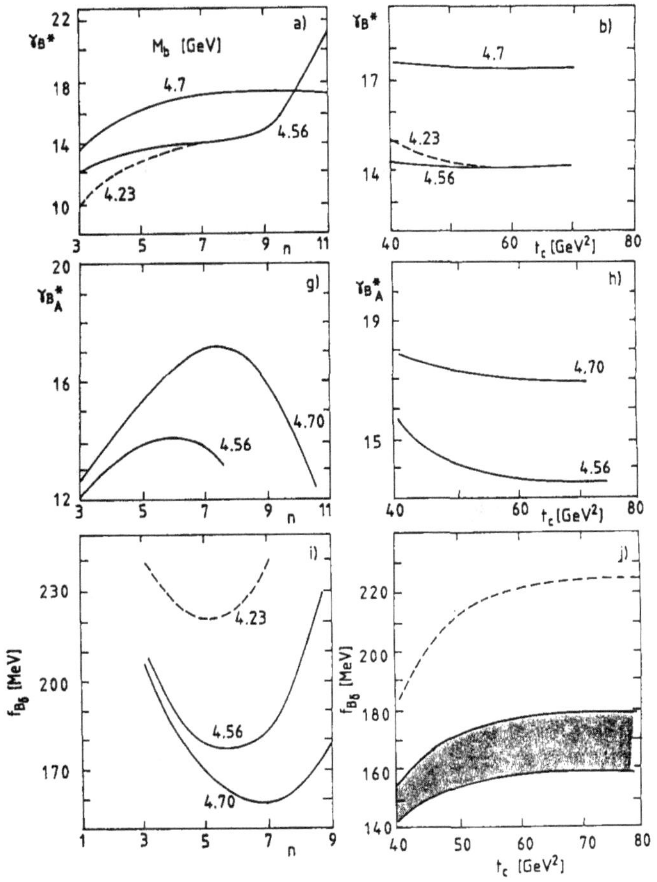

Fig. 10.6 : Coupling constants $\gamma_{B^*_{(A)}}$ *and* f_{B_δ} *versus n and* t_c.

4. $0^{-+}/0^{++}$ AND $1^{--}/1^{++}$ SPLITTINGS

We work with the double ratio of moments DMSR as the individual ones associated with the 0^{++} and 1^{++} do not present a clear inflexion point. As usual we study the n- and t_c-stabilities of the predictions. This is shown in Fig. 10.5 from which we deduce the ratio of mass squared :

$$d^n_A \simeq d^n_{\cdot} \simeq (0.86 \pm 0.06) \quad . \tag{10.20}$$

The corresponding values of the splitting are given in Table 10.2. It may be noticed that the splitting is mainly due to the chiral condensate as can be expected on general grounds. The ambiguous definition of the b quark mass to be used into the non-perturbative contributions is the main source of errors.

5. DECAY CONSTANTS OF OTHER HEAVY-LIGHT ($\bar{d}$ b) MESONS

Extending our analysis of f_B to the other channels, we also estimate the $B^*(1^{--})$ $B^*_A(1^{++})$ and $B_\delta(0^{++})$ couplings to the currents. Those of the spin-one mesons are normalized as :

$$\left\langle 0 \left| \bar{d}\ \gamma^\mu(\gamma_5)\ b \right| B^*_{(A)} \right\rangle = \sqrt{2}\ \epsilon^\mu \frac{M^2_{B^*_{(A)}}}{2\gamma_{B^*_{(A)}}} \ . \tag{10.21}$$

We give successively the behaviours of γ_{B^*}, $\gamma_{B^*_A}$ and f_{B_δ} in Fig. 10.6. Our optimal predictions are shown in Table 10.2 where it may be seen that :

$$\gamma_{B^*} \simeq \gamma_{B^*_A} \tag{10.22}$$

but f_{B_δ} is much higher than f_B.

6. $SU(3)_F$ BREAKINGS FOR THE STRANGE BEAUTIFUL MESONS

Let us now study the splittings between the B (respectively B^*) and B_s $\left(\text{respectively } B_s^*\right)$ when we use the value of the strange-quark mass and condensates obtained in Chapter 8. We do not have good control of the $SU(3)_F$ breakings for the four-quark and mixed condensates. We assume that either we have a $SU(3)_F$ symmetry or that they are broken like the $\langle \bar{s}\, s\rangle$ condensate. Owing to this uncertainty, we cannot provide a prediction for the splittings. In turn, we can give an upper bound obtained from the set 1 of QCD parameters in Eq. (10.19) and for $SU(3)_F$ symmetrical high-dimension condensates. The behaviour of the bound is given in Fig. 10.7. Using the value of $x \equiv \langle \bar{s}s\rangle / \langle \bar{u}u\rangle$ obtained previously, we obtain the result in the Table 10.2. $SU(3)_F$ breaking related to the decay constant f_{B_s} can be "roughly" deduced from the lowest moment expression given in Ref. 129) :

$$f_{B_s} \simeq f_B \left(\frac{M_{B_s}}{M_B}\right) \left\{1 + \frac{3\bar{m}_s}{m_b}\left(1 - \frac{M_B^2}{t_c}\right) - \frac{8\pi^2}{m_b^3}\langle \bar{s}s - \bar{u}u\rangle\right\}^{1/2} , \qquad (10.23)$$

where the corrections to the lowest order term amount to 8% for the B. Then, we deduce the predictions :

$$f_{D_s} \simeq 1.26\, f_D \quad \text{and} \quad f_{B_s} \simeq 1.1\, f_B \quad , \qquad (10.25)$$

if we assume for the latter the phenomenological relation :

$$M_{B_s} - M_B \simeq M_{D_s} - M_D \qquad (10.25)$$

expected from potential model arguments[175].

7. $SU(4)_F$ BREAKINGS AND THE CHARMING BEAUTIFUL MESONS

We study here the masses and couplings of the 0^{-+} and 1^{--} mesons $\bar{c}b$. We have a new phenomenon as the chiral condensate of the *charm quark vanishes as the inverse of its mass. (see Eq. 9.1)*. Therefore, we cannot repeat what we have done for the $\bar{u}b$ systems where the light quarks are associated with the Goldstone realization of chiral symmetry governed by the PCAC relation in Eq. (1.17). We analyse the n- and t_c-stabilities of the predictions in the usual way. These are shown in Fig. 10.8 for two values of $\langle \alpha_s G^2 \rangle$ (ST ≡ standard value 0.04 GeV^4 and NST ≡ 2.ST). In the ST case we have good n-stability but not often in the NST one. We have checked that moving the b quark mass inside the range given in Table 10.1 and the invariant mass $\hat{m}_c$ inside 1.8-2.14 GeV[166] does not give appreciable errors in the analysis. The curves in Fig. 10.8 correspond to M_b=4.56 GeV and $\hat{m}_c$=2.14 GeV. The main source of errors is $\langle \alpha_s G^2 \rangle$. Our result is given in Table 10.2. The strength of the mass difference is about the value of the running mass $m_c\left(p^2=M_b^2\right)$, as expected from naive arguments. The value of $\gamma_{B_c^*}$ is about that of γ_{B^*} but f_{B_c} is quite high (about $3f_B$). One should notice that flavour breakings tend to increase the value of the decay constant of pseudoscalar mesons, i.e. the case $f_K \simeq 1.2\, f_\pi$ seems not to be a particular feature of the light-meson systems ! We are not able (within our approximation) to obtain reliably the masses of the chiral partners (scalar and axial-vector mesons) owing to the fact that the associated MSR and DMSR do not present a n-stability to this approximation.

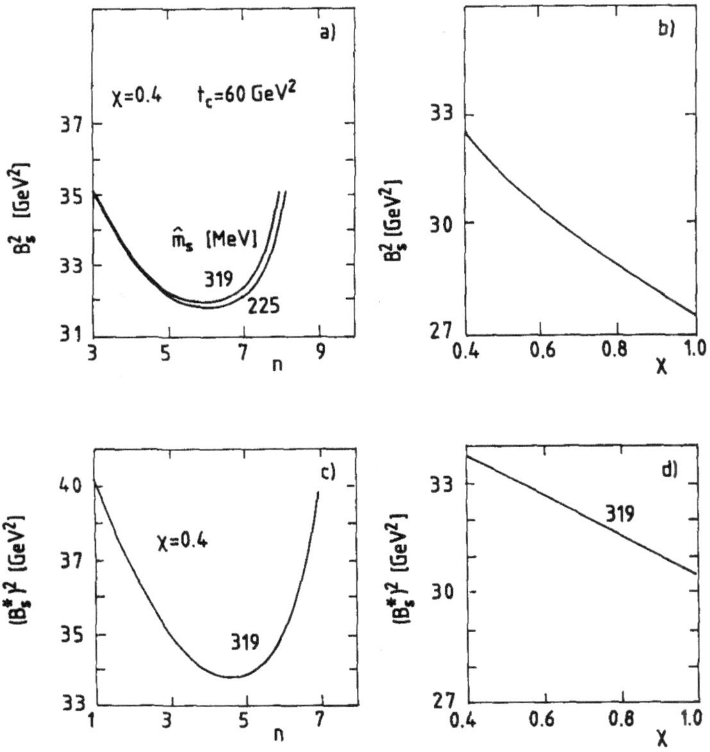

Fig. 10.7 : Upper bound on B_s^2 *and* B_s^{*2} *from moment ratios for*

$\chi \equiv \langle \bar{s}s\rangle / \langle \bar{u}u\rangle \geqslant 0.4$.

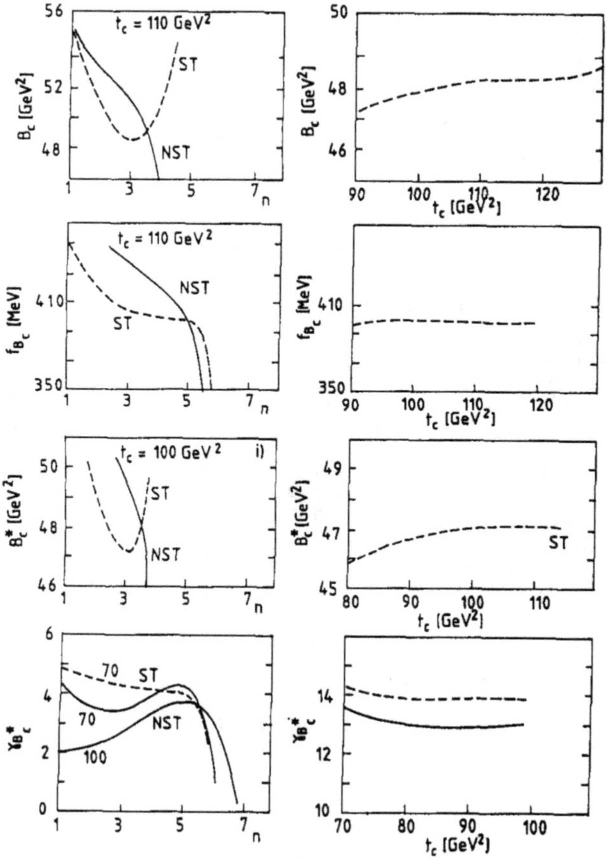

Fig. 10.8 : B_c , f_{B_c}, B_c^*, $\gamma_{B_c^*}$ *versus* n *and* t_c. *We use* M_b = 4.56 GeV, $\hat{m}_c$ = 2.14 GeV .

8. DOES f_B^2 ALREADY SCALE LIKE $\frac{1}{M_b}$?

This important question has been re-examined in Ref. 176). The main point is to understand whether the scaling law :

$$f_B^2 \sim \frac{1}{M_b} |\psi(0)|^2 \tag{10.26}$$

expected from a non-relativistic approach at $M_b \to \infty$ already applies to the b and c quark masses. $\psi(0)$ is the B-meson wave function at the origin.

We study this problem in two ways. Firstly, we examine the analytical expressions of f_B given by the different sum rules in paragraph 1 at finite M_B and then we analyze its limit when $M_b \to \infty$. Secondly, we study the validity of the extrapolation of the expression of f_B obtained for $M_b \to \infty$ down to the real world of finite quark mass M_b.

a) f_B at finite M_b and the limit $M_b \to \infty$:

Different expressions of f_B from the low-energy behaviour of the pseudoscalar correlator, the Laplace sum rule and the $Q^2 = 0$ moments have been discussed in Ref. 176). Here we limit ourselves to the $Q^2 = 0$ moment which is based on the $1/M_b^2$-mass expansion and is very analogous to the non-relativistic approximation. The expression of the moment is given in Eq. (10.7). The lowest superconvergent moment (n=1) which is valid for both the D and B mesons gives to the lowest order in α_s :

$$2f_B^2 \simeq \left(\frac{M_B}{M_b}\right)^2 \left\{\frac{1}{8\pi^2 t_c^2} \left(t_c - M_b^2\right)^3 - \frac{\langle\bar{\psi}\psi\rangle}{M_b} + \dots\right\} , \tag{10.27}$$

while higher moments which converge only for the b and higher quark mass values give the exact expression :

$$2f_B^2 \simeq \left(\frac{M_B}{M_b}\right)^{2n} \left\{\frac{3}{8\pi^2}\frac{M_b^2}{n}\left[1-\left(\frac{M_b^2}{t_c}\right)^n - \left(\frac{2n}{n+1}\right)\left(1-\left(\frac{M_b^2}{t_c}\right)^{n+1}\right)+\right.\right.$$

$$\left.\left(\frac{n}{n+2}\right)\left(1-\left(\frac{M_b^2}{t_c}\right)^{n+2}\right)\right] + \frac{1}{M_b^2}\left[- M_b\langle\bar{\psi}\psi\rangle +\right.$$

$$\left.\left.\frac{(n+1)(n+2)}{4M_b} g\left\langle\bar{\psi}\sigma^{\mu\nu}\frac{\lambda a}{2}\psi G^a_{\mu\nu}\right\rangle + \frac{(n+2)(n^2+10n+9)}{M_b^2} k\frac{\pi}{4}\alpha_s\langle\bar{\psi}\psi\rangle^2\right\}, \quad (10.28)$$

where we have included the effects of the non-perturbative high-dimension condensates in order to show why the expansion in $\frac{1}{M_b^2}$ fails for the charm quark. Let us concentrate on the perturbative contribution. It is interesting to analyze its limit when :

$$\Delta_c \equiv \frac{t_c}{M_b^2} - 1 \ll 1 \quad . \quad (10.29)$$

Then, we obtain under this condition :

$$2\, f_B^2 \simeq \left(\frac{M_B}{M_b}\right)^{2n}\left\{\frac{1}{n}\left(\frac{3}{8\pi^2}\right)\frac{(t_c-M_b^2)^3}{t_c^2} - \frac{\langle\bar{\psi}\psi\rangle}{M_b} + \ldots\right\} , \quad (10.30)$$

in good agreement with the behaviour in (10.26), if the condition in (10.29) is satisfied.

b) <u>How good is the extrapolation of</u> f_B <u>from</u> $M_b \to \infty$ <u>down to</u> M_c ?

As in Ref. 174), we can also analyze the behaviour of the spectral function for $M_b \to \infty$ and study the FESR :

$$2\, f_B^2 \simeq \frac{1}{M_b^4} \int_0^{t_c - M_b^2} dt\, \frac{1}{\pi}\, \text{Im}\, \psi_5 \left(t - M_b^2\right) - \frac{\langle \bar{\psi}\psi \rangle}{M_b} . \qquad (10.31)$$

Here we ignore the mass difference between the quark and the meson. For $M_b \to \infty$, one is tempted to do a $\frac{t}{M_b^2}$ expansion in the spectral function as in Ref. 174) but one should not forget that this is only valid under the condition in Eq(10.29). In this case, we get :

$$2\, f_B^2 \simeq \frac{3}{16\pi^2} \frac{\left(t_c - M_b^2\right)^2}{M_b^2} - \frac{\langle \bar{\psi}\psi \rangle}{M_b} , \qquad (10.32)$$

which again satisfies Eq (10.26) under the condition in Eq. (10.29). To make further program with the $M_b \to \infty$ result, we should also recall that within this limit the axial current gets an anomalous dimension[177]. In this case it is therefore much more accurate to write :

$$\left(\hat{f}_B^{\infty}\right)^2 \simeq \frac{3}{16\pi^2} \frac{\left(t_c - M_b^2\right)^2}{M_b^2} - \frac{\langle \bar{\psi}\psi \rangle}{M_b} , \qquad (10.33a)$$

where for $SU(6)_F$:

$$\hat{f}_B^{\infty} \equiv f_B^{\infty}(\nu)\, (\alpha_s(\nu))^{2/9} , \qquad (10.33b)$$

is the renormalization group invariant coupling which is related to the f_B of finite-quark mass in a non-trivial way. $f_B^{\infty}(\nu)$ is the renormalized coupling in the infinite quark mass theory. Eq. (10.33) is consistent with the finite quark mass result <u>but</u> again under the condition in Eq. (10.29).

We now test the realization of the condition in Eq. (10.29) for the D and B mesons. We use the value of $\sqrt{t_c}$ obtained from the detailed sum rule analysis of the correlator which we take at the beginning of the t_c-stability region of f_D and f_B. These values are :

$$\sqrt{t_c}\ [\text{GeV}] \simeq \begin{matrix} 2.83 \\ 6.30 \end{matrix} \quad \text{for} \quad M_b[\text{GeV}] \simeq \begin{matrix} 1.46 & \text{charm} \\ 4.6 & \text{bottom} \end{matrix} \tag{10.34a}$$

which implies :

$$t_c \simeq 2\ M_b^2 \quad , \tag{10.34b}$$

a value which is still too large for the condition in (10.29) to be applied. Analogous values of t_c have been deduced from a QSSR analysis of the B_o-$\bar{B}_o$ matrix element[147] and on the B-meson mass[166].
In order to check the "real physical meaning" of Eq. (10.34), we can identify $\sqrt{t_c}$ at about the value of the first radial excitation. Then, we deduce the mass differences :

$$M_{D'} - M_D \simeq 0.96\ \text{GeV} \qquad M_{B'} - M_B \simeq 1.05\ \text{GeV} \tag{10.35}$$

which look reasonable compared to the observed splittings in the light meson systems :

$$M_{\rho'} - M_\rho \simeq M_{\pi'} - M_\pi = \ldots \simeq 1\ \text{GeV} \quad . \tag{10.36}$$

We do not yet have data to test the value in Eq. (10.35) and findings of these radial excitations might give improvements on the estimate of the decay constants f_D and f_B . However, the agreement between the lattice [178] and QSSR results on f_D is already strong support for the reliability of the value of f_D from the QSSR within the t_c-stability criterion if one can consider the lattice result as a reference value. f_B obtained in the same way is also expected to make sense unless some new unexpected strong effects appear in the B-meson case.

9. CONCLUSIONS

The uses of the moments sum rules for the heavy-light quark systems are interesting per se, as we combine here our knowledge of the light-quark systems (Goldstone realization of chiral symmetry) with that of heavy quarks (mass expansion and Wigner-Weyl realization of chiral symmetry). Our results summarized in Table 10.2 are obtained within *n- and t_c-stability criterions ($n \simeq 5,7$) and t_c larger than* 50 GeV^2. They indicate that :

- Heavy-light quark systems can be a good laboratory for measuring the QCD parameters (M_b, mixed and four-quark condensates). The values of these two quantities come out very accurately as shown in Table 10.2.
- Mass splittings for the B_δ-B and B^*_A-B^* are roughly equal and seem to follow the phenomenological expectations from the potential models. The results quoted here, though in agreement with RRY result[29] are difficult to compare with theirs as they have not studied the dependence of their result on the mass definition in the OPE. We infer that the tendency of RRY to have larger splittings is due to the implicit use of the vacuum saturation in their analysis.
- The size of the B_c-B and B^*_c-B^* splittings are almost equal to the value of the running charm quark mass evaluated at M_b^2, as naively expected!
- The couplings of the spin-one meson to the current are almost $SU(4)_F$ symmetric. On the other hand, the decay constants of the pseudoscalar mesons exhibit a large $SU(4)_F$ breaking.

We hope that some of our results will be tested in the forthcoming B-physics experiments. One of the particular uses of our results is the prediction for the B_o-$\bar{B}_o$ mixing from which one can deduce a lower limit on the top quark mass. Following Ref. 179), we can write :

$$x_d = \frac{G_p^2}{3\pi^2} M_t^2 \tau_B B_B f_B^2 \left|V_{td} V_{tb}^*\right|^2 \frac{A(z_t)}{z_t} \eta_t \qquad (10.37a)$$

where :

$$\frac{P\left(B_o \to \bar{B}_o\right)}{P(B_o \to B_o)} \simeq \frac{x^2}{2+x^2} \quad , \qquad x \equiv \frac{M_L - M_S}{\frac{1}{2}(\Gamma_L + \Gamma_S)} \qquad (10.37b)$$

$$x_d = \frac{G_F^2}{3\pi^2} M_t^2 \tau_B B_B f_B^2 \left|V_{td} V_{tb}^*\right|^2 \frac{A(z_t)}{z_t} \eta_t \qquad (10.38a)$$

where :

$$\frac{P\left(B_o \to \bar{B}_o\right)}{P(B_o \to B_o)} \simeq \frac{x^2}{2+x^2} \quad , \qquad x \equiv \frac{M_L - M_S}{\frac{1}{2}(\Gamma_L + \Gamma_S)} \quad . \qquad (10.38b)$$

L,S refer to long and short lifetime mass eigenstates ; $\tau_B = \frac{1}{\Gamma_B}$ is the B_o-lifetime ; $\left\langle B_o \left| \left(\bar{d}_L \gamma_\mu b_L\right) \left(\bar{d}_L \gamma^\mu b_L\right) \right| \bar{B}_o \right\rangle \equiv \frac{4}{3} f_B^2 M_B^2 B_B$; V_{ij} are Cabibbo Kobayashi-Maskawa (CKM) mixing matrices and $A(z_t)$ is the short distance effect $\left(z_t \equiv M_t^2 / M_W^2\right)$:

$$\frac{A(z)}{z} = \frac{1}{4} + \frac{9}{4(1-z)} - \frac{3}{2}\frac{1}{(1-z)^2} - \frac{3}{2}\frac{z^2 \log z}{(1-z)^3} . \qquad (10.38c)$$

The value of the top-quark mass is very sensitive to the non-perturbative matrix element $f_B \sqrt{B_B}$ once one fixes the value of the CKM mixing angles. One also expects that the vacuum saturation leading to $B_B \simeq 1$ is a good approximation for the heavy quarks. This assumption can be tested from the estimate :

$$f_B \sqrt{B_B} \simeq (0.15 - 0.17) \text{ GeV} \qquad (10.39)$$

first obtained by Pich[147] using a QCD sum rule analysis of the four-quark operator to a leading order in α_s. Using our value of f_B obtained previously, we deduce :

$$B_B \simeq (1.0 - 1.3) \quad , \qquad (10.40)$$

in good agreement with the vacuum saturation assumption. With this value of the mixing angle $\left(\tau_B \ |V_{td}|^2 \simeq (0.02\text{-}6.2)\ 10^{-6} s\right)$, one can deduce a non-trivial constraint on the value of the top-quark pole mass as one can see in Fig. 10.9 from Ref. 179) where the Argus data are also shown. This bound is :

$$m_t \geqslant 50 \text{ GeV} \qquad (10.41)$$

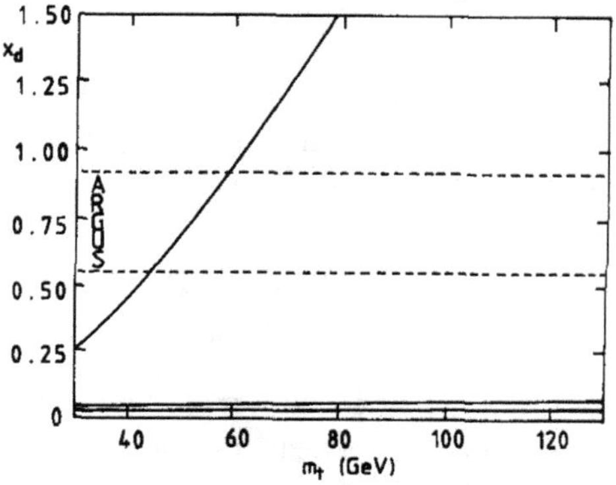

Fig. 10.9 : Predicted values of B_d^0 - $\bar{B}_d^0$ mixing in the standard model with three fermion families, as a function of the top quark mass, for different choices of the relevant parameters $r_d = x_d^2 / (2+x_d^2)$. All curves are for $\Gamma(b\to y)/\Gamma(b\to c) < 0.08$. The Argus value for r_d is also shown, which translates to $x_d = 0.73 \pm 0.18$ by the relation. In the curve we use the following values of the parameters $F = B_B^{1/2}\, f_b \simeq 1.4\, f_\pi, \tau_b |V_{bc}|^2\, 10^{-15}$ sec $\simeq 3.5$ and the largest values of the mixing angles allowed by the $|\epsilon|$ constraint.

CHAPTER 11

BARYON SYSTEMS

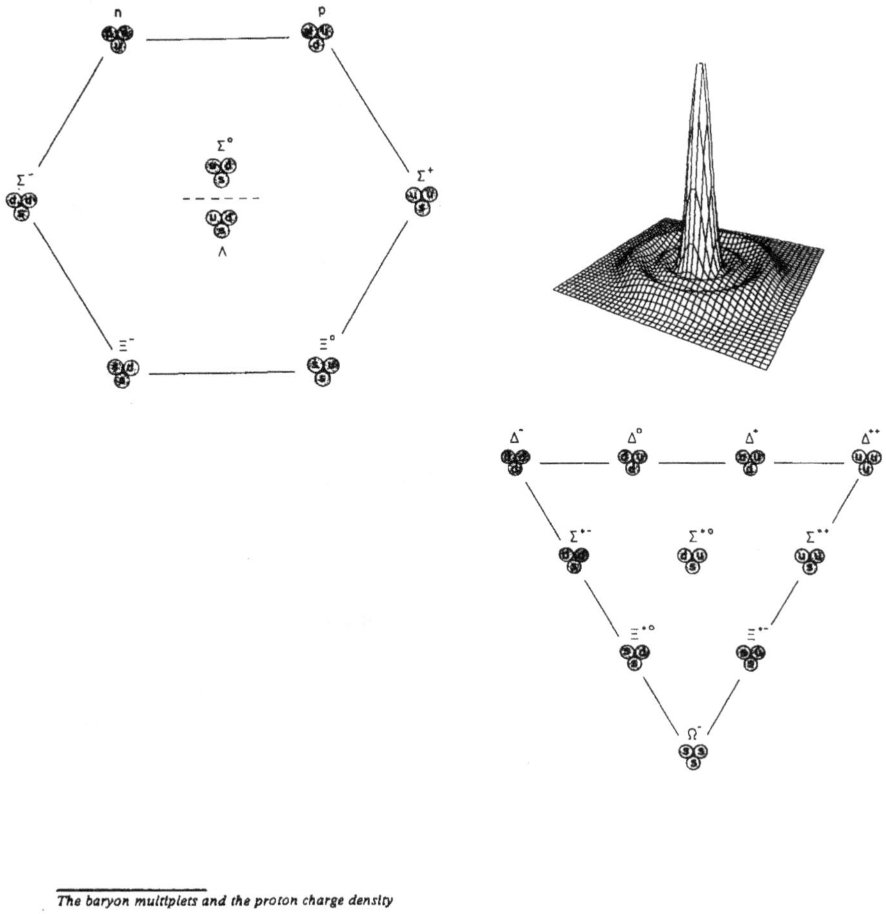

The baryon multiplets and the proton charge density

Baryon masses and decay constants have been related to the chiral quark condensate $\langle \bar{\psi} \psi \rangle$ using the QSSR approaches[127,160,181-183] following the pioneer work of Ioffe[103]. Contrary to the case of mesons, the analysis is much more involved here as we have to deal with operators built with three quark fields which are not unique for given quantum numbers. In this approach we shall only consider the valence quarks of the proton while the sea-quark effect will only appear to higher order and can safely be neglected.

1. THE DECUPLET

The lowest dimension interpolating operator having the quantum number of the $\Delta(3/2)$ is :

$$\Delta^\mu = \frac{1}{\sqrt{2}} : \psi^T C \gamma_\lambda \psi \, (g^{\mu\lambda} - \gamma^\mu \gamma^\lambda/4)\psi : \quad , \tag{11.1}$$

where C is the charge conjugation matrix and we have suppressed the colour indices. ψ is the "valence" quark field. We study the associated two-point correlator :

$$S^{\mu\nu} = i \int d^4x \, e^{iqx} \left\langle 0 \left| \mathbb{T} \, \Delta^\mu(x), \, (\Delta^\nu(0))^+ \right| 0 \right\rangle \quad , \tag{11.2}$$

which can be parametrized without loss of generalities by two invariants :

$$S^{\mu\nu} = \left(\hat{q} \, F_1 + F_2 \right) g^{\mu\nu} + \ldots \tag{11.3}$$

The QCD expressions of these form factors are known in the literature up to condensate effects of dimension nine. But the contributions of these high-dimension condensates are very uncertain owing to the introduction of further assumptions used to estimate them and their coefficients. Therefore, we shall limit ourselves to the OPE series including the dimension-six condensates. The LSR of these quantities reads in a generic notation :

$$\int_0^{t_c} dt\, e^{-t\tau} \frac{1}{\pi} \mathrm{Im}\, F_1(t) \simeq \left(\frac{\alpha_s(\tau^{-1/2})}{\alpha_s(\nu)}\right)^{\gamma/-\beta_1} \Big\{ A_1\, \tau^{-3}(1-\rho_2) + A_2 \pi \left\langle \alpha_s\, G^2 \right\rangle + A_3\, \pi^2\, m_i(\tau) \left\langle \bar{\psi}\, \psi \right\rangle \tau^{-1} (1-\rho_0) + A_4\, \pi^4 k \left\langle \bar{\psi}\, \psi \right\rangle^2 \Big\} \tag{11.4}$$

and

$$\int_0^{t_c} dt\, e^{-t\tau} \frac{1}{\pi} \mathrm{Im}\, F_2(t) \simeq \left(\frac{\alpha_s(\tau^{-1/2})}{\alpha_s(\nu)}\right)^{\gamma/-\beta_1} \Big\{ B_1\, m_i \tau^{-3} (1-\rho_2) + B_2\, \pi^2 \left\langle \bar{\psi}\, \psi \right\rangle \tau^{-2}(1-\rho_1) + B_3 \pi^2 \left\langle g\, \bar{\psi}\, \sigma^{\mu\nu} \frac{\lambda_a}{2} \psi\, G^a_{\mu\nu} \right\rangle . \tau^{-1}(1-\rho_0) \Big\} \tag{11.5}$$

where $m_i = 0$ for u and d quarks. The coefficients A_i and B_i for each channel can be obtained from Table 11.1 as obtained from previous calculations. $\gamma = 2\left(-\frac{2}{3}\right)$ is the anomalous dimension corresponding to the octet (decuplet). The quantity :

$$\rho_n = \exp(-\, t_c \tau) \sum_n \frac{(t_c \tau)^n}{n\, !} \tag{11.6}$$

takes into account the QCD continuum effects starting from the threshold t_c and deriving from the discontinuity of the Feynman diagrams.

Table 11.1 : Wilson coefficients appearing in Eqs 11.4 and 11.5

	A_1	A_2	A_3	A_4
Δ	$\frac{1}{20}$	$-\frac{5}{36}$	0	$\frac{32}{3}$
Σ^*	$\frac{1}{20}$	$-\frac{5}{36}$	$-\frac{2}{3}(4-\chi_3)$	$\frac{32}{9}(1+2\chi_3)$
Ξ^*	$\frac{1}{30}$	$-\frac{5}{36}$	$-\frac{4}{3}(2+\chi_3)$	$\frac{32}{9}\chi_3(2+\chi_3)$
Ω	$\frac{1}{20}$	$-\frac{5}{36}$	$-6\chi_3$	$\frac{32}{3}\chi_3^2$
N	$\frac{1}{256}(5+2b+5b^2)$	$\frac{1}{128}(5+2b+5b^2)$	0	$\frac{2}{6}(7-2b-5b^2)$
Λ	$\frac{1}{256}(5+2b+5b^2)$	$\frac{1}{128}(5+2b+5b^2)$	$-\frac{1}{48}[4(5-4b-b^2)$ $-3(5+2b+5b^2)\chi_3]$	$\frac{1}{9}[(11+2b-13b^2)$ $+2(5-4b-b^2)\chi_3]$
Σ	$\frac{1}{256}(5+2b+5b^2)$	$\frac{1}{128}(5+2b+5b^2)$	$-\frac{1}{16}[12(1-b^2)$ $-(5+2b+5b^2)\chi_3]$	$\frac{1}{2}[(1-b)^2$ $+6(1-b^2)\chi_3]$
Ξ	$\frac{1}{256}(5+2b+5b^2)$	$\frac{1}{128}(5+2b+5b^2)$	$-\frac{1}{8}[2(1-b^2)$ $-(1+b)^2\chi_3]$	$\frac{1}{2}\chi_3[6(1-b^2)$ $+(1-b)^2\chi_3]$

	B_1	B_2	B_3
Δ	0	$-\frac{8}{3}$	$\frac{4}{3}$
Σ^*	$\frac{1}{8}$	$-\frac{8}{9}(2+\chi_3)$	$\frac{4}{9}(2+\chi_3)$
Ξ^*	$\frac{1}{4}$	$-\frac{8}{9}(1+2\chi_3)$	$\frac{4}{9}(1+2\chi_3)$
Ω	$\frac{3}{8}$	$-\frac{8}{3}\chi_3$	$\frac{4}{3}\chi_3$
N	0	$-\frac{1}{6}(7-2b-5b^2)$	$\frac{3}{8}(1-b^2)$
Λ	$\frac{1}{192}(11+2b-13b^2)$	$-\frac{1}{24}[2(5-4b-b^2)$ $+(11+2b-13b^2)\chi_3]$	$\frac{1}{32}(1-b^2)(10+4\chi_3)$
Σ	$\frac{1}{64}(1-b)^2$	$-\frac{1}{8}[6(1-b^2)$ $+(1-b)^2\chi_3]$	$\frac{3}{4}(1-b^2)$
Ξ	$\frac{1}{32}(1-b^2)$	$-\frac{1}{8}[(1-b)^2$ $+6(1-b^2)\chi_3]$	$\frac{1}{8}(1-b^2)\chi_3$

As usual, we introduce the invariant $\hat{m}_i$ and $\hat{\mu}_i$ quark mass and condensate with the value in Eqs (7.32, 7.49). We parametrize the mixed condensate as :

$$g\left\langle \bar{\psi}\, \sigma^{\mu\nu} \frac{\lambda a}{2} \psi\, G^a_{\mu\nu} \right\rangle = -\hat{\mu}^3\, M_0^2 \left(-\log \Lambda \sqrt{\tau}\right)^{2/27} , \qquad (11.7)$$

with $M_0^2 \simeq 0.8$ GeV^2 as obtained in Table 10.2. The anomalous dimension comes from the calculation in Ref. 42). We use k = 2 as a deviation from the vacuum saturation estimate of the four-quark condensate and take $\Lambda_{\overline{MS}} \simeq 150$ MeV from τ decay analysis[165].

In principle, we can use three moments for determining the Δ mass :

$$R_{1i}(\tau) \equiv \frac{\int_0^{t_c} dt\; t e^{-t\tau}\; \mathrm{Im}\, F_i(t)}{\int_0^{t_c} dt\; e^{-t\tau}\; \mathrm{Im}\, F_i(t)} \qquad i = 1,2 \qquad (11.8)$$

and

$$R_{12}(\tau) \equiv \frac{\int_0^{t_c} dt\; e^{-t\tau}\; \mathrm{Im}\, F_2(t)}{\int_0^{t_c} dt\; e^{-t\tau}\; \mathrm{Im}\, F_1(t)} \quad . \qquad (11.9)$$

In the previous literature $R_{12}(\tau)$ has often been used with an elaborate fitting procedure[103,127,181] or by simply ignoring continuum effects in the analysis[182,183]. We have tested the reliability of this moment in Ref. 160) by studying the τ- and t_c-stabilities. Neither of the two criteria is satisfied by $R_{12}(\tau)$ at reasonable values of $t_c \leqslant 7$ GeV^2, though it appears that $R_{12}(\tau)$ gives a good result if one assumes an unrealistic lowest ground state dominance of the sum rule. The same conclusion also applies to $R_{11}(\tau)$. Working with $R_{22}(\tau)$ as in Ref. 160) gives a far greater insight into the understanding of the physics behind the sum rules. $R_{22}(\tau)$ also has the advantage that the leading lowest dimension condensates are accurately known while the radiative corrections to the leading term do not exceed 10%. The τ-stability of R_{22} starts for $t_c > 1.86$ FeV^2 while the t_c-stability of the τ minima is reached for $t_c > 4.0$ GeV^2 (see Fig. 11.1). We parametrize the spectral function using the duality ansatz "one narrow resonance" plus "QCD continuum". The resonance contribution being parametrized as :

$$\mathrm{Im}\, F_2\, (t \leqslant t_c) = M_\Delta\, |Z_\Delta|^2\; \delta\left(t-M_\Delta^2\right) \quad . \qquad (11.10)$$

For the above range (1.86-4) GeV^2 of t_c values, one obtains :

$$M_\Delta \simeq (1.15 - 1.34)\ \text{GeV} \qquad |Z_\Delta|^2 \simeq (0.97 - 1.32)\ \text{GeV}^6 \tag{11.11}$$

where we have taken the value of $|Z_\Delta|^2$ at the inflexion point (Fig. 11.2). We improve the above results with the help of the lowest dimension FESR constraint :

$$\frac{1}{\pi} M|Z_\Delta|^2 \simeq B_1\, m_s \frac{t_c^3}{6} + B_2\, \pi^2 \left\langle \bar\psi\psi \right\rangle \frac{t_c^2}{2} + B_3\, \pi^2 \left\langle g\bar\psi\, \sigma^{\mu\nu} \frac{\lambda_a}{2}\, \psi\, G^a_{\mu\nu} \right\rangle . t_c \tag{11.12}$$

The above values of M_Δ and $|Z_\Delta|^2$ lead to :

$$t_c \simeq 2.2\ \text{GeV}^2 \tag{11.13}$$

which, used in Fig. 11.1, gives the final estimate :

$$M_\Delta \simeq 1.21\ \text{GeV (exp 1.23 GeV)} \qquad |Z_\Delta|^2 \simeq 1.15\ \text{GeV}^6 \tag{11.14}$$

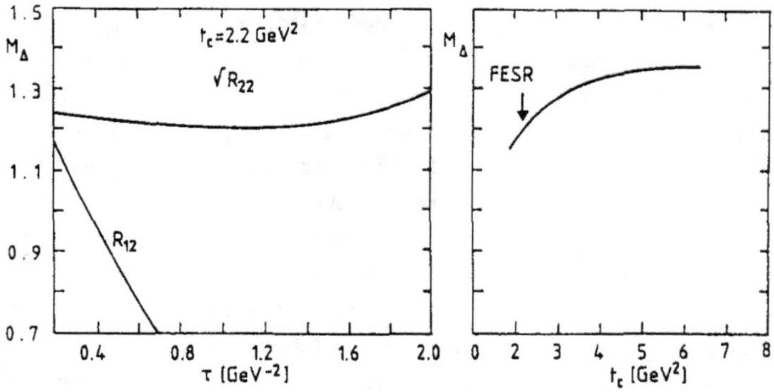

Fig. 11.1 : M_Δ versus τ and t_c from R_{22}, R_{12} and R_{11} (not shown in the scale). The arrow indicates the value of t_c from FESR.

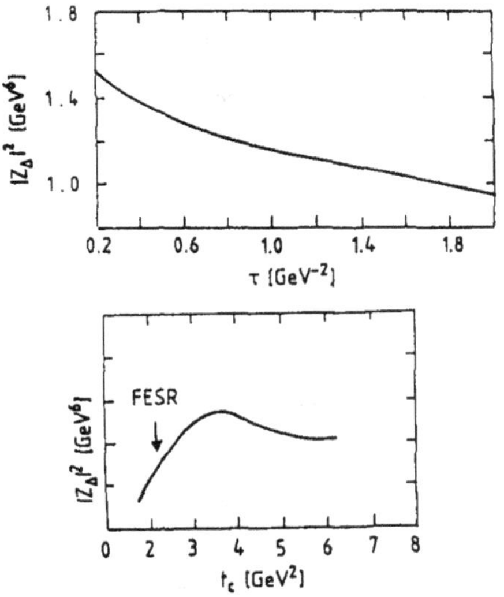

Fig. 11.2 : Z_Δ from Eq. (11.5) versus τ and t_c . The arrow indicates the value of t_c from FESR.

The agreement of the result with the data is fantastic, while the value of $\sqrt{t_c}$ almost coincides with the mass of the first radial excitation. We test the effect of the mixed condensate on the mass value. Moving M_0^2 from 0.8 to 0.9 GeV^2 leads to an increase of M_Δ by about 2% while the lower value of M_0^2 is incompatible with the data.

The other members of the decuplet can be studied in the same way after introducing the $SU(3)_F$ breaking parameters, namely the strange-quark mass and the ratio of chiral condensates :

$$\chi_3 \equiv \frac{\langle \bar{s}s \rangle}{\langle \bar{d}d \rangle} \qquad \chi_5 \equiv \frac{\langle \bar{s}\, \sigma^{\mu\nu} \lambda_a\, s\, G^a_{\mu\nu} \rangle}{\langle \bar{u}\, \sigma^{\mu\nu} \lambda_a\, u\, G^a_{\mu\nu} \rangle} \tag{11.15}$$

The values of χ_3 (0.6 ± 0.1) and m_s = (267 ± 15) MeV come from the light-meson systems (Chapter 8). We leave χ_5 as a free parameter. A better fit of the Ω mass is reached for

$$\chi_5 \quad 1.4$$

as shown in Fig. 11.3. We list our results in Table 11.2 ,

Table 11.2 : Estimates of the decuplet parameters

Names	Mass [GeV]	$\|Z\|^2$ [GeV]^6	t_c [GeV]^2
Ω	1.61	5.16	4.08
Σ^*	1.35	1.89	2.78
Ξ^*	1.48	3.07	3.39

where the method has the tendency to underestimate the absolute value of the decuplet masses. However, the mass ratio is well reproduced. Here, we should mention that the good fit of Ref. 182) obtained by ignoring the continuum effect or by taking the high value of t_c might be due to an accidental cancellation of this continuum effect. However, the corresponding output values are inconsistent with the FESR constraint in (11.12).

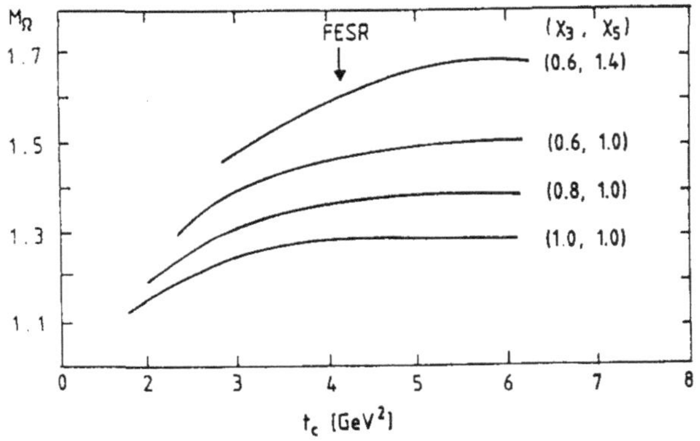

Fig. 11.3 : M_Ω versus t_c for different values of χ_3 and χ_5. The arrow indicates the value of t_c from FESR.

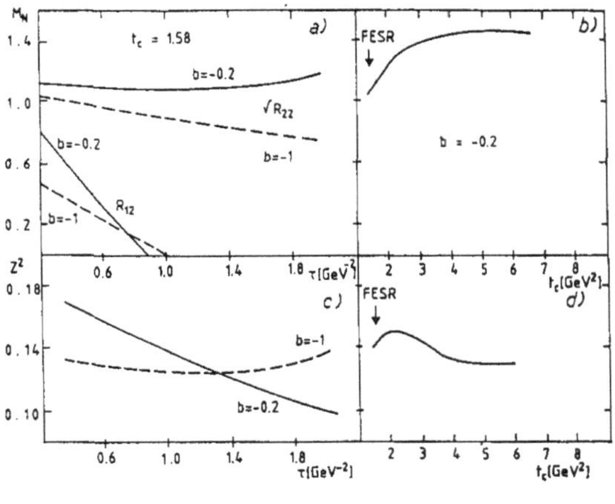

Fig. 11.4 : M_N and Z_N versus τ and t_c for different values of b. The arrow indicates the value of t_c from FESR.

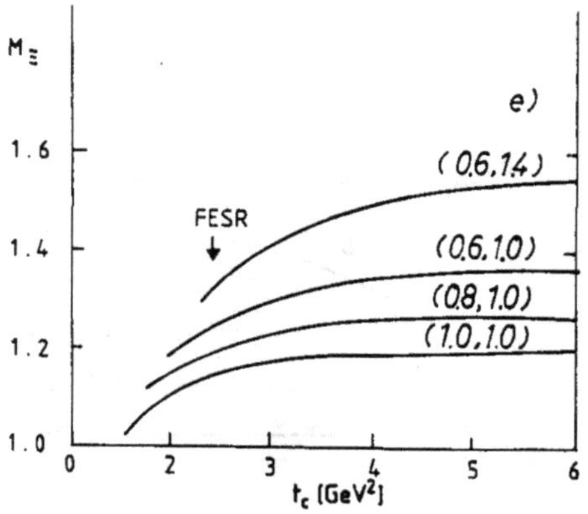

Fig. 11.5 : M_Ξ versus t_c for b = -1/5 and for different values of (χ_3, χ_8).

2. THE OCTET

Here the analysis is much more involved than for the decuplet, since the nucleon is interpolated by two lowest dimension operators[181]:

$$N = \frac{1}{\sqrt{2}} : \{(\psi\, C\lambda_5 \psi)\psi + b(\psi C\psi)\ \lambda_5 \psi\} \qquad (11.16)$$

where b is an arbitrary mixing parameter. In principle, the nucleon parameters should be independent of b. However, for truncated series such as in Eqs (11.4) and (11.5), one expects a minimal sensitivity if the leading term is stationary in b. This requirement applied to F_1 and F_2 gives in the chiral limit ($m_u = m_d = 0$) :

$$\frac{dA_1}{db} = \frac{dB_2}{db} = 0 \qquad (11.17a)$$

which corresponds to :

$$b = -\frac{1}{5} \ . \qquad (11.17b)$$

This value is that found by CDKS[181] after a painful numerical procedure. Though we are inclined to use this value, we also consider other choices of b for a safer estimate. In particular, we move b up to the value near -1, the "a priori" choice favoured by Ioffe[103].

As in the case of the decuplet, $R_{11}(\tau)$ and $R_{12}(\tau)$ have no τ-stabilities for any values of b. We use the same procedure as for the decuplet, i.e. we look for a first stage of results corresponding to the range of τ- and t_c-stabilities in the mass estimate. We insert these first results into the FESR constraints from which we deduce the final results. This analysis is shown in Figs 11.4 and 11.5. For $b = -\frac{1}{5}$ the mass estimate is summarized in Table 11.3 :

Table 11.3 : Estimates of the octet parameters

Names	Mas [GeV]	$\|Z\|^2$ [GeV]^6	t_c [GeV^2]
N	1.05	0.14	1.58
Σ	1.16	0.27	2.09
Λ	1.24	0.23	2.15
Ξ	1.33	0.31	2.42

Better agreement with the data is reached for a smaller value of $b \to -0.7$ but we have no good arguments to explain why this choice of current should be better than the one in Eq. (11.17). Here again, we have noticed that the size of the nucleon mass increases with M_o^2. A larger value of M_o^2 makes the nucleon too heavy and a smaller one makes the Δ too light. This fact indicates that the value of $M_o^2 \simeq 0.8$ GeV² from the heavy-light quark systems is a good compromise value. This value has also been previously obtained by Ioffe from baryon systems, though less accurate, while the too low value of M_o^2 found in Ref. 181) is due to a numerical error in the analysis.

3. WHAT HAVE WE LEARNED ?

The uses of QSSR for the baryon systems indicate that :

- the intuitive expectation of the baryon mass dependence on the chiral condensate value is demonstrated ;
- the approach has the tendency to overestimate the nucleon mass and underestimate that of Δ, the splitting being strongly controlled by the size of the mixed condensate $\langle g\bar\psi\sigma^{\mu\nu}\lambda_a\psi G^a_{\mu\nu}\rangle$;
- the correct SU(3) splittings among the same multiplets suggest *a large breaking of the chiral* $\langle\bar s s\rangle$ and $\langle g\bar s\sigma^{\mu\nu}\lambda_a s G^a_{\mu\nu}\rangle$ condensates, where the breaking is likely to act in opposite directions. A further test of the $SU(3)_F$ breaking of the mixed condensate is necessary to prove or disprove the previous result.

4. QSSR AT LARGE N_c AND THE SKYRME MODEL

The large N_c-behaviour of the QSSR in the baryon sector has been investigated for N_c flavour in Ref. 184) and for two flavours in Ref. 69). The latter case is much more complicated than the former owing to the large number of combinatorial possibilities in the evaluation of the two-point correlator. In our approach[69] we define the nucleon interpolating operator as the one with lowest spin and isospin I=J=1/2, while the Δ_c would correspond to the highest possible one $I = J = N_c/2$. Therefore we have successively the operators :

$$\mathcal{O}^{N}_{\alpha} = : u^{A_1}_{\alpha} \left(u^{A_2}\, C\, \gamma_5 d^{A_3}\right) \ldots \left(u^{A_{N_c-1}}\, C\, \gamma_5\, d^{A_{N_c}}\right) : \epsilon_{A_1 \ldots A_{N_c}}$$

$$\mathcal{O}^{\Delta\, \mu_1 \ldots \mu_\nu}_{\alpha} = : u^{A_1}_{\alpha} \left(u^{A_2} C\, \gamma^{\mu_1} u^{A_3}\right) \ldots \left(u^{A_{N_c-1}}\, C\, \gamma^{\mu_\nu}\, u^{A_{N_c}}\right) : \epsilon_{A_1 \ldots A_{N_c}} \qquad (11.18)$$

where $\nu \equiv \frac{N_c - 1}{2}$ with odd values of N_c ; u,d are the up and down massless quark fields ; A_j is the colour index and C denotes the charge conjugation. For $N_c = 3$, the nucleon current is close to that giving b-stability while the Δ_c current is that in Eq. (11.1). The associated two-point correlator can still be described by the two invariants F_1 and F_2. Owing to the complexity of the technical calculation, we shall limit ourselves to the evaluation of the leading term, as in any case we shall confine ourselves here to a qualitative discussion. The typical form of the two-point correlator is given in Fig. 11.6. It is most easily evaluated in position space. Using simple γ-matrix algebra, one obtains the following convenient form for the perturbation part of the two point function for the N :

$$F^{N}_{\alpha\bar{\alpha}} = \sum_{\ell_1 \ldots \ell_n} \hat{S}^{\ell_1} (\mathrm{Tr}\, \hat{S}^{\ell_2}) \ldots (\mathrm{Tr}\, \hat{S}^{\ell_n}) = \sum_{\lambda=0} \hat{S}^{2\lambda+1}\, A^{N_c}_{2\lambda+1}\, \hat{S}^{N_c - 2\lambda - 1} \qquad (11.19a)$$

where $\hat{S}$ is the propagator of a massless quark in position space.
The leading non-perturbative term is proportional to the quark condensate $\langle \bar{\psi}\psi \rangle = \sum_{A=1}^{N_c} \langle \bar{\psi}^{A}\, \psi_{A} \rangle$ which contributes to F^{N}_{2}.

The evaluation of the two-point function in this two flavours case is more complex than in the case of an infinite number of flavours considered in Ref.18) owing to the large number of combinatorial possibilities (see Fig. 11.6).

From Eq. (11.19a) one obtains immediately

$$F_1(x^2) = \sum_{\lambda=1}^{\nu} A^{N_c}_{2\lambda+1} \; (\hat{S})^{N_c - 1} \quad , \tag{11.19b}$$

and from the number of possibilities to cut fermion lines, in order to form condensates, one deduces :

$$F_2(x^2) = \sum_{\lambda=1}^{\nu} (2\lambda+1) \; A^{N_c}_{2\lambda+1} \; (\hat{S})^{N_c - 1} \; \frac{\langle \bar{\psi}\psi \rangle}{N_c} \; . \tag{11.19c}$$

The coefficients $A_{2\lambda+1}$ are determined by the decomposition of permutations into cycles. The product $\hat{S}^{\ell_1}$ in Eq. (11.19a) is due to a cycle of ℓ_1 elements containing the first one. The trace Tr $\hat{S}^{\ell_2}$ is due to a cycle of ℓ_2 elements *not containing the first one. From combinatorics,* we obtain the following recurrence relations for the coefficients A :

$$A^{N_c+2}_{2\lambda+1} = \lambda \; A^{N_c}_{2\lambda-1} \; + \left(4-\lambda + \frac{N_c - 1}{2}\right) A^{N_c}_{2\lambda+1}$$

$$\text{with} \quad A^1_1 = 1 \; , \; A^{N_c}_{-1} = A^{N_c}_{N_c+2} = 0 \; . \tag{11.19d}$$

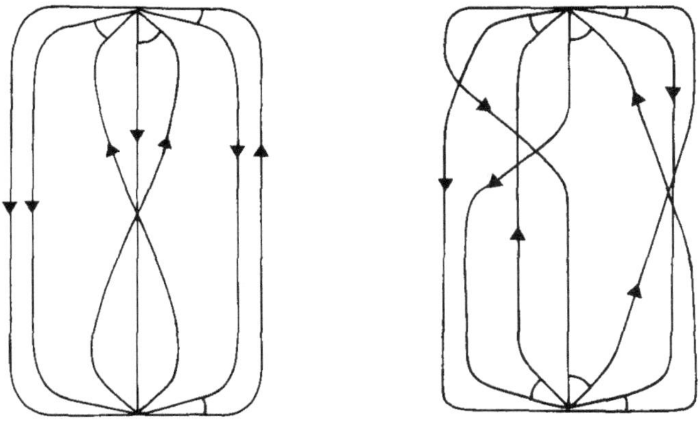

Fig. 11.6 : Some typical graphs of the baryon two-point function for $N_c = 7$. The connection between outgoing and incoming lines corresponds to $C\gamma_5$ or $C\gamma_\mu$. The first graph corresponds to a term proportional to $\hat{S}$; the second one to $\hat{S}^7$ in Eq. (11.19a).

For the Δ_c, we use a similar approach and consider the terms proportional to the Lorentz factor :

$$\sum_{\sigma\in S_\nu} g^{\mu_1\mu_{\sigma(1)}} \dots g^{\mu_\nu\mu_{\sigma(\nu)}} \quad . \tag{11.20a}$$

Then, we obtain :

$$F_1^\Delta = \sum_{\lambda=1}^{\nu} B_{2\lambda+1}^{N_c} \; (\hat{S})^{N_c-1}$$

$$F_2^{\Delta} = \sum_{\lambda=1}^{\nu} (2\lambda+1)\, B_{2\lambda+1}^{N_c}\, (\hat{S})^{N_c-1} \frac{\langle \bar{\Phi}\phi \rangle}{N_c} \tag{11.20b}$$

with :

$$B_{2\lambda+1}^{N_c} = \left[\frac{(N_c - 1)\,!!}{(N_c - 2\lambda - 1)\,!!}\right]^2 T_{N_c-2-2\lambda}$$

$$T_n = 4n \sum_{\lambda=0}^{n/2} \left[\frac{(n-2)\,!!}{(n-2-2\lambda)\,!!}\right]^2 T_{N_c-2-2\lambda} \tag{11.20c}$$

$$T_o = 1 \qquad \ell\,!! = \begin{cases} 1.3.5\ldots\,\ell & \ell \text{ odd} \\ 2.4\ldots\,\ell & \ell \text{ even} \end{cases} .$$

As, we limit ourselves here to leading order effects, the best quantity which we can work with is F_2^B/F_1^B despite the weak points of this sum rule as discussed previously. However, we shall only take advantage of it to give a good description of the spectrum in the lowest ground state dominance assumption.

From the above results, we obtain by induction :

$$F_2^B/F_1^B = C_B \frac{\langle \bar{\Phi}\,\phi \rangle}{N_c} \tag{11.21a}$$

with :

$$C_N = \frac{N_c+4}{5} \qquad C_{\Delta_c} = \frac{N_c+2}{3} . \tag{11.21b}$$

Taking the Fourier transform of the correlator and using the Laplace transform, we obtain the sum rule :

$$M_B \simeq \frac{F_2^B}{F_1^B} \simeq 4\pi^2 \frac{(3N_c-1)(3N_c-3)}{4} \left| \frac{\langle \bar\psi \psi \rangle}{N_c} \right| \tau C_B \quad . \tag{11.22}$$

We can exploit the above sum rule in two ways :
Assuming that the sum rule scale τ behaves like M_B^{-2}, one obtains the mass value :

$$M_B \simeq \left\{4\pi^2 \frac{(3N_c-1)(3N_c-3)}{4} C_B\right\}^{1/3} \left|\frac{\langle \bar\psi \psi\rangle}{N_c}\right|^{1/3} \quad , \tag{11.23}$$

leading to the mass ratio :

$$M_{\Delta_c}/M_N \simeq \left(\frac{5}{3}\right)^{1/3} \quad . \tag{11.24}$$

We compare it with the expression given by the Skyrme model at large N_c [64] :

$$M_N \simeq \frac{\mu f_\pi \sqrt{N_c}}{e\sqrt{3}} + \frac{3\, f_\pi \sqrt{N_c}}{8L\, e^3 \sqrt{3}}$$

$$M_{\Delta_c} \simeq \frac{\mu\, f_\pi \sqrt{N_c}}{e\sqrt{3}} + \frac{N_c\,(N_c+2)\, f_\pi \sqrt{N_c}}{8\, L\, e^3 \sqrt{3}} \tag{11.25}$$

where f_π is the pion decay constant equal to 93.3 MeV for $N_c = 3$. We have used the fact that the decay constant increases like $\sqrt{N_c}$. The numbers $L = 51$ and $\mu = 73$ come from radial integrals over the soliton distribution. By evaluating the quantity :

$$\frac{M_{\Delta_c}}{M_N} - 1 \qquad (11.26)$$

within the two approaches, one obtains a value of the skyrme parameter :

$$e \simeq \frac{12}{\sqrt{N_c}} \quad . \qquad (11.27)$$

One can also assume in Eq. (11.22) that the sum-rule scale τ is the same for the N and Δ_c. This leads to :

$$e \simeq \frac{9}{\sqrt{N_c}} \quad . \qquad (11.28)$$

Our result is consistent with the expectations that the Skyrme parameter decreases as $1/\sqrt{N_c}$ for large N_c. The extrapolation of our result to the real $N_c = 3$ case gives :

$$e \simeq (3.5 - 4.7) \qquad (11.29)$$

compared with the fitted value[64] of 5.45. Bearing in mind the crude approximation which we have used to get this number in Eq. (11.29) one can consider that the agreement between the two approaches is fantastic. One can do[69] much better work for $N_c = 3$, by using the $N_c = 3$ version of the previous sum rules where the non-leading condensate effects are known. Therefore from the ratio F_2/F_1, one can deduce (see previous paragraphs)

$$M_N \simeq - 24\pi^2 \langle \bar{u} u \rangle Z_N \tau_N \qquad (11.30)$$

$$M_\Delta \simeq -\frac{80\ \pi^2}{3}\left\langle \bar{u}\ u\right\rangle Z_\Delta\ \tau_\Delta \qquad (11.31)$$

with :

$$Z_N \simeq 1 - \frac{2}{5} M_o^2\ \tau_N - \frac{23}{30}\ \pi\left\langle \alpha_s\ G^2\right\rangle \tau_N^2 - 64\ \pi^4 \left\langle \bar{u}\ u\right\rangle^2 \tau_N^3$$

$$Z_\Delta \simeq 1 + \frac{M_o^2}{z}\ \tau_\Delta + \frac{11}{9}\ \pi\left\langle \alpha_s\ G^2\right\rangle \tau_\Delta^2 - \frac{320}{3}\ \pi^4 \left\langle \bar{u}\ u\right\rangle^2 \tau_\Delta^3 \ . \qquad (11.32)$$

Assuming that $\tau_N \simeq \tau_\Delta \equiv \tau_o$ one can deduce the sum rule :

$$e^4 \simeq \frac{8\pi\mu L}{27}\left[\frac{1 + \frac{43}{5} M_o^2\ \tau_o + \frac{1721}{90}\ \pi\left\langle \alpha_s G^2\right\rangle \tau_o^2 - \frac{1472}{9}\ \pi^4\left\langle \bar{u}u\right\rangle^2 \tau_o^3}{1 - \frac{23}{55} M_o^2\ \tau_o - \frac{2523}{1890}\ \pi\left\langle \alpha_s G^2\right\rangle \tau_o^2 - \frac{1088}{24}\ \pi^4 \left\langle \bar{u}u\right\rangle^2 \tau_o^3}\right] \qquad (11.33)$$

which gives

$$e \simeq 5 - 7 \quad , \qquad (11.34)$$

in good agreement with the numerical fit[64]. Another way[69] of exploiting the $N_c = 3$ sum rule is to combine it with the pseudoscalar sum rule discussed in Chapter 7, which contributes via the sum of quark masses (m_u+m_d), the latter being related to the chiral condensate through the PCAC relation. The sum of quark masses reads :

$$(m_u+m_d) \simeq \frac{4\pi}{\sqrt{3}}\ m_\pi^2\ f_\pi\ \tau_\pi\ Z_\pi^{1/2} \qquad (11.35a)$$

where :

$$Z_\pi^{1/2} \simeq 1 - \frac{\pi}{6}\left\langle \alpha_s G^2\right\rangle \tau_\pi^2 - \frac{448\pi^3}{81}\ \alpha_s\left\langle \bar{u}\ u\right\rangle^2 \tau_\pi^3 \ . \qquad (11.35b)$$

Using the approximate relation :

$$\tau_\pi / \tau_N \simeq \left(\frac{324\pi}{35\ \alpha_s(\tau)} \right)^{1/3} \simeq 3 \tag{11.36}$$

at the optimization scale within our parametrization, we again deduce from Eq. (11.30) :

$$e \simeq 5 - 8 \ . \tag{11.37}$$

The agreement between the numerical values of e from two independent methods is interesting. In fact, the QSSR and the Skyrme model are two aspects of non-perturbative QCD. It should also be noticed that our determination of e is independent to a leading order of the chiral symmetry breaking parameters. This property is a non-trivial result.

CHAPTER 12

GLUONIA SUM RULES AND LOW-ENERGY THEOREMS

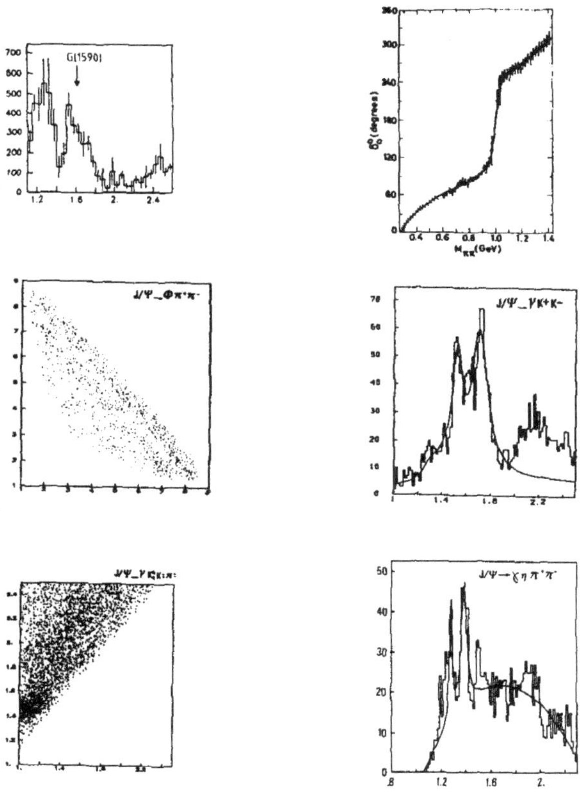

Typical events indicating the possible existence of gluonia

The existence of gluonia or glueballs as consequences of QCD was first advocated by Harald Fritzsch and Murray Gell-Mann[1]. Later, systematics of glueball properties were discussed[185]. Gluonia spectra were studied within the bag model[71-73], while calculations of gluonia masses in Lattice Gauge Theories (LGT) are progressing[53,186], although the most fundamental of the non-perturbative QCD approaches, LGT, is still affected (at present) by uncertainties arising from limited statistics, from finite size and from the neglect of dynamical fermions. The latter limitation is particularly troublesome for the kind of problems we wish to address here : the possible existence of relatively "pure" gluonium states. We are thus forced to have recourse to more phenomenological tools.

Of these, two have attracted considerable interest. The first one is the approach based on current algebra and effective Lagrangians. Predictivity can be further enhanced by combining current algebra with the non-perturbative expansions, such as the $1/N_c$ or the quark loop n_f expansions.

The second method is the one based on QCD spectral sum rules. QSSR has been originally applied to gluonia by NSVZ[187]. Their results are quite interesting but only at the qualitative level, as they are strongly based on the "direct" instanton contributions which are not under good control. Some other works[188,191] based on the standard OPE governed by the few lowest dimension condensates of QCD look much more attractive. In this chapter, we shall review the latter and some results from the so-called low-energy theorems based on current algebra.

1. CLASSIFICATION OF THE GLUONIA CURRENTS

We have given a classification of the gluonia current with the help of Landau-Yang's theorem. They have proved that a vector 1^- gluonium state cannot exist as two-gluon states. The lowest dimension operators associated with each gluonium state for $SU(n)_F$ are :

$$\hat{J}_+ = :\beta(\alpha_s)\, G_{\mu\nu}\, G^{\mu\nu}: \qquad 0^{++}$$
$$\hat{J}_{\mu\nu} = :-G_{\mu\alpha}\, G^{\alpha}_{\nu} + \frac{1}{4} g_{\mu\nu}\, G_{\alpha\beta}\, G^{\alpha\beta}: \qquad 2^{++}$$
$$\hat{J}_- = :\left(\frac{n\alpha_s}{4\pi}\right) G_{\mu\nu}\, \tilde{G}^{\mu\nu}: \qquad 0^{-+}$$
$$\hat{J}_3 = :g^3 f_{abc}\, G^a_{\mu\nu}\, G^b_{\nu\rho}\, G^c_{\rho\mu}: \qquad 0^{++}$$
$$\hat{J}^{\alpha}_{V} = :\mathrm{Tr}\, G_{\mu\nu}\left(D^{\alpha}\, G_{\nu\rho}\right) G_{\rho\mu}: \qquad 1^{--}$$
$$\cdot \quad \cdot \quad \cdot$$
$$\cdot \quad \cdot \quad \cdot$$
$$\cdot \quad \cdot \quad \cdot \qquad (12.1)$$

$\hat{J}_+$ is the current associated with the anomalous trace of the energy momentum tensor. $\hat{J}_-$ is that corresponding to the QCD $U(1)_A$ anomaly. The renormalization of the tensor current $\hat{J}_{\mu\nu}$, to which the anomalous dimension corresponds : $\gamma_T = \frac{n}{3}\left(\frac{\alpha_s}{\pi}\right)$ has been studied in Chapter 1. In this chapter we shall limit ourselves to studying the gluonia associated with the 0^{++}, 0^{-+}, and 2^{++} .

2. SCALAR GLUONIA

a) Low-energy theorems (LET) for the correlator

A classical LET is the one derived by Witten and Veneziano[84] in Eq. (2.17) which relates the η' mass to the θ derivative of the QCD vacuum energy and which indicates that the η' gets its mass through its gluon component.

A quite analogous result has been derived by NSVZ[187] for the scalar gluonium. It consists in the evaluation of the scalar gluonium two-point correlator at zero momentum transfer :

$$\psi(q^2) = 16\, i \int d^4 x\, e^{iqx} \left\langle 0 \left| T\, \theta^{\mu}_{\mu}(x) \left(\theta^{\nu}_{\nu}(0)\right)^{+} \right| 0 \right\rangle , \qquad (12.2)$$

where :

$$\theta^{\mu}_{\mu} = \frac{1}{4}\, \beta(\alpha_s)\, G^2 + (1 + \gamma_m(\alpha_s)) \sum_{u,d,s} m_i\, \bar{\psi}_i\, \psi_i \qquad (12.3)$$

is the anomalous trace of the energy momentum tensor. $\beta(\alpha_s)$ and $\gamma_m(\alpha_s)$ are respectively the usual β function and the mass anomalous dimension already defined in Chapter 1. NSVZ demonstrate that for a local operator $\mathcal{O}(x)$ built from light quarks and/or gluon fields and with a canonical dimension d, one has :

$$\psi_{\mathcal{O}}(0) = \lim_{q\to 0} i \int d^4 x\, e^{iqx} \left\langle 0 \left| \mathbb{T}\, \mathcal{O}(x), \frac{1}{\alpha_s}\, \theta^{\mu}_{\mu}(0) \right| 0 \right\rangle = - d\, \langle \mathcal{O} \rangle + \mathcal{O}(m_i) \Big\} \quad . \qquad (12.4)$$

The proof comes from the dimensional counting where one can write :

$$\langle \mathcal{O} \rangle \sim \left[-\nu \exp\left(\frac{-8\pi^2}{b g_B^2} \right) \right]^d \quad , \qquad (12.5a)$$

where ν is the ultra-violet subtraction scale, the only available scale in the chiral limit ($m_i \to 0$). g_B is the bare coupling. By rescaling the gluon-field strengths as :

$$\bar{G}_{\mu\nu} \equiv g_B\, G_{\mu\nu} \qquad (12.5b)$$

the g_B dependence in the Lagrangian reduces to $\left(-\frac{1}{4g_B^2}\right) \bar{G}_{\mu\nu} \bar{G}_{\mu\nu}$ and then :

$$\psi_{\mathcal{O}}(0) \equiv - \frac{d}{d\left(1/4g_B^2\right)} \langle \mathcal{O} \rangle \quad . \qquad (12.6)$$

One can iterate the result in Eq. (12.4) to get the LET for the vertex :

$$i^2 \int d^4 x \, d^4 y \left\langle 0 \left| \mathbb{T} \, \theta^\mu_\mu(x) \; \theta^\nu_\nu(y) \; \mathcal{O}(0) \right| 0 \right\rangle = (-d)^2 \langle \mathcal{O} \rangle \quad . \qquad (12.7)$$

Applying the previous result to the two-point correlator in Eq. (12.2), one can easily obtain the important result :

$$\psi(0) = - \frac{16}{\pi} \beta_1 \left\langle \alpha_s \, G^2 \right\rangle \qquad (12.8)$$

which will be useful later on.

b) The two sum rules*)

We shall successively discuss the two sum rules obeyed by $\psi(q^2)$. From its expected asymptotic behaviour we can write a twice-subtracted dispersion relation for ψ :

$$\psi(q^2) = \psi(0) + \frac{q^4}{\pi} \int_0^\infty \frac{dt \; \mathrm{Im}\, \psi(t)}{t^2(t - q^2 - i\epsilon)} + \psi'(0) \, q^2 \qquad (12.9)$$

Using standard techniques, one can derive two sum rules from (12.9)

*) The following discussion is mainly based on the works in Ref. 190).

and from its known QCD expression. The first[189], hereinafter referred to as the unsubtracted sum rule (USR), is independent of $\psi(0)$. After renormalization group improvement it reads :

$$\frac{1}{\pi}\int_0^\infty dt\; e^{-\tau t}\;\mathrm{Im}\,\psi(t) \simeq$$

$$\simeq \left(\frac{4}{\pi^2}\right)\beta_1^2\left(\frac{\bar\alpha_s}{\pi}\right)\left\{\left(\frac{\bar\alpha_s}{\pi}\right)+\frac{1}{2}\langle g^3G^3\rangle\tau^3+\frac{9\pi^2}{8}\left\langle\alpha_s G^2\right\rangle^2\tau^4\right\}\,, \qquad (12.10a)$$

where

$$\frac{\bar\alpha_s}{\pi}\equiv\left(\beta_1\log\sqrt{\tau}\,\Lambda\right)^{-1}\quad;\quad\beta_1=-\frac{9}{2}\quad(n_F=3)$$

$$\langle g^3G^3\rangle\equiv\left\langle g^3 f_{abc}\,G^a_{\mu\nu}\,G^b_{\nu\rho}\,G^c_{\rho\mu}\right\rangle\,. \qquad (12.10b)$$

As usual, the left-hand side of the sum rule is supposed to be saturated by low-energy resonances and, above a certain t_c, by a continuum identified with the discontinuity of the QCD diagrams.

The second, subtracted sum rule[187] (SSR) depends crucially on $\psi(0)$ and reads :

$$\frac{1}{\pi}\int_0^\infty\frac{dt}{t}\,e^{-\tau t}\;\mathrm{Im}\,\psi(t)=$$

$$=\frac{2}{\pi^2}\beta_1^2\left(\frac{\bar\alpha_s}{\pi}\right)+\psi(0)-\frac{4}{\pi}\beta_1^2\left(\frac{\bar\alpha_s}{\pi}\right)\left[\left\langle\alpha_s G^2\right\rangle+\frac{\tau}{2\pi}\langle g^3G^3\rangle\right]\,. \qquad (12.11)$$

We shall discuss the two sum rules in succession.

a) The USR (12.10) stabilizes at large values of τ ($\tau\simeq 0.8$ GeV^{-2})

and is consequently dominated by the low-energy spectrum of 0^{++} mesons. Saturating the USR with the σ contribution plus a QCD continuum above a threshold t_c, one obtains in a narrow width approximation :

$$2 f_\sigma^2 M_\sigma^4 e^{-\tau M_\sigma^2} + \frac{4}{\pi^2} \beta_1^2 \left(\frac{\alpha_s}{\pi}\right)^2 \tau^{-3} e^{-t_c \tau} \left(1 + t_c\tau + (t_c\tau)^2 / 2\right) \simeq$$

$$\simeq \frac{4}{\pi^2} \beta_1^2 \left(\frac{\alpha_s}{\pi}\right) \tau^{-3} \left\{\frac{\alpha_s}{\pi} + \frac{1}{2} \langle g^3 G^3 \rangle \tau^3 + \frac{9\pi^2}{8} \langle \alpha_s G^2 \rangle^2 \tau^4 \right\}, \qquad (12.12)$$

where fermionic condensates can be neglected. Using the standard condensate values

$$\langle \alpha_s G^2 \rangle \simeq 0.04 \text{ GeV}^4$$

and

$$\langle g^3 G^3 \rangle \simeq (1 \text{ GeV}^2) \langle \alpha_s G^2 \rangle , \qquad (12.13)$$

$\Lambda \simeq 0.18$ MeV and $M_\sigma \simeq 1$ GeV*) as input, we can study the stability of f_σ given by (12.12) with respect to τ (Fig. 12.1a). We find a reasonable minimum at $\tau \simeq 0.8$ GeV^{-2} for $\sqrt{t_c} \geqslant 2.1$ GeV. In Fig. 12.1b, we consider the effects of $\sqrt{t_c}$ at the optimal value of τ obtained in Fig. 12.1a for two values of Λ. The stability of f_σ versus the changes of t_c is reached for $\sqrt{t_c} \geqslant 2.9$ GeV. For $M_\sigma \simeq 1$ GeV and $\Lambda \simeq 100 \sim 180$ MeV, one thus obtains :

$$f_\sigma \sim (546 \div 677) \text{ MeV} \qquad (12.14a)$$

*) In the case where one has multistates with degenerate masses, one should interpret M_σ as an effective resonance.

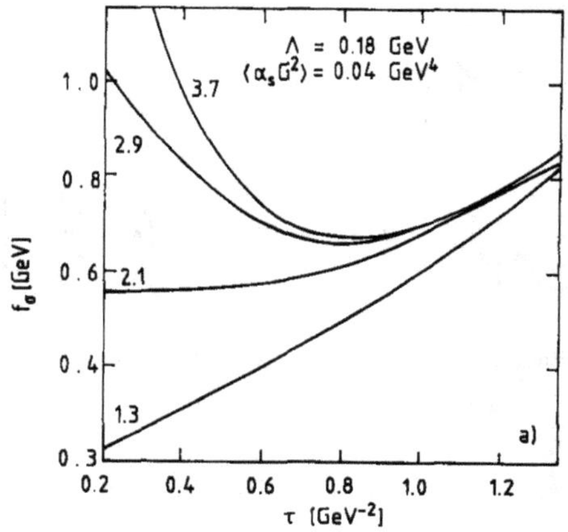

Fig. 12.1a : Behaviour of f_σ versus τ

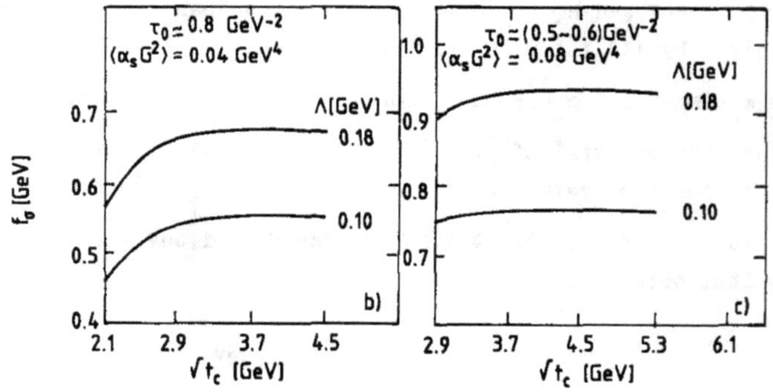

Fig. 12.1b : f_σ versus $\sqrt{t_c}$ for two values of $\langle \alpha_s\ G^2 \rangle$.

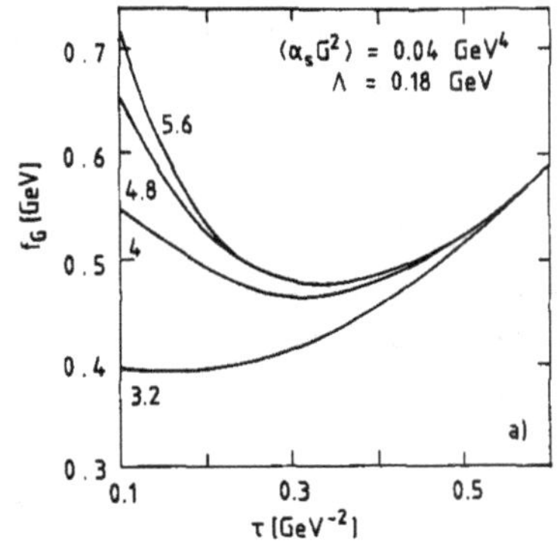

Fig. 12.2a : f_G versus τ

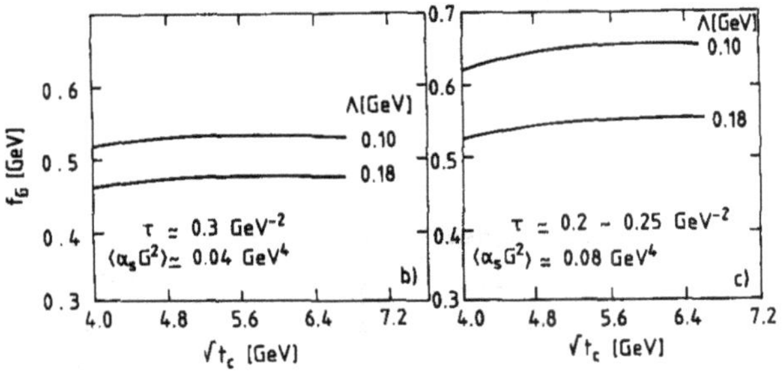

Fig. 12.2b : f_G versus $\sqrt{t_c}$ for two values of $\langle\alpha_s G^2\rangle$.

with the definition

$$\left\langle 0 \left| 4\ \theta^\mu_\mu \right| \sigma \right\rangle = \sqrt{2}\ f_\sigma\ M_\sigma^2 \quad . \tag{12.14b}$$

The result in (12.14) is consistent with that obtained from the ratio of moments analysis[189] where M_σ and f_σ have been left as free parameters. If, instead, one allowed a deviation by a factor two from the standard value of the condensates in (12.13) one would obtain :

$$f_\sigma \simeq (769 \div 931)\ \mathrm{MeV}, \qquad (12.14c)$$

as shown in Fig. 12.1 .

b) Turning now to the SSR, Eq.(12.11), let us first recall that the subtraction constant $\psi(0)$ is actually known in terms of the gluon condensate via the low-energy theorem in Eq. (12.8).
Thus, no new input parameter is needed and the SSR reads :

$$2 \sum_{i=\sigma,\ldots} f_i^2\, M_i^2\, e^{-\tau M_i^2} \simeq \frac{2}{\pi^2} \frac{\tau^{-2}}{\log^2 \sqrt{2}\,\Lambda} \left(1 - e^{-\tau t_c} (1+\tau t_c)\right) +$$

$$+ \psi(0) - \frac{4\beta_1^2}{\pi^2} \bar{\alpha}_s \left(\langle \alpha_s G^2 \rangle + \frac{\tau}{2\pi} \langle g^3 G^3 \rangle\right) \qquad (12.15)$$

A saturation of Eq. (12.15) by only the σ(1 GeV) gives results which are contradictory with those of the USR. We thus conclude that at least one more 0^{++} meson is needed and we shall test hereafter the hypothesis that this is precisely the recently observed G(1.6) state.

The sum rule (12.15) shows a stability region at a value of $\tau(\tau \sim 0.3\ \mathrm{GeV}^{-2})$ which is much smaller than that at which the USR stabilizes, owing to the important contribution of $\psi(0)$ relative to the unit operator. This is consistent with the idea that the G contributes appreciably only to the SSR. The dominance of the σ in the USR and the appreciable contribution of the G in the SSR might explain the sparsity of previous results obtained within a one-resonance + QCD continuum parametrization of the sum rules[187-191].

Saturating the SSR by the σ and the G, and using standard condensate values, one obtains (Fig. 12.2a,b) :

$$f_G \simeq (478 \div 533)\ \mathrm{MeV} \quad , \qquad (12.16a)$$

whilst taking twice the standard values :

$$f_G \simeq (554 \div 656) \text{ MeV} \quad . \qquad (12.16b)$$

One can double-check at this point that the G contribution to the SSR is about the same as the σ contribution while it is negligible in the USR. Adding also the $f_0(1.3)$ contribution, we deduce by matching the continuum of the USR with the f_0 peak :

$$f_{f_0} \lesssim (139 \sim 224) \text{ MeV} \qquad (12.16c)$$

which only slightly affects the result in (12.16).

c) *σ coupling to pairs of Goldstone bosons*

At this point we use vertex sum rules to obtain further constraints and/or predictions. Consider the vertex :

$$V(q^2) = \left\langle \pi_1 \left| \theta^\mu_\mu \right| \pi_2 \right\rangle \quad , \quad q = p_1 - p_2 \qquad (12.17a)$$

where

$$V(0) = O\left(m^2_\pi\right) \longrightarrow 0 \quad . \qquad (12.17b)$$

In the chiral limit $\left(m^2_\pi \simeq 0\right)$, we have :

$$V(q^2) = q^2 \int_{4m^2_\pi}^{\infty} \frac{dt}{t - q^2 - i\epsilon} \frac{1}{\pi} \text{Im } V(t)/t \quad . \qquad (12.17c)$$

Using the fact that $V'(0) = 1$, one obtains :

$$\frac{1}{4} \sum_{i=\sigma, G} g_{i\pi\pi} \sqrt{2}\, f_i \Big/ M_i^2 = 1 \qquad (12.18)$$

Since the G coupling to ππ is bound from above by the GAMS data[192,193], one finds that (12.18) is dominated by the σ, with the result that :

$$g_{\sigma\pi^+\pi^-} \simeq (4.6 \pm 0.5)\ \text{GeV} \qquad (12.19a)$$

for the standard values of the condensates, while :

$$g_{\sigma\pi^+\pi^-} \simeq (3.3 \pm 0.3)\ \text{GeV} \qquad (12.19b)$$

if the condensates are twice the standard values. The inclusion of the $f_0(1.3)$ contribution with the sign required by $V(0) = 0$ slightly increases the result in (12.19) by an amount less than 16%. Considering this effect as another source of error, we deduce for the standard (ST) and non-standard (NST) values of the condensates :

$$\Gamma(\sigma \to \pi^+\pi^- + 2\pi^0) \simeq \begin{matrix} \nearrow(798 \pm 429)\ \text{MeV} \\ \searrow(410 \pm 220)\ \text{MeV} \end{matrix} \cdot \left(\frac{M\sigma}{1\ \text{GeV}}\right)^3 \begin{matrix} :\ \text{ST} \\ :\text{NST.} \end{matrix} \qquad (12.20)$$

We have repeated the derivation of f_σ in (12.14) by taking into account finite width corrections. This leads to an increase of f_σ which is compensated for by the propagator effects in the estimate of $g_{\sigma\pi\pi}$, i.e., the result in (12.19) remains almost unchanged.

It is interesting to compare this prediction with the known ππ scattering [194] and J/ψ data [195]. Indeed the σ(900) with a width of about 700 GeV and the $f_0(975)$ with a width of 40 MeV are good gluonia candidates. Both are revealed from a coupled channel analysis of the I=0 s-wave ππ and $K\bar{K}$ final states and are coupled almost universally to ππ and $K\bar{K}$ pairs. In addition, the latter is produced in $J/\psi \to \phi\pi\pi$, $\phi K\bar{K}$, ωππ and γππ. Our result in (12.20) supports the presence of gluons inside the $f_0(975)$ and σ(900) wave functions. Their relative amounts might only be fixed after a complete mixing scheme analysis, as we shall see later on.

d) Current algebra constraints

In order to compute couplings of G to η and η', we use the approach of VW[54]. Consider the three-point function :

$$\tilde{V}_{\mu\nu}(q_1,q_2) \equiv \int d^4x_1 d^4x_2 \; e^{i(q_1x_1+q_2x_2)} \langle T[Q(x_1)Q(x_2)\theta_{\mu\nu}(0)]\rangle , \qquad (12.21)$$

where $\theta_{\mu\nu}$ is the energy-momentum tensor of QCD with three light quarks and

$$Q(x) = \frac{\alpha_s}{16\pi} \epsilon_{\mu\nu\rho\sigma} G^a_{\mu\nu} G^a_{\rho\sigma} \qquad (12.22)$$

is the topological charge density. We recall that from the large N_c (or the quenched) solution to the U(1) problem VW find :

$$\Gamma_2(q) \equiv i \int d^4x \, e^{iqx} \langle T[Q(x)\, Q(0)]\rangle \xrightarrow[\text{NO QUARK LOOPS}]{} \Gamma_2^{YM}(q)$$

$$\Gamma_2^{YM}(q) \underset{q \ll \Lambda}{\sim} \Gamma_2^{YM}(0) \simeq (180 \text{ MeV})^4 \quad . \qquad (12.23a)$$

Including the quark loops present[196], $\Gamma_2(q)$ reads :

$$\Gamma_2(q) \simeq \Gamma_2^{YM} \left(1 - \sum_{i=1}^{3} \frac{f_i(m_i)}{p^2 - m_i^2}\right) , \qquad (12.23b)$$

where m_i are the physical pseudoscalar nonet masses and the known (but complicated) constants f_i ensure the vanishing of $\Gamma_2(0)$ whenever a quark mass goes to zero.

A consistent formula for $\tilde{V}_{\mu\nu}$, which satisfies Ward identities as well as large N_c, and small quark-mass limits, turns out to be :

$$\tilde{V}_{\mu\nu}(q_1,q_2) = \Gamma_2^{YM}\left[\frac{1}{2}\, g_{\mu\nu}\, (q_1+q_2)^2\ )\ q_{1\mu}\, q_{2\nu} - q_{2\mu}\, q_{1\nu}\right] .$$

$$.\ \sum_i \frac{f_i\ (m_i)}{\left(q_1^2 - m_i^2\right)\left(q_2^2 - m_i^2\right)} + \frac{1}{2}\, g_{\mu\nu}\, (\Gamma_2(q_1) + \Gamma_2(q_2))\ . \qquad (12.24)$$

From this formula it is easy to argue that

$$\left\langle \eta_1 \left| \theta^\mu_\mu \right|_{n_f=3} |\eta_1\right\rangle \simeq \frac{9}{11}\left\langle \eta_1 \left| \theta^\mu_\mu \right|_{Y.M.} |\eta_1\right\rangle = \frac{9.12}{11}\, f_\pi^{-2}\Gamma_2^{YM} \simeq 1.15\ \text{GeV}^2 . \quad (12.25)$$

It should be noted that we have taken seriously the factor 9/11 between $\beta_1^{n_f=3}$ and $\beta_1^{n_f=0}$, although this is a higher order effect (both in $1/N_c$ and in n_f).

Saturating a dispersion relation in $q^2 = (q_1+q_2)^2$ with σ, f_o (1.3) and G yields :

$$\frac{1}{4} \sum_{i=\sigma, f_o, G} g_{i\eta_1\eta_1} \sqrt{2} f_i = 1.15\ \text{GeV}^2 \quad . \qquad (12.26)$$

Picking up the singlet component in the physical η, η' via $\eta \sim \sin\theta_p \eta_1 + \ldots$, $\eta' \sim \cos\theta_p \eta_1 + \ldots (\theta_p \simeq = (18\pm2)^\circ)$ being the pseudoscalar mixing angle[197] estimated from $\eta, \eta' \to 2\gamma)$, we find :

$$\sum_{i=\sigma, f_o, G} g_{i\eta\eta'}\, f_i = \frac{8}{\sqrt{2}} M^2_{\eta_1} \sin\theta_p \longrightarrow \frac{4}{\sqrt{2}} (1.15) \sin\theta_p\ \text{GeV}^2 \quad . \ (12.27)$$

The $\sigma\eta\eta'$ and $f_o\eta'\eta$ couplings are limited by the observed $\pi\pi$ spectrum in $\eta' \to \eta\pi\pi$ which is known to go mainly through $a_o\pi$. Assuming that the 25% allowed by the experimental errors is due either to σ of f_o exchanges, one deduces using a Gell-Mann-Sharp-Wagner-type model :

$$g_{\sigma\eta\eta'} \leq 0.75 \text{ GeV} \quad ,$$

$$g_{f_o\eta\eta'} \leq 2 \text{ GeV} \quad . \qquad (12.28)$$

This allows us to obtain for the previous two values of the gluon condensates :

$$g_{G\eta\eta'} \leq (3.6 \pm 0.3) \text{ GeV} \quad : \quad (ST)$$

$$g_{G\eta\eta'} \leq (3.3 \pm 0?3) \text{ GeV} \quad : \quad (NST) \quad , \qquad (12.29a)$$

which becomes in terms of the width :

$$\Gamma_{G\eta'\eta}(\text{MeV}) \leq 52 \pm 9 \quad : \quad (ST)$$

$$\Gamma_{G\eta'\eta}(\text{MeV}) \leq 44 \pm 8 \quad : \quad (NST) \quad . \qquad (12.29b)$$

The result in (12.29) is in good agreement with the GAMS data. The scheme is also known to predict :

$$r \equiv \Gamma_{G\eta\eta}/\Gamma_{G\eta\eta'} \simeq 0.26 \quad ; \quad g_{G\eta\eta} \simeq \sin\theta_p \; g_{G\eta\eta'} \qquad (12.30)$$

compared with the data[192,193] $r \simeq 0.34 \pm 0.13$ and with the one in Ref. 198). In other words, what we have found appears to be consistent with interpreting the G meson as an almost "pure" gluonium state. The coupling of G to pseudoscalar pairs could just reflect the amount of glue in each pseudoscalar and will be maximal for the η', minimal for the pion and intermediate for the η which has a small glue component. Within our analysis, one could also interpret the tendency of the G to decay copiously into $4\pi^0$ as due to exchange of virtual σ-like pairs. This could be checked from a reconstruction of the $2\pi^0$ invariant mass.

We are thus led to conclude that, while both the σ and the G mesons couple to glue, only the latter is, to a good approximation, a pure glue state, while the σ has a large $q\bar{q}$ admixture. This fact can be related to the improving validity of the OZI rule with increasing q^2.

e) Couplings to photons and to heavy quarkonia

Let us now study the $\gamma\gamma$ widths of these gluonium candidates and their productions in J/ψ and Υ radiative decays. As usual, we start from the Euler-Heisenberg Lagrangian which controls the low-energy behaviour of the $\gamma\gamma gg$ box diagrams in Fig. 12.3 :

$$\mathcal{L}_{\gamma g} = \frac{\alpha\, \alpha_s\, Q_H^2}{180\, M_H^4} \Big\{ 28\, F_{\mu\nu} F_{\nu\lambda} G_{\lambda\sigma} G_{\sigma\mu} + 14\, F_{\mu\nu} G_{\nu\lambda} F_{\lambda\sigma} G_{\sigma\mu}$$

$$- 10\, F_{\mu\nu} G_{\mu\nu} F_{\alpha\beta} G_{\alpha\beta} - 5\, F_{\mu\nu} F_{\mu\nu} G_{\alpha\beta} G_{\alpha\beta} \Big\} , \qquad (12.31)$$

where M_H is the mass of the heavy quark ($M_c \simeq 1.46$ GeV ; $M_b \simeq 4.6$ GeV for $\Lambda \simeq 0.1$-0.18 GeV). The $J/\psi \to \gamma X$ ($X \equiv \sigma, f_0, G$) process can be estimated using standard dispersion relation techniques[187] where a spectral function is saturated by the J/ψ plus a continuum. The glue part of the amplitude can be converted into a physical matrix element in terms of $\langle 0 | \alpha_s G^2 | X \rangle$ which is known from our previous analysis. If the continuum contribution is small, as argued in Ref. 187), we get :

$$\Gamma(J/\psi \to \gamma X) \simeq \frac{\alpha^3 \pi}{\beta_1^2\ 656100} \left(\frac{M_{J/\psi}}{M_H}\right)^4 \left(\frac{M_X}{M_H}\right)^4 \frac{f_X^2 \left(1 - M_X^2 / M_{J/\psi}^2\right)^3}{\Gamma(J/\psi \to e^+ e^-)} , \qquad (12.32a)$$

where we take $-\beta_1 = 7/2$ for six flavours. This leads to the rough estimates :

$$B(J/\psi \to \gamma\sigma) \ . \ B(\sigma \to \pi\pi) \simeq (6 \div 16)\ 1^{-4} \quad ,$$

$$B(J/\psi \to \gamma_{o})\ B(f_{o} \to \pi\pi) \simeq (0.7 \div 1.9)\ 10^{-4} \quad ,$$

$$B(J/\psi \to \gamma G)\ B(G \to \eta\eta' + 4\pi^{o}) \simeq (13 \div 25)\ 10^{-4} \ . \qquad (12.32b)$$

These branching ratios can be compared with the observed $B(J/\psi \to \eta'\psi)$ and $B(J/\psi \to f_2\gamma)$ ones which are 4.10^{-3} and $1.6.10^{-3}$. The extension of this analysis to the Υ is straightforward. One gets :

$$B(\Upsilon \to \gamma\sigma) \quad B(\sigma \to \pi\pi) \simeq (2 - 6)\ 10^{-6} \quad ,$$

$$B(\Upsilon \to \gamma\ f_{o})\quad B(f_{o} \to \pi\pi) \simeq (0.5 \sim 1)\ 10^{-6} \quad ,$$

$$B(\Upsilon \to \gamma G)\ B(G \to \eta\eta' + 4\pi^{o}) \simeq (9.5 - 17.5)\ 10^{-6} \ . \qquad (12.33)$$

The two-photon widths of these gluonia can be estimated from the identification of the scalar $\gamma\gamma$ Lagrangian :

$$\mathcal{L}_{\sigma\gamma\gamma} = g_{\sigma\gamma\gamma}\ \sigma(x)\ F^{(1)}_{\mu\nu}\ F^{(2)}_{\mu\nu} \qquad (12.34)$$

with the previous one in (12.31) where light quarks have to be added. This gives :

$$g_{\sigma\gamma\gamma} \simeq \frac{\alpha}{60}\sqrt{2}\ f_{\sigma}\ M^2_{\sigma}\left(\frac{\pi}{-\beta_1}\right)\sum_{i=u,d,s} Q_i^2\Big/m_i^4 \quad , \qquad (12.35a)$$

where Q_i is the quark charge and m_i the constituent light-quark masses which we take to be :

$$m_u \simeq m_d \simeq M_\rho\Big/2 \quad , \quad m_s \simeq M_\phi\Big/2 \ . \qquad (12.35b)$$

Therefore, we obtain

$$\Gamma(\sigma \to \gamma\gamma) \simeq (0.03 - 0.08) \text{ KeV} \quad ,$$

$$\Gamma(f_o \to \gamma\gamma) \simeq (0.01 \sim 0.03) \text{ KeV} \quad ,$$

$$\Gamma(G \to \gamma\gamma) \simeq (0.3 \sim 0.6) \text{ KeV} \quad . \qquad (12.36)$$

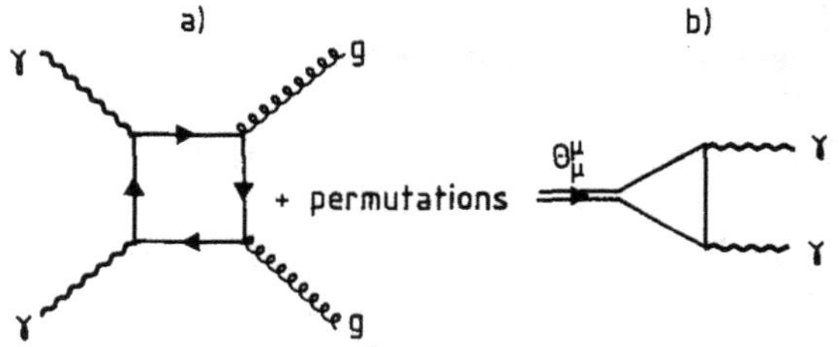

Figure 12.3 : Box and anomaly diagrams controlling the decay $G \to \gamma\gamma$.

Now let us check the approximate validity of the results in (12.36). Alternatively we can estimate the $\gamma\gamma$ width from the trace anomaly (Fig. 12.3b) :

$$\left\langle 0 \left| \theta^\mu_\mu \right| \gamma_1 \gamma_2 \right\rangle = \left\langle 0 \left| \frac{1}{4} \beta(\alpha_s) G^2 + \frac{\alpha R}{3\pi} F_1^{\mu\nu} F_2^{\mu\nu} \right| \gamma_1 \gamma_2 \right\rangle \qquad (12.37)$$

where $F^{\mu\nu}$ is the photon field strength and $R \equiv 3\Sigma Q_i^2$. The fact that the left-hand side of (12.37) is $O(k^4)$ whilst its right-hand side is $O(k^2)$ implies the sum rule[199,187] :

$$\left\langle 0 \left| \frac{1}{4}\, \beta(\alpha_s)\, G^2 \right| \gamma_1 \gamma_2 \right\rangle = - \left\langle 0 \left| \frac{\alpha R}{3\pi}\, F_1^{\mu\nu}\, F_2^{\mu\nu} \right| \gamma_1 \gamma_2 \right\rangle \qquad (12.38)$$

One can deduce from (12.38) the couplings :

$$\frac{\sqrt{2}}{4} \sum_{i=\sigma,G} f_i\, g_{i\,\gamma\gamma} \simeq \frac{\alpha R}{3\pi} \quad . \qquad (12.39)$$

There are two ways of exploiting (12.39). Using the σ and f_0 couplings from Eq. (12.35) we deduce from (12.39)

$$\Gamma_{G\to\gamma\gamma} \simeq (1\text{-}6) \text{ KeV} \quad , \qquad (12.40)$$

which looks too high compared to (12.36). This appreciable G contribution to the anomaly constraint in (12.39) might explain why the authors in Ref. 200) obtain an unexpectedly high $\sigma \to \gamma\gamma$ width by using a σ-dominance in (12.39) or might indicate that the anomaly approach in (12.39) is a very rough approximation for the estimate of the $\gamma\gamma$ gluonia width. However, we expect that the approximate use of the single-meson dominance in order to get Eqs (12.35) and (12.40) is less accurate for higher mass mesons like the G(1.6) owing to the proximity of the radial excitations in this energy region. Therefore, Eqs (12.35) and (12.40) might give an overestimate of the real value of the $\gamma\gamma$ widths.

The $\gamma\gamma$ widths in (12.36) can be compared with the well-known quarkonia widths : $\Gamma_{\eta\to\gamma\gamma} \simeq 0.56$ keV and $\Gamma_{f_2\to\gamma\gamma} \simeq 2.64$ keV.

The conclusions stemming from the above analysis of the couplings of 0^{++} states to heavy quarkonia + γ and $\gamma\gamma$ are as follows :

i) The absence of a G signal in $J/\psi \to \gamma\pi\pi$ is due mainly to its weak coupling to $\pi\pi$.

ii) One should stand a better chance of observing the G in the $\gamma 4\pi^0$ or $\gamma\eta(\eta')\eta$ decay modes of J/ψ.

iii) The fact that the f(975) has been seen in J/ψ decays and the $\sigma(0.9)$ has not may be due to the large width of the latter.

From the alternative way of detecting gluonium candidates from the study of inclusive ($J/\psi \to \gamma X$) or exclusive ($J/\psi \to \gamma\gamma\gamma$) γ spectra we conclude that :

iv) Effects of gluonia in the 1 GeV region are difficult to isolate, being swamped by the η' contribution (which is one to two orders of magnitude wider than σ in the $\gamma\gamma$ mode). Moreover, the absence of a $\gamma\gamma \to 2\pi^0$ signal is consistent with the small width found Eq (12.36).

v) A G(1.6) signal is also difficult to disentangle from so many other candidates in this energy region.

Finally, as far as $\gamma\gamma$ scattering data are concerned :

vi) The weak couplings of the σ and f_o to $\gamma\gamma$ could explain their absence there.

vii) $\gamma\gamma$ production of G(1.6) selecting $\eta'\eta$ or $4\pi^0$ final states can provide a good way of confirming its properties.

f) Conclusions

The G(1.6) meson passes all our QCD tests well and can be identified with a (relatively pure) gluonium state, within experimental and theoretical errors. Moreover, the existence of an object with the observed properties is welcome for fulfilling QCD sum rules in the 0^{++} channel. Of course, more experimental and theoretical work is needed in order to confirm the above interpretation of G(1.6). Experimentally, its isolation in inclusive γ spectra and in the $\gamma\eta\eta(\eta')$, $\gamma 4\pi^0$ channels of J/ψ and Υ decays, as well as its formation in $\gamma\gamma$ scattering, looks like the necessary complements to its confirmation in hadronic reactions. Theoretically, a mixing scheme for the σ, the f_0, the G and other 0^{++} states explaining simultaneously the rich body of data for all mesons in this channel should be worked out.

3. SCALAR MESON-GLUONIUM MIXING FOR THE $f_0(975)$*)

Our previous discussions can be used to study the nature of the $f_0(975)$ seen in $\psi \to \phi\pi\pi$, $\phi K\bar{K}$ experiments[195] having the widths [150,202]:

$$\Gamma(f_0 \to \pi\pi) \simeq 26 \text{ MeV} \quad ,$$

$$\Gamma(f_0 \to \gamma\gamma) \simeq \begin{matrix} (0.24 \pm 0.06 \pm 0.15) \text{ keV} \quad , \\ (0.31 \pm 0.14 \pm 0.11) \text{ keV} \quad . \end{matrix} \tag{12.41}$$

For this purpose, we use the minimal two-component mixing scheme :

$$|f_0\rangle = -\sin\theta\,|\sigma_B\rangle + \cos\theta\,|S_2\rangle \quad ,$$

$$|\sigma\rangle = \cos\theta\,|\sigma_B\rangle + \sin\theta\,|S_2\rangle \quad , \tag{12.42}$$

where σ_B is the gluonium discussed previously while $|S_2\rangle \equiv \frac{1}{\sqrt{2}}\left(\bar{u}u+\bar{d}d\right)$ is the SU(2) singlet scalar meson. The hypothetical decay width of the $|S_2\rangle$ can be estimated from the well controlled width ratio :

$$\Gamma(S_2 \to \gamma\gamma) \simeq \frac{25}{9}\,\Gamma(a_0 \to \gamma\gamma) \simeq 0.67 \text{ keV} \quad ,$$

$$\Gamma(S_2 \to \pi\pi) \simeq \frac{3}{2}\left(\sqrt{\frac{2}{3}}\right)^2 \frac{\vec{P}_\pi}{P_{\vec{\eta}}}\,\Gamma(a_0 \to \eta\pi) \simeq 180 \text{ MeV} \quad , \tag{12.43}$$

deduced from the quark content and/or vertex sum rules[146], where the

*) Based on the work in Ref. 201).

$|a_o\rangle \equiv \frac{1}{\sqrt{2}}\left(uu - \bar{d}d\right)$ is the isovector partner of the $|S_2\rangle$ with the observed width*):

$$\Gamma(a_o \to \gamma\gamma)\ .\ B(a_o \to \eta\pi) \simeq \begin{array}{l}\left(0.19 \pm 0.07 \pm {0.10 \atop 0.07}\right) \text{keV} \\ (0.29 \pm 0.05 \pm 0.14)\ \text{keV}\end{array}$$

$$\Gamma(a_o \to \eta\pi) \simeq (57 \pm 7)\ \text{MeV} \qquad B(a_o \to \eta\pi) \simeq 1. \qquad (12.44)$$

We fix the mixing angle θ using Eqs (12.41) to (12.43) and the predicted small $\gamma\gamma$ width of the σ_B (Eq 12.36), from which we deduce :

$$|\theta| \simeq 45^\circ. \qquad (12.45)$$

Using this value within a destructive interference, the hadronic width of the σ_B in Eq. (12.20) and the one of S_2 in Eq (12.43), one deduces :

$$\Gamma(f_o \to \pi\pi) \simeq (22 - 85)\ \text{MeV} \qquad (12.46)$$

in perfect agreement with the data. It is encouraging that this "naive" mixing scheme is able to save the weakness of previous $\bar{q}q$ model predictions for the f_o.

4. THE SCALAR TRIGLUONIUM SUM RULE[203]

a) Mass and coupling

Let us study the gluonium associated with the $g^3 f_{abc} G^a G^b G^c$ current. The QCD expression of the two-point correlator is[203]:

$$\psi_3 = -\alpha_s^2 \left\{\frac{3}{10\pi}\alpha_s . (q^2)^4 \log\frac{-q^2}{\nu^2} + 18\pi\ q^4\left\langle\alpha_s G^2\right\rangle\right.$$

*) We note that an estimate[201] of these widths within a $\bar{q}q$ scheme for the a_0 agrees very well with the data.

$$- \frac{27}{2} \left(q^2 \log \frac{-q^2}{\nu^2} \right) \left\langle g^3 f_{abc} G^a G^b G^c \right\rangle$$

$$\left. + \alpha_s \pi^3 \, 36.64 \, (\phi_7 - \phi_5) \right\} ,$$

with :

$$\phi_5 = \frac{1}{16} \mathrm{Tr} \left\langle G_{\nu\mu} G^{\mu\rho} G_{\rho\tau} G^{\tau\nu} \right\rangle ,$$

$$\phi_7 = \frac{1}{16} \mathrm{Tr} \left\langle G_{\nu\mu} G^{\nu\rho} G^{\mu\rho} G_{\rho\tau} \right\rangle . \qquad (12.47)$$

We introduce the resonance effect to the sum rule via :

$$\left\langle 0 \left| \hat{J}_3 \right| G \right\rangle = \sqrt{2} \, M_3^4 \, f_3 \qquad (12.48)$$

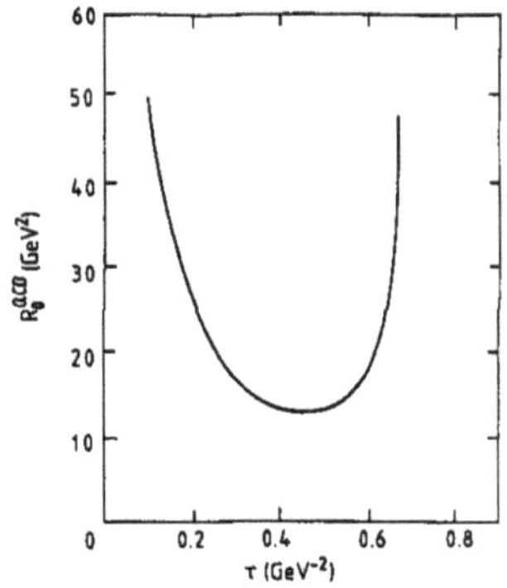

Fig. 12.4 : Behaviour of the QCD moments $R_o(\tau)$ versus the sum-rule scale τ for $g^3 f_{abc}\langle G^3 \rangle = 0.04$ GeV^6.

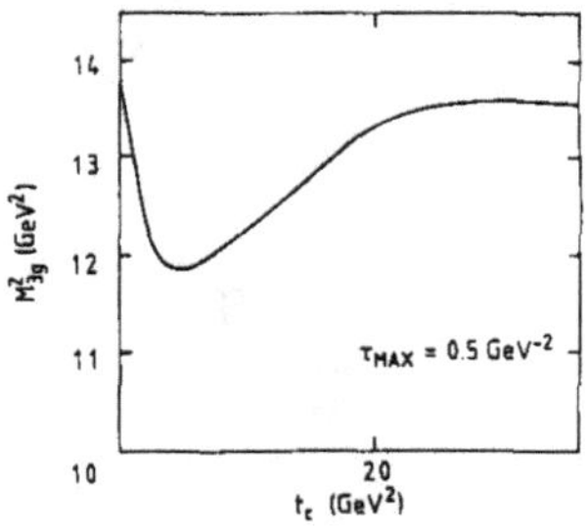

Fig. 12.5 : Behaviour of the trigluonium mass versus t_c for a given value of the fit interval $[0, \tau_{MAX}]$.

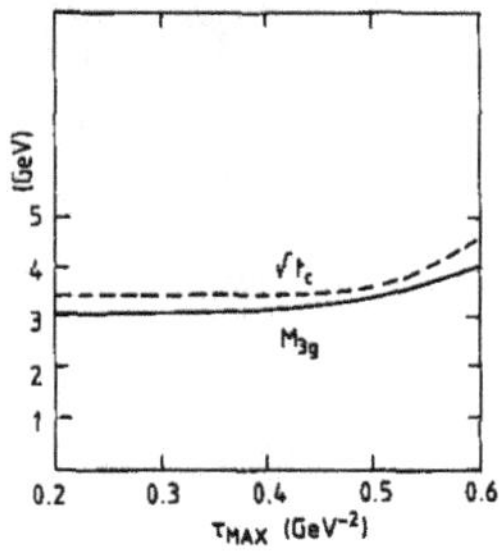

Fig. 12.6 : Behaviour of the set $\left(M_3, \sqrt{t_c}\right)$ of the optimal values versus the changes of τ_{MAX}.

and we study the trigluonium mass with the moment ratio :

$$R_0(\tau) = -\frac{d}{d\tau} \log \int_0^\infty dt\, e^{-t\tau} \frac{1}{\pi} \text{Im}\, \psi_3(t) \tag{12.49}$$

or its variant in Eq. (1.41). $R_0(\tau)$ is free of the unknown subtraction constant $\psi_3(0)$ which disappears in the sum-rule procedure. We show in Fig. 12.4 the τ-behaviour of $R_0(\tau)$ where one can extract an upper bound on the resonance mass at the τ minima :

$$M_3 \leq 3.7 \text{ GeV} , \qquad (12.50)$$

for the standard values of the condensates. We use the FESR constraint :

$$2 f_3^2 M_3^8 \simeq \frac{3}{10} \frac{\alpha_s^3}{\pi} \frac{t_c^5}{5} + \frac{27}{2} \alpha_s^2 \left\langle g^3 f_{abc} G^a G^b G^c \right\rangle \qquad (12.51)$$

in order to eliminate f_3 in the fitting procedure which corresponds to a least square two-parameter fit of the two sides of the sum rules for the range $[0, \tau_{MAX}]$. The output for M_3 versus the τ_{MAX} and t_c values are given in Figs 12.5,6. The optimal prediction is :

$$M_3 \simeq 3.1 \text{ GeV} \qquad \sqrt{t_c} \simeq 3.4 \text{ GeV} . \qquad (12.52a)$$

The associated value of the coupling is :

$$f_3 \simeq 62 \text{ MeV} . \qquad (12.52b)$$

It is interesting to note that the value of M_3 is relatively high compared to that of the scalar gluonia built with two gluons. We may therefore expect that the mixing between the di and tri scalar gluonia will be minimal.

b) Di and trigluonia mixing

We test this intuitive expectation by evaluating dynamically[203] the mixing angle through the off-diagonal correlator :

$$\psi_{23}(q^2) = i \int d^4x \, e^{iqx} \left\langle 0 \left| \mathbb{T} \, \hat{J}_+(x) \left(\hat{J}_3(0)\right)^+ \right| 0 \right\rangle \qquad (12.53)$$

Its QCD expression reads :

$$\psi_{23}(q^2) = \alpha_s^2 \left(\log - \frac{q^2}{\nu^2}\right) \left[\frac{9}{4\pi^3} g^2 (q^2)^3 - \frac{9}{4} g^2 \langle G^2 \rangle . q^2\right]$$

$$- 24 \pi g \left\langle f_{abc} G^a G^b G^c \right\rangle . \qquad (12.54)$$

The associated Laplace transform sum rule is :

$$\int_0^{t_c} dt \, e^{-t\tau} \frac{1}{\pi} \operatorname{Im} \psi_{23}(t) \simeq \tau^{-3} \alpha_s^2 \beta_1 \left(\frac{54 \alpha_s}{\pi^2} (1-\rho_3) - \right.$$

$$\left. - 9 \left\langle \alpha_s G^2 \right\rangle \tau^2 (1-\rho_1)\right) \qquad (12.55)$$

where $\rho_i \equiv e^{-t_c\tau} \sum_{i=o} 1 \frac{(t_c\tau)^i}{i\ !}$ is the continuum effect.

Within a two-component mixing formalism :

$$|G_3\rangle \equiv |2\rangle \sin\theta + |3\rangle \cos\theta$$

$$|G_2\rangle \equiv |2\rangle \cos\theta - |3\rangle \sin\theta \ , \qquad (12.56)$$

(where the indices 2 and 3 refer to two- and three-gluon bound states) the spectral part of the sum rule reads :

$$\int_0^{c} dt \, e^{-t\tau} \frac{1}{\pi} \operatorname{Im} \psi_{23}(t) \simeq \sin 2\theta \, M_2^2 \, f_2 M_3^4 f_3$$

$$\left(e^{-M_2^2\tau} - e^{-M_3^2\tau}\right) . \qquad (12.57)$$

We use the values $M_2 \simeq 1.6$ GeV, $M_3 \simeq 3.1$ GeV and the associated couplings $f_2 \simeq 5.76$ MeV, $f_3 \simeq 62$ MeV. τ-stability of the predicted value of θ is reached for $t_c \geqslant 3$ GeV^2 but the value of these extremes increases strongly with t_c. As $t_c \simeq 3$ GeV^2 is the continuum threshold associated with the M_2 gluonium and expecting that in the off-diagonal is larger than this value, we can deduce the lower bound :

$$\theta \geqslant 4^\circ \quad . \tag{12.58}$$

Further phenomenological consequences of this result should be studied once detailed data in the energy range 1.6 - 3.1 GeV are available.

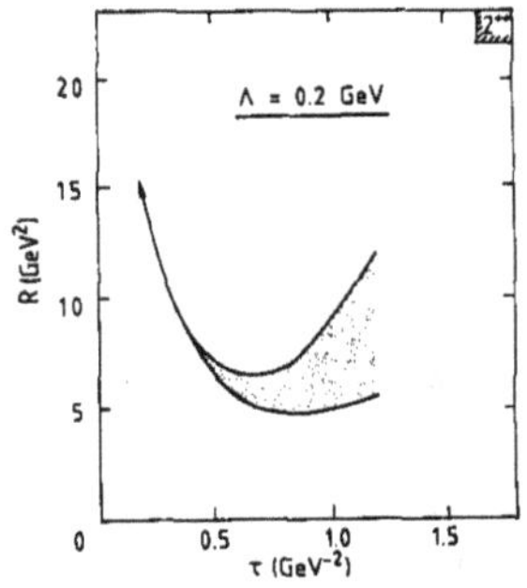

Fig. 12.7 : Behaviour of the QCD expression of the moments R of the 2^{++} current for various values of τ. The uncertainties come from the determination of $\langle \alpha_s G^2 \rangle$.

5. TENSOR GLUONIUM

a) Mass and coupling

We shall be concerned with the two-point correlator associated with the current $\hat{J}_{\mu\nu}$ defined in (12.1). It reads :

$$T_{\mu\nu,\rho\sigma} = i \int d^4x \; e^{iqx} \left\langle 0 \left| T \; \hat{J}_{\mu\nu}(x) \; \left(\hat{J}_{\rho\sigma}(0)\right)^+ \right| 0 \right\rangle$$

$$\equiv \left(\eta_{\mu\rho}\eta_{\nu\sigma} + \eta_{\mu\sigma}\eta_{\rho\nu} - \frac{2}{3}\eta_{\mu\nu}\eta_{\rho\sigma} \right) \psi_T(q^2\ ) \qquad (12.59)$$

with

$$\eta_{\mu\nu} \equiv \left(g_{\mu\nu} - q_\mu q_\nu / q^2 \right) .$$

Its QCD expression is known in Ref. 187). NSVZ have studied the moment ratio $R_T(\tau)$:

$$R_T(\tau) \equiv -\frac{d}{d\tau} \log \int_0^\infty dt\ e^{-t\tau} \frac{1}{\pi} \operatorname{Im} \psi_T(t)$$

$$\simeq 3\tau^{-1} \left\{ 1 + \frac{200}{9} \pi^3 \alpha_s \left\langle 0 \right| (f_{abc}\ G_{\mu\nu} G_{\alpha\beta})^2 \right.$$

$$\left. - 2\ (f_{abc}\ G_{\mu\alpha} G_{\nu\alpha})^2 \left| 0 \right\rangle \tau^4 + \ldots \right\} . \qquad (12.60)$$

One can estimate the condensates using the factorization or its variant. Using the former :

$$\left\langle 0 \left| (f_{abc} G_{\mu\nu} G_{\alpha\beta})^2 - 2 (f_{abc} G_{\mu\alpha} G_{\nu\alpha})^2 \right| 0 \right\rangle \simeq \frac{3}{16} \langle G^2 \rangle^2 , \qquad (12.61)$$

one can deduce from the minimum of $R(\tau)$ the optimal upper bound[187,189] (see Fig. 12.7) :

$$M_T \leqslant (2.4 \pm 0.3)\ \text{GeV} \quad ,$$

where the error bar is induced by an assumed 50% uncertainty on the estimate of the condensate and by a 20% error due to the high dimension ones.

Now, using a much more involved fitting procedure[189], we eliminate (as usual) the coupling constant of the gluonium via the lowest dimension FESR :

$$\frac{t_c^3}{20\pi^2} \simeq 6\ M_T^4\ f_T^2 \quad , \tag{12.62}$$

with the normalization :

$$\frac{1}{\pi}\ \text{Im}\ \psi_T(t) = 2\ M_T^4\ f_T^2\ \delta\left(t - M_R^2\right) + \theta(t-t_c)\ .\ \text{"QCD continuum"} \quad .$$

A comparison of the two sides of the moments in the region $[0, \tau_{MAX}]$ gives :

$$M_T \simeq (1.73 \pm 0.10)\ \text{GeV}$$

$$t_c \simeq (1.80 \pm 0.15)\ \text{GeV} \quad , \tag{12.63a}$$

from which we deduce the coupling to the vacuum :

$$g_T \equiv \frac{f_T}{M_T} \simeq (47 \pm 13)\ 10^{-3} \quad . \tag{12.63b}$$

The above analysis can be alternatively checked by using the FESR-like moment in Eq. (1.41) within τ-stability. Another check of the previous result is the FESR in Ref. 191). These different methods give results analogous to that in Eq. (12.63) and increase our confidence on the solidity of the previous estimate.

It is tempting to identify the gluonium obtained earlier with the observed θ meson seen in ψ-decay experiments[204)].

It may be noticed that the mass ratio :

$$r_{0T} \equiv \frac{M_{0^{++}}}{M_T} \leqslant 1 \quad , \tag{12.64}$$

where the exact value depends on whether the σ or the G(1.6) is taken as a scalar gluonium. A comparison with LGT results obtained without dynamical fermions should be done carefully.

b) Tensor meson-gluonium mixing

Meson-gluonium mixing has been studied in Ref. 48a) from the off-diagonal correlator :

$$\psi^{\mu\nu\rho\sigma}_{gq} = i \int d^4x \; e^{iqx} \left\langle 0 \left| \mathrm{T}\; J^{\mu\nu}(x) \left(J^{\rho\sigma}(0)\right)^+ \right| 0 \right\rangle \tag{12.65a}$$

with the Lorentz decomposition in Eq. (12.59). Its QCD expression is :

$$\Psi_{gq} \simeq \frac{q^4}{15\pi^2}\left(\frac{\alpha_s}{\pi}\right)\left(\log^2\frac{-q^2}{\nu^2} - \frac{91}{15}\log\frac{-q^2}{\nu^2}\right) - \frac{7}{36\pi}\left(\log\frac{-q^2}{\nu^2}\right)\left\langle \alpha_s G^2 \right\rangle \quad , \tag{12.65b}$$

from which one can derive the FESR :

$$\int_0^{t_c} dt \, \frac{1}{\pi} \, \mathrm{Im}\, \psi_{gq}(t) \simeq - \frac{101}{675\pi^2}\left(\frac{\alpha_s}{\pi}\right) t_c^3 \left(1 + \frac{525\pi^2}{404}\frac{\langle G^2 \rangle}{t_c^2}\right) \quad , \tag{12.66}$$

where we use $t_c \simeq 2.8$ GeV^2 as the mean value of t_c from the diagonal quark and gluon correlators. We can improve this t_c value by using the moment ratio :

$$\left\langle t_c^{eff} \right\rangle \simeq \frac{\int_0^{t_c} dt\, t\, \mathrm{Im}\, \psi_{gq}(t)}{\int_0^{t_c} dt\, \mathrm{Im}\, \psi_{gq}(t)} \simeq \frac{591}{808} t_c \simeq 2\ \mathrm{GeV}^2 \quad . \tag{12.67}$$

We use a standard two-component mixing formalism for parametrizing the spectral function. As we have to deal with mesons having degenerate masses, we factorize out the spurious meson mass difference. In this way, one obtains :

$$\frac{1}{\pi}\, \mathrm{Im}\, \psi_{gq}(t \leqslant t_c) \simeq \sin 2\theta\, M_T^2\, f_T\, M_F^2\, f_F\, \delta(t-M^2) \quad , \tag{12.68}$$

where T and F denote the tensor gluonium and quarkonium. M is the average value of the meson mass. Therefore, one deduces :

$$\theta \simeq -10^\circ \quad . \tag{12.69}$$

The result is quite interesting but one should be aware of the crude approximation used for its derivation. Therefore, Eq. (12.69) should only be considered as a very rough estimate. A phenomenological consequence of this result can be obtained in the parametrization of the tensor f_2, f'_2 couplings to Goldstone boson pairs :

$$g^2_{f\pi\pi} \simeq \frac{9}{10} g^2 \left(1+2\sin\theta r + O(\theta^2)\right)$$

$$g^2_{fK\bar{K}} \simeq \frac{3}{10} g^2 \left(1+4\sin\theta r + \ldots\right)$$

$$g^2_{f'K\bar{K}} \simeq \frac{3}{5} g^2 \left(1+2\sqrt{2}\sin\theta r + \ldots\right) \quad . \tag{12.70}$$

$g^2 \simeq (205 \pm 10)\,GeV^{-2}$ is the unmixed coupling estimated from $a_2 \to K\bar{K}$ and $K^{**} \to K\pi$ data. We have assumed a universal coupling of gluonium to pairs of pseudoscalars. $r \equiv g_{\theta\pi\pi}/g_{f\pi\pi} \simeq -(0.68 \pm 0.23)$ is the amount of glue in the meson wave function and is fixed from the $f \to \pi\pi$, $K\bar{K}$ data. These results are used for obtaining :

$$\Gamma\left(\theta \to \pi\pi + K\bar{K} + \eta\eta\right) \leqslant (200 \pm 60)\ \text{MeV} \quad , \qquad (12.71)$$

which is a non-trivial constraint in view of our ignorance of the true value of the width of the gluonium. From Eq(12.71) one can speculate about the origin of the suppression of the $\theta \to \pi\pi$ versus the $K\bar{K}$ width. Ref. 48a) has qualitatively discussed that this suppression is due to a destructive interference between the gluonium and the radial excitation of the f_2 or f'_2. Some consequences of this assumption have also been discussed in Ref. 48a).

c) **The vertex sum rule and the θ coupling to Goldstone pairs**[205)]

One could also assume that the inclusion of the gluonium saturates Eq. (8.49)[206)]. Therefore, we have :

$$\sum_{f,f',\theta} g_T\, g_{TP\bar{P}} \simeq 2 \qquad P \equiv \pi,\ K,\ \eta \qquad (12.72)$$

which is very analogous to that in the 0^{++} meson channel[190)]. We use the previous estimated decay constants g_T associated with the matrix element $\langle 0\ |\theta^{\mu\nu}|\ T\rangle$ deduced in the previous section (Eqs 8.45 and 12.63). Solving Eq. (12.72) for each channel, we obtain :

$$g_{\theta\pi\pi} \simeq (3 \pm 4)\,GeV^{-1} \quad ; \quad g_{\theta K\bar{K}} \simeq (12.4 \pm 4.1)\ GeV^{-1} \ , \qquad (12.73)$$

where the difference between $g_{\theta\pi\pi}$ and $g_{\theta K\bar{K}}$ is due to the $SU(3)_F$ breaking effects in the value of $g_{f'}$ and $g_{f'K\bar{K}}$. Eq. (12.73) leads to :

$$\Gamma(\theta \to \pi^+\pi^- + \pi^0\pi^0) = (g_{\theta\pi\pi})^2 \frac{M_\theta}{320\pi} \left(1 - \frac{4m_\pi^2}{M_\theta^2}\right) \lesssim 89 \text{ MeV}$$

$$\Gamma\left(\theta \to K^+K^- + K^0\bar{K}^0\right) \lesssim 326 \text{ MeV} \quad . \qquad (12.74)$$

The results in Eq. (12.74) can provide an alternative explanation for the enhancement of the $K\bar{K}$ production versus the $\pi\pi$ one in the ψ decay experiment. The result in Eq. (12.74) can raise some doubts regarding the naive expectation based on the SU(3) singlet counting rule. Properties of the $\theta(1.72)$ observed in some other hadronic processes might need a much more involved mechanism like soft behaviour of the form factor assumed to be related to the non-locality of the $\theta\pi\pi$ interaction. However, a true quantitative QCD analysis is needed for a better understanding of such a phenomenological assumption.

6. PSEUDOSCALAR GLUONIUM

a) Mass and coupling

The use of the sum rule in this channel is much more subtle owing to the "strange" nature of the η' and to its relation to the subtraction constant $\psi_-(0)$ of the two-point correlator. In fact, according to the analysis of Witten and Veneziano[54] reviewed in Section 2, we should be aware of the behaviour of $\psi_-(0)$ in the world with or without quarks. We shall be concerned with the correlator built from the anomalous current $Q(x)$ where its QCD expression reads in the $\overline{MS}$-scheme :

$$\psi_-(q^2) \simeq \left(\frac{\alpha_s}{8\pi}\right)^2 q^4 \left\{ -\frac{2}{\pi^2} \log \frac{-q^2}{\nu^2} - 4 \frac{\langle G^2 \rangle}{q^4} + \frac{8}{(q^2)^3} \left\langle g_s f_{abc} G^a G^b G^c \right\rangle \right.$$

$$\left. - \frac{g_s^2}{(q^2)^4} f_{abc} f_{ade} \left\langle 0 \left| 2\, G^b_{\mu\nu} G^c_{\alpha\beta} G^d_{\mu\nu} G^e_{\alpha\beta} + 20\, G^b_{\mu\alpha} G^c_{\alpha\nu} G^d_{\mu\beta} G^e_{\beta\nu} \right| 0 \right\rangle \right\} + \ldots \quad (12.75)$$

where the radiative corrections obtained in Ref. 207) are renormalization scheme dependent and have not been included here. We shall use the normalization of the current

$$\hat{J}^- \equiv 2n\, Q(x) \quad . \quad (12.76)$$

From the above QCD expression one can derive the ratio of moments in pure Yang-Mills :

$$R_-(\tau) = -\frac{d}{d\tau} \log \int_0^\infty dt\; e^{-t\tau} \frac{1}{\pi} \operatorname{Im} \psi_-(t)$$

$$\simeq 3\, \tau^{-1} \left(1 + 2\, \pi^2 g\, \langle G^3 \rangle\, \tau^3 \right) \quad (12.77)$$

which is independent of the subtraction constant $\psi(o)$ and its slope $\psi'(o)$. $R_-(\tau)$ has the nice feature of having a minimum at which one can derive an upper bound on the value of the lowest gluonium mass :

$$M_- \lesssim (1.9 \pm 0.4)\ \text{GeV} \quad , \quad (12.78)$$

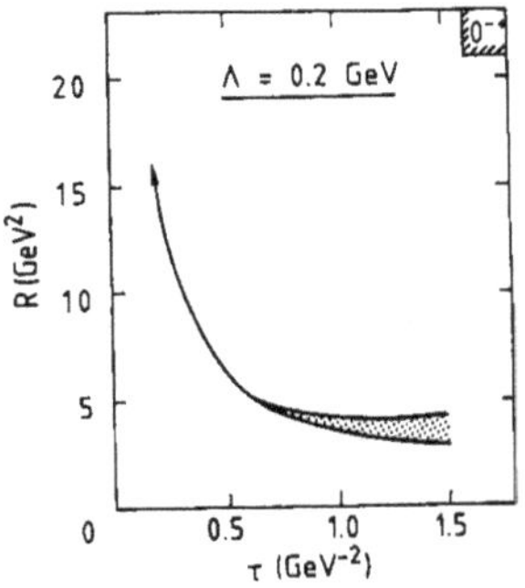

Fig. 12.8 : Behaviour of $R(\tau)$ versus τ where one has used $\langle g^3 G^3 \rangle = (1.1 \pm 0.2)\,GeV^2 \left\langle \alpha_s G^2 \right\rangle$

where the error is due to $g^3\langle G^3 \rangle$ condensate and to a 25% effect of the higher-dimension non-perturbative condensates in the moments. We now do a much more elaborate fitting procedure by using a least χ^2 two -parameter fit and by studying the τ-stability of the analysis. The resulting value without radiative corrections is[189)*)] :

$$M_- \simeq (1.46 - 1.66)\ \text{GeV} \simeq \sqrt{t_c}$$

$$f_- \simeq 3\ \text{MeV} \quad . \tag{12.79}$$

This value is confirmed by the FESR-like moment $R^c_-(\tau)$ where we show in Fig. 12.9 the τ- and t_c-stabilities of M^2_- and that of f_- from the un-subtracted sum rule :

$$\int_0^{t_c} dt\ e^{-t\tau}\ \frac{1}{\pi}\ \text{Im}\ \psi_-(t) \quad . \tag{12.80}$$

One obtains :

*) See also S. Narison, a talk given at the Moriond Conference, Hadronic Interactions (Les Arcs, March 1989) CERN preprint TH 5353/89 and S. Narison and G. Veneziano (in preparation).

$$M_- \simeq (1.36 \sim 1.55)\ \text{GeV}$$

$$f_- \simeq 2\ \text{MeV} \quad . \qquad (12.81)$$

It is clear that this gluonium cannot be identified with the η'. Our result suggests the presence of a new state other than the η'. This new state might be the η (1.43) seen in ψ decay and hadronic experiments. A further investigation of the properties of this pseudoscalar gluonium should be worked out for a much better understanding of the rich data around this energy region.

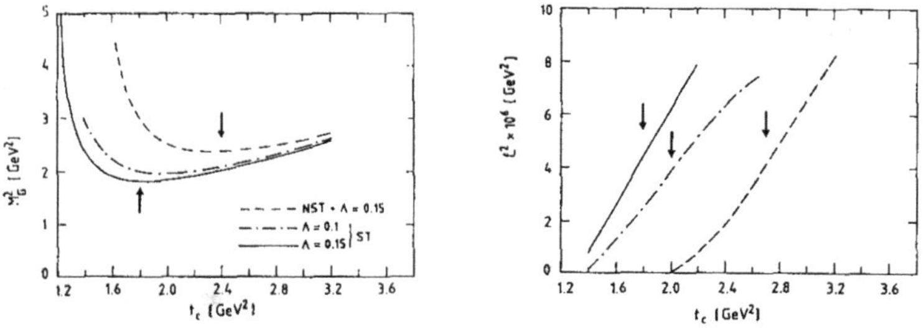

Fig. 12.9 : M_-^2 and f_-^2 respectively from $R_c(\tau)$ and the unsubtracted sum rule for two values of the gluon condensates :
— $\langle \alpha_s G^2 \rangle = 0.04\ \text{GeV}^4$ --- $\langle \alpha_s G^2 \rangle = 2.(0.04)\ \text{GeV}^4$.

b) The subtracted sum rule and the topological susceptibility of the $U(1)_A$ sector

One can try to evaluate the topological susceptibility $\psi_-(o)$ by the subtracted sum rule (SSR) which reads

$$\int_0^\infty \frac{dt}{t} e^{-t\tau} \frac{1}{\pi} \operatorname{Im} \psi_-(t) \simeq \psi_-(o) + \tau^{-2} \left(\frac{\alpha_s}{8\pi}\right)^2 \frac{2}{\pi^2} \cdot$$

$$\cdot \left\{1 + 2\pi^2\tau^2 \langle G^2 \rangle + 4\pi^2\tau^2 \langle G^3 \rangle\right\} \quad , \qquad (12.82)$$

or its modified form obtained by combining (12.80) and (12.82)[189]:

$$\int_0 \frac{dt}{t} e^{-t\tau} \left(1 - \frac{t\tau}{2}\right) \frac{1}{\pi} \operatorname{Im} \psi_-(t) \simeq \psi_-(o)\,|_{\text{NO-quarks}}$$

$$+ \left(\frac{\alpha_s}{8\pi}\right)^2 \frac{2}{\pi^2} \tau^{-2} .\ 2\pi^2 \left\{\langle G^2 \rangle\tau^2 + 3g \langle G^3 \rangle \tau^3\right\} \quad , \qquad (12.83)$$

where we have parametrized the spectral function by the gluonium obtained previously plus the QCD continuum which contributes as :

$$\left(\frac{\alpha_s}{8\pi}\right)^2 \left(\frac{2}{\pi^2}\right) \tau^{-2} e^{-t_c\tau} (1 + t_c\tau) \frac{1}{\log \tau\Lambda^2} \quad . \qquad (12.84)$$

Using either (12.82) or (12.83) and the resonance parameters obtained previously, one obtains a value of $\psi_-(o)$ about a factor 5 lower than the current algebra estimate[54] :

$$\psi_-(o)_{C.A} \simeq -(180 \text{ MeV})^4 \quad . \qquad (12.85)$$

We realize that owing to the large value of $\psi_-(o)$ compared to the perturbative contribution, the SSR stabilizes at small τ values and is much more weighted by the higher mass gluonium (a situation quite simi-

lar to the 0^{++} channel) so that the resonance in (12.81) alone does not saturate the sum rule. This fact agrees with the qualitative counting of Ref. 187) on the relative strength of each QCD contribution. After subtracting the effects in (12.81) to the SSR in (12.82) and using as input the current algebra estimate in (12.85), one needs for consistency a value of about :

$$2 f'^2 M'^2 \simeq 1.14\ 10^{-3}\ \text{GeV}^4 \tag{12.86}$$

which could also be deduced from the FESR :

$$2 \sum_i f_-^2 M_-^2 \simeq \psi_-(o) + \left(\frac{\alpha_s}{8\pi}\right)^2 \frac{2}{\pi^2} \frac{t_c^2}{2} + \dots \tag{12.87}$$

This high-mass state has not the standard property of a radial resonance as its coupling to the current might be larger than the ground state one in (12.81).

The extension of the previous analysis for the estimate of the slope of $\psi_-(o)$ and on the $SU(3)_F$-breaking effects has been done in Ref. 187) but without taking into account the effect of that radial excitation. An extensive and careful analysis of various possible corrections in this pseudoscalar channel is under way.

c) Pseudoscalar meson-gluonium mixing and the $\iota \to \gamma\gamma, \rho\gamma$ widths :

The mixing is analyzed through the off-diagonal correlator built with quark and gluonic currents :

$$\psi^-_{gq}(q^2) = i \int d^4x\ e^{iqx} \left\langle 0 \left| J^-_g(x) \left(J^-_q(o)\right)^+ \right| 0 \right\rangle \tag{12.88}$$

with :

$$J^-_g(x) = :\alpha_s G\tilde{G}:$$

$$J_q^-(x) = : 2\, i m\, \bar{\psi}\, \gamma_5 \psi \; : \qquad (12.89)$$

Its QCD expression reads[208] :

$$\psi_{gq}^-(q^2) = \left(\frac{\alpha_s}{\pi}\right)^2 \frac{3}{2}\, m_s^2\, q^2 \log \frac{-q^2}{\nu^2} \left\{\log \frac{-q^2}{\nu^2} - \frac{2}{3}\left(\frac{11}{4} - 3\gamma_E\right)\right\}$$

$$- 8\, \alpha_s \left(\frac{\alpha_s}{\pi}\right) m_s \left\langle \bar{s}s \right\rangle \log \frac{-q^2}{\nu^2} + 2 \left(\frac{\alpha_s}{\pi}\right) \left\langle \alpha_s G^2 \right\rangle \;.$$

$$. \left(\frac{m_s^2}{q^2}\right) \log \frac{-q^2}{m_j^2} \;. \qquad (12.90)$$

Using a two-component mixing formalism, we can derive the sum rule :

$$\sin 2\theta\, M_{\eta'}^2 f_{\eta'} .\, M_G^2 f_G \left(e^{-M_{\eta'}^2 \tau} - e^{-M_G^2 \tau}\right) \simeq$$

$$\left(\frac{3}{4\pi}\right) \left(\frac{\alpha_s}{\pi}\right)^2 \frac{3}{\pi}\, \tau^{-2} \left\{\bar{m}_s^2 \left(\frac{1}{6} + 2\, \gamma_E\right) (1 - \rho_1) - \frac{8\pi^2}{3}\, m_s \left\langle \bar{s}s \right\rangle .\, \tau\right.$$

$$\left. - \frac{2\pi^2}{3}\, \bar{m}_s^2 \log \bar{m}_s^2\; \tau\, \langle G^2 \rangle\, \tau^2\, (1-\rho_0)\right\} \;, \qquad (12.91)$$

where ρ_1 as usual includes the continuum contribution. We have identified the η' and G with the meson and gluonium states. We show in Fig. 12.10, the τ- and t_c-behaviour of the mixing angle for two values of the gluon condensate $\langle \alpha_s G^2 \rangle \simeq 0.04,\ 0.08\ \mathrm{GeV}^4$.

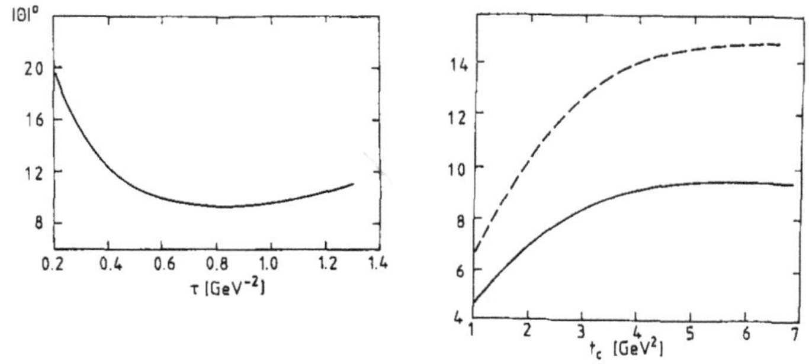

Fig. 12.10 : τ- and t_c-behaviour of the 0^{-+} meson-gluonium mixing angle.

We use $\Lambda = 0.15$ GeV and the value of the invariant strange-quark mass and condensate determined in previous sections. The results are quite sensitive to $\langle \alpha_s G^2 \rangle$ where θ increases with it. We deduce :

$$|\theta| \simeq (8.5 - 14.5)^\circ \quad , \tag{12.92}$$

a value which is compatible with previous findings[208] but is better controlled here. We can use the result in (12.92) to predict the gluonium electromagnetic widths. Assuming that the "physical" gluonium is around the ι mass, we can deduce :

$$\Gamma(\iota \to \gamma\gamma) \simeq \Gamma(\eta' \to \gamma\gamma) \, . \, \mathrm{tg}^2\theta \left(\frac{M_\iota}{M_{\eta'}}\right)^3 \simeq 0.4 \sim 1\ \mathrm{keV}$$

$$\Gamma(\iota \to \rho\gamma) \simeq \Gamma(\eta' \to \rho\gamma) \, . \, \mathrm{tg}^2\theta \left(\frac{\vec{p}_\iota}{\vec{p}_{\eta'}}\right)^3 \simeq 54.7 \sim 130.4\ \mathrm{keV} \quad , \tag{12.93}$$

where we have used $\Gamma(\eta' \to \gamma\gamma) \simeq 4.3$ keV and $\Gamma(\eta' \to \rho\gamma) \simeq 72$ keV. Measurements of these widths can help in testing the amount of glue inside the ι wave function. The corresponding hadronic width should also be investigated. However, a comparison of the previous results with the experimental ones shows that though the $\gamma\gamma$ width is consistent with the data, the $\rho\gamma$ one is too small. This inconsistency should stimulate a better understanding of the nature of the ι.

CHAPTER 13

EXOTICS : HYBRIDS OR HERMAPHRODITE MESONS AND FOUR QUARK STATES

The "exotic" supraconducting palm

In this Chapter we shall be concerned with the mesons associated with the "hybrid" colourless and local gauge invariant operators :

$$\mathcal{O}^{\mu}_{V}(x) \equiv : g \bar{\psi} \lambda_a \gamma_\nu \psi G^{\mu\nu}_a :$$
$$\mathcal{O}^{\mu}_{A}(x) \equiv : g \bar{\psi} \lambda_a \gamma_\nu \gamma_5 \psi G^{\mu\nu}_a(x) : \qquad (13.1)$$

which are the only lowest-dimension operators that can be used to study the quantum numbers of exotic mesons 1^{-+} and 0^{--}. It is worthwhile studying these states experimentally as they are expected not to mix with ordinary mesons and gluonia so that their identification might be easier. However, we should bear in mind that we do not have any "theorem" predicting their existence and even the system might not bind to form a resonance. Within QSSR, this question can only be indirectly answered by testing the consistency of the usual duality "one resonance" plus "QCD continuum" ansatz for parametrizing the spectral function and the QCD expression of the two-point correlator :

$$\Pi^{\mu\nu}_{i(q)} \equiv i \int d^4x\, e^{iqx} \langle 0 | \mathbb{T}\, \mathcal{O}^{\mu}_{i}(x) \left(\mathcal{O}^{\nu}_{i}(0)\right)^{+} | 0 \rangle$$
$$= -(g^{\mu\nu}q^2 - q^\mu q^\nu)\, \Pi^{(1)}_{i}(q^2) + q^\mu q^\nu\, \Pi^{(0)}_{i}(q^2) \quad . \qquad (13.2)$$

As may already have been noticed, the QCD evaluation of the correlator is cumbersome (see Table 13.1). These technical difficulties have led to some controversies [209,210] in the QCD expression and then in the mass predictions. After an effort of communication between different groups, agreement has finally been reached on the correctness of the latest QCD expressions of LNP[211] in Table 13.1 for light hybrids.

Table 13.1 : QCD contributions to the two-point function in Eq. 13.2 [a] *:*

lowest order

(I)

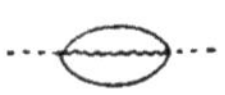

$$\Pi_V^{\mu\nu}(q) = \Pi_A^{\mu\nu}(q) = (\alpha_s/120\pi^3)q^4\{[1/\hat{\epsilon} + \gamma + \ln(-q^2/4\pi\nu^2) - 117/20]q^2g^{\mu\nu}$$
$$- (3/2)[1/\hat{\epsilon} + \gamma + \ln(-q^2/4\pi\nu^2) - 331/60]q^\mu q^\nu\}.$$

$1/\hat{\epsilon} \equiv 1/\epsilon + \gamma + \ln(-q^2/4\pi\nu^2)$, $\alpha_s = g^2/4\pi$, $D = 4 + 2\epsilon$ space-time dimensions.

dimension-four:

(II)

$$\Pi_V^{\mu\nu}(q) = (4\alpha_s/9\pi)m\langle\bar{\psi}\psi\rangle[(2/\hat{\epsilon} - 7/3)q^2g^{\mu\nu} + (1/\hat{\epsilon} - 13/6)q^\mu q^\nu].$$
$$\Pi_A^{\mu\nu}(q) = (4\alpha_s/9\pi)m\langle\bar{\psi}\psi\rangle[(4/\hat{\epsilon} - 17/3)q^2g^{\mu\nu} + (-7/\hat{\epsilon} + 43/6)q^\mu q^\nu].$$

(III)

$$\Pi_V^{\mu\nu}(q) = \Pi_A^{\mu\nu}(q) = (\alpha_s/18\pi)\langle F^2\rangle[(2/\hat{\epsilon} - 4/3)q^2g^{\mu\nu} + (1/\hat{\epsilon} - 5/3)q^\mu q^\nu].$$
$$\langle F^2\rangle \equiv \langle 0|:F_a^{\mu\nu}(0)F_{\mu\nu}^a(0):|0\rangle.$$

dimension-six

(a) four-fermion:

(IV)

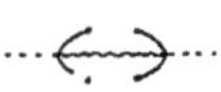

$$\Pi_V^{\mu\nu}(q) = -\Pi_A^{\mu\nu}(q) = -(16\pi\alpha_s/9q^2)\langle\bar{\psi}\psi\rangle^2(q^2g^{\mu\nu} + 2q^\mu q^\nu).$$

(b) triple gluon:

(V)

$$\Pi_V^{\mu\nu}(q) = \Pi_A^{\mu\nu}(q) = (1/36\pi^2q^2)\{[(-9/\hat{\epsilon} + 15/2)q^2g^{\mu\nu} - 6q^\mu q^\nu]O_1$$
$$+ [(-3/\hat{\epsilon} + 5/2)q^2g^{\mu\nu} - 6q^\mu q^\nu]O_2\}.$$
$$O_1 \equiv \langle 0|:\mathrm{Tr}[F_{\sigma\tau}(0)F^{\tau\rho}(0)F_\rho{}^\sigma(0)]:|0\rangle \equiv (1/4)g^3f_{abc}\langle F_{abc}^3\rangle.$$
$$O_2 \equiv \langle 0|:\mathrm{Tr}\{[D^\sigma(0), F_{\sigma\rho}(0)][D_\tau(0), F^{\tau\rho}(0)]\}:|0\rangle.$$
$$F_{\mu\nu} \equiv (i/2)g\lambda_aF_{\mu\nu}^a, \quad D_\mu \equiv \partial_\mu - ig(\lambda_a/2)B_\mu^a$$

(VI)

$$\Pi_V^{\mu\nu}(q) = \Pi_A^{\mu\nu}(q) = (1/24\pi^2q^2)[(6/\hat{\epsilon} - 7)q^2g^{\mu\nu} + 4q^\mu q^\nu]O_1.$$

(c) mixed condensate:

(VII)

$$\Pi_V^{\mu\nu}(q) = -\Pi_A^{\mu\nu}(q) = (\alpha_s/8\pi q^2)mg\langle\bar{\psi}F\psi\rangle[(3/2\hat{\epsilon} - 5/4)q^2g^{\mu\nu} - q^\mu q^\nu].$$
$$\langle\bar{\psi}F\psi\rangle \equiv \langle 0|:\bar{\psi}(0)\sigma^{\mu\nu}\lambda_aF_{\mu\nu}^a(0)\psi(0):|0\rangle.$$

(VIII)

$$\Pi_V^{\mu\nu}(q) = (\alpha_s/8\pi q^2)mg\langle\bar{\psi}F\psi\rangle[(-3/\hat{\epsilon} + 7)q^2g^{\mu\nu} + 2q^\mu q^\nu].$$
$$\Pi_A^{\mu\nu}(q) = (\alpha_s/8\pi q^2)mg\langle\bar{\psi}F\psi\rangle[(6/\hat{\epsilon} - 17/2)q^2g^{\mu\nu} - 8q^\mu q^\nu].$$

(IX)

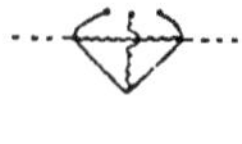

$$\Pi_V^{\mu\nu}(q) = (\alpha_s/144\pi q^2)mg\langle\bar{\psi}F\psi\rangle[(-3/\hat{\epsilon} + 1/2)q^2g^{\mu\nu} - 14q^\mu q^\nu].$$
$$\Pi_A^{\mu\nu}(q) = (\alpha_s/144\pi q^2)mg\langle\bar{\psi}F\psi\rangle[(-3/\hat{\epsilon} - 11/2)q^2g^{\mu\nu} + 10q^\mu q^\nu]$$

(X)

$$\Pi_V^{\mu\nu}(q) = (\alpha_s/24\pi q^2)mg\langle\bar{\psi}F\psi\rangle[(1/\hat{\epsilon} - 1/2)q^2g^{\mu\nu} + 2q^\mu q^\nu].$$
$$\Pi_A^{\mu\nu}(q) = (\alpha_s/24\pi q^2)mg\langle\bar{\psi}F\psi\rangle[(3/\hat{\epsilon} - 13/6)q^2g^{\mu\nu} + (2/3)q^\mu q^\nu].$$

(XI)

$$\Pi_V^{\mu\nu}(q) = \Pi_A^{\mu\nu}(q) = 0.$$

(XII)

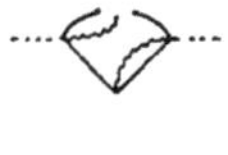

$$\Pi_V^{\mu\nu}(q) = (\alpha_s/3\pi q^2)mg\langle\bar{\psi}F\psi\rangle[(1/\hat{\epsilon} - 29/18)q^2g^{\mu\nu} + (4/3\hat{\epsilon} - 28/9)q^\mu q^\nu].$$
$$\Pi_A^{\mu\nu}(q) = (\alpha_s/3\pi q^2)mg\langle\bar{\psi}F\psi\rangle[(5/3\hat{\epsilon} - 55/18)q^2g^{\mu\nu} + (-4/3\hat{\epsilon} + 4/3)q^\mu q^\nu].$$

(XIII)

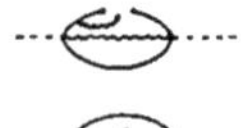

$$\Pi_V^{\mu\nu}(q) = (\alpha_s/6\pi q^2)mg\langle\bar{\psi}F\psi\rangle[(-1/\hat{\epsilon} + 1/6)q^2g^{\mu\nu} - 2q^\mu q^\nu].$$
$$\Pi_A^{\mu\nu}(q) = (\alpha_s/6\pi q^2)mg\langle\bar{\psi}F\psi\rangle[(-1/\hat{\epsilon} - 11/6)q^2g^{\mu\nu} + 6q^\mu q^\nu].$$

(XIV)

$$\Pi_V^{\mu\nu}(q) = \Pi_A^{\mu\nu}(q) = 0$$

[a] $F_a^{\mu\nu}$ throughout this table is identical to $G_a^{\mu\nu}$ in the text

1. MASS AND COUPLINGS OF THE LIGHT HYBRIDS

a) $\tilde{\rho}(1^{-+})$

In the chiral limit $m_u^2 = m_d^2 = 0$, one obtains from LNP[211]:

$$\Pi_V^{(1)}(q^2) \simeq - \left\{ \frac{\alpha_s}{60\pi^3} q^4 + \frac{1}{9\pi} \left(\left\langle \alpha_s G^2 \right\rangle + 8\alpha_s \, m \left\langle \bar{\psi} \psi \right\rangle \right) \right\} \log \frac{-q^2}{\nu^2}$$

$$+ \frac{1}{q^2} \left[\frac{16\pi}{9} \alpha_s \left\langle \bar{\psi} \psi \right\rangle^2 + \frac{1}{48\pi^2} \langle g^3 G^3 \rangle - \frac{83\alpha_s}{432\pi} mg \left\langle \bar{\psi} G\psi \right\rangle \right] . \qquad (13.3)$$

For greater convenience, we shall use the generic notation :

$$\Pi(q^2) = - \left\{ C_o q^4 + C_2 \, m^2 q^2 + C_4 \langle O_4 \rangle + \frac{C_6 \langle O_6 \rangle}{q^2} \right\} \log \frac{-q^2}{\nu^2}$$

$$- \frac{1}{q^2} C_6^F \langle O_6 \rangle . \qquad (13.4)$$

In principle, one can work with three LSR :

$$\mathcal{F}_o(\tau) \equiv \int_o^{t_c} dt e^{-t\tau} \frac{1}{\pi} \mathrm{Im}\, \Pi(t) = 2! \; C_o \tau^{-3}(1-\rho_2) + C_2 m^2 \tau^{-2}(1-\rho_1) +$$

$$\tau^{-1} C_4 \langle O_4 \rangle (1-\rho_o) + C_6^F \langle O_6 \rangle$$

$$\mathcal{F}_1(\tau) \equiv \int_o^{t_c} dt \; t e^{-t\tau} \frac{1}{\pi} \mathrm{Im}\, \Pi(t) = 3! \; C_o \tau^{-4}(1-\rho_3) + 2! \; C_2 m^2 \tau^{-3} (1-\rho_2)$$

$$+ \tau^{-2} C_4\langle O_4\rangle (1-\rho_1) + C_6\langle O_6\rangle \tau^{-1} (1-\rho_0)$$

$$\mathcal{F}_2(\tau) \equiv \int_0^{t_c} dt\ t^2 e^{-t\tau} \frac{1}{\pi} \text{Im}\,\Pi(t) = 4!\ C_0\tau^{-5}(1-\rho_4) + 3!\ C_2 m^2\tau^{-4}(1-\rho_3)$$

$$+ 2!\ \tau^{-3}\ C_4\ \langle O_4\rangle (1-\rho_2) + C_6\langle O_6\rangle \tau^{-2}\ (1-\rho_1) \qquad (13.5)$$

and the associated moment ratios :

$$R_0^c(\tau) = \frac{\mathcal{F}_1(\tau)}{\mathcal{F}_0(\tau)} \quad ; \quad R_1^c(\tau) = \frac{\mathcal{F}_2(\tau)}{\mathcal{F}_1(\tau)} \qquad (13.6)$$

or their Bell-Bertlmann variants.

If one relies on the low-energy estimate of the subtraction constant of Ref. 209), then the $\mathcal{F}_0(\tau)$ sum rule will be affected by this contribution :

$$q^2\ \Pi_V^{(1)}(q^2)\Big|_{q^2=0} \simeq -q^2\ \Pi_A^{(0)}(q^2)\Big|_{q^2=0} \simeq -\frac{16\pi}{9}\alpha_s \left\langle\bar\psi\psi\right\rangle^2 , \qquad (13.7)$$

which would cancel the four-quark condensate contribution given in (13.3). The lowest dimension FESR constraints are :

$$2f_H^2\ M_H^4 \simeq C_0\ \frac{t_c^3}{3} + C_2\ m^2\ \frac{t_c^2}{2} + t_c\ C_4\ \langle O_4\rangle$$

$$2f_H^2\ M_H^6 \simeq C_0\ \frac{t_c^4}{4} + C_2\ m^2\ \frac{t_c^3}{3} + \frac{t_c^2}{2}\ C_4\ \langle O_4\rangle\ , \qquad (13.8)$$

if one parametrizes the spectral function within the usual duality ansatz :

$$\frac{1}{\pi} \operatorname{Im} \Pi(t \leqslant t_c) \simeq 2f_H^2 M_H^4 \, \delta\left(t - M_H^2\right) \quad . \qquad (13.9)$$

One can see that neither $R_o^c(\tau)$ nor $R_1^c(\tau)$ present τ-stability if one takes into account (13.7). This fact can indicate that the reliability of the estimates for the masses from these moments might depend strongly on the validity of Eq. (13.7). We have also checked the τ-stability of the LSR in Eq. (13.5).

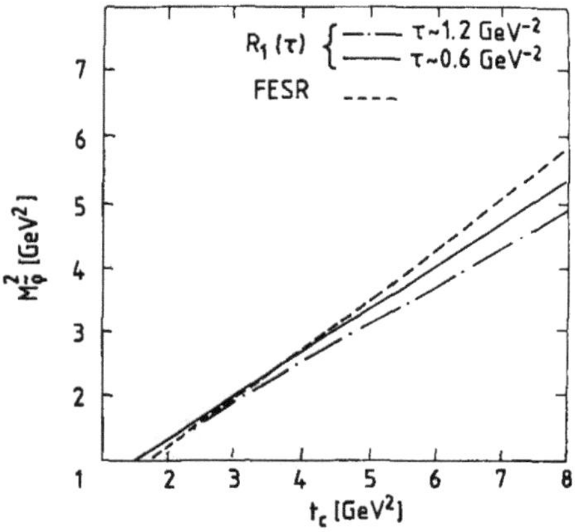

Fig. 13.1 : Behaviour of the results of M_ρ^2 versus t_c at fixed value of τ. For comparison, we give the results obtained from the analysis of R_1 for two expected extreme values of τ and of the FESR associated to $R_o(\tau)$.

We reach the same negative conclusions as for the moments. Therefore, the only available method for exploiting the previous sum rules is the least square fit one. As in Ref.211), we do the fitting procedure of $R_1(\tau)$ which is independent of (13.7). We work inside the interval $[0, \tau_{MAX}]$ where we took $\tau_{MAX} \simeq 0.6 \sim 1.2$ GeV^{-2}. The resulting value of M_ρ^2 is almost insensitive to this choice. In Fig. 13.1, we show the predicted value of $M^2_{\tilde{\rho}}$ versus the choice of the continuum threshold t_c from $R_1(\tau)$ and from the ratio of the two FESR in (13.8). Though the results have no t_c-stability, it indicates the common solution from each different method :

$$M_{\tilde{\rho}} \simeq (1.4 \sim 1.6)\ \text{GeV} \tag{13.10}$$

$$\sqrt{t_c} \simeq (1.7 \sim 1.9)\ \text{GeV}\ ,$$

to which one might add the typical 10% estimated accuracy of the method. The FESR in Eq. (13.8) helps to obtain the decay amplitude :

$$f_{\tilde{\rho}} \simeq (24 \sim 26)\ \text{MeV}\ . \tag{13.11}$$

The high value of $M_{\tilde{\rho}}$ obtained in Ref. 211) is due to the use of the moment $R_0(\tau)$ and to the neglect of the low-energy theorem in Eq. (13.7). Previous predictions for $M_{\tilde{\rho}}$ lower than 1.3 GeV are due to the incorrect QCD expressions used in the sum rules. Therefore, it is misleading to quote these results alongside those corresponding to the range in Eq. (13.10).

b) $\tilde{\eta}\ (0^{--})$

We shall be concerned here with the longitudinal part of the two-point correlator. Its QCD expression is :

$$\Pi^{(0)}_A(q^2) \simeq -\left\{\frac{\alpha_s}{120\pi^3}\,q^4 - \frac{1}{6\pi}\left(\langle\alpha_s\, G^2\rangle - 8\alpha_s\, m\,\langle\bar{\psi}\,\psi\rangle\right) - \right.$$

$$\left. - \frac{1}{q^2}\,\frac{11}{18}\left(\frac{\alpha_s}{\pi}\right) mg\,\langle\bar{\psi}\, G\,\psi\rangle\right\}\ \log\left(\frac{-q^2}{\nu^2}\right)\ . \tag{13.12}$$

In this case, the ν-dependence in the dimension-six operators can be eliminated if one works with the $R_1(\tau)$ moment. At the same time, $R_1(\tau)$ is free of the value of the correlator at $q^2=0$. Its QCD behaviour is shown in Fig. 13.2 where there is a minimum in τ. At this minimum, we deduce the upper bound on the $\tilde{\eta}$ mass :

$$M_{\tilde{\eta}} \leq 4.2 \text{ GeV} \quad . \tag{13.13}$$

A two-parameter fit $\left(M^2_{\tilde{\eta}}, t_c\right)$ plus the FESR-like lowest dimension constraint provides the result[211] :

$$M_{\tilde{\eta}} \simeq 3.8 \text{ GeV} \quad t_c \simeq 16.5 \text{ GeV}^2 \tag{13.14}$$

corresponding to the τ-stability of the set $\left(M^2_{\tilde{\eta}}, t_c\right)$ as shown in Fig. 13.2. The presence of this τ-stability can indicate that the result in Eq. (13.14) is much more accurate than the one for the 1^{-+} case in Eq. (13.10). One should also notice that the high value of the $\tilde{\eta}$ mass is "dual" to the low value of τ at which the minimum occurs and is mainly due to the relative importance of the condensate effects compared to that of the perturbative term.

c) $SU(3)_F$ breaking and the $\tilde{\phi}(1^{-+})$

The leading $SU(3)_F$-breaking terms are due to the strange quark fields in Eq. (13.1). We give these effects in Table 13.2 for equal mass in the quark loops.

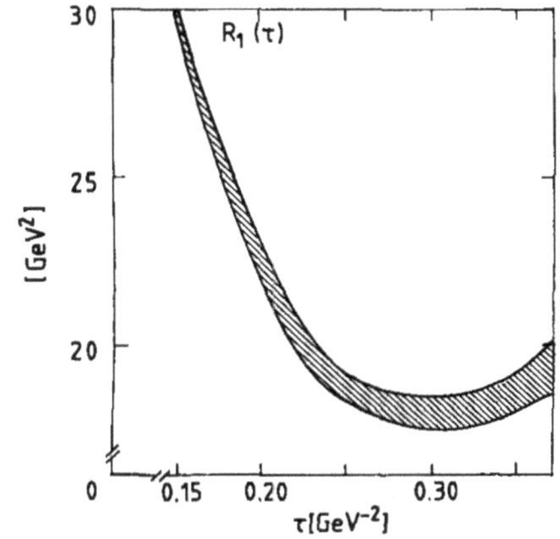

Fig. 13.2 : Variation of the moments $R_1(\tau)$ in the $J^{PC} = 0^{--}$ case. The error bars are due to the uncertainties on Λ and $\langle \alpha_s G^2 \rangle$.

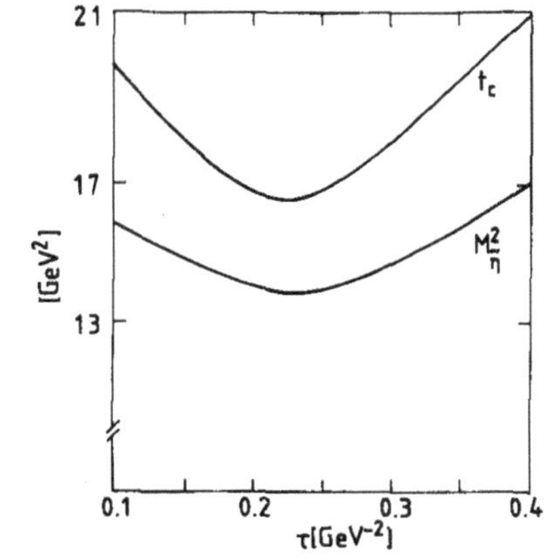

Fig. 13.3 : Set of $\left(M^2_{\tilde{\eta}}, t_c\right)$ coming from the fit procedure versus τ.

Table 13.2 : Leading $SU(3)_F$-breaking effects to the two-point function

	$\frac{m^2}{q^2}\left(\frac{\alpha_s}{\pi^3}\right)q^4 \log\frac{-q^2}{\nu^2}\left(-\frac{1}{3}g^{\nu\mu}q^2+\frac{1}{3}q^\mu q^\nu\right)$
	$O\left(\frac{m^2}{q^2}\alpha_s\langle G^2\rangle\right)$
	$-\frac{16}{3}\frac{\alpha_s}{\pi}m\langle\bar\psi\psi\rangle\frac{m^2}{q^2}q^\mu q^\nu \log -\frac{q^2}{\nu^2}$
	$O\left(\frac{m^2}{q^4}\langle G^2\rangle\right)$
	$\frac{m^2}{q^4}\frac{\langle G^3\rangle}{96\pi^2}(4g^{\mu\nu}q^2+8q^\mu q^\nu)\log -\frac{q^2}{\nu^2}$

Then, for the $\tilde\phi(1^{-+})$ channel, the correlator reads[211] :

$$\Pi_{\tilde\phi}(q^2) = \left[-\frac{\alpha_s}{60\pi^3}q^4 + \frac{\alpha_s}{3\pi^3}q^2 m_s^2 - \frac{1}{9\pi}\left(\langle\alpha_s G^2\rangle 8\alpha m_s\langle\bar s s\rangle\right)\right] \log -\frac{q^2}{\nu^2}$$

$$- \frac{1}{q^2} \left[\frac{16\pi}{9} \alpha_s \left\langle \bar{\psi}\psi \right\rangle^2 + \frac{1}{48} g^3 \langle G^3 \rangle - \frac{83}{42} \left(\frac{\alpha_s}{2\pi} \right) m_s \, g \left\langle \bar{s} \, G \, s \right\rangle \right] . \quad (13.15)$$

One should notice that there are some notorious cancellations for the coefficients of $m_s^3 \left\langle \bar{s}s \right\rangle$ and $\left\langle \alpha_s \; G^2 \right\rangle m_s^2$.

There are various ways of studying the splittings between the $\tilde{\phi}$ and $\tilde{\rho}$ hybrids. The most informative method can be the ratio $R_{\tilde{\phi}}/R_{\tilde{\rho}}$ or the difference $R_{\tilde{\phi}} - R_{\tilde{\rho}}$. If one approximately neglects the $SU(3)_F$ breaking of the continuum threshold, one obtains the GMO-like mass formula from the R_1 moments :

$$R_{\tilde{\phi}} - R_{\tilde{\rho}} \simeq \frac{20}{3} \bar{m}_s^2 - \frac{160\pi^2}{9} m_s \left\langle \bar{s} \, s \right\rangle \tau \qquad (13.16a)$$

which for $\tau \simeq 0.5 \sim 1$ GeV^2 gives :

$$M^2_{\tilde{\phi}} \approx M^2_{\tilde{\rho}} + (0.24 \sim 0.33) \text{ GeV}^2 \quad . \qquad (13.16b)$$

This result indicates that it can be inaccurate to extrapolate the observed ϕ-ρ splitting into the hybrid channel as here the mass difference can be much smaller.

2. HADRONIC AND RADIATIVE WIDTHS OF THE $\tilde{\rho}(1^{-+})$

a) $\tilde{\rho} \longrightarrow \rho\pi$ and K^*K

It has been suggested[73] that these decay modes are among the most important decays of the $\tilde{\rho}$ QSSR predictions on these widths are based on three-point function sum rules where the accuracy is not under good control (see Chapter 14). The $\tilde{\rho} \longrightarrow \rho\pi$ width was originally evaluated in Ref. 212) but the result is sensitive on the resonance parameters. We shall be concerned in this case with the three-point function :

$$T^{\mu\nu}(p,q) = \int d^4x\, d^4y\, e^{i(qx+py)} \left\langle 0 \left| T\, \partial_\alpha A^\alpha(x)\, V^\mu(y)\, O^\nu(o) \right| 0 \right\rangle$$

$$\equiv i\, \epsilon^{\mu\nu\rho\sigma} p_\rho q_\sigma\, T(p,q) \quad , \qquad (13.17)$$

where :

$$\partial_\alpha A^\alpha(x) = (m_u+m_d) : \bar{u}\,(i\,\gamma_5)\, d(x) :$$

$$V^\mu(x) = \frac{1}{\sqrt{2}} : \left(\bar{u}\,\gamma^\mu\, u - \bar{d}\,\gamma^\mu d\right) :$$

$$O^\nu(x) = g : \bar{u}\,\lambda_a\,\gamma^\alpha d\, G_a^{\nu\alpha} : \quad . \qquad (13.18)$$

Here one exploits the duality between the triangle QCD diagram (LHS) and the "Mercedes" diagram of the meson (RHS) :

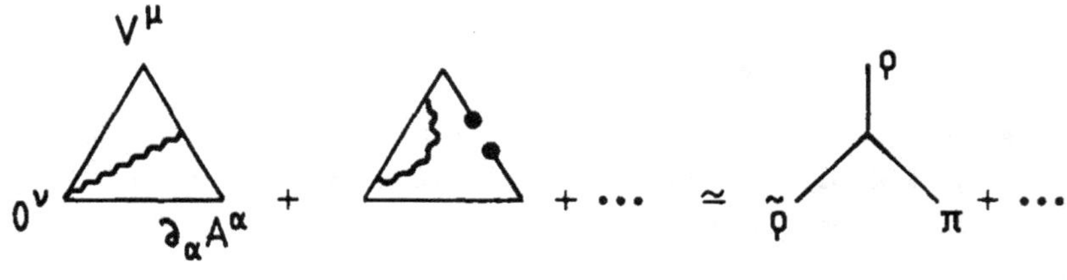

Fig. 13.4 : Quark-hadron duality in the vertex sum rules.

The OPE of the LHS reads at the symmetric point $\left(p^2=q^2=(p+q)^2 \equiv -(Q^2>0)\right)^{211)}$

$$T_{QCD}(p,q) = 2\sqrt{2}\,(m_u+m_d)\,\frac{1}{Q^2}\left\{-\frac{\alpha_s}{3\pi}\log\frac{Q^2}{\nu^2}\left\langle\bar{u}u+\bar{d}d\right\rangle+\frac{g}{48Q^2}\left\langle\bar{u}Gu+\bar{d}Gd\right\rangle\right.$$

$$\left.+\frac{\pi}{18Q^4}\left(1-\frac{2}{3}\right)\left\langle\alpha_s G^2\right\rangle\left\langle\bar{u}u+\bar{d}d\right\rangle\right\} \quad , \qquad (13.19)$$

where the first two terms agree with the original calculation[212] but in the third term the authors have missed some diagrams associated with diagrams with local condensation.

Parametrizing the phenomenological side of the sum rules by resonance poles via :

$$\left\langle 0\left|\partial_\alpha A^\alpha(o)\right|\pi\right\rangle = \sqrt{2}\,f_\pi\,m_\pi^2 \quad : \quad f_\pi \simeq 93 \text{ MeV}$$

$$\left\langle 0\left|V^\mu(o)\right|\rho\right\rangle = \sqrt{2}\,\frac{M_\rho^2}{2\gamma_\rho}\,\epsilon^\mu \quad : \quad \gamma_\rho \simeq 2.56$$

$$\left\langle 0\left|O^\nu(o)\right|\tilde{\rho}\right\rangle = \sqrt{2}\,f_{\tilde{\rho}}\,M_{\tilde{\rho}}^3\,\epsilon^\nu \qquad (13.20)$$

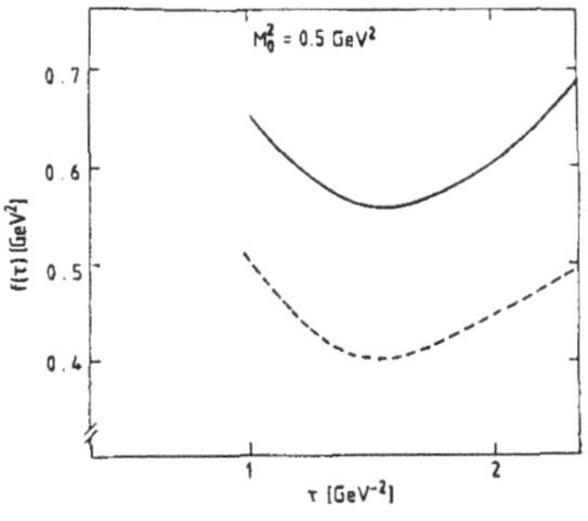

Fig. 13.5 : Behaviour of the function f(τ) which controls the $\tilde{\rho}\rho\pi$ coupling constant for two values of M_o^2.

one obtains the amplitude :

$$T_{phe}(Q^2) = g_{\tilde{\rho}\rho\pi} \frac{\sqrt{2}\, f_\pi m_\pi^2}{Q^2+m_\pi^2} \frac{\sqrt{2}\, M_\rho^2}{2\gamma_\rho \left(Q^2+M_\rho^2\right)} \frac{\sqrt{2}\, f_{\tilde{\rho}} M_{\tilde{\rho}}^3}{\left(Q^2 + M_{\tilde{\rho}}^2\right)} \quad . \tag{13.21}$$

The use of the pion PCAC relation plus the Laplace transform sum rule gives the constraint :

$$\left|g_{\tilde{\rho}\rho\pi}\right| \simeq \frac{2f_\pi}{f_{\tilde{\rho}}} \left(\frac{M_{\tilde{\rho}}^2 - M_\rho^2}{M_{\tilde{\rho}}^3}\right) \frac{\gamma_\rho}{M_\rho^2} f(\tau) \quad ,$$

with :

$$f(\tau) = \frac{\tau^{-1}}{e^{-M_\rho^2\tau} - e^{-M_{\tilde{\rho}}^2\tau}} \left\{\frac{2}{3}\frac{\bar{\alpha}_s}{\pi} + \frac{M_0^2\tau}{12} + \frac{\pi}{27}\left\langle \alpha_s G^2\right\rangle \tau^2\right\} \quad , \tag{13.22}$$

where M_0^2 is the mixed condensate scale. The function $f(\tau)$ is shown in Fig. 13.5 and has a τ-stability at $\tau \simeq 1.5$ GeV^{-2} where the $\left\langle \alpha_s G^2\right\rangle$ effect is still 17% of the leading term. At this τ value one has :

$$f(\tau_M) \simeq 0.5 \text{ GeV}^2 \Longrightarrow \left|g_{\tilde{\rho}\rho\pi}\right| \simeq 7.7 \text{ GeV}^{-1} \tag{13.23}$$

The corresponding decay width is :

$$\Gamma\left(\tilde{\rho} \longrightarrow \rho\pi\right) = \left|g_{\tilde{\rho}\rho\pi}\right|^2 \left(\frac{M^2_{\tilde{\rho}}}{96\pi}\right) \left[1 - \left(\frac{M_\rho + m_\pi}{M_{\tilde{\rho}}}\right)^2\right]^{3/2} \left[1 - \left(\frac{M_\rho - m_\pi}{M_{\tilde{\rho}}}\right)^2\right]^{3/2} \quad (13.24)$$

which gives :

$$\Gamma\left(\tilde{\rho} \longrightarrow \rho\pi\right) \simeq 274 \text{ MeV} \quad \text{for} \quad M_{\tilde{\rho}} \simeq 1.5 \text{ GeV} \quad . \quad (13.25)$$

It should again be noticed that the value of the width is very sensitive to the value of the resonance parameters. Larger value of the width in Ref. 211) is associated with larger $M_{\tilde{\rho}}$-mass and smaller $f_{\tilde{\rho}}$. One should also mention that this prediction can fail if the single vector meson dominance used to derive it is not justified. Improvement of this prediction can only be done after good control of the higher resonance effects.

It is easy to estimate in the same way the $\tilde{\rho} \longrightarrow K^*K$ decay. It is more suppressed than the $\tilde{\rho} \longrightarrow \rho\pi$ one owing to phase-space but is still appreciable :

$$\Gamma\left(\tilde{\rho} \longrightarrow K^*K\right) \simeq 8 \text{ MeV} \quad . \quad (13.26)$$

b) $\tilde{\rho} \longrightarrow \gamma\pi$

One can estimate this width using vector meson dominance. One obtains :

$$g_{\tilde{\rho}\pi\gamma} \simeq g_{\tilde{\rho}\pi\rho} \left(\frac{e}{2\gamma_\rho}\right) \simeq 0.46 \text{ GeV}^{-1} \quad , \quad (13.27a)$$

which implies quite a large radiative width compared to ordinary mesons :

$$\Gamma\left(\tilde{\rho} \longrightarrow \gamma\pi\right) \simeq \left|g_{\tilde{\rho}\pi\gamma}\right|^2 \frac{M^3_{\tilde{\rho}}}{96\pi} \simeq 2.4 \text{ MeV} \quad . \tag{13.27b}$$

The importance of the $\tilde{\rho} \longrightarrow \rho\pi$ and $\gamma\pi$ decays has stimulated a $\tilde{\rho}$ search in the Primakoff production experiment [213]. Narrow $\Gamma_{\tilde{\rho}} \leqslant 200$ MeV and light $M_{\tilde{\rho}} \leqslant 1.5$ GeV appear to be excluded from this experiment. This bound is still consistent with the previous QSSR results and should be checked in some other experiments.

c) $\tilde{\rho} \longrightarrow \pi\pi$, $K\bar{K}$ and $\pi\eta_8$ decay

The pseudoscalar π, K and η_8 mesons are correctly described by the quark bilinears :

$$\partial_\mu A^\mu_{(x)} = :(m_i + m_j)\,\bar{\Psi}_i\,(i\gamma_5)\,\psi_j: \quad , \tag{13.28}$$

which is the well-known divergence of the octet axial current and there is no rational way in QCD to add extra terms to it. This point already amends the result of Ref. 214 (hereafter denoted FT). We shall be concerned with the vertex sum rules :

$$T^\mu(p,q) = \int d^4x\, d^4y\, e^{i(qx+py)} \left\langle 0 \left| \mathbb{T}\, \partial_\alpha A^\alpha(x)\ \partial_\alpha A^\alpha(y) . \mathcal{O}^\mu(o) \right| 0 \right\rangle \tag{13.29a}$$

which have the Lorentz decomposition :

$$T^\mu(p,q) = T_S\,(p+q)^\mu + T_A\,(p-q)^\mu \quad . \tag{13.29b}$$

This amplitude has been correctly evaluated in Ref. 212). In QCD only the T_S part of it contributes. This result was later confirmed by LNP [211] using the symmetry properties of the amplitude in the exchange p into q. Thus, one can expect that the decays of the $\tilde{\rho}$ into these pseudoscalar pairs are suppressed.

d) $\tilde{\rho} \longrightarrow \eta'\pi$ and $\eta\pi$ via the QCD $U(1)_A$ anomaly

Therefore, one might expect that the most relevant decay of the $\tilde{\rho}$ into a pair of pseudoscalars occurs via the QCD $U(1)_A$ anomaly :

$$\partial_\beta \; S^\beta(x) = -\left(\frac{3\alpha_s}{4\pi}\right) G^{\mu\nu}\, \tilde{G}_{\mu\nu} + 2 \sum_{u,d,s} m_i \; \bar{\psi}_i (i\gamma_5)\, \psi_i \qquad . \tag{13.30}$$

The relevant *QCD* vertex is the one in *Table 13.3 where LNP choose to* evaluate at the symmetric Euclidian point. A correct evaluation of the perturbative graph shows that the leading contribution is proportional to the chiral breaking $\left(\frac{m^2}{q^2}\right)$ term. This result can be understood as there is no *QCD* anomaly *in the fermion loop. Indeed* $\Gamma(\pi^0 \longrightarrow gg)$ is well known to be trivially zero. FT "wrongly" add an anomalous term to describe the π^0 field using arguments based on mysterious and incorrect τ-matrix counting rules[*]. The relevant QCD contribution to the amplitude is :

$$T_A(p,q) = \alpha_s (m_u + m_d) \left\{ \frac{3}{2\pi^2} \left(\alpha_s \; \log \frac{Q^2}{\nu^2} \right) \left\langle \bar{u}u + \bar{d}d \right\rangle \right.$$
$$+ \frac{1}{4\pi Q^2}\, g \left\langle \bar{u}\; Gu + \bar{d}\; Gd \right\rangle + \frac{13}{18Q^4} \left\langle \bar{u}u + \bar{d}d \right\rangle .$$
$$\left. . \left\langle \alpha_s \; G^2 \right\rangle \right\} \tag{13.31}$$

---

*) *Analogous manipulation applied to the evaluation of the $\omega\rho\pi$ coupling using QSSR leads to* obsolete results. The cut-off dependence of the T result can already indicate that their result has no physical relevance.

Table 13.3 : QCD contribution to the $\tilde{\rho} \longrightarrow \eta'\pi$ amplitude normalized to $\alpha_s/2\pi(m_u+m_d)$

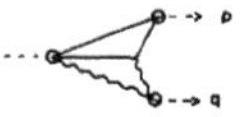	$O\left(\frac{m^2}{q^2}\right)$
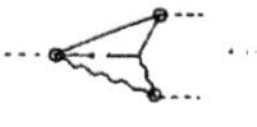	$\frac{g}{2\pi^2}\log-\frac{q^2}{\nu^2}(2p-q)^\mu\langle\bar{u}u+\bar{d}d\rangle$
	$\frac{g}{2\pi^2}(q-p)^\mu\langle\bar{u}Gu+\bar{d}Gd\rangle$
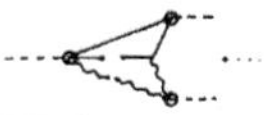	$\frac{g^2}{6q^4}\{-(2p+7q)^\mu-\frac{1}{3}(p-q)^\mu\}\langle\bar{u}u+\bar{d}d\rangle\langle G^2\rangle$

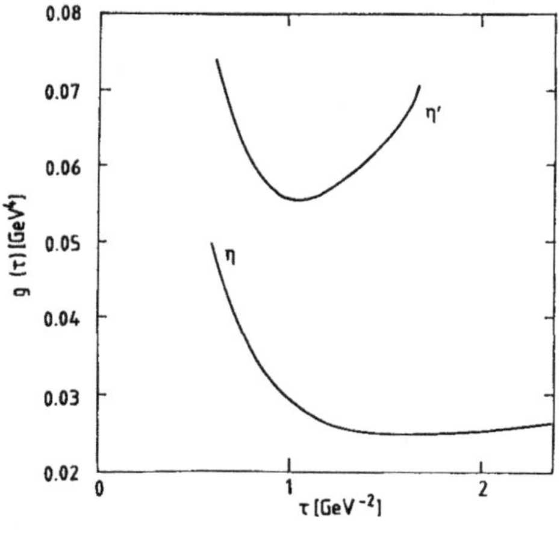

Fig. 13.6 : Behaviour of the function $g(\tau)$ which controls the $\tilde{\rho}\eta'\pi$ and $\tilde{\rho}\eta\pi$ couplings versus τ.

The η_1 effect is introduced as :

$$\left\langle 0 \left| \partial_\alpha S^\alpha(0) \right| \eta' \right\rangle = \sqrt{2}\, f_{\eta_1} M_\eta \quad , \tag{13.32a}$$

where :

$$f_{\eta_1} \simeq \sqrt{3}\, f_\pi \quad . \tag{13.32b}$$

One can take the Laplace transform of T_A or $Q^2 T_A$. We have tried both and they give almost the same numerical results. The Laplace transform of $Q^2 T_A(Q^2)$ gives the sum rule :

$$g_{\tilde{\rho}\eta_1\pi} \simeq \frac{f_\pi}{f_{\eta_1}} \left(\frac{3}{2\sqrt{2}}\right) \left(\frac{M^2_{\tilde{\rho}} - M^2_{\eta_1}}{M^2_{\eta_1}}\right) \frac{1}{f_{\tilde{\rho}}} \cdot \frac{1}{M^3_{\tilde{\rho}}}\, g(\tau) \tag{13.33a}$$

with the QCD function :

$$g(\tau) \simeq \left(\frac{\tau^{-2}}{e^{-M^2_{\eta'}\tau} - e^{-M^2_{\tilde{\rho}}\tau}}\right) \bar{\alpha}_s(\tau) \left\{\frac{\bar{\alpha}_s}{\pi^2} + \frac{13}{27}\left\langle \alpha_s G^2 \right\rangle \tau^2\right\} \quad . \tag{13.33b}$$

$g(\tau)$ is shown in Fig. 13.6 where at the τ-stability :

$$g(\tau) \simeq 0.055 \text{ GeV}^4 \quad . \tag{13.34}$$

Therefore, one obtains :

$$g_{\tilde{\rho}\eta_1\pi} \simeq 1 \quad . \tag{13.35a}$$

By introducing the mixing between the singlet η_1 and octet η_8, one deduces :

$$\Gamma_{\tilde{\rho}\to\eta'\pi} \simeq \cos^2\theta \left| g_{\tilde{\rho}\eta_1\pi} \right|^2 \frac{M_{\tilde{\rho}}}{48\pi} \left(1 - \frac{M^2_{\eta'}}{M^2_{\tilde{\rho}}}\right)^3 \simeq 3 \text{ MeV} \quad (13.35b)$$

with $|\theta| \simeq 18°$.

A decay of $\tilde{\rho} \to \eta'\pi$ is a clean signature of the $\tilde{\rho}$. The approach also predicts :

$$\frac{\Gamma\left(\tilde{\rho} \to \eta\pi\right)}{\Gamma\left(\tilde{\rho} \to \eta'\pi\right)} \simeq 3.1 \, tg^2\theta \ . \quad (13.36)$$

Previous results would indicate that the 1^{-+} decay into pseudoscalar pairs are not among the dominant ones. Thus the GAMS 1^{-+} candidate at 1420 MeV found from $\eta\pi$ mode with a width of about 150 MeV[215] could not be a pure hybrid state. Taking into account all possible decay modes of the $\tilde{\rho}$ also including some possible large decay of the $\tilde{\rho}$ into $b_1\pi$ and $f_1\pi$ one can safely deduce the bound

$$\Gamma\left(\tilde{\rho} \to \text{all}\right) \geq \Gamma\left(\tilde{\rho} \to \rho\pi + \eta'\pi\right) \simeq 280 \text{ MeV} \quad . \quad (13.37)$$

This result would indicate that the $\tilde{\rho}$ is too wide to be observed and presumably its experimental search might be harder than naively expected.

It would be worthwhile for experiments such as LEAR to test the existing data and theoretical ideas on the hybrids in the future[216].

3. MASS AND COUPLINGS OF "HEAVY" HYBRIDS

We shall now discuss the mesons associated with the operators in (13.1) where the quark fields are the charm and bottom ones. The correlator has been evaluated by one group [217] and an independent calculation will be useful for a further check (remember the divergences of previous results for the light hybrid case). As usual, the masses of the resonances have been estimated from the ratio of moments. The predicted values show a τ-(sum rule variable) stability but no t_c-(continuum threshold) stability (M_R increases with t_c). In order to have a much better control of the t_c value, one should investigate a two-parameter fit $\left(M_R^2, t_c\right)$ or use a FESR constraint. This investigation should complement that of Ref. 217). According to this later analysis, one might expect that only the sum rules corresponding to the channels shown in the Table 13.4 have τ-stability. Their results also indicate that the splitting between two opposite C-parity states is typically 300 $\sim$ 500 MeV. The spin zero states are often heavier than the spin one states, a feature which is similar to the case of the light hybrids.

Table 13.4 :
Predicted value of heavy hybrid masses from the τ-stable sum rule.

	$\bar{b}bg$	$\bar{c}cg$	$\bar{b}ug$	$\bar{c}ug$
0^{++}	10.9	5	6.8	4
0^{--}	11.4	5.4	7.7	4.5
1^{+-}	10.6	4.1	-	-
1^{++}	10.9	4.7	6.5	3.4

The corresponding values of $\sqrt{t_c}$ have been chosen by the authors to be 0.6-0.7 GeV and 0.3-0.4 GeV above the meson masses for the $\bar{b}bg$ and for $\bar{c}cg$. The $\sqrt{t_c}$ values for the $\bar{b}ug$ and $\bar{c}ug$ are taken to be 0.3-0.4 GeV and 0.2 GeV. The reasons for these choices should be checked.

4. FOUR-QUARK NATURE OF THE $a_o(980)$?

a) The mass and coupling

The existence of four-quark states has been conjectured from the bag model approach[218,71] in order to explain degeneracy of the $a_o(980)$ and $f_o(975)$ and at the same time their possible large couplings to $K\bar{K}$ pairs. In this scheme, their natural quark content would be :

$$|a_o\rangle = \frac{1}{\sqrt{2}}\,\bar{s}s\left(\bar{u}u - \bar{d}d\right)$$

$$|f_o\rangle = \frac{1}{\sqrt{2}}\,\bar{s}s\left(\bar{u}u + \bar{d}d\right) \quad . \qquad (13.38)$$

On the other hand, this scheme is unlikely as it never explains why the normal $\bar{q}q$ scalar mesons are absent and in addition it leads to a proliferation of states (too many (not yet seen) cryptoexotic states ...). The notion of chiral and flavour symmetries is also not obvious in this scheme.

Latorre and Pascual [219] were the first to study the two-point correlator associated with the colour singlet operators :

$$\mathcal{O}_1^{\pm} = \frac{1}{\sqrt{2}} \sum_{\Gamma=\mathbb{1},\ \pm\gamma_5} \bar{s}\,\Gamma\, s\left(\bar{u}\Gamma u - \bar{d}\Gamma d\right) \qquad (13.39)$$

within QSSR.

They obtained the QCD expression of the associated correlator :

$$\psi(q^2) = q^8 \log - \frac{q^2}{\nu^2} \left\{ - \frac{1}{40960\pi^6} + \frac{m_s^2}{q^2 1020\pi^6} \left(\frac{m_s \langle \bar{s} s \rangle}{128\pi^4} + \frac{\langle \alpha_s G^2 \rangle}{64\pi^5} \right) / q^4 \right.$$

$$\left. - \left(m_s \langle \bar{s} G s \rangle \frac{1}{64\pi^4} + \frac{3}{8\pi^2} \left(\langle \bar{u} u \rangle^2 + \langle \bar{s} s \rangle^2 \right) \right) / q^6 \right\}$$

$$- \frac{8}{3q^2} m_s \langle \bar{s} s \rangle \langle \bar{u} u \rangle^2 , \qquad (13.40)$$

which is free of the $\frac{1}{\epsilon} \log - \frac{q^2}{\nu^2}$ non-local pole. Factorization hypothesis has been used for the estimate of the high-dimension quark condensates. Introducing the resonance as :

$$\frac{1}{\pi} \mathrm{Im}\, \psi(t \leqslant t_c) = 2 f_E^2 M_E^8 \delta\left(t - M_E^2\right) , \qquad (13.41)$$

a usual LSR approach leads them to the mass estimate :

$$M_E \simeq 1 \text{ GeV} \qquad f_E \simeq 2.5 \text{ MeV} , \qquad (13.42)$$

(E denotes exotic !), i.e. the mass is around that conjectured for the a_0 and f_0 mesons. Pursuing this approach by the inclusion of the operator[146]:

$$\mathcal{O}_2^{\pm} = \frac{1}{\sqrt{2}} \sum \bar{s}\, \Gamma\, \lambda_a s \left(\bar{u}\, \Gamma\, \lambda^a u - \bar{d}\, \Gamma\, \lambda^a d \right) , \qquad (13.43)$$

one can study the two-point correlator associated with the combination of operators :

$$\mathcal{O}_E^{+} = \mathcal{O}_1^{+} + t\ \mathcal{O}_2^{+} + \ldots \tag{13.44}$$

where in principle t is an arbitrary mixing parameter.

b) Coupling to $\eta\pi$ and $\bar{K}K$

Ref.146) fixes the value of t from the estimate of the $g_{a_o\eta\pi}$ coupling using vertex sum rules. In this case, one has to evaluate the diagram in Fig. 13.7.

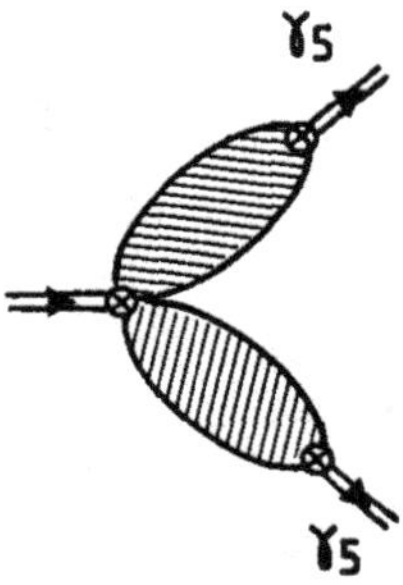

Fig. 13.7 : Vertex diagrams relevant for $a_o \rightarrow \eta\pi$

where the pseudoscalar currents describe the η_8 and π fields :

$$J_{\eta_8} = \sum_{u,d} 2\ m_u\ \bar{u}(i\gamma_5)u - 4\ m_s\ \bar{s}(i\gamma_5)s \tag{13.45a}$$

with :

$$\left\langle 0 \left| J_{\eta_8} \right| \eta_8 \right\rangle = \sqrt{2}\left(\sqrt{6}\ f_\pi\right) \cdot M_\eta^2 \ . \tag{13.45b}$$

This calculation gives the QCD expression of the amplitude at the symmetric Euclidian point :

$$T_{\eta\pi} \simeq (m_u+m_d)\, m_s \left\langle \bar{s}\, s \right\rangle \left(\frac{3}{2\pi^2}\right) \left\{ m_s \log \frac{Q^2}{\nu^2} + \frac{8\pi^2}{3Q^2} \left\langle \bar{u}\, u \right\rangle \right\} \qquad (13.46)$$

where the operator $O_2^{\pm}$ cannot contribute to a leading order. Introducing narrow resonance poles and taking the Laplace transform, one deduces[146]:

$$g_{a_0\eta\pi} \simeq \cos\theta \frac{1}{f_E M_E^4} \left(1 - \frac{M_\eta^2}{M_E^2}\right) \frac{\sqrt{3}}{8\pi^2} \frac{\left\langle \bar{s}\, s \right\rangle}{\left\langle \bar{u}\, u \right\rangle} f(\tau) \;, \qquad (13.47a)$$

where :

$$f_E \simeq 2.5 \text{ MeV} \left(1 + \frac{32}{9} t^2\right)^{1/2} \;, \qquad (13.47b)$$

is the a_0-decay amplitude rescaled by the inclusion of the $O_2^{\pm}$ operator into the two-point function analysis of Ref.219).

$$f(\tau) \equiv \frac{\tau^{-1}\, \bar{m}_s \left(\bar{m}_s - \frac{8\pi^2}{3} \tau \left\langle \bar{u}\, u \right\rangle \right)}{\left[\left(1 - e^{-M_\eta^2 \tau}\right) - \frac{M_\eta^2}{M_E^2}\left(1 - e^{-M_E^2 \tau}\right)\right]} \qquad (13.48)$$

is the sum rule function which presents τ-stability for $\tau \simeq 0.7\text{-}1.1$ GeV^{-2}. By assuming 60-90% effects of the higher radial excitation, one deduces

$$\Gamma(a_0 \rightarrow \eta\pi) \simeq (52 - 88) \text{ MeV} \frac{1}{\left(1 + \frac{32t^2}{9}\right)} \tag{13.49}$$

which for the data $\Gamma(a_a \rightarrow \eta\pi) \simeq 57$ MeV corresponds to :

$$|t| \simeq 0.2 \sim 0.4 . \tag{13.50}$$

Some other operators such as $\bar{s}\gamma^\mu s\, \bar{u}\gamma_\mu u, \ldots$ can be added into the interpolating operators but at the same time the numbers of parameters also increase, which decreases the predictive power of the approach. Within the previous minimal choice of interpolating fields, one also expects :

$$g_{a_0 \bar{K}K^0} \simeq \sqrt{\frac{3}{2}}\, g_{a_0\eta\pi} \cdot \left(1 + \frac{16}{3} t\right) \simeq 5 \sim 8 \text{ GeV} , \tag{13.51}$$

where here the O_2^- operator also contributes (Fig. 13.8). Eq.(13.51) suggests that the coupling of a four-quark state to $K\bar{K}$ can deviate strongly for the $SU(3)_F$ expectation owing to the extra $\left(1 + \frac{16}{3} t\right)$ factor. This deviation is one of the characteristic features of the four-quark nature of the a_0.

c) $\gamma\gamma$ width.

The $\gamma\gamma$-coupling of the a_0 has been found[146] to be dominated by the O_2 operator responsible for the QCD diagram with one-gluon exchange. (Fig. 13.9). Working within the chiral limit ($m_i = 0$), one obtains the leading contribution to the amplitude :

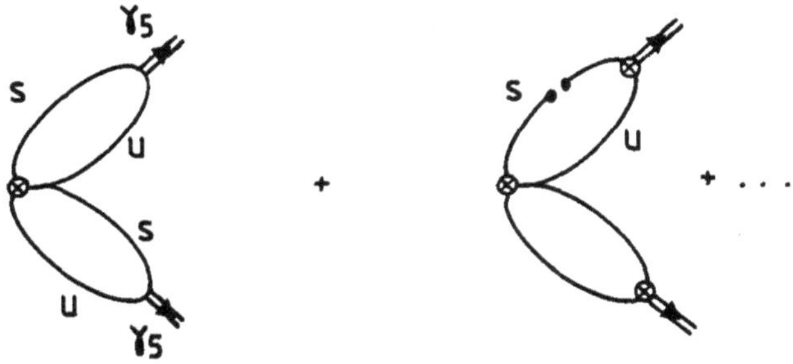

Fig. 13.8. QCD diagrams for $a_0 \rightarrow \bar{K}K$ from the O_2 operator

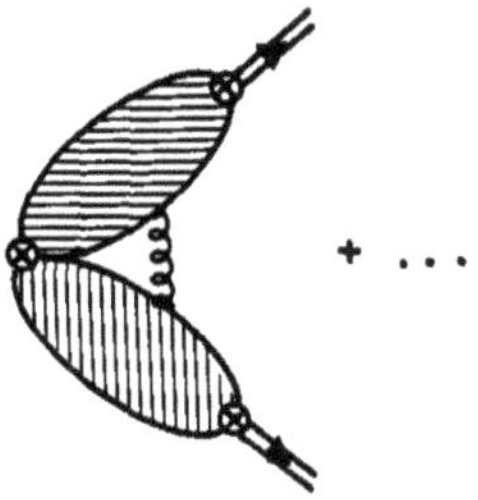

Fig. 13.9. QCD diagrams relevant for $a_0 \rightarrow \gamma\gamma$ from the O_2 operator in Eq.(13.43)

$$T^{\mu\nu}_{\gamma\gamma} = (g^{\mu\nu}(p.q) - q^{\mu}p^{\nu}) \frac{8}{27\sqrt{2}} \left(\frac{\alpha_s}{\pi}\right) \langle\bar{s}s\rangle\langle\bar{u}u\rangle \frac{t}{(Q^2)^2} \log \frac{Q^2}{\nu^2} \qquad (13.52a)$$

where one has normalized the current as :

$$J^\mu_\phi = \frac{1}{3}\bar{s}\,\gamma^\mu s \quad : \quad \left\langle 0 \left| J^\mu_\phi \right| \phi \right\rangle = \epsilon^\mu \frac{M^2_\phi}{2\gamma_\phi} \quad . \tag{13.52b}$$

Using for instance the Laplace transform of $(Q^2)^2\, T^{\mu\nu}_{\gamma\gamma}$ in order to eliminate the unwanted subtraction scale and taking its ratio with the $\gamma\gamma$ amplitude of the $\bar{q}q$ state (see next Chapter 14), one obtains :

$$\frac{g_{E\gamma\gamma}}{g_{a_o\gamma\gamma}} \simeq \frac{4t}{27}\left(\frac{\alpha_s}{\pi}\right)\frac{\left\langle \bar{s}\, s\right\rangle}{(m_d-m_u)\, M^2_E}\,\frac{f_{a_o}}{f_E}\left(1 - 2\, M^2_\nu\tau + \frac{M^4_\nu\, \tau^2}{2}\right) \tag{13.53}$$

where we have assumed for convenience $M_V \equiv M_{a_o} \simeq M_\rho \simeq M_\phi \simeq \frac{1}{2}\left(M_{a_o}+M_\rho+M_\phi\right)$. Eq. (13.53) stabilizes in τ for $\tau \simeq 2\, M_V^{-2}$ at which values one obtains :

$$\Gamma(E\to\gamma\gamma) \simeq (2-5)\, 10^{-4}\ \text{keV} \quad , \tag{13.54}$$

which is too small (even assuming a factor ten error) compared to the $\Gamma(a_o \to \gamma\gamma)$ data of about 0.3 keV[202]. It is interesting to notice that Eq. (13.53) is very similar to the successful (!) four-quark estimate[220]:

$$\Gamma(E\to\gamma\gamma)_{\left(\bar{q}q\right)^2} \simeq 0.24\, \alpha_s^2\, \Gamma(a_o \to \gamma\gamma)_{\bar{q}q} \quad . \tag{13.55}$$

We understand Eq. (13.55) as a very qualitative relation where dynamical loop-factor suppression such as $\frac{1}{16\pi^2}$ has not have been included correctly. The too small $\gamma\gamma$ width obtained in Eq. (13.54) does not favour the $\left(\bar{q}q\right)^2$ interpretation of the a_o.

CHAPTER 14

VERTEX SUM RULES, FORM FACTORS AND WAVE FUNCTIONS

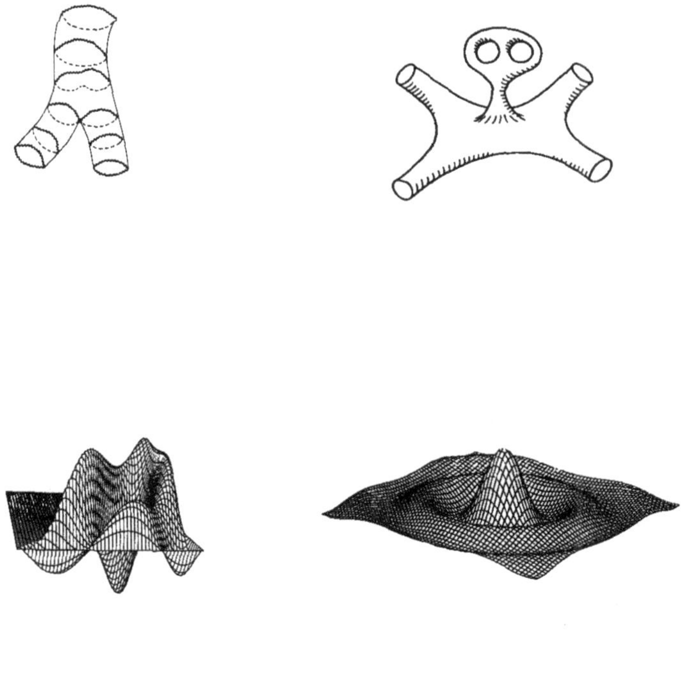

Some types of vertices and wave functions

In previous sections we have anticipated the uses of the vertex sum rules in order to estimate some hadronic trilinear couplings. The extensions of the QSSR analysis into three-point functions have been discussed by many authors [221-225]. In most of them, the vertex (three-point function) is parametrized phenomenologically by the simplest polar form, assuming the dominance of the lowest hadronic states in the different channels. The QCD expressions in the Euclidian region are evaluated in the configuration best suited to the processes considered.

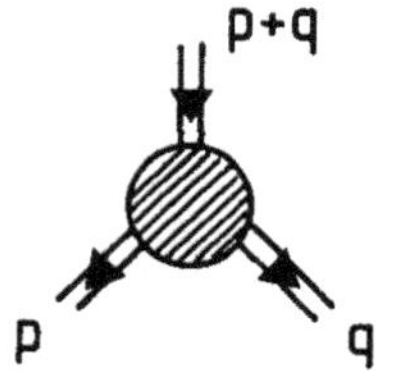

Fig. 14.1 : Hadronic vertex.

1. SPECTRAL REPRESENTATION AND CHOICE OF THE CONFIGURATION

Let us consider the vertex shown in Fig. 14.1. One can generally write its spectral representation as :

$$T(p^2, q^2, (p+q)^2) = \int dt_1\, dt_2\, dt_3 \frac{\rho(t_1, t_2, t_3)}{\left(t_1 - p^2\right)\left(t_2 - q^2\right)\left(t_3 - (p+q)^2\right)} \quad (14.1)$$

The validity of this representation has been studied in Ref. 226) within perturbation theory. However, Eq. (14.1) is not a very convenient expression in practice, but it can simplify in the case of a narrow resonance pole (Fig. 14.2) and in a given choice of configuration for the vertex. Under the symmetrical choice of configuration[224]:

Fig. 14.2 : Duality between a QCD vertex and a hadronic vertex in a narrow width approximation.

$$p^2 = q^2 = (p+q)^2 \equiv -(Q^2 \gg \Lambda^2) \qquad (14.2)$$

where some possible QCD mass singularities[227] and some possible anomalous thresholds are avoided, the vertex takes the nice form :

$$T(Q^2) = 2\int_0^1 x dx \int_0^1 dy \int_0^\infty dt_1 dt_2 dt_3 \frac{\rho(t_1,t_2,t_3)}{\left[Q^2+(t_1-t_2)xy + (t_2-t_3)x + t_3\right]^3} \qquad (14.3)$$

after a Feynman parametrization of the propagators. One can ideed easily apply the usual Laplace transform operator to Eq. (14.3) in order to deduce the sum rule :

$$\mathcal{F}(\tau) = \hat{\mathbb{L}}\, T(Q^2)$$

$$= \tau^3 \int_0^1 x dx \int_0^1 dy \int_0^\infty dt_1 dt_2 dt_3 \; e^{-[(t_1-t_2)xy + (t_2-t_3)x + t_3]\tau}$$

$$\cdot\, \rho(t_1,t_2,t_3) \quad , \tag{14.4}$$

where the depressive role of the exponential only manifests itself if the mass of the higher states is much higher that any of the ground states involved in the three channels, the depressive role of the exponential being less evident if the mass of some of the excited states are about that of the ground states.

Another alternative choice of vertex configuration is that where one or two of the square of the external momenta or the momenta themselves are kept small or at zero. This choice is suitable in the case of form factors. In this case, the assumed spectral representation is :

$$T(p^2,q^2,0) = \frac{1}{\pi^2} \int \frac{\rho(t,t')\, dtdt'}{(t-p^2)(t'-q^2)} + \ldots \quad . \tag{14.5}$$

One assumes double "Borel" transformation for each variable p^2 and q^2, which might be justified if there are no terms of the form[134]:

$$\mathrm{Poly}(p^2) \int \frac{\Delta(t)dt}{t-q^2} \quad \text{or} \quad \mathrm{Poly}(q^2) \int \frac{\Delta(t')dt'}{t-p^2} \quad , \tag{14.6}$$

which would induce non-controllable contributions to the sum rules. However, another criticism against the double independent Laplace transform is that of Craigie and Stern[221] where they noticed that the variables (p^2,q^2) are correlated, implying that the correct dispersion representation of the vertex should be done in the (p^2,q^2) plane along a straight line, which is a linear combination of the p^2 and q^2 variables.

2. PHENOMENOLOGICAL APPLICATIONS

There are various applications of the vertex sum rules in the literature, such as the determination of form factors and hadronic couplings. A comparison of the estimates with the data in most cases indicates good agreement. Here however, one could not claim the real accuracy of the QSSR predictions owing to the peculiar points discussed previously and mainly to the fact that the "continuum" and higher state effects are out of control. Let us discuss a few examples :

a) The $g_{\omega\rho\pi}$ coupling

The relevant three-point function is :

$$T_{\mu\nu}(p,q) = (i)^2 \int d^4x d^4y \; e^{-ipx} \; e^{i(p+q)y} \left\langle 0 \left| \mathbb{T} \; J^{\rho}_{\mu}(x) J^{\pi}(y) J^{\omega}_{\nu}(o) \right| 0 \right\rangle$$

$$\equiv i \, \epsilon_{\mu\nu\rho\sigma} \, p^{\rho} \, q^{\sigma} \, T(p,q) \qquad (14.7)$$

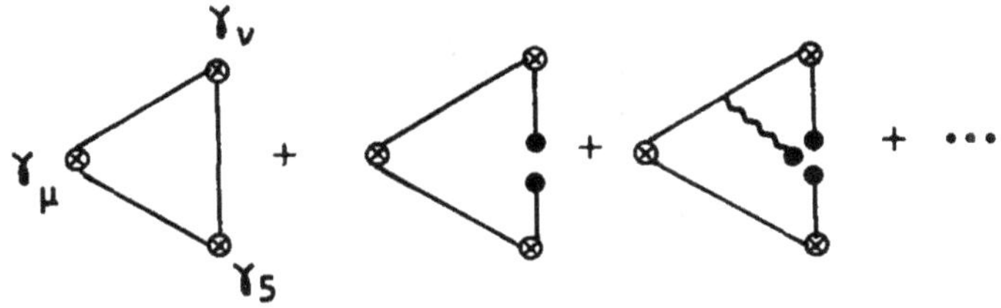

Fig. 14.3 : Leading QCD vertices for the $\omega\rho\pi$ coupling.

where the quark currents are normalized as :

$$J^{\rho}_{\mu} = : \bar{u} \gamma_{\mu} d :$$

$$J^{\omega}_{\nu} = \frac{1}{6} : \bar{u} \gamma_{\nu} u + \bar{d} \gamma_{\nu} d :$$

$$J^{\pi} = (m_u + m_d) : \bar{d}(i \gamma_5) u : \quad . \tag{14.8}$$

The QCD expression of the vertex is in the symmetrical point[224] (see Figure 14.3) :

$$T(p,q) = \frac{1}{16\pi^2} \frac{(m_u+m_d)}{Q^2} \left\{ (m_u+m_d) I_{x,y} - \frac{1}{Q^2} \langle \bar{u}u \rangle \right\} , \tag{14.9}$$

where :

$$I_{x,y} = \int_0^1 dx \int_0^{1-x} dy \frac{1}{x(1-x) + y(1-y) - xy} \simeq 2.34 \quad , \tag{14.10}$$

is a typical Feynman integral from a loop diagram. The phenomenological side of the vertex reads in a narrow resonance approximation :

$$T_{exp} \simeq \left(\frac{\sqrt{2} f_\pi m^2_\pi}{Q^2 + m^2_\pi} \right) \left(\frac{\sqrt{2} M^2_\rho}{2 \gamma_\rho} \right) \left(\frac{M^2_\omega}{2 \gamma_\omega} \right) \cdot g_{\omega\rho\pi} \quad , \tag{14.11}$$

with the usual normalizations :

$$\langle 0 | J^{\pi} | \pi \rangle = \sqrt{2} m^2_\pi f_\pi \quad ,$$

$$\langle 0 | J^{\rho}_{\mu} | \rho \rangle = \sqrt{2} \frac{M^2_\rho}{2\gamma_\rho} \epsilon_\mu \quad ,$$

$$\left\langle 0 \left| J_\nu^\omega \right| \omega \right\rangle = \frac{M_\omega^2}{2\gamma_\omega} \epsilon_\nu \quad ,$$

$$\langle \omega\, (p_1, \epsilon_1)\; | \rho(p_2, \epsilon_2)\; \pi(p_3) \rangle = g_{\omega\rho\pi} \epsilon_{\mu\nu\rho\sigma} \epsilon_1^\mu \epsilon_2^\nu p_1^\rho p_2^\sigma \quad . \qquad (14.12)$$

We assume the duality between the two sides of the vertex (Fig. 14.2) and take their Laplace transform. Then, one obtains the sum rule for $\gamma_\omega \simeq 3\gamma_\rho$, $M_\rho \simeq M_\omega$ and $m_\pi^2 \simeq m_u^2 \simeq 0$[224,146]:

$$g_{\omega\rho\pi} \simeq \left(\frac{6\ \gamma_\rho^2}{M_\rho^4}\right) \cdot \left(\frac{1}{f_\pi m_\pi^2}\right) \cdot (m_u + m_d) \left\langle \bar{u}u \right\rangle \tau^{-1} e^{M_\rho^2 \tau} \quad . \qquad (14.13)$$

The sum rule stabilizes for $\tau \simeq 1\text{-}2\ \mathrm{GeV}^{-2}$. Using pion PCAC in order to eliminate the quark parameters, one obtains the leading order result[146]:

$$g_{\omega\rho\pi} \simeq 19.2\ \gamma_\rho^2\ f_\pi / \mathrm{GeV}^2 \simeq 12\ \mathrm{GeV}^{-1} \quad , \qquad (14.14)$$

where one should not take literally the analytical dependence in f_π. The expression is only a numerical one owing to the unknown dependence of the sum rule scale on the decay constant f_π.
Possible corrections to this leading order result have been discussed in Ref. 224) :

- The mixed quark condensate contributes as $(5\ M_0^2/36\tau)$ relative to the $\left\langle \bar{u}u \right\rangle$ one. The stability region is about 11 to 20% of the latter and remains a negligible correction.
- The most important corrections are due to the higher ω', ρ' and π' states. These can be larger than 30%, and indicate that the value in (14.14) should only be understood as order of magnitude indications

rather than as accurate estimates.

b) $\pi^0 \to \gamma\gamma$ and $a_0 \to \gamma\gamma$ widths

The previous estimate of $g_{\omega\rho\pi}$ can be used via vector-meson dominance for predicting the $\pi_0 \to \gamma\gamma$ width. One obtains :

$$g_{\pi\gamma\gamma} \simeq 2e^2 \, g_{\omega\rho\pi} \left(\frac{1}{2\gamma_\rho}\right) \left(\frac{1}{2\gamma_\omega}\right) \simeq 2.5 \; 10^{-2} \; \mathrm{GeV}^{-1} \qquad (14.15a)$$

leading to :

$$\Gamma(\pi_0 \to \gamma\gamma) \simeq |g_{\pi\gamma\gamma}|^2 \, \frac{m_\pi^3}{64\pi} \simeq 8.5 \; \mathrm{eV} \quad , \qquad (14.15b)$$

in good agreement with the data ((7.6 ± 0.3)eV). One can study in the same way the $a_0\left(u\bar{u} - \bar{d}d\right) \to \gamma\gamma$ width. Using the a_0 current as :

$$J_{a_0} = \frac{i}{\sqrt{2}} (m_d - m_u) : \left(\bar{d}d - \bar{u}u\right) : \quad , \qquad (14.16)$$

where we have taken the quark mass factor in order to have a renormalization group invariant operator and to show an explicit $SU(2)_F$ breaking, we obtain the QCD expression of the amplitude[146]:

$$T^{\mu\nu} = (p^\nu q^\mu - g^{\mu\nu}(pq)) \, (m_d - m_u) \, \frac{\langle \bar{u}u \rangle}{\sqrt{2} \, (Q^2)^2} \left\{1 + O\left(\frac{m^2}{Q^2}\right)\right\} \quad , \qquad (14.17)$$

where p and q are the momenta of the outgoing photon. We parametrize the spectral function by the a_0, ω and ρ poles. Then, taking the ratio of the π and a_0 amplitudes, we have[146]:

$$\frac{g_{a\omega\rho}}{g_{\pi\omega\rho}} \simeq \left(\frac{Q^2 + M_a^2}{Q^2 + m_\pi^2}\right) \left(\frac{m_\pi}{M_a}\right)^2 \left(\frac{f_\pi}{f_a}\right) \left(\frac{m_d = m_u}{m_d + m_u}\right) \tag{14.18}$$

at the symmetric Euclidian point. The value of f_a is known from the previous section to be between 0.5-1.8 MeV while owing to the absence of Q^2-stability in this leading order expression, we use $Q^2 \geqslant M_a^2$ where the OPE is expected to make sense. Combining these different sources of uncertainties leads to[201]:

$$\Gamma(a_0 \to \gamma\gamma) \simeq (0.3 \sim 1.5) \text{ keV} \quad , \tag{14.19}$$

where the upper value comes from the ratio $g_{a\omega\rho}/g_{a\eta\pi}$ studied in Ref. 201). The result is not quite precise but still agrees with the data[202]:

$$\Gamma(a_0 \to \gamma\gamma)\, B(a_0 \to \eta\pi) \simeq \begin{matrix} \left(0.19 \pm 0.07 \begin{matrix} + 0.10 \\ - 0.07 \end{matrix}\right) \text{ keV} \\ (0.19 \pm 0.05 \pm 0.14) \text{ keV} \quad . \end{matrix} \tag{14.20}$$

We should notice that owing to these large uncertainties, we are not surprised that the prediction[228] $\psi \to \eta_c \gamma$ obtained within a similar context, though with a different choice of vertex configuration, is higher than the data by a factor 2-3.

Finally, we can also derive the $\gamma\gamma$ width of the singlet $(m_u + m_d) \frac{1}{\sqrt{2}} \left(\bar{u}u + \bar{d}d\right)$ meson, which we call S_2. Using a similar technique, we obtain for $M_2 \simeq M_a$ [201]:

$$\frac{g_{a_0\gamma\gamma}}{g_{S_2\gamma\gamma}} \simeq \left(\frac{m_d - m_u}{m_d + m_u}\right) \left(\frac{f_2}{f_u}\right) \frac{2\,\gamma_\rho \gamma_\omega}{\left(\gamma_\rho^2 + \gamma_\omega^2\right)} \simeq \frac{3}{5} \quad , \tag{14.21}$$

using the fact that $f_a \sim (m_d - m_u)$ and $f_2 \sim (m_d + m_u)$ from the two-point function sum rule. Eq. (14.21) is in nice agreement with the quark charge counting rule factor circulating inside the triangle :

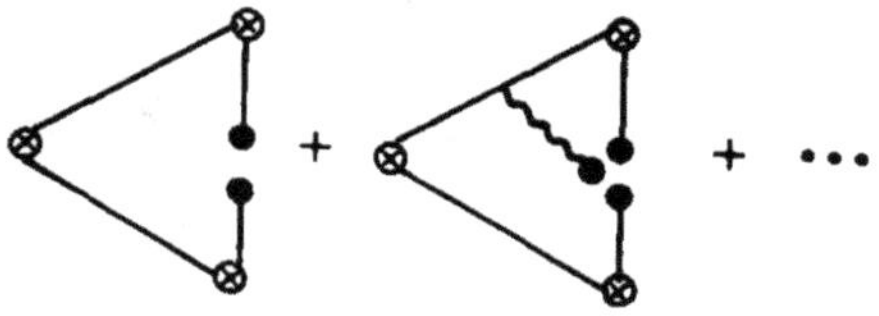

Fig. 14.4 : Leading QCD diagrams controlling the a_o $K\bar{K}$ coupling.

$$\Gamma(S_2 \to \gamma\gamma) \simeq \frac{25}{9} \Gamma(a_o \to \gamma\gamma) \simeq 0.67 \text{ keV} \quad . \qquad (14.22)$$

This agreement might indicate that vertex sum rules can provide a precise prediction for the ratio of width of the mesons having equal masses. A further check on this result should be worked out.

c) $a_o \to \eta\pi$ and $S_2 \to \pi^+\pi^-$ widths

For the former, we shall estimate the $a_o K\bar{K}$ coupling and deduce the $a_o\eta\pi$ one from a SU(3) relation :

$$g_{a_o\eta\pi} \simeq \sqrt{\frac{2}{3}}\, g_{a_o K^+ K^o} \qquad (14.23)$$

between the two couplings. The quark amplitude for the $a_o K\bar{K}$ reads[224] (Fig. 14.4) :

$$T(Q^2) \simeq - m_s^2 (m_d - m_u) \left\{ \frac{3}{4\pi^2} m_s \log \frac{Q^2}{\nu^2} + \frac{1}{2Q^2} \left\langle \bar{s}s - 2\bar{d}d \right\rangle \right\} . \quad (14.24)$$

Introducing as usual the lowest mass real poles, we obtain the Laplace sum rule :

$$g_{aK^+K^0} \simeq \left(\frac{M_a^2 - M_K^2}{2f_K^2 M_K^4 \sqrt{2}\, f_a M_a^2} \right) \frac{\tau^{-4}}{e^{-M_K^2 \tau} - \frac{\tau^{-1}}{M_a^2 - M_K^2} \left(e^{-M_K^2 \tau} - e^{-M_a^2 \tau} \right)} .$$

$$. \; m_s^2 \; (m_d - m_u) \left\{ \frac{3}{4\pi^2} m_s - \left\langle \bar{s}s - 2\bar{d}d \right\rangle \tau \right\} . \quad (14.25)$$

It is instructive to use the two-point function sum rules between the quark masses and the decay amplitudes. To leading order, one obtains (see Chapter 7) :

$$m_d - m_u \simeq \frac{4\pi}{\sqrt{3}} f_a \, M_a^2 \, \tau \, e^{-M_a^2 \tau/2}$$

$$m_s^2 \simeq \frac{16\pi^2}{3} f_K^2 M_K^4 \, \tau^2 \, e^{-M_K^2 \tau} \quad (14.26)$$

and therefore[224]:

$$g_{a_o K\bar{K}} \simeq \frac{8\pi^2}{3\sqrt{2}} \frac{m_s \left\langle \bar{s}s - 2\bar{d}d \right\rangle}{M_K^2 \, f_K} \simeq 2 \text{ GeV} \quad (14.27a)$$

which gives :

$$\Gamma(a_o \to \eta\pi) \simeq \frac{1}{16\pi} |g_{a\eta\pi}|^2 \frac{1}{M_a} \left(1 - \frac{M_\eta^2}{M_a^2}\right) \simeq 37 \text{ MeV} \quad . \qquad (14.27b)$$

The $S_2 \to \pi^+\pi^-$ can be studied in a similar way. The amplitude reads[229]:

$$T(Q^2) = \frac{\sqrt{2}}{Q^2} (m_u + m_d)^3 \left\langle \bar{u}u \right\rangle \quad . \qquad (14.28)$$

Taking the Laplace transform and using the two-point function sum rule :

$$(m_d + m_u) \simeq \frac{4\pi}{\sqrt{2}} f_2 M_2^2 \tau e^{-M_2^2 \tau/2} \qquad (14.29)$$

in order to eliminate the S_2-decay amplitude, one obtains[201]:

$$g_{S_2\pi^+\pi^-} \simeq \frac{16\pi^3}{3\sqrt{3}} \left\langle \bar{u}u \right\rangle \tau \exp\left(M_2^2 \tau/2\right) \quad , \qquad (14.30)$$

which for the value of $\tau \simeq 1$ GeV^{-2} leads to a width :

$$\Gamma\left(S_2 \to \pi^+\pi^- + \pi^0\pi^0\right) \simeq 180 \text{ MeV} \quad , \qquad (14.31)$$

in perfect agreement with that deduced from the non-relativistic quark model :

$$\Gamma(S_2 \to \pi\pi) \simeq \frac{3}{2} \left(\sqrt{\frac{3}{2}}\right)^2 \frac{\vec{P}_\pi}{\vec{P}_\eta} \Gamma(a_o \to \eta\pi) \simeq 180 \text{ MeV} \quad . \qquad (14.32)$$

or from an improved linear σ model Lagrangian.

d) $SU(3)_F$ breakings and some other uses of the vertex sum rules

Ratios of some vertex sum rules have also been used[230] for measuring the $SU(3)_F$ breaking of the hadronic couplings despite the large uncertainties for each individual sum rule involved. The evaluation of the vertex at the symmetric point leads to :

$$T_{\rho\pi\pi}(Q^2) = -(m_u+m_d)^2 \left\{ \frac{3}{8\pi^2} \log \frac{Q^2}{\nu^2} - \frac{(m_u+2m_d)\langle \bar{u}u\rangle + (m_d+2m_u)\langle \bar{d}d\rangle}{4\,Q^4} - \frac{\langle \alpha_s G^2\rangle}{48\pi\,Q^4} \right\}$$

$$T_{\rho\bar{K}K} \simeq -(m_u+m_s)^2 \left\{ \frac{3}{16\pi^2} \log \frac{Q^2}{\nu^2} + I_{x,y} \frac{m_s^2}{f_\pi^2\,Q^2} - \frac{(2m_s-m_u)\langle \bar{s}s\rangle - 2m_u\langle \bar{u}u\rangle}{4\,Q^4} - \frac{\langle \alpha_s G^2\rangle}{96\,Q^4} \right\} . \tag{14.33}$$

Taking the Laplace transform of each vertex and working with the ratio of the two transforms leads to the prediction for :

$$r \equiv \frac{2g_{\rho K\bar{K}}}{g_{\rho\pi\pi}} \simeq \left(\frac{f_\pi}{f_K}\right)^2 \left(\frac{m_\pi}{M_K}\right)^4 \frac{\left(M_\rho^2 - M_K^2\right)^2}{\left(M_\rho^2 - m_\pi^2\right)^2} \left(\frac{m_u + m_s}{m_u + m_d}\right)^2 \cdot \frac{F_{\rho\pi\pi}(\tau)}{F_{\rho\bar{K}K}(\tau)} .$$

$$\cdot \frac{18\tau^{-2} - 12\, m_s^2 \tau^{-1}\, I_{x,y} + 24\pi^2 \left[2(m_u - m_s)\left\langle \bar{s}s \right\rangle + 2m_u \left\langle \bar{u}u \right\rangle\right] + \pi \left\langle \alpha_s G^2 \right\rangle}{18\tau^{-2} + 36\pi^2 (m_u + m_d) \left\langle \bar{u}u \right\rangle + \pi \left\langle \alpha_s G^2 \right\rangle}$$

with :

$$F_{VPP} \equiv \tau^{-1}\, e^{-M_V^2 \tau} - \left(\tau^{-1} - M_V^2 + m_p^2\right) e^{-m_p^2 \tau} \quad . \qquad (14.34)$$

Analogous sum rules have also been derived for determining the ratio of the ϕKK over $\rho\pi\pi$ and $K^{*0}K^0\gamma$ over $K^{*+}K^+\gamma$ couplings. All these results exhibit nice τ-stabilities and are not quite sensitive to the values of $\bar{m}_s$ and $\left\langle \bar{s}s \right\rangle$ given in the range in Chapter 8. The success of these predictions leads one to imagine that there are some cancellations among the higher-state contributions in the ratios of sum rules, but a rigourous proof of this observation is still missing. The estimates[222,224] of the $\pi NN, \eta NN, \rho NN$... couplings and the proton wave function is much more subtle than in the meson cases. In addition to the various sources of uncertainties encountered for the mesons, we have to be aware of the arbitrariness of the choice of the nucleon operators. In fact, the optimal choice obtained in the two-point function sum rules might not be the same for three-point function[224,231]. Therefore, one should consider these estimates as qualitative results. It is for instance important to notice that the ηNN coupling is suppressed[224] compared to the πNN one. This result is relevant for the present crisis of the proton spin.

e) Pion form factors

For the study of the pion form factor, the symmetric configuration in (14.2) cannot help, since the rho momentum should be kept at moderate energies. We shall be concerned with the three-point function :

$$T_{\alpha\beta\mu}(p,q) = -\int d^4x d^4y\, e^{-ipx}\, e^{i(p+q)y} \left\langle 0 \left| \mathbb{T} A_\alpha(x)\; J^\rho_\mu(y) (A_\beta(o))^+ \right| 0 \right\rangle \quad (14.35)$$

where $A_\alpha(x)$ is the axial-vector current but one could also have started from the pseudoscalar current where the associated vertex is related to the former by Ward identities (obscure instanton arguments unfavourable to the choice of the pseudoscalar current are unclear to us). The pion form factor is introduced as :

$$\left\langle 0 \; |J^\pi| \; p \right\rangle \left\langle p \left| J^\rho_\mu \right| q \right\rangle \left\langle q \; |J^\pi| \; 0 \right\rangle = 2f^2_\pi\, F_\pi(Q^2)\, p^\alpha q^\beta\, (p+q)^\mu . \quad (14.36)$$

A QCD evaluation of the relevant amplitude plus the use of a double Laplace transform would lead to the sum rules at equal sum-rule scale $(\tau_1 = \tau_2 \equiv \tau)$ 222) :

$$2f^2_\pi F_\pi(Q^2,\tau) = \frac{1}{\pi^2} \int_0^{t_c} dt_1 \int_0^{t_c} dt_2\; \rho\left(t_1, t_2, Q^2\right) \exp(-\,(t_1+t_2)\,\tau)$$

$$+ \frac{\left\langle \alpha_s G^2 \right\rangle}{12\pi}\,\tau + \frac{208\pi\, \alpha_s \left\langle \bar{u}u \right\rangle^2}{81}\,\tau^4 \left(1 + \frac{2}{13}\, Q^2 \tau\right) + \ldots , \quad (14.37)$$

where :

$$\rho = \frac{3Q^4}{4} \left\{ \left(\frac{d}{dQ^2}\right)^2 + \frac{Q^2}{3} \left(\frac{d}{dQ^2}\right)^3 \right\} \frac{1}{\left[\left(t_1+t_2+Q^2\right)^2 - 4t_1 t_2\right]^{1/2}} \quad (14.38)$$

is the spectral function associated with the perturbative graph. It should be noticed that at given τ where the OPE is expected to work, one can see that the role of power terms is much greater than the perturbative contributions previously obtained in Ref. 232) :

$$F_\pi(Q^2)\Big|_{Pert.} \simeq \frac{4\pi\ \alpha_s(Q^2)}{Q^2} \quad . \tag{14.39}$$

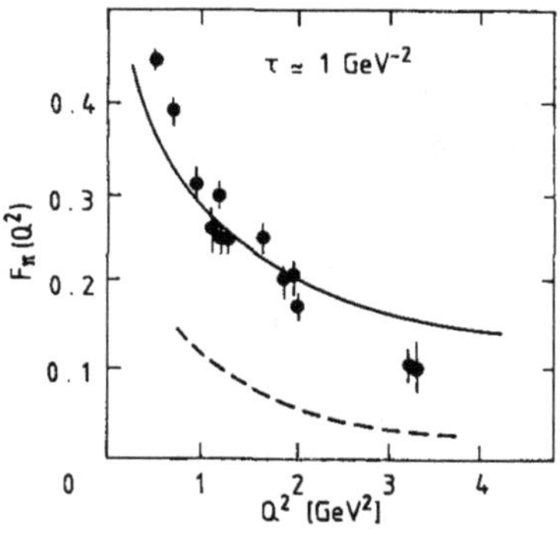

Fig. 14.5 : Pion form factor as a function of the Q^2: --- the perturbative QCD contribution , ____ the QCD prediction including power corrections.

For large Q^2 and small τ, one also recovers the $\frac{1}{Q^2}$ behaviour intuitively expected. The behaviour of the form factor is given in Fig.14.5 for $\tau \simeq 1\ GeV^{-2}$. There is good qualitative agreement with the data, which indicates that the role of power corrections is great in the intermediate region of QCD. Unfortunately, the lack of control for other high mass state effects in the sum rules renders the previous results no more than a qualitative estimate.

f) Meson-wave functions

Let us discuss the estimate of the meson-wave functions according to the original works of Chernyak and Zhitnitsky[222] (hereafter denoted CZ). For the vector and axial currents, we shall be concerned with the correlator :

$$I^{(5)}_{no}(z,q) = i \int d^4x\, e^{iqx} \langle 0 \,|T\, O_n(x)\, O_o(o)|\, 0\rangle$$

$$= (z.q)^{n+2}\, I^{(5)}_{no}\,(q^2\,) \quad , \tag{14.40a}$$

where the operator $O^{(5)}_n(x)$ is :

$$O^{(5)}_n(x) = \bar{d}\, \hat{z}(\gamma_5)\, (iz\, \overleftrightarrow{D})^n\, u \quad , \tag{14.40b}$$

where $\overleftrightarrow{D}_\mu \equiv \overrightarrow{D}_\mu - \overleftarrow{D}_\mu$ is the covariant derivative and z is a light-like vector ($z^2=0$, $z.q \equiv q^+$).

i) π and A_1 wave functions : in the axial-vector channel, the correlator reads (CZ) :

$$I^5_{no}(q^2\,) = -\frac{1}{4\pi^2}\log\frac{-q^2}{\nu^2}\,\frac{3}{(n+1)(n+3)} + \frac{\left\langle \alpha_s G^2 \right\rangle}{12\pi\, q^4}$$

$$-\frac{32\pi}{81}(11-4n)\;\frac{\alpha_s\left\langle \bar{u}u \right\rangle^2}{q^6} + \ldots \quad , \tag{14.41}$$

in the chiral limit $m_u = m_d = m^2_\pi = 0$ and where the factorization of the four-quark condensate has been used. Radiative corrections to it have been evaluated by Gorskii[222] but they are small. The corresponding spectral function can be parametrized as :

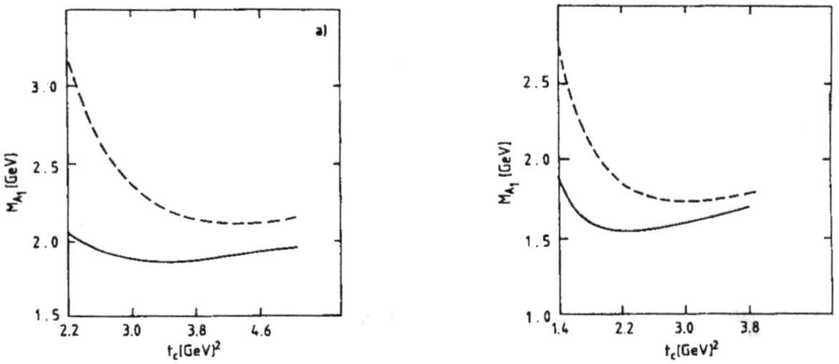

Fig. 14.6 a and b : t_c behaviour of the A_1 mass from FESR for n=2 and 4 for two different values of the QCD condensates : — $\langle \alpha_s G^2 \rangle$ = 0.04 GeV^4, $\alpha_s \langle \bar{u}u \rangle^2$ = 3.5 10^{-4} GeV^6 ; --- twice the previous values.

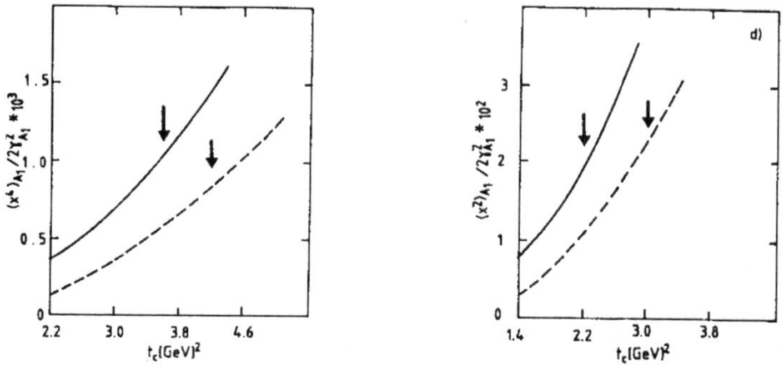

Fig. 14.6 c and d : n = 2 and 4 A_1 moments versus t_c from FESR.

$$\frac{1}{\pi} \, \mathrm{Im} \, I_{no}(t) = 2 \, f_\pi^2 \, \langle x^n \rangle_\pi \, \delta\left(t - m_\pi^2\right) + \frac{M_{A_1}^2}{2\gamma_{A_1}^2} \, \langle x^n \rangle_{A_1} \, .$$

$$\cdot\ \delta\left(t-M^2_{A_1}\right) + \text{"QCD continuum"}\ .\ \theta(t-t_c)\quad . \qquad (14.42)$$

Using the standard techniques discussed previously, one can derive a set of FESR constraints[222g)]:

$$2\ f^2_\pi\ \langle x^n\rangle_\pi + \frac{M^2_{A_1}}{2\gamma^2_{A_1}}\langle x^n\rangle_{A_1} \simeq C_{1}\, t_c^{(n)}\ , \qquad (14.43)$$

$$\frac{M^4_{A_1}}{2\gamma^2_{A_1}}\langle x^n\rangle_{A_1} \simeq \frac{C_{\mathbb{1}}}{2}\, t_c^{(n)^2} - C_4\langle O_4\rangle\ , \qquad (14.44)$$

$$\frac{M^6_{A_1}}{2\gamma^2_{A_1}}\langle x^n\rangle_{A_1} \simeq \frac{C_{\mathbb{1}}}{3}\, t_c^{(n)^3} - C_6\langle O_6\rangle\ , \qquad (14.45)$$

with the notation :

$$I^5_{no} \equiv -\ C_{\mathbb{1}}\ \log\frac{-q^2}{\nu^2} + C_4\langle O_4\rangle/q^4 - C_6\langle O_6\rangle/q^6\ , \qquad (14.46)$$

and f_π = 93.3 MeV, $\gamma_{A_1} \simeq 4.2$ (Chapter 6).

The ratio of (14.45) over (14.44) gives the A_1 mass squared where the optimal prediction is reached at the t_c-stability (Fig. 14.6 a and b for n=2 and 4). We have used two values of the QCD condensates :

$$\left\langle\alpha_s G^2\right\rangle \simeq 0.04\ (\text{resp. } 0.08)\ \text{GeV}^4\ ,$$

$$\alpha_s\left\langle\bar{u}u\right\rangle^2 \simeq 3.5\ (\text{resp. } 7)\ 10^{-4}\ \text{GeV}^6\quad . \qquad (14.47)$$

It can be seen that the prediction for the A_1 mass and the corresponding value of t_c increase with n. These t_c values do not coincide with the (ad hoc) choice of CZ. At these t_c values one obtains the moments (Fig. 14.6 c and d):

$$\langle x^2 \rangle_{A_1} \simeq 0.07 - 0.08 \qquad \langle x^4 \rangle_{A_1} \simeq 0.03 - 0.04 \ , \tag{14.48}$$

which are not sensitive to the values of the QCD condensates.

In order to deduce the π moments, we use (14.48) in (14.43) (see Fig. 14.7 a and b)). One can check these FESR predictions with that from LSR :

$$\int_0^{t_c} dt\, e^{-t\tau} \frac{1}{\pi} \operatorname{Im} I^5_{n0}(t) = C_{\mathbb{1}}\left(1 - e^{-t_c^{(n)}\tau}\right) \tau^{-1} + C_4\langle O_4 \rangle \tau + C_6 \langle O_6 \rangle \frac{\tau^2}{2} + \dots \tag{14.49}$$

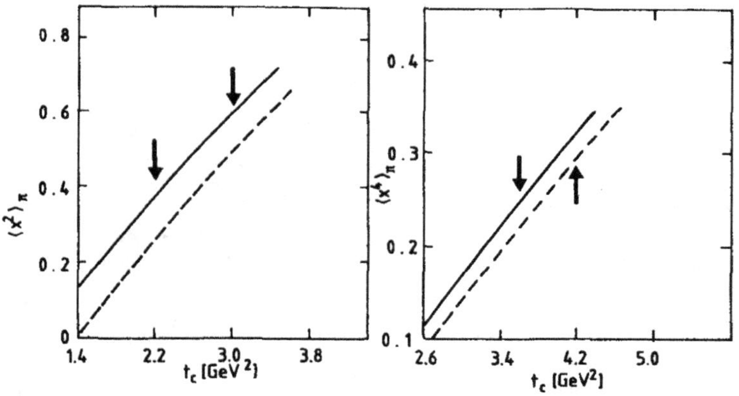

Fig. 14.7 a and b : $\langle x^n \rangle_\pi$ (n=2 and 4) from FESR versus t_c.

These LSR predictions are shown versus τ and t_c in Fig. 14.7 c and f. Both methods lead to the results :

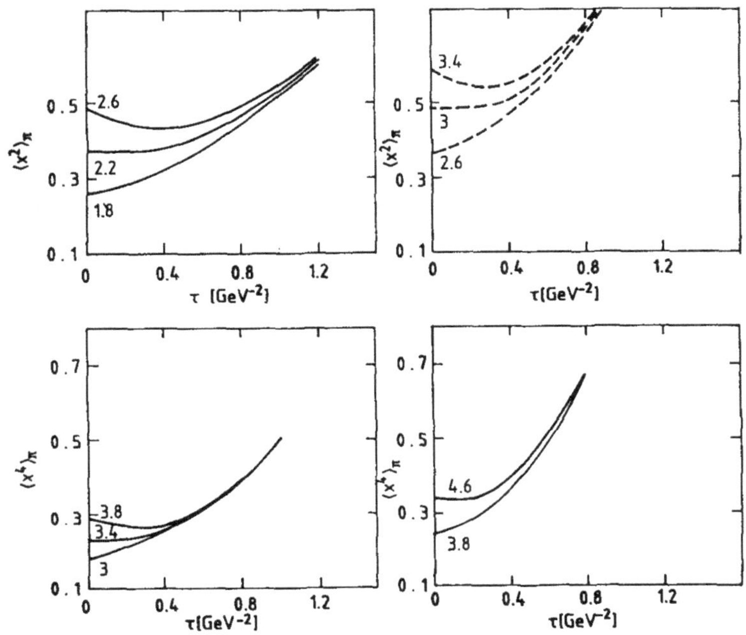

Fig. 14.7 c-f : $\langle x^n \rangle_\pi$ from LSR versus τ and t_c.

$$\langle x^2 \rangle_\pi = 0.38 - 0.60 \qquad \langle x^4 \rangle_\pi = 0.22 - 0.35 \quad , \tag{14.50}$$

where there is agreement among Refs 222 a,e and g). The errors in (14.50) are due to the QCD condensate values. Applying the same methods, one obtains[222g)] :

$$\langle x^6 \rangle_{A_1} \simeq 0.02 \qquad \langle x^6 \rangle_\pi \simeq 0.17 - 0.22 \quad . \tag{14.51}$$

It is interesting to notice that the n-behaviour of the A_1 moment can be qualitatively understood from (14.44) where :

$$\langle x^n \rangle_{A_1} \simeq \frac{15}{(n+1)(n+3)} \langle x^2 \rangle_{A_1} , \qquad (14.52)$$

while that for the pion is difficult to understand owing to the partial compensation of the increase of $t_c^{(n)}$ and n suppression at large n.

ii) ρ **wave functions:** In the vector channel, the correlator associated with the operator (14.40b) can be deduced from (14.41) by changing the coefficient (11+4n) of the four-quark operator by (4n-7) owing to the γ_5-chirality flip (CZ). The moments associated with the transverse ρ_T can be analyzed from the correlator associated with the operator :

$$\mathcal{O}_\mu^{(n)}(x) = \bar{d}\, \sigma^{\mu\nu} z^\nu (iz \overleftrightarrow{D})^n u \quad . \qquad (14.53)$$

The QCD expression of this correlator is (CZ) :

$$I_{no}(q^2) = -\frac{1}{4\pi^2}\left(\log \frac{-q^2}{\nu^2}\right)\frac{3}{(n+1)(n+3)}$$

$$+ \left(\frac{n-1}{n+1}\right)\frac{\langle \alpha_s G^2 \rangle}{12\pi\, q^4} - \frac{128\pi}{81}(1-n)\frac{\alpha_s \langle \bar{u}u \rangle^2}{q^6} \quad . \qquad (14.54)$$

While the first correlator is only affected by the longitudinal ρ_L, the second one in practice involves ρ_T and b_1 mesons, where for the later mesons we only use the positivity of its contribution.
We shall study the ρ_L and ρ_T in the same way, i.e., we shall use FESR constraints analogous to these in (14.43) and (14.45) and the LSR similar to (14.49). First, we study the ratio (14.44) over (14.43) but this quantity does not provide useful information owing to the absence

of t_c-stability,in contrast to the analogue (14.45 over 14.44) which is t_c-stable in both cases. (Fig. 14.8a,b and 14.9a,b). We should point out that the value of the $t_c^{(n)}$ and the ρ-mass are almost equal for ρ_T and ρ_L.

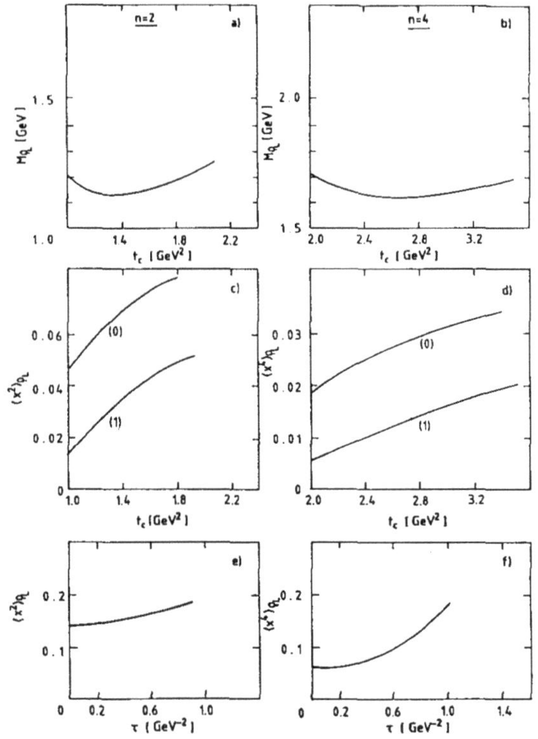

Fig. 14.8 a and b : ρ_T mass from FESR Eq (14.44) over (14.43)
c and d : n = 2 and 4 ρ_T moments from FESR versus t_c
e and f : n = 4 ρ_T moments from LSR versus τ and t_c

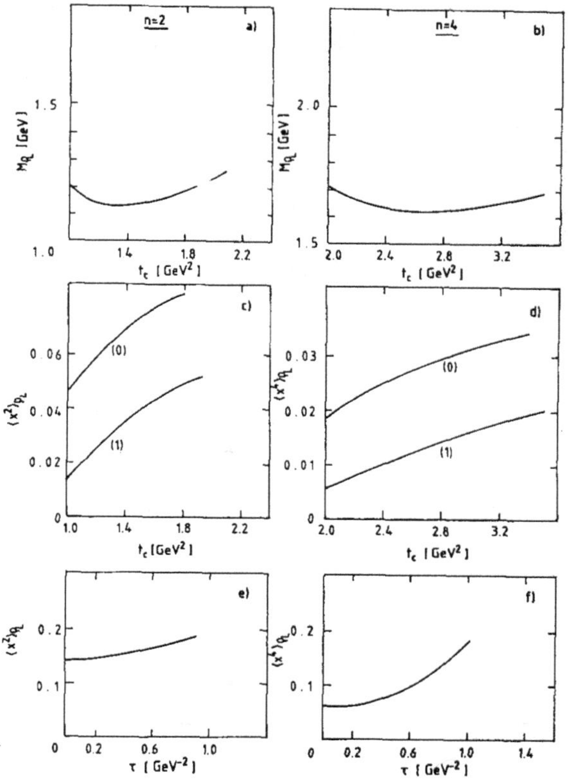

Fig. 14.9 : analogous to ρ_T but for ρ_L.

Next we study the t_c-behaviour of the moments from Figs 14.8 c and d, 14.9 c,d by using the 0th and 1st constraints in (14.43) and (14.44). As can be seen, they give almost the same predictions for ρ_L and ρ_T but do not show a common solution. From this last weak point, the prediction for $\langle x^n \rangle_\rho$ from the FESR cannot be accurate. The only reliable information from FESR is the value of the ratio of the moment distributions :

$$\langle x^n \rangle_{\rho_T} \Big/ \langle x^n \rangle_{\rho_L} \lesssim 1 - 1.2 \quad n = 2,4 \tag{14.55}$$

from (14.43) and (14.44) where we have used the positivity of the b_1-contribution in order to get the inequality.

Now, let us turn to the LSR. In contrast to the ρ_T ones (Fig. 14.8 e and f), the ρ_L moments do not have τ-stability (Fig. 14.9 e and f) owing to the sign of the four-quark contribution. Therefore the values of the ρ_L moments from CZ taken at $\tau^{-1} \simeq 1.2 - 1.5$ GeV^2 are difficult to justify and cannot be accurate. Therefore the most reliable estimate we can extract from the ρ-sum rule is that for ρ_T at the stability point :

$$\langle x^2 \rangle_{\rho_T} \lesssim 0.14 \qquad \langle x^4 \rangle_{\rho_T} \lesssim 0.06 \quad . \tag{14.56}$$

These inequalities agree with the estimate of Ref. 222a) obtained by neglecting the b_1 contribution. They also indicate that the ρ_T is narrower than the naive asymptotic estimate $\left(\langle x^2 \rangle^{as}_{\rho_T} \simeq 0.2\right)$.

Finally, we test the method independence of the result in (14.56) by working with the LSR ratio of the ρ_L and ρ_T. This ratio does not stabilize as expected but for the τ values in the τ-stability of ρ_T ($\tau \leqslant 0.6$ GeV^{-2}), one obtains :

$$\langle x^n \rangle_{\rho_T} \Big/ \langle x^n \rangle_{\rho_L} \lesssim 0.80 - 0.85 \quad : \quad n = 2,4 \tag{14.57}$$

which is comparable in strength with that in (14.55).
A better comparison with the result of CZ can be done by neglecting as they do the effect of b_1. In this case, the inequalities in (14.55) to (14.57) become equalities. Therefore our results suggest that the ρ_L-moments obtained in Ref. 222a) have been overestimated by a factor more than 1.7 for the n = 2 moment and 2.3 for the n=4 one. Their results also indicate that like the ρ_T, the ρ_L is narrower than the asymptotic estimate. We are aware of the consequences of our results for a realistic model building of the ρ-meson wave functions and for ρ-pair

decays of heavy quarkonia.

iii) B meson wave function. Before concluding, we also test the reliability of the value of the n=1 mean momentum fraction of the light quark in the B meson by using the non-relativistic sum rule (CZ) :

$$2 f_B^2 M_B^2 \langle x_q \rangle e^{-E_R/m} \simeq \left[\frac{1}{6}\int_o^{\infty} \frac{dx\ x^3\ e^{-x}}{(1+\beta x)^4}\right] \frac{24}{\pi^2} m^4 \ .$$

$$\cdot \left(1 - e^{-E_c/m} \sum_0^3 \left(\frac{E_c}{m}\right)^i \frac{1}{i!}\right) + \frac{g}{12m} \left\langle \bar{u}\ \sigma^{\mu\nu}\ G^a_{\mu\nu}\ \lambda_a\ u \right\rangle +$$

$$+ \frac{\pi}{144} \frac{1}{m^2} \alpha_s \left\langle \bar{u}\ u \right\rangle^2 + \dots \ , \qquad (14.58)$$

where $E_R \simeq 0.4$ GeV and E_c are respectively is the energy of the B meson and of the QCD continuum ; $M_b \simeq M_B \simeq (4.56 - 5.28)$ GeV are the quark and meson masses in this large mass limit : $\beta = \frac{2m}{M_b}$; m is the non-relativistic sum rule variable ; $g \left\langle \bar{u}\ \sigma^{\mu\nu}\ \lambda_a\ G^a_{\mu\nu}\ u \right\rangle \simeq 2\ M_o^2 \left\langle \bar{u}\ u \right\rangle$ where $M_o^2 \simeq 0.8\ GeV^2$ from the B-B* sum rule (see Chapter 10). The LSR prediction of $\langle x_q \rangle$ has m- but not E_c-stability. The LSR ratio of moments cannot help for fixing E_c*). Therefore, we work with the FESR version of (14.58) and of the n = o moment obtained in Ref. 171). Working with their ratio one can give a prediction independent of f_B :

$$M_B \langle x_q \rangle \simeq \frac{E_c^{(1)}}{1 - \left\langle \bar{q}\ q \right\rangle \frac{\pi^2}{\left(E_c^{(o)}\right)^3}} \qquad (14.59)$$

*) We cannot find a good argument for justifying the choice $E_c = 1.6$ GeV of (CZ).

where the indices 0 and 1 correspond to the n=0 and 1 moments.
A clean dependence of $\langle x_q \rangle$ on M_b can only be deduced if one knows the scaling behaviour of E_c.
Assuming that $E_c^{(1)} \simeq E_c^{(0)} \simeq 1.2$ GeV[171], one obtains :

$$\langle x_q \rangle \simeq 0.21 - 0.24 \tag{14.60}$$

a value which is a factor two higher than the one in CZ. However, it is difficult to appreciate the real accuracy of the approach based on (14.58) in our case where the b-quark mass value is not too high for accurate non-relativistic treatment.

iv) **Conclusions.** Our analysis has shown that QSSR can provide an improved determination of the meson wave functions compared to the naive perturbative QCD expectations. However, the accuracy reached is lesser than in the case of the n = 0 moments discussed in previous chapters, with the exception of the pion moments. One should note that the applications of this method to the nucleon moments can be affected by the arbitrariness of the value of the mixing between the two nucleon operators. In this way, we do not expect that the accuracy of the predictions in this channel is better than in the case of mesons.

CHAPTER 15

WEAK DECAYS - CP VIOLATION

$T(4S) \to B^0 \bar{B}^0$
$\quad \hookrightarrow B^0$

$B_1^0 \to D_1^- \mu_1^+ \nu_1$
$\quad \hookrightarrow \pi_1^- D^0$
$\qquad \hookrightarrow K_1^+ \pi_1^-$

$B_2^0 \to D_2^{*+} \mu_2^- \nu_2$
$\quad \hookrightarrow \pi^+ D^-$
$\qquad \hookrightarrow K_2^+ \pi_2^- \pi_2^-$
$\qquad\quad \hookrightarrow \gamma\gamma$

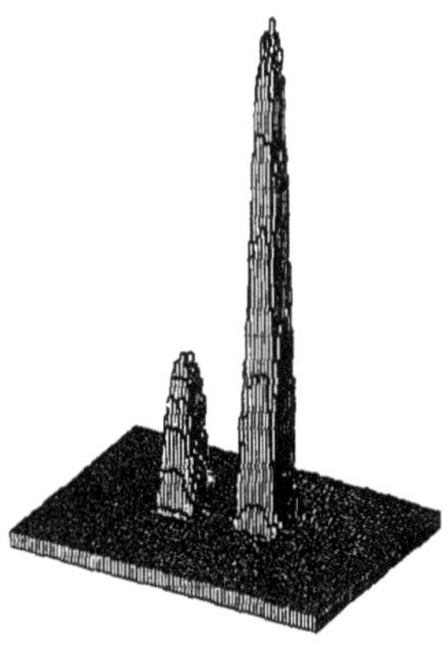

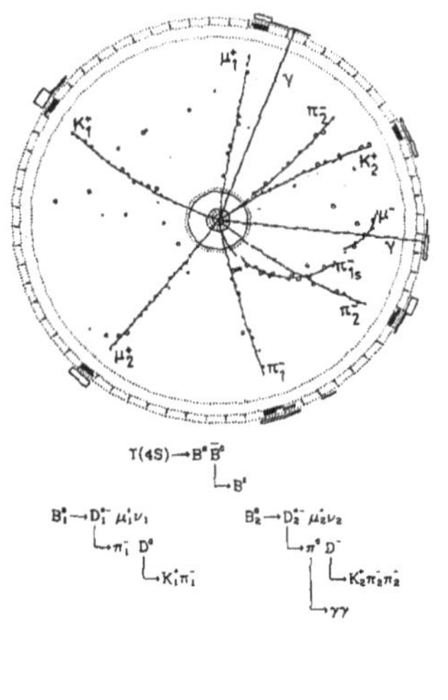

Typical CP-violating events in K- and B-decays

1. $\Delta S = 2$ K_o-$\bar{K}_o$ MASS DIFFERENCE AND THE B PARAMETER

The QSSR in this channel was first used in Ref. 233) and later improved by Pich and de Rafael (PR)[94].

The $\Delta S = 2$ $K_o \longrightarrow \bar{K}_o$ transition is allowed via the so-called box diagram with two W exchanges. Using a perturbative approach and removal of the heavy-quark fields, this off-diagonal matrix element can be expressed as[234]:

$$M_{12} = \frac{G_F^2 M_W^2}{4\pi^2} \left\{ \lambda_c^2 \eta_1 \frac{m_c^2}{M_W^2} + \lambda_t^2 \eta_2 \frac{m_t^2}{M_W^2} + 2 \lambda_c \lambda_t \eta_3 \frac{m_c^2}{M_W^2} \log \frac{m_t^2}{M_W^2} \right\} \cdot$$

$$\cdot \left(\frac{\alpha_s \left(m_c^2 \right)}{\alpha_s (\nu^2)} \right)^{2/9} \frac{1}{2M_K} \left\langle \bar{K}_o \; | \mathcal{O}_{\Delta S=2} | \; K_o \right\rangle \qquad (15.1a)$$

where the K_S^o - K_L^o mass difference is :

$$M_L - M_S \simeq 2 \text{ Re } M_{12} \quad . \qquad (15.1b)$$

The four-quark operator $\mathcal{O}_{\Delta S=2}$ is defined as :

$$\mathcal{O}_{\Delta S=2} = \left(\bar{s}_L \gamma^\mu d_L \right) \left(\bar{s}_L \gamma_\mu d_L \right) \; : \; s_L \equiv \frac{1}{2} (1 - \gamma_5) s \qquad (15.2)$$

G_F is the Fermi coupling ; $\lambda_i \equiv V_{id} V_{is}^*$ $i \equiv u,c$ and d with $\lambda_u + \lambda_c + \lambda_t = 0$ are products of two-matrix elements of the Cabibbo-Maiani-Kobayashi-Waskawa mixing matrix [235] ; η_i are QCD correction Wilson coefficients. The $(\alpha_s(\nu))^{2/9}$ appears to be due to the fact that $\mathcal{O}_{\Delta S=2}$ has an anomalous dimension.

We shall be concerned here with the estimate of the matrix element in (15.2) using QSSR. Using a vacuum saturation assumption, Gaillard and Lee found[236]:

$$\left\langle \bar{K}_o \, | O_{\Delta S=2} | \, K_o \right\rangle = \frac{1}{2} \left(1 + \frac{1}{3} \right) 2 \, f_K^2 \, M_K^2 \qquad (15.3)$$

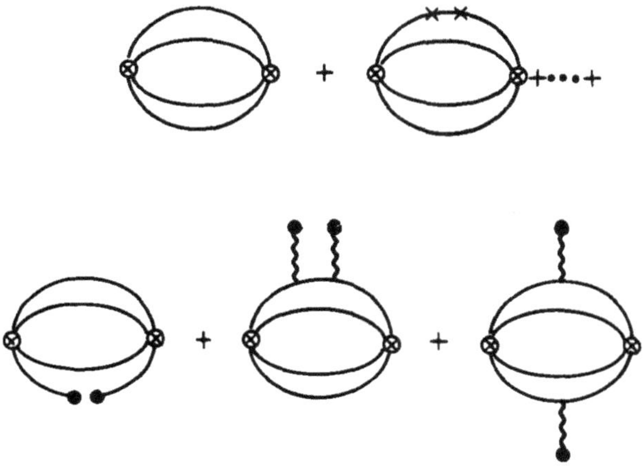

Fig.15.1

where $f_K \simeq 1.2 \, f_\pi$: f_π = 93.3 MeV. The bulk of QSSR is to test the validity of the vacuum saturation estimate by measuring the so-called B parameter defined as :

$$\left\langle \bar{K}_o \, | O_{\Delta S=2} | \, K_o \right\rangle = \frac{4}{3} f_K^2 \, M_K^2 \, B \, . \qquad (15.4)$$

The relevant correlator is (see Fig. 15.1)

$$\psi(q^2) = i \int d^4x \, e^{iqx} \left\langle 0 \left| \mathbb{T} \, O_{\Delta S=2}(x) \, \left(O_{\Delta S=2}(0) \right)^+ \right| 0 \right\rangle \qquad (15.5)$$

The QCD expression of its spectral function is[94]:

$$\frac{1}{\pi}\,\mathrm{Im}\,\psi_{QCD}(t) = \frac{1}{(16\pi^2)^3}\,\frac{1}{10}\left(1+\frac{1}{3}\right) t^4 \left(\frac{\alpha_s(\nu)}{\alpha_s(t)}\right)^{4/9} \cdot$$

$$\cdot\left\{1 - \frac{28}{15}\frac{\alpha_s}{\pi} - \frac{40\,\bar{m}_s^2}{t} - \frac{200\,m_s^4}{t^2} + \frac{20\pi^2}{t^2}\left(16\pi\, m_s\langle\bar{s}s\rangle - \langle\alpha_s G^2\rangle\right)\right\} \qquad (15.6)$$

where a $m_s^4 \log m_s$ term due to four-mass insertion has been absorbed by that of the $m_s\langle\bar{s}s\rangle$ term in the way defined in Eq. (3.37). It should be noted that the mass corrections and leading non-perturbative terms have large coefficients which force us to work at very short distances in order to justify the validity of the OPE. PR[94] derive two FESR :

$$\int_{4M_K^2}^{t_c} dt\,\frac{1}{\pi}\,\mathrm{Im}\,\psi(t) = \frac{1}{(16\pi^2)^3}\left(\frac{\alpha_s(\nu^2)}{\alpha_s(t_c)}\right)^{4/9}\frac{2}{75}\,t_c^5$$

$$\left\{1 - \frac{28}{15}\frac{\alpha_s}{\pi} - \frac{50\,\bar{m}_s^2}{t_c} - \frac{5}{3}\langle\Omega\rangle\Big/t_c^2\right\} \qquad (15.7)$$

$$\int_{4M_K^2}^{t_c} dt\,\frac{t}{\pi}\,\mathrm{Im}\,\psi(t) = \frac{1}{(16\pi^2)^3}\left(\frac{\alpha_s(\nu^2)}{\alpha_s(t_c)}\right)^{4/9}\frac{1}{45}\,t_c^6$$

$$\left\{1 - \frac{28}{15}\frac{\alpha_s}{\pi} - \frac{48\,\bar{m}_s^2}{t_c} - \frac{3}{2}\langle\Omega\rangle\Big/t_c^2\right\} \qquad (15.8)$$

where $\langle\Omega\rangle \equiv 200\ \bar{m}_s^4 - 20\pi\left[16\pi\ m_s\left\langle\bar{s}\ s\right\rangle - \left\langle\alpha_s G^2\right\rangle\right]$. QCD corrections to the FESR are smaller than 100% for $t_c \geq 2.3\ GeV^2$. These imply that the lowest K_o-$\bar{K}_o$ dominance of the spectral function is a very poor approximation. The situation is not improved by working with the Laplace transform[237] as at this scale, the depressive factor due to the exponential is not efficient. The best which one can do is to improve the parametrization of the spectral function by including higher states than K_o-$\bar{K}_o$. For that purpose, PR analyze the low-energy behaviour of the spectral function within chiral perturbation theory. In this way, the quark operator $\mathcal{O}_{\Delta S=2}$ has the low-energy realization :

$$\mathcal{L}_{\Delta S=2} \equiv g_2\ \frac{1}{3}\left(f_K^2/f_\pi^2\right)\ i\left(f_\pi^2\ U\ \partial_\mu U^+\right)\ \left(i\ f_\pi^2\ U\ \partial^\mu U^+\right) \qquad (15.9a)$$

where U(x) is a 3×3 unitary matrix incorporating the Goldstone fields :

$$U(x) = \exp\left[i\ \sqrt{2}\ /f_\pi\ \phi(x)\right] \qquad (15.9b)$$

with :

$$\phi(x) = \begin{bmatrix} \pi^0/\sqrt{2} + \eta/\sqrt{6} & \pi^+ & K^+ \\ \pi^- & \pi^0/\sqrt{2} + \eta/\sqrt{6} & K_o \\ K^- & \bar{K}_o & -2\eta/\sqrt{6} \end{bmatrix} . \qquad (15.9c)$$

The overall coupling g_2 has been normalized in such a way that it is equal to B. Then one has to study the low-energy behaviour of the two-point correlator :

$$\psi(q^2) = i \int d^4x \; e^{iqx} \left\langle 0 \left| \mathbb{T} \, \mathcal{L}_{\Delta S=2}(x) (\mathcal{L}_{\Delta S=2}(0))^{+} \right| 0 \right\rangle \quad (15.10)$$

in chiral perturbation theory in the presence of the strong interactions effective Lagrangian :

$$\mathcal{L}_S = \frac{1}{4} f_\pi^2 \, \mathrm{Tr} \, \partial_\mu U \left(\partial_\mu U^{+}\right) + 2 M_K^2 \left(U(x)_{33} + U^{+}(x)_{33}\right) \quad (15.11)$$

and final-state interactions between pseudo-Goldstone particles leading to the formation of the 1^- $K^*(892)$, 0^+ $K_0^*(1350)$ ($K-\pi$ subchannel); 1^+ $Q(1.27)$ ($K-\pi\pi$) $K_0^*(1350)\pi$ and $K^*(892)\pi$... The spectral function can be parametrized as[94]:

$$\frac{1}{\pi} \mathrm{Im}\, \psi(t) = g_2^2 \, \frac{1}{16\pi^2} \, \frac{1}{18} \left(\frac{f_K}{f_\pi}\right)^4 \left[\int_{t_{10}}^{\left(\sqrt{t}-\sqrt{t_{20}}\right)^2} dt_1 \int_{t_{20}}^{\left(\sqrt{t}-\sqrt{t_1}\right)} dt_2 \; \cdot \right.$$

$$\cdot \; \lambda^{1/2}\left(1, \frac{t_1}{t}, \frac{t_2}{t}\right) \left\{ (t_1+t_2-t)^2 \, \frac{1}{\pi} \mathrm{Im}\, \Pi^{(0)}(t_1) \, \frac{1}{\pi} \mathrm{Im}\, \Pi^{(0)}(t_2) \; + \right.$$

$$2\lambda\,(t, t_1, t_2) \, \frac{1}{\pi} \mathrm{Im}\, \Pi^{(0)}(t_1) \, \frac{1}{\pi} \mathrm{Im}\, \Pi^{(1)}(t_2) \; +$$

$$\left. \left. \left[(t_1+t_2-t)^2 + 8 t_1 t_2\right] \frac{1}{\pi} \mathrm{Im}\, \Pi^{(1)}(t_1) \, \frac{1}{\pi} \mathrm{Im}\, \Pi^{(1)}(t_2) \right\} + \ldots \right] \quad (15.12a)$$

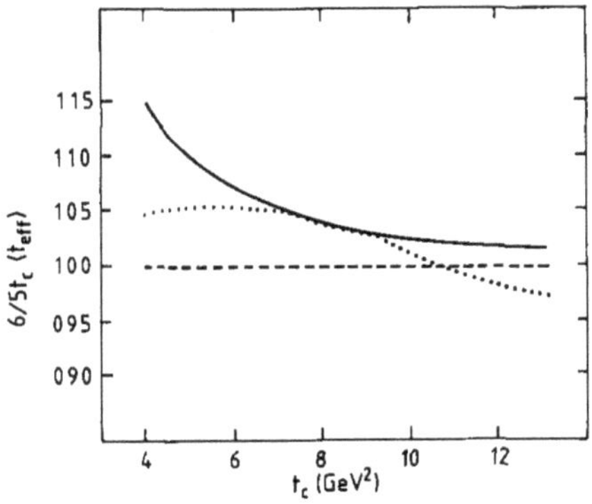

Fig. 15.2 : The ratio $6\langle t_{eff}\rangle/5t_c$ defined in Eq. (15.13) is plotted against t_c in GeV^2. The continuous line represents the QCD prediction as obtained from the ratio on the RHS of Eqs (15.8) and (15.7). It approaches asymptotically the value one - the dashed line in the Figure - at large t_c. The line of dots are the values obtained using the effective spectral function defined by Eqs (15.9) to (15.12).

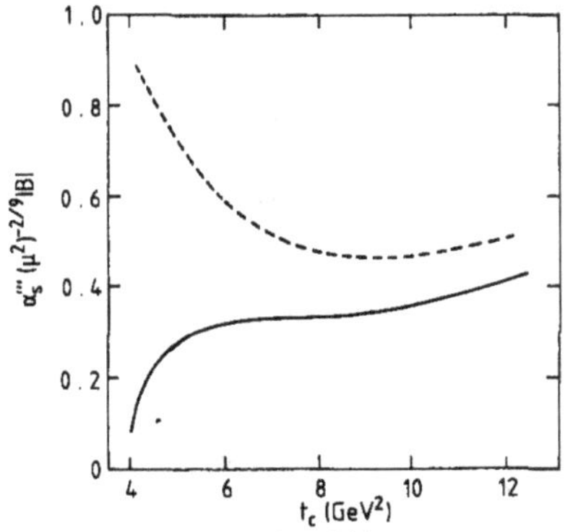

Fig. 15.3 : The quantity $\alpha_s''' (\mu^2)^{-2/9}$ |B| obtained from Eq. (15.7) is plotted against t_c in GeV^2. The continuous line is the result obtained after inclusion of the QCD corrections shown in Eq. (15.7), the dashed line is the result with the asymptotic freedom term only.

where :

$$\frac{1}{\pi} \operatorname{Im} \Pi^{(0,1)} \equiv \frac{1}{\pi}\left(\operatorname{Im} \Pi^{(0,1)}_{AA} + \operatorname{Im} \Pi^{(0,1)}_{VV}\right) \quad . \qquad (15.12b)$$

The indices 0,1 refer to spin 0 and 1 mesons, AA and VV to spectral functions which are built from two axial and two vector currents. The...denotes the higher states continuum such as $KK\pi$, $KK\eta$... The resulting expressions of the spectral function are quite ugly but quite trivial. The validity of the spectral parametrization has been controlled by the Weinberg sum rules obeyed by the spectral function (see Chapter 1). For much better information, we show in Fig. 15.2 the t_c-behaviour of the ratio of FESR in Eqs (15.7,8) given by QCD and by the above spectral parametrization :

$$\langle t_c \rangle \equiv \frac{\int_{t_o}^{t_c} dt\ t \operatorname{Im} \psi(t)}{\int_{t_o}^{t_c} dt \operatorname{Im} \psi(t)} \quad . \qquad (15.13)$$

Consistency of the two sides (the so-called eigenvalue solution) is only reached at very high t_c values compared to the kaon mass :

$$t_c \simeq 6.5 - 9.5 \text{ GeV}^2 \qquad (15.14)$$

where one expects to have optimal information from the sum rules. The corresponding value of $g_2 \equiv |B|$ is given in Fig. 15.3. $|B|$ has t_c-stability at the same range of t_c value as the one in Eq. (15.14). Its t_c-behaviour can easily be understood. For $t_c \leqslant 6.5$ GeV^2, the QCD corrections are large but negative while for $t_c \geqslant 9.5$ GeV^2 the increase of B signals the breaking of the spectral function parametrization owing to the importance of higher states.

At this t_c-stability region, one obtains the renormalization group invariant B parameter :

$$\hat{B} \equiv B(\nu) \;.\; \left(\alpha_s(\nu^2)\right)^{-2/9} = (0.33 \pm 0.09) \qquad (15.15)$$

which is about 1/3 of the usual vacuum saturation estimate[236] ($B \simeq 1$).

2. COMMENTS ON THE DETERMINATION OF B FROM VERTEX SUM RULES

The B parameter has also been estimated from the three-point function[238,239]:

$$\Gamma_{\mu\nu}(p,q) = - \int d^4x \; d^4y \; e^{i(px - qy)} .$$

$$. \left\langle 0 \left| \mathbb{T} \; \partial_\mu A^\mu(x)^d_s \; \mathcal{O}_{\Delta S=2}(0) \; \partial_\mu A^\mu(y)^d_o \right| 0 \right\rangle \qquad (15.16)$$

where : $\partial_\mu A^\mu(x)^d_s = (m_d + m_s) : \bar{s}(i\gamma_5)d$ is the pseudoscalar current coupled to the kaon. The $\Delta S=2$ operator has been used as a zero momentum squared operator insertion (see Fig. 15.4).

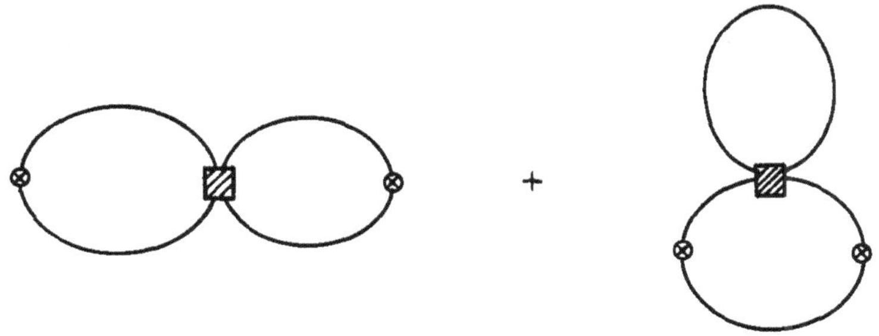

Fig. 15.4 : Vertex diagrams representing Eq. (15.16).

We have tried[240] to evaluate this diagram and encountered a serious difficulty. There is no straightforward way to absorb the $\frac{1}{\epsilon}\log\frac{-q^2}{\nu^2}$ non-local pole emerging from this kind of two-loop calculation except at a particular point $p^2 = -m_s^2$ where the effects of the two diagrams in Fig. 15.4 compensate. This point is however too low to justify the OPE. To our surprise, this crucial point has been overlooked by all authors in Refs 238) and 239) where the non-local pole has been removed without any explanation. If one forgets this technical point, one can pursue the QSSR analysis using standard techniques. Each author in Refs 238) gives his own predictions which depend on the parametrization of the spectral function and on the way of treating the vertex. The resulting value of B ranges from[238,239]:

$$B \simeq 0.2 - 1.3 \tag{15.17}$$

and reflects the typical uncertainties related to the uses of vertex sum rules. Improvement of these results is not straightforward and Eq. (15.17) should only be considered as an order of magnitude estimate which is not comparable in terms of accuracy with that in Eq. (15.15). The result in Eq. (15.15) which we consider as a definite QSSR prediction is comparable with that in Ref. 241) using kaon PCAC and chiral perturbation theory. Eq. (15.15) is however lower than the $1/N_c$-expansion estimate[242] :

$$B_{N_c} \simeq (0.7 \pm 0.1) \quad , \tag{15.18a}$$

while the lattice estimate is[178]

$$B_{\text{lattice}} \simeq (0.87 \pm 0.20) \quad . \tag{15.18b}$$

The apparent discrepancy between previous estimates can be solved once the systematic uncertainties of each different approach have been properly included.

3. $\Delta S = 1$ EFFECTIVE NON-LEPTONIC HAMILTONIAN AND $\Delta I = 1/2$ RULE

Relative enhancement of the $\Delta I = 1/2$ processes has been observed in different processes :

The $K_s \to \pi^+\pi^-$ width is observed to be about two orders of magnitude larger than the pure $|\Delta I| = 3/2$ $K^+ \to \pi^+\pi^0$ process :

$$\Gamma\left(K_s \to \pi^+\pi^-\right) \simeq (21.2)^2 \ \Gamma(K^+ \to \pi^+\pi^0) \quad . \tag{15.19a}$$

The $\Delta I = 1/2$ non-leptonic rates are found to be systematically larger than the corresponding (semi) leptonic amplitudes :

$$\Gamma(K_s \to 2\pi) \simeq (12.2)^2 \ \Gamma(K^+ \to \mu\nu) \quad ,$$

$$\Gamma(\Lambda \to p\pi) \simeq (27.7)^2 \ \Gamma(\Lambda \to pe\nu) \quad , \tag{15.19b}$$

while the $|\Delta I| = 3/2$ rates are of the same order of magnitude as the corresponding semileptonic ones :

$$\Gamma(K^+ \to \pi^+\pi^0) \simeq (4.4) \ \Gamma(K^+ \to \pi^0 e^+\nu) \quad ,$$

$$\Gamma(K^+ \to \pi^+\pi^+\pi^-) \simeq (1.2) \ \Gamma(K^+ \to \pi^0 e^+\nu) \quad . \tag{15.19c}$$

The advent of QCD has made it possible to study a Wilson expansion of the weak hadronic currents. Therefore, the effective $\Delta S = 1$ Hamiltonian can be reduced as a sum of local four-quark operators constructed with light u, d and s quark fields times the Wilson coefficients $C_n(\nu)$ which are functions of the heavy masses of the W and c,b and t quark fields. It is then found[243] that the effect of the leading QCD gluonic corrections provides an enhancement of the $\Delta I = 1/2$ transition by a factor 2 to 3 in the Wilson coefficients if one of the four-quark operators transforms like a $SU(3)_c$ octet. However, this gain is not enough to explain the observed enhancement. Thus, the $\Delta I = 1/2$ rule has remained a major puzzle in weak interaction phenomenology.

The lowest dimension four-quark operators which govern the $\Delta S=1$ effective Hamiltonian are[234]:

$$Q_1 = 4\left(\bar{s}_L \gamma^\mu d_L\right)\left(\bar{u}_L \gamma_\mu u_L\right)$$

$$Q_2 = 4\left(\bar{s}_L^\alpha \gamma^\mu d_L^\beta \bar{u}_L^\beta \gamma_\mu u_L^\alpha\right)$$

$$Q_3 = 4\left(\bar{s}_L \gamma^\mu d_L\right) \sum_{u,d,s} \bar{u}_L \gamma_\mu u_L$$

$$Q_4 = 4\, \bar{s}_L^\alpha \gamma^\mu d_L^\beta \sum_{u,d,s} \bar{u}_L^\beta \gamma_\mu u_L^\alpha$$

$$Q_5 = 4\left(\bar{s}_L \gamma^\mu d_L\right) \sum_{u,d,s} \bar{u}_R \gamma_\mu u_R$$

$$Q_6 = 4\left(\bar{s}_L^\alpha \gamma^\mu d_L^\beta\right) \sum_{u,d,s} \bar{u}_R^\alpha \gamma_\mu u_R^\beta \quad , \qquad (15.20a)$$

with :

$$\mathcal{H}_{\Delta S=1} = \Sigma\, C_n(\nu)\, Q_n(\nu) \quad . \qquad (15.20b)$$

$C_n(\nu)$ being the Wilson coefficients ; $\psi_{L(R)} \equiv \frac{1}{2}(1 - (+)\, \gamma_5)\psi$; α and β are colour indices and $\bar{\psi}_1 \gamma^\mu \psi_1 \equiv \sum_\alpha \bar{\psi}_1^\alpha \gamma^\mu \psi_1^\alpha$. Only five of these operators are independent since :

$$Q_1 + Q_4 = Q_2 + Q_3 \tag{15.21}$$

and the usual chosen basis is Q_1, Q_2, Q_3, Q_5 and Q_6.
From the point of view of $SU(3)_L \times SU(3)_R$ chiral symmetry,

$$\tilde{Q}_4 = Q_1 + 2/3\; Q_2 - 1/3\; Q_3 \tag{15.22}$$

transforms like $(27_L, 1_R)$ and induces both $|\Delta I| = 1/2$ and $|\Delta I| = 3/2$ transitions via its components

$$\tilde{Q}_4 = \frac{4}{9}\left(O_4^{1/2} + 5\; O_4^{3/2}\right) \quad , \tag{15.23a}$$

with :

$$O_4^{1/2} = \left(\bar{s}_L \gamma^\mu d_L\right)\left(\bar{u}_L \gamma_\mu u_L\right) + \left(\bar{s}_L \gamma^\mu u_L\right)\left(\bar{u}_L \gamma_\mu d_L\right)$$
$$+ 2\left(\bar{s}_L \gamma_\mu d_L\right)\left(\bar{d}_L \gamma_\mu d_L\right) - 3\left(\bar{s}_L \gamma^\mu d_L\right)\left(\bar{s}_L \gamma_\mu s_L\right)$$
$$O_4^{3/2} = \left(\bar{s}_L \gamma^\mu d_L\right)\left(\bar{u}_L \gamma_\mu u_L\right) - \left(\bar{s}_L \gamma^\mu d_L\right)\left(\bar{d}_L \gamma_\mu d_L\right)$$
$$+ \left(\bar{s}_L \gamma^\mu u_L\right)\left(\bar{u}_L \gamma_\mu d_L\right) \quad . \tag{15.23b}$$

The operator $\tilde{Q}_4$ is multiplicatively renormalizable and does not mix with the remaining octet $(8_L, 1_R)$ operators :

$$Q_2 - Q_1 \;,\; Q_3 \;,\; Q_5 \text{ and } Q_6 \quad , \tag{15.24}$$

which induce pure $|\Delta I| = 1/2$ transitions. These $(8_L, 1_R)$ operators mix under renormalizations.

Therefore, for convenience, one can divide the $\Delta S = 1$ short-distance Hamiltonian into three pieces with definite transformation properties :

$$\mathcal{H}_{\Delta S=1} = -\frac{G_F}{\sqrt{2}} s_1 c_2 c_3 \left\{\mathcal{H}_{27}^{(1/2)} + \mathcal{H}_{27}^{(3/2)} + \mathcal{H}_{8}^{1/2}\right\} \qquad (15.25)$$

where s_i, c_i are C-K-M rotation matrices.
The associated chiral Lagrangian can be written as :

$$\mathcal{L}_{\Delta S=1} = g_8^{(1/2)} \mathcal{L}_8^{(1/2)} + g_{27}^{(1/2)} \mathcal{L}_{27}^{(1/2)} + g_{27}^{(3/2)} \mathcal{L}_{27}^{(3/2)} \qquad (15.26a)$$

where :

$$\mathcal{L}_8^{(1/2)} = \left(L_\mu^+ L^\mu\right)_{23}$$

$$\mathcal{L}_{27}^{(1/2)} = L_{\mu 13}^+ L_{21}^\mu + \frac{1}{2} L_{\mu 23}^+ \left(9 L_{11+22}^\mu - L_{11-22}^\mu\right)$$

$$\mathcal{L}_{27}^{(3/2)} = L_{\mu 13}^+ L_{21}^\mu + L_{\mu 23}^+ L_{11-22}^\mu \quad , \qquad (15.26b)$$

with :

$$L_\mu \equiv i f_\pi^2 U \partial_\mu U^+ \quad , \qquad (15.26c)$$

is associated with the 3×3 unitary matrices describing the Goldstone fields (Eq. 15.9). The couplings g_i are not fixed by chiral symmetry requirements alone. Their values can be extracted from the observed $K \longrightarrow \pi\pi$ rates :

$$\left|g_8^{(1/2)} + g_{27}^{(1/2)}\right|_{EXP} \simeq 5.1$$

$$\left|g_{27}^{3/2}\right|_{EXP} \simeq 0.16 \qquad (15.27)$$

which indicate the large enhancement of the $|\Delta I = 1/2|$ transitions.

The aim of QSSR is to provide an estimate of these coupling constants in the same way as the one done for the B parameter in the $\Delta S=2$ transition. The key object with which one will work is the correlator :

$$\psi_R^{(I)}(q^2) \equiv i \int d^4x\, e^{iqx} \left\langle 0 \left| T\, \mathcal{H}_R^{(I)}(x) \left(\mathcal{H}_R^{(I)}(0)\right)^+ \right| 0 \right\rangle , \qquad (15.28)$$

where the high-energy behaviour will be studied within the usual OPE. The associated spectral function will be parametrized using its low-energy behaviour in chiral perturbation theory and by incorporating final-state interactions of pseudo-Goldstone bosons ($K\pi$, $K\pi\pi$...) leading to the formation of resonances with definite quantum numbers.

4. ESTIMATE OF THE $g_{27}^{(I)}$ COUPLINGS FROM QSSR

The QCD expressions of the sum rules for the $\Delta I = 3/2$ transitions are :

$$\int_{M_K^2}^{t_c} dt\, \frac{1}{\pi}\, \mathrm{Im}\, \psi(t) \simeq \frac{1}{(16\pi^2)^3} \left(\frac{\alpha_s(\nu^2)}{\alpha_s(t_c)}\right)^{4/9} \frac{1}{25}\, t_c^5 \,.$$

$$\cdot \left\{1 - \frac{28}{15}\left(\frac{\alpha_s}{\pi}\right) - \frac{25\, \bar{m}_s^2}{t_c} - \frac{5}{3}\, \langle\Omega\rangle \Big/ t_c^2 \right\}$$

$$\int_{M_K^2}^{t_c} dt\, t\, \frac{1}{\pi}\, \mathrm{Im}\, \psi(t) \simeq \frac{1}{(16\pi^2)^3} \left(\frac{\alpha_s(\nu^2)}{\alpha_s(t_c)}\right)^{4/9} \frac{1}{30}\, t_c^6 \,.$$

$$\cdot \left\{1 - \frac{28}{15}\, \frac{\alpha_s(t_c)}{\pi} - \frac{24\, \bar{m}_s^2}{t_c} - \frac{3}{2}\, \Omega \Big/ t_c^2 \right\} , \qquad (15.29a)$$

with :

$$\langle \Omega \rangle \equiv - 20\pi^2 \left(8\, m_s \langle \bar{s}\, s \rangle - \frac{\langle \alpha_s\, G^2 \rangle}{\pi} \right) . \qquad (15.29b)$$

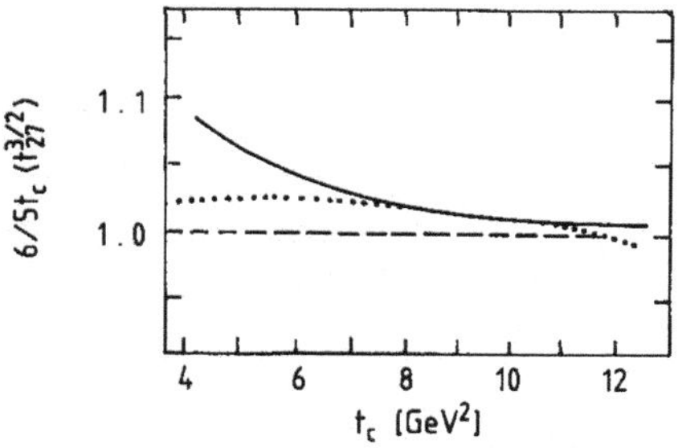

Fig. 15.5 : $\langle t_{eff} \rangle$ versus t_c.

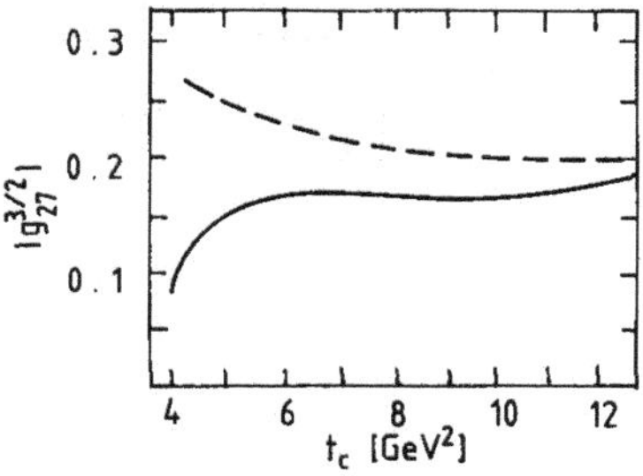

Fig. 15.6 : $g_{27}^{3/2}$ against t_c.

The same sum rules strategy as for the B parameter will be used, i.e. one is looking for the t_c values at which the QCD and spectral sides of the sum rules have a common solution and then one determines the coupling constant from the lowest FESR. This analysis is shown in Figs 15.5 and 15.6 from which one deduces[244]:

$$\left| g_{27}^{(3/2)} \right| \simeq 0.16 \tag{15.30}$$

in perfect agreement with the data. In the same way, one also deduces[245]:

$$\left| g_{27}^{(1/2)} \right| \simeq 3.2\ 10^{-2} \tag{15.31}$$

in nice agreement with the $SU(3)_F$ limit prediction :

$$\left| g_{27}^{(1/2)} \right| \simeq \frac{1}{5}\, g_{27}^{(3/2)} \tag{15.32}$$

However, the result for $g_{27}^{(1/2)}$ indicates that the $\Delta I = 1/2$ observed enhancement cannot come from the $(27_L, 1_R)$ operators.

5. $g_8^{(1/2)}$ AND THE $\Delta I = 1/2$ ENHANCEMENT

In this section we shall concern ourselves with the two-point correlator :

$$\psi_{ij}(q^2) \equiv i \int d^4x\ e^{iqx} \left\langle 0 \left| \mathbb{T}\ Q_i(x)\ \ (Q_j(o))^+ \right| 0 \right\rangle \tag{15.33}$$

built from the octet operators in (15.24) which mix under renormalizations :

$$Q_i^B = \sum_j Z_{ij}\ Q_j^R \tag{15.34}$$

where B and R refer to bare and renormalized operators. As an educated guess, Ref. 246) have examined the results for the renormalization of the operators [242,247] and the two-point correlator in the large N_c limit. They realize that only the Q_6 operator gets renormalized to a leading order in $1/N_c$ with the anomalous dimension :

$$\gamma_{66} \equiv \frac{\nu}{Z}\frac{dZ}{d\nu} = \left(\frac{\alpha_s}{\pi}\right)\left(-\frac{9}{2}\right) . \tag{15.35}$$

Ignoring for a first approximation the contribution of other non-leading $1/N_c$ operators, one then obtains[246] :

$$\alpha_s(\nu)^{(2\gamma_{66}/\beta_1)} \frac{1}{\pi} \operatorname{Im} \psi_{66}(t) \simeq (\alpha_s(\nu))^{18/11} \frac{12}{5} \frac{t^4}{(16\pi^2)^3} .$$

$$\cdot \left\{1 + \frac{\alpha_s(\nu)}{\pi}\left(-\frac{9}{2}\log\frac{t}{\nu^2} + \frac{423}{20}\right) + \ldots\right\} \tag{15.36}$$

where the factor $(\alpha_s(\nu))^{18/11}$ comes from the Wilson coefficient. The log-coefficient is equal to γ_{66} while the constant term is very large. Even at $t = 10$ GeV^2, it amounts to about 120% of the leading term and explains the failure of previous analysis[245] carried out without this effect. In order to study the correlators associated with the operators

$$Q_{\pm} \equiv Q_1 \pm Q_2$$

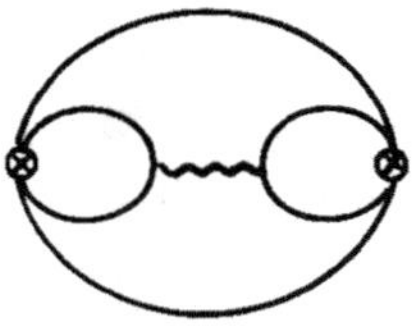

Fig. 15.7 : Non-leading in $1/N_c$ Penguin-like diagrams for the $\Delta I = 1/2$ process.

the authors select some dominant diagrams. By ignoring for a first approximation the penguin-like diagram in Fig. 15.7, they obtain[246]:

$$(\alpha_s(\nu))^{2\gamma_-/\beta_1} \frac{1}{\pi} \operatorname{Im} \psi_{--}(t) \Big|_{\text{NO Penguin}} \simeq (\alpha_s(\nu))^{8/9} \left(\frac{16}{15}\right) \frac{t^4}{(16\pi^2)^3} \cdot$$

$$\cdot \left\{1 + \left(\frac{\alpha_s(\nu)}{\pi}\right) \left(-2 \log \frac{t}{\nu^2} + \frac{45}{5}\right)\right\}$$

$$(\alpha_s(\nu))^{2\gamma_+/\beta_1} \frac{1}{\pi} \operatorname{Im} \psi_{++}(t) \Big|_{\text{NO-Penguin}} \simeq (\alpha_s(\nu))^{-4/9} \left(\frac{32}{15}\right) \frac{t^4}{(16\pi^2)^3}$$

$$\cdot \left\{1 + \left(\frac{\alpha_s(\nu)}{\pi}\right) \left(\log \frac{t}{\nu^2} = \frac{49}{20}\right)\right\} , \qquad (15.37a)$$

where the anomalous dimensions :

$$\gamma_- = -2 \qquad \gamma_+ = +1 \qquad (15.37b)$$

are equal, as they should be, to the coefficients of the corresponding log term. $\psi_{--}(q^2)$ has a large correction with the same sign as $\psi_{66}(q^2)$ whilst $\psi_{++}(q^2)$ has a moderate and negative effect which is the same as for the $(27_L, 1_R)$ correlator. A QSSR analysis within these preliminary results [246] does indeed indicate that the huge radiative corrections may provide an enhancement of the $|\Delta I| = 1/2$ transition and may stimulate the computation of the next-to leading term.

6. THE B_B PARAMETER FOR THE $B_0 - \bar{B}_0$ MIXING

An estimate of the B parameter associated with the B_0-$\bar{B}$ matrix element has been done by Pich[147] along the same lines as that of the

K_o-$\bar{K}_o$ matrix elements but using the $Q_o^2 = 0$ moments :

$$M_n = \int_{4m_b^2}^{t_c} \frac{dt}{t^n} \frac{1}{\pi} \mathrm{Im}\, \psi(t) \quad . \tag{15.38}$$

The OPE of the spectral function gives :

$$\frac{1}{\pi} \mathrm{Im}\, \psi_{\Delta B=2}(t) \simeq \frac{t^4}{(16\pi^2)^3} \left\{ A(t) + 16\pi \frac{\langle \alpha_s G^2 \rangle}{t^2} B(t) + \mathcal{O}\left(\frac{\alpha_s}{\pi}, \frac{1}{t^3}\right) \right\} .$$

$$. \ \theta\left(t - 4\, m_b^2\right) \quad , \tag{15.39a}$$

where

$$A(t) = 2\left(1 + \frac{1}{3}\right) \int_{\delta \equiv \frac{m_b^2}{t}}^{(1-\sqrt{\delta})^2} dz \int_{\delta}^{(1-\sqrt{z})^2} du\ \lambda^{1/2}(1,z,u) \ .$$

$$\left(1 - \frac{\delta}{z}\right)^2 \left(1 - \frac{\delta}{u}\right)^2 \left\{ (1-z-u)^2 \left(1 + \frac{2\delta}{z}\right)\left(1 + \frac{2\delta}{u}\right) + 8zu - 2\delta(z+u) - 4\delta^2 \right\} \tag{15.39b}$$

$$B(t) = \int_{x_o}^{1} dx \int_{y_-}^{y_+} dy \left\{ -\frac{\Delta}{2y^2} [\Delta - y(1-y)]\, [2xy + (1-x)^2(1-y)] + \right.$$

$$\left. \frac{\delta x}{3y^3} \left((1-x)^2\ (1-y)^3\ (2\Delta - y(1-y)) \right) \right\}$$

$$- \int_{\delta}^{(1 - \sqrt{\delta})^2} dz \; z(1 - \delta/z)^2 \; \lambda^{1/2}(1,z,\delta) \quad , \qquad (15.39c)$$

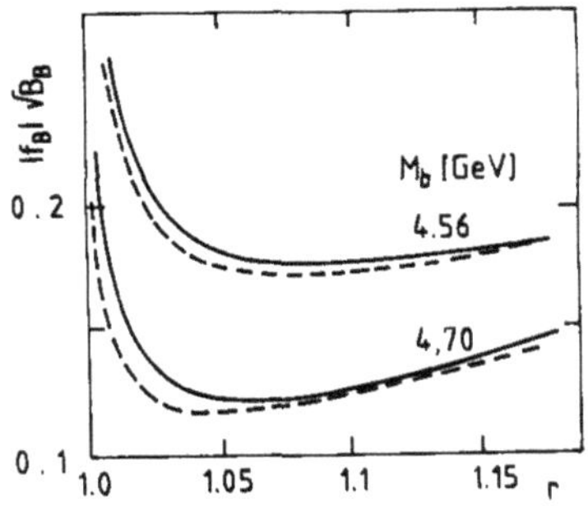

Fig. 15.8 : t_c-behaviour of $f_B\sqrt{B}$.

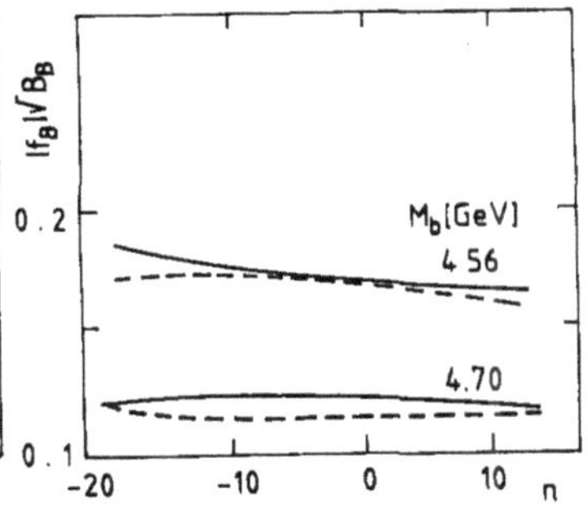

Fig. 15.9 : n-behaviour of $f_B \sqrt{B}$.

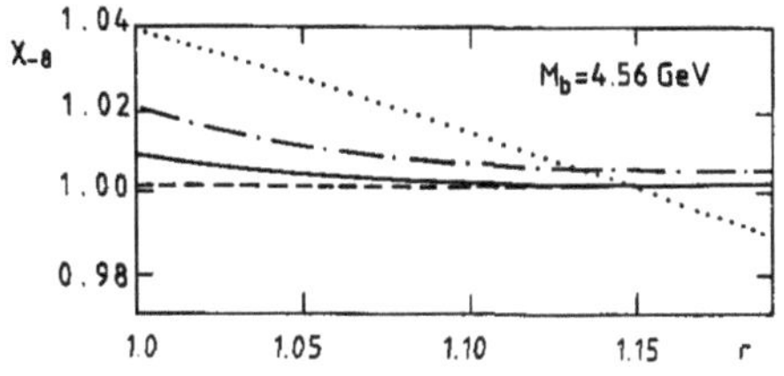

Fig. 15.10 : eigenvalue solution of the ratio of moments.

where :

$$\lambda(x,y,z) \equiv x^2 + y^2 + z^2 - 2xy - 2xz - 2yz$$

$$\Delta \equiv \delta(y/x + 1-y) - y(1-y)$$

$$x_o \equiv \delta\Big/(1 - \sqrt{\delta})^2$$

$$y_{\pm} = \frac{1}{2}\,\{1 + \delta(1 - 1/x) \pm \lambda^{1/2}\,(1,\,\delta,\,\delta/x)\}\quad , \qquad (15.39d)$$

while the phenomenological parametrization leads to :

$$\frac{1}{\pi}\,\mathrm{Im}\,\psi\,(t \leqslant t_c) \simeq \frac{2}{9}\,.\,f_B^2\,B_B^2\left(\frac{1}{16\pi^2}\right)t^2\left(1 - \frac{2M_B^2}{t}\right)\sqrt{1 - \frac{4M_B^2}{t}}\;.$$

$$.\,\theta\left(t - 4M_B^2\right) \qquad (15.40)$$

where $M_B = 5.275$ GeV is the B-meson mass. A QSSR analysis of the moments indicates that the predicted quantity $f_B\sqrt{B}$ has t_c- (Fig. 15.8) and n- (Fig. 15.9) stabilities. The values of the t_c-stabilities coincide with the eigenvalue solution of the ratio of two moments (Fig. 15.10).

Within the range of values of the pole mass given previously in Chapters 9 and 10, one obtains :

$$f_B\sqrt{B_B} \simeq (0.15 - 0.17)\ \mathrm{GeV} \qquad (15.41)$$

which gives

$$B_B \simeq 1.0 - 1.3 \qquad (15.42)$$

when one combines this result with the independent estimate of f_B in

Chapter 10. The B parameter agrees with the naive vacuum saturation estimate which is expected to be a good approximation for heavy quarks.

7. OTHER WEAK PROCESSES : CHARM NON-LEPTONIC DECAYS AND THE DIQUARK DECAY CONSTANT

QSSR have also been applied in Ref. 248) for the study of the charm non-leptonic decays via four-point functions. The agreements of the predictions with the data are impressive but as in the case of the vertex sum rules, one cannot claim to control the accuracy of these predictions.

QSSR have also been used in order to estimate[249] the decay constant of a hypothetical diquark $J^{PC} = 0^{++}$ $I = 0$ state which may be expected to make an important contribution in weak decays[250]. In this case we shall be concerned with the operator :

$$J^D_i(x) = \epsilon_{ijk}\, u^T_j\, C\, \gamma_5\, d_k \quad , \tag{15.43}$$

where C is the charge conjugation matrix.
With the normalization :

$$\langle 0\ |J_i|\ D_{3\ell}\rangle = \sqrt{\frac{2}{3}}\, g^D_3\, \delta_{i\ell} \quad , \tag{15.44}$$

the estimated value of the decay constant for $M_D \simeq 0.4 - 0.7$ GeV[249] is :

$$g^D_3 \simeq 0.16 - 0.22 \text{ GeV}^2 \tag{15.45}$$

in the τ- and t_c-stability regions. This value is about that of the pion :

$$g_\pi \equiv f_\pi \frac{m_\pi^2}{(m_u+m_d)} \simeq 0.17 \text{ GeV}^2 \quad . \tag{15.46}$$

8. *CONCLUSION*

The readers may have noticed that we have not tried to emphasize discussion of weak-interaction phenomenology which is not the aim of this report. We have presented only those quantities which have been estimated from QSSR methods and which are useful for further phenomenological discussions.

CHAPTER 16

DETERMINATION OF LAMDA TO FOUR-LOOPS AND QCD FORMULATION OF THE TAU DECAY

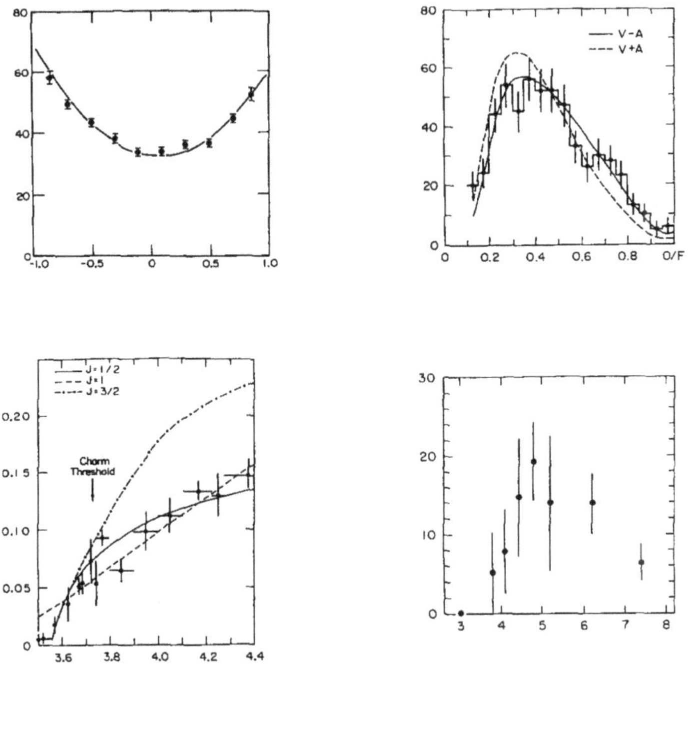

Typical experimental proofs for the tau being a sequential lepton

1. FORMULATION

Considering the present progress for determining the properties of the tau (τ) lepton, and at the same time for our understanding of the weak currents within QCD, one can make predictions[165] on the semi-inclusive and exclusive tau-decay properties. We shall be concerned with the ratio :

$$R_H \equiv \frac{\Gamma(\tau \to \nu_\tau + \text{Hadrons})}{\Gamma\left(\tau \to \nu_\tau\, e\, \bar{\nu}_e\right)} , \qquad (16.1)$$

which can be expressed as an integral over the invariant mass of the final hadrons :

$$R_H = 12\pi \int_0^{M_\tau^2} \frac{dt}{M_\tau^2} \left(1 - \frac{t}{M_\tau^2}\right)^2 \left\{\left(1 + 2\frac{t}{M_\tau^2}\right) \text{Im}\, \Pi^{(1)}(t) + \text{Im}\, \Pi^{(0)}(t)\right\} , \qquad (16.2)$$

where the spectral functions are associated with spin 1 and 0 mesons respectively. These can be decomposed as :

$$\text{Im}\, \Pi^{(k)}(t) \equiv \cos^2\theta_c \left\{\text{Im}\, \Pi_V^{(k)}(t)^{12} + \text{Im}\, \Pi_A^{(k)}(t)^{12}\right\}$$

$$+ \sin^2\theta_c \left\{\text{Im}\, \Pi_V^{(k)}(t)^{13} + \text{Im}\, \Pi_A^{(k)}(t)^{13}\right\} , \qquad (16.3)$$

i.e. into Cabibbo favoured and suppressed channels while the indices A and V refer to axial and vector currents.

These spectral functions are normalized as :

$$\left(\Pi^{\mu\nu}_{V(A)}\right)^{ij} = i \int d^4x\, e^{iqx} \left\langle 0 \left| T\, J^{\mu}_{V(A)}(x)^{ij} \left(J^{\nu}_{V(A)}(o)^{ij}\right)^{+} \right| 0\right\rangle$$

$$\equiv -(g^{\mu\nu}q^2 - q^\mu q^\nu)\, \Pi^{(1)}_{V(A)}\,(q^2)^{ij} + q^\mu q^\nu\, \Pi^{(o)}_{V(A)}(q^2)^{ij} \quad , \qquad (16.4)$$

where : $J^\mu_{V(A)}(x)^{ij} \equiv \bar{\psi}_i\, \gamma^\mu\, (\gamma_5)\, \psi_j$ are the charged vector (axial) currents and i,j denote quark flavours. We shall deal here with the QCD expression of the moments :

$$\mathcal{M}^{(k)}_n \left(M^2_\tau\right)^{ij} = \int_0^{M^2_\tau} dt\ t^n\ \mathrm{Im}\ \Pi^{(k)}_{V(A)}\ (t)^{ij} \qquad (16.5)$$

for n = 0,1,2 and 3 and k = 0 and 1.

2. 1^{--} VECTOR CHANNEL

Using the derivation of these moments as in Ref. 23), one obtains to four loops in the $\overline{MS}$ scheme[165a]:

$$R_H(1^-)^{ij} = \xi^{ij}\, \frac{3}{2} \left\{ 1 + a_s + a_s^2 \left(F_3 - \frac{19}{24}\beta_1 - \frac{\beta_2}{\beta_1} L \right) \right.$$

$$+ a_s^3 \left[F_4 - \frac{19}{12} F_3\, \beta_1 - \frac{19}{24}\beta_2 + \frac{265}{288}\beta_1^2 - 2\frac{\beta_2}{\beta_1} L \left(F_3 - \frac{19}{24}\beta_1 \right) \right.$$

$$\left. + \left(\frac{\beta_2}{\beta_1}\right)^2 (L^2 - L - 1) + \frac{\beta_3}{\beta_1} \right]$$

$$\left. + 2\, \frac{C_2\langle O_2\rangle_V}{M^2_\tau} - 6\, \frac{C_6\langle O_6\rangle_V}{M^6_\tau} - 4\, \frac{C_8\langle O_8\rangle}{M^8_\tau} \right\} \qquad (16.6)$$

where, for $SU(3)_c \times SU(n)_F$:

$$\beta_1 = -\frac{11}{2} + \frac{n}{3} \quad , \quad \beta_2 = -\frac{51}{4} + \frac{19}{12} n \quad ,$$

$$\beta_3 = \frac{1}{32} \left(-\frac{2857}{2} + \frac{5033n}{18} - \frac{325}{54} n^2 \right) \quad ,$$

$$F_3 = 1.986 - 0.115n \quad \text{[124]} \quad ,$$

$$F_4 = 70.985 - 1.200n - 0.005n^2 \quad \text{[251]} \quad ,$$

$$a_s \equiv \frac{1}{-\beta_1 \log M_\tau \big/ \Lambda_{\overline{MS}}} \qquad L \equiv \log \log M_\tau^2 \big/ \Lambda^2_{\overline{MS}} \quad . \tag{16.7}$$

The factor ξ^{ij} takes into account the different Cabibbo mixing structures of the $|\Delta S| = 0$ and 1 channels :

$$\xi^{12} \equiv \cos^2 \theta_c \qquad \xi^{13} \equiv \sin^2 \theta_c \quad . \tag{16.8}$$

The leading quark mass corrections come from the QCD expression from Ref. 34) in Chapter 1 :

$$C_2\langle O_2 \rangle = -3\left(\bar{m}_i^2 + \bar{m}_j^2\right) + \frac{17}{4}\left(\bar{m}_i - \bar{m}_j\right)^2 \quad , \tag{16.9a}$$

where :

$$\bar{m}_i(M_\tau) \equiv \hat{m}_i \Big/ \left(\log M_\tau \big/ \Lambda_{\overline{MS}}\right)^{2/-\beta_1} \quad , \tag{16.9b}$$

with the invariant light-quark masses determined in previous sections. One should note that only $C_6\langle O_6 \rangle$ and $C_8\langle O_8 \rangle$ non-perturbative effects due to dimension-six and -eight vacuum condensates enter Eq. (16.6) and there is no contribution by dimension-four condensates to this order. Therefore, unlike the $e^+e^- \to$ Hadrons data, one cannot determine these dimension-four condensates from the real τ data. The effects of $C_6\langle O_6 \rangle$

and $C_8\langle O_8\rangle$ are also known and can be estimated to be :

$$C_6\langle O_6\rangle_V \simeq -\frac{896}{81}\pi^3\alpha_s\left\langle\bar{u}u\right\rangle^2.\rho \quad ,$$

$$C_8\langle O_8\rangle_V \simeq \frac{\pi^2}{648}(26 \sim 39)\left\langle\alpha_s G^2\right\rangle^2 \quad , \qquad (16.10)$$

where $\rho = 1$ corresponds to the usual vacuum saturation estimate of the four-quark operators. We will use $\rho \simeq 2$ to 4 as suggested in previous sections. We estimate the dimension-eight operator effect by using the result of Ref. 125) and the modified factorization of Ref. 95). For the $\Delta S = 0$ case, the dimension-six condensates contribute between 2.2% and 4.5% for the above range of ρ values. The effects of the dimension-eight condensates are negligible while the quark mass corrections amount to 2-3%. The most important corrections are due to the radiative corrections and therefore one might expect that this process would provide a strong constraint on the size of the value of $\Lambda_{\overline{MS}}$. For $\Lambda_{\overline{MS}} = 100$ MeV the a_s, a_s^2 and a_s^3 terms contribute by 7.7, 1.2 and 3.5% of the lowest order term. For larger values of $\Lambda_{\overline{MS}}$, say 300 MeV, it may be noticed that the a_s^3 term provides the leading correction. These effects amount to 12.5, 4.6 and 16.0% respectively for the a_s, a_s^2 and a_s^3 terms. Therefore, one finds the value :

$$R_H(1^-) \simeq \begin{pmatrix} 1.72 - 1.75 \\ 1.85 - 1.88 \\ 2.03 - 2.06 \end{pmatrix} \quad \text{for} \quad \Lambda_{\overline{MS}} \simeq \begin{pmatrix} 0.1 \\ 0.2 \\ 0.3 \end{pmatrix} \text{ GeV} \qquad (16.11)$$

in the chiral limit and to four loops. The value of $R(1^-)^{ij}$ for massive quarks is given in the Table 16.1 for each Cabibbo-favoured

*Table 16.1 : QCD predictions and experimental (exclusive) values of the τ-hadronic widths in units of $\Gamma\left(\tau \to \nu_\tau\, e\, \bar{\nu}_e\right)$**)

$\Lambda_{\overline{MS}}$ (MeV	$\|\Delta S\| = 0$ 1^-	$1^+ + 0^-$	$\|\Delta S\| = 0$ 1^-	$1^+ + 0^-$	Total Width
100	1.64 - 1.67	1.50 - 1.55	0.086 - 0.089	0.070 - 0.076	3.22 - 3.35
200	1.76 - 1.79	1.63 - 1.68	0.092 - 0.096	0.074 - 0.082	3.58 - 3.60
300	1.93 - 1.96	1.80 - 1.85	0.101 - 0.105	0.081 - 0.090	3.93 - 3.95
EXPERIMENT	1.62 ± 0.06	1.44 ± 0.08	0.102 ± 0.017	0.062 ± 0.026	3.22 ± 0.10

and suppressed channel and is compared with the data of exclusive modes : the $(1^-)^{12}$ channel is estimated phenomenologically as a source of even number of pions [252-254] where we have used the CVC argument [255] for the estimate of the 4π final states by using the $e^+e^- \to 4\pi$ data. Better agreement between QCD predictions and the data is obtained for :

$$\Lambda_{\overline{MS}} \simeq 100 \text{ MeV} \quad . \tag{16.12}$$

The $R(1^-)^{13}$ ratio is estimated from $\tau \to K\,\nu_\tau$ + pions data[252,256] but these data are quite inaccurate for a good determination of $\Lambda_{\overline{MS}}$.

3. 0^{++} SCALAR CHANNEL

The corresponding mesons are associated with the divergence of the vector current which is proportional to the light-quark mass differences :

---

*) We have not included in the Table the 0^{++} contribution, which are negligible[257].

$$\partial_\mu V^\mu(x)_{ij} = (m_j - m_j) : \bar{\psi}_i(i)\, \psi_j : \qquad (16.13)$$

The tau decays into these modes are expected to be dominated by the a_o (980) in the $\Delta S = 0$ and by the K^*_o (1350) in the $|\Delta S| = 1$ channels. These widths are unobservable[257] owing to their smallness which is due to the vanishing value of the a_o and K^*_o decay constants with the quark mass difference (see Chapter 7).

4. $1^{++} + 0^{-+}$ CHANNELS

These two channels are responsible for odd numbers of pions in the tau-decay experiments so that their separation is quite difficult. Moreover, from the theoretical side, a separate QCD estimate of the 0^{-+} channels requires theoretical knowledge of the slope $\frac{\partial \Pi^{(o)}}{\partial q^2}$ at $q^2 = 0$ which is not under control at present. In the chiral limit $m_i = 0$, only one invariant function $\Pi_A(q^2)$ appears in the Lorentz decomposition of the axial-axial correlator :

$$\Pi^{\mu\nu}_A(q^2) \equiv -(g^{\mu\nu}q^2 - q^\mu q^\nu)\left[\Pi_A(q^2) \equiv \left(\Pi^{(1)}_A(q^2) + \Pi^{(o)}_A(q^2)\right)\right] , \qquad (16.14)$$

which corresponds to the sum of the 1^{++} and 0^{-+} invariants. Chiral invariance allows us to use Eq. (16.6) in the chiral $m_i = 0$ limit with the change due to γ_5-flip :

$$C_6\langle O_6\rangle_{1+0} \simeq -\frac{11}{7}\, C_6\langle O_6\rangle_V \quad . \qquad (16.15)$$

Then, one obtains in the chiral limit :

$$R(1^{++} + 0^{-+}) = \begin{pmatrix} 1.58 - 1.64 \\ 1.71 - 1.77 \\ 1.89 - 1.94 \end{pmatrix} \quad \text{for} \quad \Lambda \simeq \begin{pmatrix} 0.1 \\ 0.2 \\ 0.3 \end{pmatrix} \text{ GeV} , \qquad (16.16)$$

where we have used $F_4|_V = F_4|_A$. The mass corrections read :

$$C_2\langle O_2\rangle_{1+0} \simeq -3\left(m_i^2 + m_j^2\right) - (m_i + m_j)^2 \quad . \qquad (16.17)$$

The mass-corrected ratio is given in the Table 16.1 together with the data from Refs 252-254, 256 and 258). These data are $\tau \to \nu_\tau + \left(\pi, 3\pi, K\bar{K}\pi, 5\pi\right)$ and $\tau \to \nu_\tau + (K, K\pi\pi)$ for the $\Delta S = 0$ and $\Delta S = 1$ channels respectively. Again, these data favour a low value of $\Lambda_{\overline{MS}}$ around 100 MeV.

5. **ELECTRONIC AND TOTAL WIDTHS AND τ LIFETIME**

By adding the branching ratio in each channel, we predict the tau-hadronic width in the chiral limit :

$$R_H \equiv R(1^-) + R(1^+ + 0^-) = \begin{pmatrix} 3.34 - 3.36 \\ 3.60 - 3.62 \\ 3.95 - 3.97 \end{pmatrix} \quad \text{for} \quad \Lambda = \begin{pmatrix} 0.1 \\ 0.2 \\ 0.3 \end{pmatrix} \text{ GeV} \quad . \qquad (16.18)$$

Mass corrections decrease (16.18) by about 0.5% after adding the different widths in the Table 16.1. R_H is related to the total tau width through :

$$B_e\left(\tau \to \nu_\tau\, e\, \bar{\nu}_e\right) \cdot \left\{1 + f\left(\left(\frac{m_\mu}{M_\tau}\right)^2\right) + R_H\right\} = 1 \quad ,$$

where :

$$f(x) = 1 - 8x + 8x^3 - x^4 - 12x^2 \log x \quad . \qquad (16.19)$$

Using $B_e|_{exp} \simeq (17.7 \pm 0.4)\%$ one can then deduce the inclusive determination :

$$R_H^{EXP} \simeq (3.68 \pm 0.13) \quad , \qquad (16.20)$$

while the sum of the experimental exclusive widths in the Table leads to :

$$R_H^{EXP} \simeq (3.22 \pm 0.10) \quad . \qquad (16.21)$$

Comparing the prediction in Eq. (16.18) with the average of these experimental values in Eqs (16.20) and (16.21), one can deduce the range :

$$\Lambda_{\overline{MS}} \simeq (100 - 200)\ \mathrm{MeV} \quad . \qquad (16.22)$$

Finally, one can insert the theoretical expression of the electronic width :

$$\Gamma\left(\tau \to \nu_\tau\, e\, \bar{\nu}_e\right) \simeq \frac{G_F^2}{192\pi^2} M_\tau^5 \qquad (16.23)$$

and use the prediction of R_H in Eq. (16.18) in (16.19). In this way, one can predict the τ lifetime :

$$\tau_\tau \simeq \begin{pmatrix} 2.99 - 3.01 \\ 2.86 - 2.87 \\ 2.69 - 2.70 \end{pmatrix} 10^{-13}\ \mathrm{sec} \quad \mathrm{for} \quad \Lambda_{\overline{MS}} \simeq \begin{pmatrix} 0.1 \\ 0.2 \\ 0.3 \end{pmatrix} \mathrm{GeV} \qquad (16.24)$$

compared to the world average :

$$\tau_\tau \simeq (3.06 \pm 0.09)\ 10^{-13}\ \mathrm{sec} \quad , \qquad (16.25)$$

while using the experimental value of the electronic width in Eq. (16.19) gives :

$$\tau_\tau \simeq (2.82 \pm 0.06)\ 10^{-13}\ \mathrm{sec} \quad . \qquad (16.26)$$

Taking again the average of these two data leads, by comparison with the prediction in (16.24), to :

$$\Lambda_{\overline{MS}} \simeq (100 - 200) \text{ MeV} \quad . \tag{16.27}$$

Our analysis has shown that the τ-decay experiments can provide a value of $\Lambda_{\overline{MS}}$ in the range given by Eq. (16.27). This range of values can be sharpened by the inclusion of the electroweak radiative corrections[165b] which reads to leading order in α/π :

$$\sqrt{R_H} = \sqrt{R_H^0}\left(1 + \frac{\alpha}{\pi}\log\frac{M_Z}{M_\tau}\right) \tag{16.28}$$

and after a renormalization group improvement :

$$\sqrt{R_H} = \sqrt{R_H^0} \cdot \left(\frac{\alpha(m_b)}{\alpha(M_\tau)}\right)^{9/38}\left(\frac{\alpha(m_t)}{\alpha(m_b)}\right)^{9/40}\left(\frac{\alpha(M_W)}{\alpha(m_t)}\right)^{3/16} \cdot \left(\frac{\alpha(M_Z)}{\alpha(M_W)}\right)^{6/11} , \tag{16.29}$$

leading to a correction of an approximately 2% increase on R_H.
Taking this result with the experimental average :

$$R_H^{ave} = (3.52 \pm 0.08) \quad , \tag{16.30}$$

one then obtains[165b]:

$$\Lambda_{\overline{MS}} \simeq (150 \pm 30) \text{ MeV} \tag{16.31}$$

where the error is due to the experimental one.
This value of $\Lambda_{\overline{MS}}$ can be considered to be the most accurate available today but relies on the accuracy of the measurement of R_H.

We again stress that the derivation of the previous result is based on a duality between the resonance physics and the QCD expressions of the hadronic correlator evaluated at the τ-mass value. In this way, it is quite clear that in the QCD evaluation of Eq. (16.2), we should not take into account simultaneously the resonance effects, otherwise we are counting double in the analysis. By duality, the resonance effects are equivalent to the non-perturbative condensates which we have allowed to vary in a larger range than the ones obtained in the previous sections. In fact, our results are quite independent of these condensate values as at the τ-mass scale, these non-perturbative effects do not exceed 4.5% of the lowest order term in R_H. One should also note that, within the SVZ OPE framework, one should use the value of the running light-quark mass defined in Eq. (16.9) instead of the constituent- quark mass. Tests of the QSSR approach within Weinberg sum rules and using the exclusive τ-decay modes have also been done in the work of Ref. 131) which we have reviewed in Chapter 6. We might conclude that the theoretical uncertainties coming from these "non-perturbative effects" are small within our approach and the improvment of Eq. (16.31) can only be done after an evaluation of the a_s^4 term.

CHAPTER 17

FURTHER USES OF THE SPECTRAL SUM RULES IN QCD

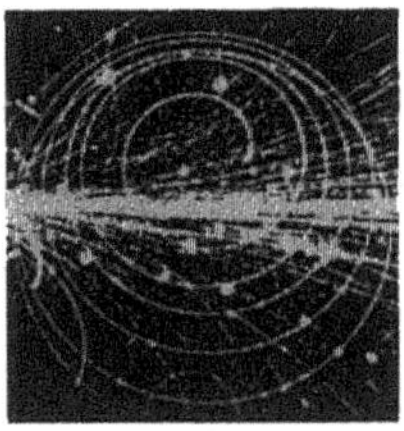

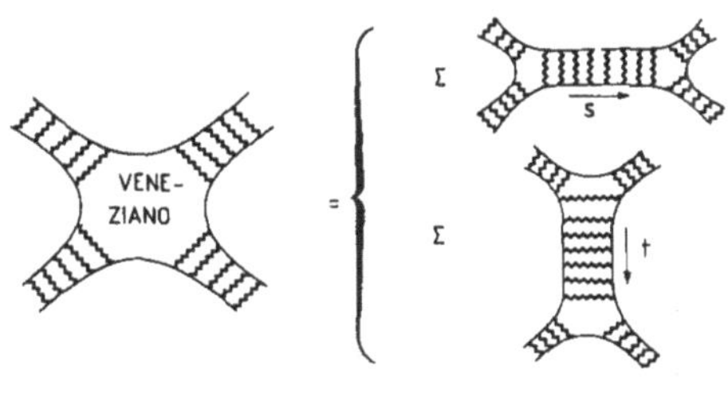

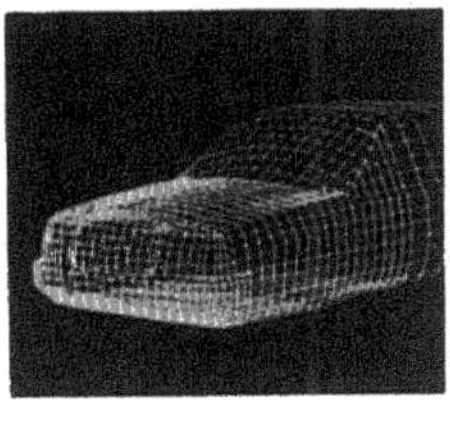

Search for quark-gluon plasma in oxygen-ion collision, typical Veneziano-string amplitudes as seen by A. De Rujula and example of a computer simulation

1. ρ MESON PROPERTIES AT FINITE TEMPERATURE

There seems general agreement from lattice Monte-Carlo calculations that the Yang-Mills sector of QCD exhibits a phase transition of the first kind at $T \simeq 220$ MeV[259] but the influence of dynamical fermions on the phase transition is less understood[260].
QSSR techniques have been extended by Bochkarev and Shaposhnikov[261] (BS) to finite temperatures where they evaluated the retarded commutator of the vector current known to have appropriate analytical properties [262] using the Matzubara formalism[263]. This analysis has been tested by Ref. 264) using conventional τ- and t_c-stability criteria. Following BS, we study the longitudinal part of the retarded commutator which is non-zero at finite temperature T. Using the usual zero-width resonance plus QCD continuum parametrization of the spectral function, one obtains the finite energy Laplace sum rule :

$$\mathcal{F}(\tau) \simeq 4\pi^2 \frac{M_\rho^2}{\gamma_\ell^2} e^{-M_\rho^2\tau} \simeq \int_0^{t_c} dt\, e^{-t\tau}\, \mathrm{th}(\sqrt{t/4T}) +$$

$$2\int_0^{\infty} dt \left[n_F\, (\sqrt{t/2T}) - \frac{1}{3}\, n_B\, (\sqrt{t/2T}) \right] +$$

$$\frac{\pi}{3}\left\langle \alpha_s\, G^2 \right\rangle \tau - 2\pi^3\, \alpha_s \left\langle \bar{\psi}\, \psi \right\rangle^2 \tau^2 \qquad (17.1)$$

for $m_u = m_d = m_\pi^2 = 0$. γ_ℓ is the ρ coupling to the vacuum ; $n_{F,B}$ are respectively the number of fermions and bosons inside the plasma. τ is as usual the Laplace SR variable. The Wilson coefficients of the non-perturbative condensates are not affected by T to leading order and can be obtained in the usual way. We shall assume that these condensates vary smoothly with the temperature[265,261] range around 150 MeV. We shall work with the ratio of moments :

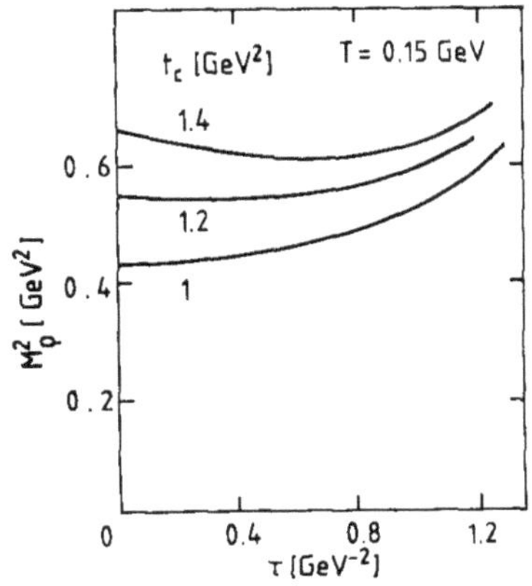

Fig. 17.1 : τ-behaviour of M_ρ^2 for different values of t_c.

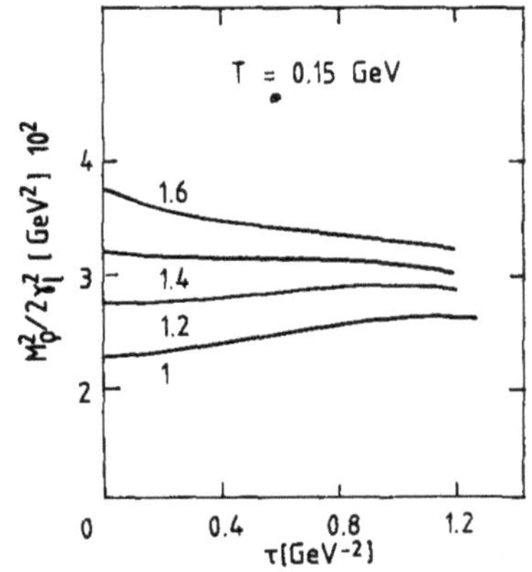

Fig. 17.2 : τ-behaviour of the coupling γ_ℓ^2 for different values of t_c.

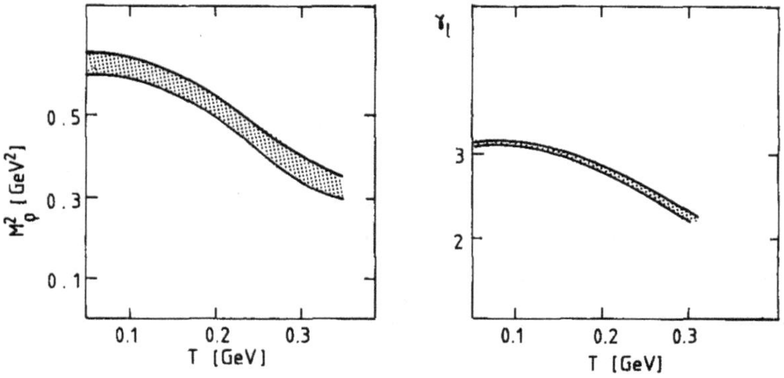

Fig. 17.3 : Temperature effect on the values of the ρ meson mass and coupling.

$$R(\tau) \equiv -\frac{d}{d\tau}\log F(\tau) \tag{17.2}$$

which gives the ρ meson mass squared. We give in Fig. 17.1 the τ behaviour of the ρ meson mass for various values of the continuum threshold t_c and for the values of the condensates

$$\left\langle \alpha_s G^2 \right\rangle \simeq 0.04 \text{ GeV}^4 \qquad \text{(canonical)}$$

$$\alpha_s \left\langle \bar\psi \psi \right\rangle^2 \simeq 3.5\ 10^{-4} \text{ GeV}^6 \qquad \text{(twice factorization)}\ . \tag{17.3}$$

The qualitative behaviour of the prediction is not affected by the changes of the condensate values in Eq. (17.3)[264]. The t_c-stability for the mass prediction is reached at $t_c \geqslant 4$ GeV ^{2} but this value is too high to be realistic. In turn, we look for t_c values where the coupling $M_\rho^2/2\gamma_\ell^2$ from Eq. (17.1) has a better τ-stability. This is reached (see Fig. 17.2) for :

$$t_c \simeq 1.2 - 1.4 \text{ GeV}^2 \quad . \tag{17.4}$$

For T $\simeq$ 50 - 300 MeV, the onset of the continuum is a very smooth function of the temperature as noticed in Ref. 264). Fixing the t_c values at this range, we look for the effect of the temperature on the values of the mass and coupling at the τ-stability. This effect is shown in Fig. 17.3 where one can see that the two parameters of the ρ meson decrease smoothly with the temperature. This fact increases our confidence that the hot Fermi gas does not lead to a drastic change of the spectra in the region T $\leqslant$ 150 MeV in agreement with the conclusion reached in Ref. 264) although a higher value of t_c (in the t_c-stability regime) has been used there for the analysis. The main difference with Ref. 261) is that the sharp transition observed there at T $\simeq$ 150 MeV is connected with an ad hoc choice of low t_c values ranging from 0.8-1.1 GeV^2 where the moments do not yet present τ stability.

FESR without exponential have also been studied in Ref. 264). They lead to the set of constraints :

$$4\pi^2 \frac{M_\rho^2}{\gamma_\ell^2} \simeq \int_0^{t_c} dt \; th\,(\sqrt{t}/4T) + 2\int_0^{\infty} dt \left(n_F - \frac{1}{3}\, n_B\right) \quad , \tag{17.5}$$

$$4\pi^2 \frac{M_\rho^4}{\gamma_\ell^2} \simeq \int_0^{t_c} dt\; t\; th\,(\sqrt{t}/4T) - \frac{\pi}{3}\left\langle \alpha_s\, G^2 \right\rangle \quad , \tag{17.6}$$

$$4^2 \frac{M_\rho^6}{\gamma_\ell^2} \simeq \int_0^{t_c} dt\; t^2\; th\,(\sqrt{t}/4T) - \frac{896}{81}\,\pi^3\, \alpha_s \left\langle \bar{\psi}\, \psi \right\rangle^2 \quad , \tag{17.7}$$

where the accuracy of the constraints decreases strongly for higher moments. For an educated guess we consider the accuracy of the ratios :

$$R^{(1)} \equiv \frac{\text{rhs Eq(6)}}{\text{rhs Eq(5)}} \qquad R^{(2)} \equiv \frac{\text{rhs Eq(7)}}{\text{rhs Eq(6)}} \tag{17.8}$$

to be respectively 20% and 40% and for our numerical analysis we use the values :

$$\langle \alpha_s G^2 \rangle \simeq 0.14 \text{ GeV}^4 \qquad \alpha_s \langle \bar{\psi} \psi \rangle^2 \simeq 12.25 \; 10^{-4} \text{ GeV}^6 \tag{17.9}$$

favoured by FESR analysis. The results are shown in Fig. 17.4 for two values of T. Previous values of t_c and M_ρ^2 obtained from Laplace sum rules are within the common corridors. These results also indicate that there is no decrease of t_c when the temperature increases and does not therefore favour the choice of t_c used by BR.

We again conclude that the effect of the temperature on (dynamical) fermions leads only to very smooth changes in the mass and coupling of the ρ meson. We also do not find any indication of the need for a drastic change in the continuum threshold t_c with temperature. Thus, QSSR indicates that there is not a drastic effect of dynamical fermions on phase transitions in QCD.*

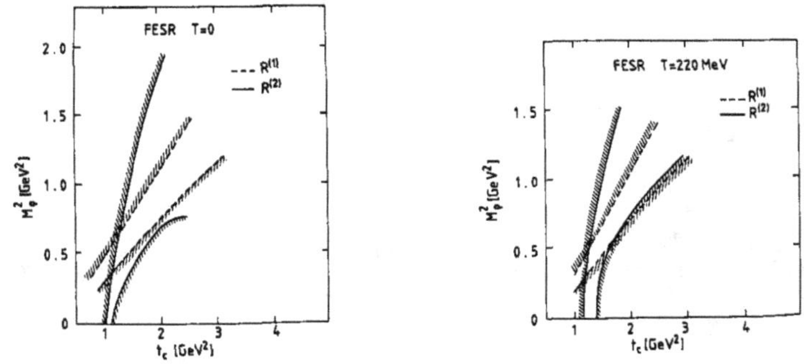

Fig. 17.4 : FESR predictions of the ratios in Eq. 17.8 versus t_c.

* The drastic change on the phase transition could only be simulated if one modifies the QCD ansatz for the spectral function at high temperature.

2. STRING TENSION FROM WILSON LOOPS

The authors in Refs 266) and 267) have extended the applicability domain of QSSR in order to study the vacuum expectation values of the Wilson loop defined in Euclidian space-time :

$$W_c = \frac{1}{N}\left\langle \mathrm{Tr}\ P \exp\left\{ - ig \oint_C dx_\mu\ A^a_\mu(x)\ \frac{\lambda^a}{2}\right\}\right\rangle \quad , \tag{17.10}$$

using the OPE including vacuum condensates. C is a closed contour, P indicates path ordering and λ_a are Gell-Mann colour matrices. The QCD perturbative and leading non-perturbative contributions to Eq. (17.10) are respectively, for $SU(N)_c \times SU(n)_F$:

$$W_C^{Pert} \simeq 1 - \left(\frac{\alpha_c}{4\pi}\right)\left(\frac{N^2-1}{2N}\right) \oint_C dx_\mu \oint dy_\nu\ \frac{\delta_{\mu\nu}}{(x-y)^2} \quad ,$$

$$W_C^{Non.Pert} \simeq - \frac{\pi}{48N}\left\langle \alpha_s G^2\right\rangle \oint_C dx_\mu\ x_\alpha \oint_C dy_\nu\ y_\beta(\delta_{\alpha\beta}\delta_{\mu\nu} - \delta_{\alpha\nu}\delta_{\mu\beta}) \ . \tag{17.11}$$

The phenomenological parametrization of the Wilson loops is done using the exponential decay corresponding to a linear static quark-antiquark potential :

$$\exp\{-\sigma.\mathrm{area}\} \quad , \tag{17.12}$$

as suggested from strong-coupling arguments. σ is the string tension. The area is that of the minimal surface containing the contour C.

The previous general results valid for any contour have been applied in the particular case of planar and non-planar Wilson loops. In the former case, a rectangular contour of size L × T was chosen. Writing $T \equiv \lambda L$, one then obtains[267)] in 4-ϵ dimension space-time :

$$W_{rect}(L,\lambda) = 1 + \left(\frac{\alpha_s}{\pi}\right)\left(\frac{N^2-1}{N}\right)\left\{\frac{2}{\epsilon} + 2\log \nu L + \log\lambda + 2 + f(\lambda) + f\left(\frac{1}{\lambda}\right)\right\} - \frac{\pi}{12N}\left\langle\alpha_s\, G^2\right\rangle \lambda^2 L^4 +$$

$$\frac{1}{12N}\left[\frac{\langle g^3 f\, G^3\rangle}{96} + \frac{1}{9}\left(\alpha_s\, \pi \left\langle\bar{\psi}\psi\right\rangle\right)^2 n_F\, \frac{N^2-1}{N^2}\right] \lambda^2\,(1+\lambda^2)\, L^6 \;, \qquad (17.136)$$

where : $f(x) = x \operatorname{atan} x + \frac{1}{2}\log(1+x^2)$ and ν is an arbitrary subtraction scale. The $\frac{1}{\epsilon}$ pole comes from the cusps of the rectangle. In order to get rid of a perimeter law, the authors in Ref. 267) work with the ratio of Wilson loops :

$$-\log\frac{W_{rect}(L,\lambda)}{W_{square}(L')} = \left(\frac{\alpha_s}{\pi}\right)\left(\frac{N^2-1}{N}\right)\left\{\log\frac{(1+\lambda)^2}{4\lambda} - f(\lambda) - f(1/\lambda) + 2\,f(1)\right\} \frac{\pi}{12N}\left\langle\alpha_s G^2\right\rangle L^4\,\lambda^2\left(1 - \left(\frac{L'}{L}\right)^4\right) -$$

$$-\frac{1}{12N}\left[\frac{\langle g^3 f\, G^3\rangle}{96} + \frac{1}{9}\left(\alpha_s \pi\left\langle\bar{\psi}\psi\right\rangle\right)^2 n_F\,\frac{N^2-1}{N^2}\right].L^6\left[\lambda^2(1+\lambda^2) - 2\left(\frac{L'}{L}\right)^6\right] \;, \qquad (17.14)$$

where $L' = (L+T)/2 = L(1+\lambda)/2$, i.e. the two Wilson loops have the same perimeters. This normalization also eliminates the $\frac{1}{\epsilon}$ and ν terms. Eq. (17.14) is studied at fixed λ as a function of L^2, where it should behave as a straight line passing through the origin as :

$$\sigma(S_{Rect} - S_{square}) = -\frac{1}{4}\,\sigma\, L^2\,(1-\lambda)^2 \;. \qquad (17.15)$$

A confrontation of the QCD side in Eq. (17.14) and the phenomenological one in Eq. (17.15) gives the best fit :

$$\sqrt{\sigma} \simeq (0.50 \pm 0.05)\ \text{GeV} \quad ; \quad \text{intercept} \simeq 0 \qquad (17.16)$$

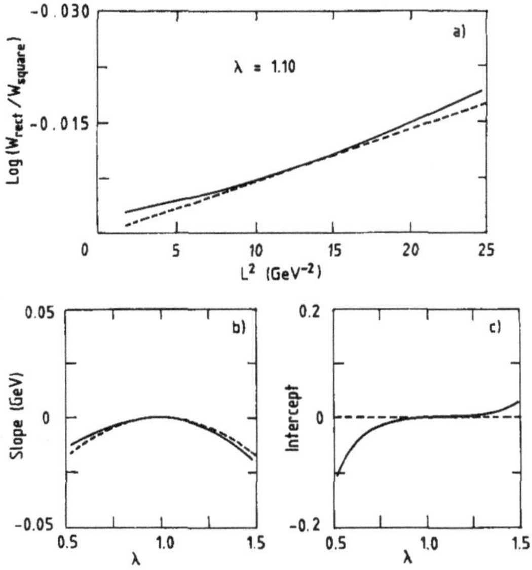

Fig. 17.5 : In (a) we plot $-\ln(W_{rect.}/W_{square})$ *(solid line) and its best phenomenological fit (dashed straight line) at* $\lambda = 1.10$. *The solid curves in (b) and (c) are obtained by comparing, for a fixed* λ, *the slope* $\times L^2$ + *intercept with* $-\ln(W_{rect.}/W_{square})$. *Again, dashed lines are the best phenomenological fit.*

for $2.7 \leqslant L[\text{GeV}^{-1}] \leqslant 4.3$ and $0.8 \leqslant \lambda \leqslant 1.2$ (see Fig. 17.5). This value corresponds to the canonical value of $\langle \alpha_s G^2 \rangle$ and tends to decrease with increasing values of the gluon condensate. This result is independent of the shape of the analyzed contour (rectangular, elliptical,...). It is comparable with the phenomenological $\sigma \simeq 0.42$ GeV value expected from Regge slopes from the data and with lattice estimates[53].

3. SOME OTHER APPLICATIONS IN QCD

Among some other applications in QCD, we mention :

a) Lattice correlators

An attempt to relate lattice correlators of meson operators at small time separation and the quark condensates which spontaneously break chiral symmetry [268]. The aim is that the degree of accuracy of such relations should provide a set of consistency checks on the ability of lattice Monte Carlo simulations to reproduce the observed spontaneous chiral symmetry breaking of the continuum limit.

b) Two-dimensional QCD

The uses of QSSR in two-dimensional QCD[269] in addition to the other solvable models which have been discussed in Chapter 3. The aim is to test the accuracy of the different sum rules approaches used in the charmonium systems where the authors conclude that the exponential and expanded ratio of moments is best suited for this channel owing to its insensitivity on the quark-mass definition... This fact has indeed been verified explicitly in the Bell-Bertlmann analysis[19].

CHAPTER 18

QHD SPECTRAL SUM RULES FOR SOME COMPOSITE MODELS OF ELECTRO-WEAK INTERACTIONS

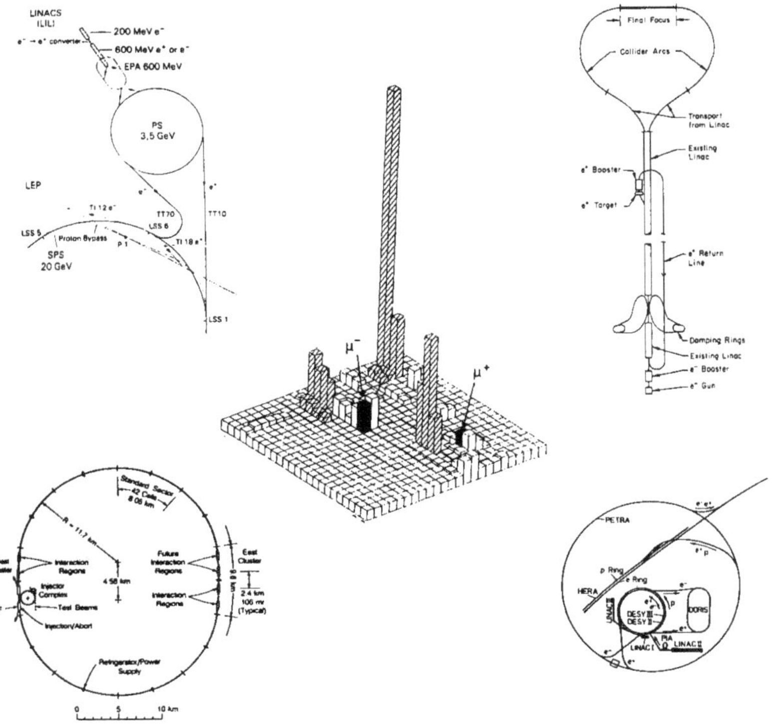

The Z-peak seen by the UA1,2 groups surrounded by the new accelerators (LEP, SLC, HERA and SSC) dedicated for further tests of the standard model and for detecting new physics

The last (but by no means the least important) applications of QSSR which we shall discuss are those in the composite models of electroweak interactions where the aim is to test the consistency of the assumed composite nature of the W and Z bosons and the quarks.

1. THE UNDERLYING IDEA OF COMPOSITENESS

The possibility of having a direct γ-Z coupling[270] becomes natural if the W and Z are bound states of much more fundamental constituents such as preons or haplons[271], in much the same way as the ρ meson of QCD is a bound state of up and down quarks.
Within this framework, weak interactions appear to be an effective theory with a $SU(n \geq 2)$ global symmetry which is a residue of preon dynamics supposed to be described at the TeV scale by a hyperstrong QHD theory based on a $SU(N)_H$ hypercolour gauge group. The $SU(n \geq 2)$ global symmetry is broken by electromagnetism through a γ-Z coupling which makes the W and Z splittings :

$$M_Z \simeq M_W \; (1 - \lambda^2)^{-1/2} \tag{18.1}$$

where :

$$\lambda \simeq g \sin^2 \theta_W / e \tag{18.2}$$

is the mixing parameter.
One of the "beyond the Standard Model" consequences of this model is the expectation of a rich spectrum of composite particles at the TeV scale while the hope is to have an understanding of the origin of the replication of the fermion family, of the hierarchy of masses and to avoid the inelegant Higgs mechanism. Tests of the effects of these non-standard weak bosons have been suggested so far[272]. In this Chapter, we propose to test the self-consistency of the idea behind these composite models using QHD (Quantum Haplo Dynamics) spectral sum rules[273-280].

2. CAN THE W BE MUCH LIGHTER THAN THE COMPOSITENESS SCALE ?

Let us study the minimal $SU(2)_L$ global symmetry scheme. The W and Z boson fields can be represented as :

$$W^- \equiv \bar{\alpha}\,\beta$$

$$Z \equiv \frac{1}{\sqrt{2}}\left(\bar{\alpha}\,\alpha - \bar{\beta}\,\beta\right) \qquad (18.3)$$

where the haplons α and β can be fermion or boson fields or both in some supersymmetric composite models.

a) spin 1/2 haplons and the Fritzsch-Mandelbaum model[281)]

The QHD Lagrangian density is very similar to the QCD one in the chiral limit $m_i = 0$ and with chiral fields modulo the presence of anomalous terms not relevant for our discussion here.
In terms of haplon fields, the charged weak current :

$$J^\mu_W = \bar{\alpha}\,\gamma^\mu\,(1 - \gamma_5)\,\beta \qquad (18.4)$$

is the Noether current.
The associated two-point correlator can be studied using an OPE including the haplon and hypergluon condensates. Due to the chirality of the fields, the $\langle\bar{\alpha}\,\alpha\rangle$ condensate cannot form. We can use either the Laplace [273,275-279)] or FESR[274,280)] sum rules for the analysis. However, the use of the Laplace transform sum rules can be very inaccurate here due to the too low mass of the W compared to the continuum threshold (about a few TeV) and to the uncertainties for higher dimension condensates. In fact, the localization of the so-called sum rule window might be impossible to achieve unless one provides a much more

involved parametrization of the continuum [276] or one proposes a new criterion to obtain the optimal result[277]. Instead, one can expect that duality constraints from FESR can be much more informative if one uses the standard parametrization ("one resonance" plus "QHD continuum") of the spectral function. The W "resonance" is introduced as :

$$\left\langle 0 \left| V^{\mu}(x)^{\alpha}_{\beta} \right| W \right\rangle = \epsilon^{\mu} \sqrt{2}\, M_W\, F_W \quad , \tag{18.5a}$$

where :

$$F_W \simeq \frac{M_W}{g} \simeq 1.6\, M_W \quad , \tag{18.5b}$$

is the decay constant of the W which has been estimated using a W dominance of the W-$f\bar{f}$ form factor at zero momentum. $V^{\mu}(x)^{\alpha}_{\beta} = \bar{\alpha}\, \gamma_{\mu} \beta$ is the vector part of the weak current and we shall study the OPE of the correlator :

$$\Pi_{\mu\nu}(q) = i \int d^4x\, e^{iqx} \left\langle 0 \left| \mathbb{T}\, V_{\mu}(x) \, (V_{\nu}(o))^{+} \right| 0 \right\rangle$$

$$\equiv - (g^{\mu\nu} q^2 - q^{\mu} q^{\nu})\, \Pi(q^2) \quad . \tag{18.6}$$

Using the moments :

$$\int_0^{t_c} dt\, t^n \frac{1}{\pi} \operatorname{Im} \Pi(t) \tag{18.7}$$

and the known QHD expressions analogous to the QCD one (see Chapter 5), one can derive the FESR constraints :

$$F_W^2 \simeq \frac{t_c}{8\pi^2} \quad , \tag{18.8}$$

$$F_W^2\, M_W^2 \simeq \frac{t_c^2}{16\pi^2} - \frac{\left\langle \alpha_H\, F^2 \right\rangle}{24\pi} \quad , \tag{18.9}$$

$$F_W^2\, M_W^4 = \frac{t_c^3}{24\pi^2} \left(\frac{1}{8\pi^2}\right) C_6\, \langle 0_6 \rangle \quad , \tag{18.10}$$

where $\left\langle \alpha_H\, F^2 \right\rangle$ and $C_6\langle 0_6 \rangle$ are respectively the hypergluon and four-fermion condensates. The accuracy of these constraints is expected to be bad for increasing dimensions.

Let us first concentrate on the relation in Eq. (18.8) which is very similar to that of the ρ-meson sum rule :

$$\frac{M_\rho^2}{\gamma_\rho^2} \simeq \frac{t_c}{2\pi^2} \quad . \tag{18.11}$$

Eqs (18.8) and (18.5) indicate that indeed the t_c value of the haplon pairs can be large :

$$\sqrt{t_c} \simeq 14\ M_W \simeq 1.15\ \text{TeV} \quad , \tag{18.12}$$

and the gain compared to the ρ meson case in Eq. (18.11) $\left(\sqrt{t_c} \simeq 1.7\ M_\rho\right)$ is mainly due to the smallness of the W coupling :

$$\gamma_W \equiv \frac{M_W}{2F_W} = \frac{g}{2} \simeq 0.3 \quad , \tag{18.13}$$

while $\gamma_\rho \simeq 2.57$.

Eq. (18.9) will be satisfied if the value of the hypergluon condensate is large. This constraint can be better seen if one takes the ratio of moments (18.9) and (18.8) :

$$M_W^2 \simeq \frac{t_c}{2}\left\{1 - \frac{2\pi}{3}\frac{\langle \alpha_H F^2 \rangle}{t_c^2}\right\} , \tag{18.14}$$

where the smallness of the W mass is dual to the large value of the hypergluon condensate as t_c is quite large from the first sum rule in Eq. (18.12). This condition is not needed in QCD as there the t_c value is quite low. An analogous conclusion can be derived for the four-quark operator appearing in Eq. (18.10) although one expects very poor accuracy from this constraint. FESR give the estimate :

$$\langle \alpha_H F^2 \rangle^{1/4} \simeq 12\ M_W \simeq (-\ C_6 \langle O_6 \rangle)^{1/6} . \tag{18.15}$$

The value of $C_6\langle O_6\rangle$ is about $(360\ \mathrm{GeV})^6$ and can spontaneously break the $SU(2)_L$ chiral symmetry as its vacuum expectation is not protected by chirality arguments. This effect can be measured by studying the Weinberg sum rules, in the chiral limit, which involve the difference between the axial and vector currents. In this way, the WSR is spontaneously broken as $\Delta \simeq \frac{(360\ \mathrm{GeV})^6}{t_c^3} \simeq 2.10^{-3}$, i.e. this breaking is quite negligible. Therefore, one expects a good realization of the preon chiral symmetry.

Previous discussions indicate that the low value of the W mass compared to the compositeness scale can be understood within the sum rules approach. They also indicate that the vacuum structure of the QHD theory even in the spin 1/2 case can be very different from the QCD one.

b) Spin 0 preons and the Abbott-Farhi model[282)]

In the case where the constituents of the W are spinless preon fields, one can start from the QHD Lagrangian density :

$$\mathcal{L}_{QHD}(x) = -\frac{1}{4} F^a_{\mu\nu} F^{\mu\nu}_a + \left(D^*_\mu \phi^+_i\right)\left(D^\mu \phi_i\right) + \text{scalar potential} . \quad (18.16a)$$

The charged current is the Noether current :

$$J^W_\mu = \phi^+_\alpha \overleftrightarrow{D}_\mu \phi_\beta \quad . \quad (18.16b)$$

The FESR corresponding to the correlator associated with this current has been derived in Ref. 275b) and later on in Refs 274) and 280). For $SU(3)_H$ these FESR constraints are :

$$2 F^2_W \simeq \frac{t_c}{8\pi^2} + 4 \langle \phi^+ \phi \rangle \ , \quad (18.17)$$

$$2 F^2_W M^2_W \simeq \frac{t^2_c}{16\pi^2} - \frac{\left\langle \alpha_H F^2 \right\rangle}{6\pi} + \frac{64\pi}{9} \alpha_H \langle \phi^+ \phi \rangle^2 \ , \quad (18.18)$$

and are derived from the QHD expression of the correlator for $SU(N)_H$:

$$\Pi_W(q^2) \simeq \frac{N}{24\pi^2} \log \frac{Q^2}{\nu^2} + 4 \frac{\langle \phi^+ \phi \rangle}{Q^2} + \frac{\left\langle \alpha_H F^2 \right\rangle}{6\pi Q^4} - 8\left(\frac{N^2-1}{N^2}\right)\pi \frac{\alpha_H \langle \phi^+ \phi \rangle}{Q^4} . \quad (18.19)$$

One should notice that the presence of the $\langle \phi^+ \phi \rangle$ condensate in Eq. (18.17) leads to a big difference from the case of spin 1/2 preons. One can exploit Eq. (18.17) using the positivity of t_c and assuming $\langle \phi^+ \phi \rangle \geqslant 0$. Then, we deduce :

$$M_W \gtrsim 2 g \langle \phi^+ \phi \rangle^{1/2} \quad . \quad (18.20)$$

The Glashow-Salam-Weinberg (GSW) mass relation would correspond to the equality sign in Eq. (18.20). This GSW mass relation is difficult to obtain from the conventional FESR in Eq. (18.17) unless the continuum effect is much smaller than any of the two contributions in Eq. (18.17). An unlikely possibility is $t_c \simeq 0$ and indicates that for a composite model with scalar preons to reproduce correctly the GSW, one should modify the usual sum rule approach. In so doing, Refs 275a) and 277) give different arguments supporting that one can neglect the perturbative terms responsible for the "unwanted" t_c factor by working with the Laplace sum rules. Within this assumption, (take $t_c = 0$ in Eqs (18.17) and 18.18)), they obtain a composite model which mimics quite well the scalar sector of the standard model but with the condition that the hypergluon condensate is also negligible. Ref. 276) also shows that the role of the low-mass continuum can be important in the spectral function. This result might be more consistent with the tendency to have $t_c \simeq 0$. One should be aware that results obtained in these ways are in contradiction with the case of the spin 1/2 preons. It is also difficult to be convinced that the neglecting of the perturbative contribution which forces the model for reproducing the GSW mass relation, is consistent with the conventional framework of QSSR.

c) Models with fermion and scalar preons and supersymmetry

Without loss of generalities, the model can be described by the Lagrangian density :

$$\mathcal{L}_{QHD}(x) = -\frac{1}{4} F^a_{\mu\nu} F^{\mu\nu}_a + \bar{\psi}_1 \hat{D} \psi_1 + \left(D^*_\mu \phi^+_1\right)(D_\mu \phi_1) + g \bar{\psi}_1 \phi_1$$

$$+ \text{ scalar potential } . \tag{18.21}$$

The term $g \bar{\psi}_1 \phi_1$ can describe models having composite quarks and leptons while for sypersymmetric models we have the additionnal Yukawa $g \bar{\psi} \lambda \phi$ term where λ is the gaugino spin 1/2 partner of the hypergluon. We shall be concerned with the Noether current :

$$J_\mu(x) = \bar{\psi}_1 \gamma_\mu \psi_1 + \phi^+_1 \overleftrightarrow{D}_\mu \phi_1 . \tag{18.22}$$

For some non-supersymmetric models, the two point correlator reads[274]:

$$\Pi(q^2) \simeq -\frac{5N}{48\pi^2}\log\frac{Q^2}{\nu^2} + \frac{2\langle\phi^+\phi\rangle}{Q^2} - \frac{8(N^2-1)}{N^2}\pi\alpha_H\frac{\langle\phi^+\phi\rangle^2}{Q^4} + \frac{\left\langle\alpha_H F^2\right\rangle}{6\pi Q^4} . \quad (18.23)$$

For some supersymmetric models, it is convenient to work within a two-component notation. In this way, the current reads :

$$J^\mu_W(x) = \tilde{\psi}\,\sigma_\mu\bar{\tilde{\psi}} - \psi\,\sigma^\mu\,\bar{\psi} + i(\phi^+\,\overleftrightarrow{D}^\mu\phi + \tilde{\phi}^+\,\overleftrightarrow{D}^\mu\,\tilde{\phi}) \qquad (18.24)$$

with :

$$\Phi \equiv \phi + \sqrt{2}\,\theta\,\psi + \ldots$$

$$\tilde{\Phi} \equiv \tilde{\phi} + \sqrt{2}\,\theta\,\tilde{\psi} + \ldots$$

$$\psi_D \equiv \begin{pmatrix}\psi_a\\ \bar{\tilde{\psi}}_\alpha\end{pmatrix} \quad \bar{\psi}_D \equiv \left(\tilde{\psi}_\alpha,\ \bar{\psi}_\alpha\right) \quad , \qquad (18.25)$$

where ψ and $\tilde{\psi}$ are chiral and anti-chiral matter superfields and θ is an anti-commuting variable. We have used the Weyl notation for the spinors.

The expression of the correlator is :

$$\Pi(-Q^2) = -\frac{N}{8\pi^2}\log\frac{Q^2}{\nu^2} + 2\,\langle\phi^+\phi + \tilde{\phi}^+\,\tilde{\phi}\rangle/Q^2$$

$$- 2\sqrt{2}\,g_H\,\langle\phi^+\lambda\,\psi + \tilde{\phi}\,\lambda\,\tilde{\phi}\rangle/Q^4 + \frac{\left\langle\alpha_H\,F^2\right\rangle}{4\pi\,Q^4} -$$

$$- 2\left(\frac{N^2-1}{N^2}\right) \pi\, \alpha_H \, \langle \phi^+\phi + \tilde{\phi}^+ \, \tilde{\phi}\rangle^2/Q^4$$

$$- 6\, \frac{N^2-1}{N^2}\, \pi\, \alpha_H \, \langle \phi\, \tilde{\phi} + \phi^+ \, \tilde{\phi}^+ \rangle^2/Q^4 \quad , \qquad (18.26a)$$

with :

$$\langle \phi^+ \; \lambda\, \psi + \tilde{\phi}\; \lambda\, \tilde{\bar{\psi}}\rangle \equiv - i \left\langle \phi^+ \; t_a \; \lambda^a \psi + \tilde{\phi}\; t_a \; \kappa^a \; \tilde{\bar{\psi}}\right\rangle \qquad (18.26b)$$

and t_a are the $SU(N)_H$ Gell-Mann matrices. We can use the following relation among condensates :

$$\langle \phi^+\phi\rangle \simeq \langle \tilde{\phi}^+ \; \tilde{\phi}\rangle \quad ,$$

$$\sqrt{2}\; g_H \; \langle \phi^+ \; \lambda\; \psi + \tilde{\phi}\; \lambda\, \tilde{\bar{\psi}}\rangle = \left(\frac{N^2-1}{N^2}\right) \pi\, \alpha_H \left\{\langle \phi^+ \; \phi + \tilde{\phi}^+ \; \tilde{\phi}\rangle^2 - \langle \phi\; \tilde{\phi} + \phi^+ \; \tilde{\phi}^+ \rangle^2 \right\} \quad , \qquad (18.27)$$

where the former is due to the fact that the condensate does not violate parity while the second comes from the fact that sypersymmetry is unbroken by the condensates.

From Eqs (18.23) and (18.26) it is easy to derive the FESR constraints which are very analogous to that in Eqs (18.17, 18). Therefore, previous discussions for scalar preons are also valid here.

3. SPIN-ZERO PARTNERS OF "COMPOSITE" W AND Z BOSONS AND MASS BEHAVIOUR OF THE CONDENSATES IN SUPERSYMMETRIC QCD

Properties of the spin-zero $\tilde{\pi}$ partners of the W and Z have been studied in Ref. 279) in the case of spin 1/2 haplons and in Ref. 278) in the case of supersymmetric QCD (SQCD). In the former case, the correlator associated with the divergence of the charged weak current for vector-like theories for spin 1/2 preons was studied. This current reads :

$$\partial_\mu J^\mu_W = i(m_\alpha + m_\beta)\,\bar{\alpha}\,\gamma_5\,\beta \quad . \tag{18.28}$$

Using the QCD techniques applied to the pion, one can derive the lowest dimension FESR constraint :

$$2\, M^4_{\tilde{\pi}}\, F^2_{\tilde{\pi}} \simeq (m_\alpha + m_\beta)^2 \left\{ \frac{3\, t_c^2}{16\pi^2} + \frac{\langle \alpha_H\, F^2 \rangle}{8\pi} \right\} , \tag{18.29}$$

which combined with the PCAC relation :

$$M^2_{\tilde{\pi}}\, F^2_{\tilde{\pi}} \simeq (m_\alpha + m_\beta) \langle \bar{\alpha}\alpha \simeq \bar{\beta}\beta \rangle \quad , \tag{18.30}$$

provides for $t_c \simeq 1\ \mathrm{TeV}^2$ and for the condensate values in (18.15) :

$$F_{\tilde{\pi}} \simeq -7\, \frac{\langle \bar{\alpha}\alpha \rangle}{t_c} \quad . \tag{18.31}$$

For vector-like theories, one can estimate $\langle \bar{\alpha}\alpha \rangle$ using the approximate vacuum saturation applied to $C_6 \langle O_6 \rangle$ obtained in (18.15) :

$$C_6 \langle O_6 \rangle \simeq \frac{896}{81}\, \pi^3\, \alpha_H\, \langle \bar{\alpha}\alpha \rangle^2 \quad , \tag{18.32a}$$

which gives :

$$\left(- \sqrt{\alpha_H} \left\langle \bar{\alpha}\alpha \right\rangle\right)^{1/3} \simeq 4.5 \ M_W \quad , \tag{18.32b}$$

and :

$$F_{\tilde{\pi}} \simeq 270 \text{ GeV} \quad , \tag{18.33}$$

which is a reasonable value compared to the weak interaction scale $G_F^{-1/2}$.

In the case of SQCD, the analysis of the correlator associated with currents built from scalar and/or fermion fields leads to constraints involving the $\langle \phi^+ \phi \rangle$ condensate[278]. Here, one can work with the currents :

$$J_f = 2m \ (\psi \ \tilde{\Phi} + \phi \ \tilde{\psi}) \quad , \tag{18.34}$$

$$J_\phi = 2m \ \phi \ \tilde{\Phi} \quad , \tag{18.35}$$

$$J_\pi = 2m \ \bar{\Psi} \ (i \ \gamma_5) \ \psi \quad , \tag{18.36}$$

associated with the θ, θ = 0 and θ^2 components of the superpotential.

The correlator associated with J_f has no perturbative part because of the absence of the $\phi \ \tilde{\Phi}$ propagator. Therefore the lowest dimension FESR constraint reads[278]:

$$2 \ M^3_{\tilde{\pi}} \ F^2_{\tilde{\pi}} \simeq 8 \ m^3 \ \langle \phi \ \tilde{\Phi} \rangle \quad , \tag{18.37a}$$

where :

$$\left\langle 0 \ |J_f| \ \tilde{\pi} \right\rangle = \sqrt{2} \ F_{\tilde{\pi}} \ M_{\tilde{\pi}} \ u(q) \quad . \tag{18.37b}$$

The evaluation of the correlator associated with J_ϕ gives the SQCD expression[278]:

$$\psi_\phi(q^2) = \frac{N}{4\pi^2} m^2 \left\{ -\log \frac{q^2}{\nu^2} + \frac{16\pi^2}{N_c} \frac{\langle \phi^+ \phi + \tilde{\phi}^+ \tilde{\phi} \rangle}{Q^2} + \dots \right\} . \tag{18.38}$$

The $\tilde{\pi}$ is introduced as :

$$\langle 0 \, | J_\phi | \, \tilde{\pi} \rangle = \sqrt{2} \, F_{\tilde{\pi}} M_{\tilde{\pi}} . \tag{18.39}$$

One can derive the FESR constraint :

$$2 \, F^2_{\tilde{\pi}} M^2_{\tilde{\pi}} \simeq N \, m^2 \left\{ \frac{t_c}{4\pi^2} + \frac{4}{N} \langle \phi^+ \phi + \tilde{\phi}^+ \tilde{\phi} \rangle \right\} . \tag{18.40}$$

The SQCD expression of the correlator associated with the current J_π reads[278]:

$$\psi_\pi(q^2) = \left(\frac{N}{2\pi^2} \right) m^2 \left\{ Q^2 \log \frac{Q^2}{\nu^2} - \frac{64\pi}{3N} \alpha_H \langle \phi^+ \phi + \tilde{\phi}^+ \tilde{\phi} \rangle \right\} . \tag{18.41}$$

Introducing the $\tilde{\pi}$ as :

$$\langle 0 \, | J_\pi | \, \tilde{\pi} \rangle = \sqrt{2} \, F_{\tilde{\pi}} M^2_{\tilde{\pi}} , \tag{18.42}$$

one can derive the FESR constraint :

$$2 \, F^2_{\tilde{\pi}} M^4_{\tilde{\pi}} \simeq \frac{N}{4\pi^2} m^2 \, t_c^2 . \tag{18.43}$$

Let us now study the relevance of the results in Eqs (18.37), (18.40) and (18.43). Combining Eq. (18.37) with the generalized Dashen formula[283]:

$$\left\langle 0 \left| \left[Q_5 , \left[Q_5 , \mathcal{L}_m^{\theta=0} \right] \right] \right| 0 \right\rangle = \frac{1}{2} M_{\tilde{\pi}} F_{\tilde{\pi}}^2 \simeq 2m \langle \tilde{\Phi} \phi \rangle \quad , \qquad (18.44a)$$

where Q_5 is the axial charge and

$$\mathcal{L}_m(x,\theta) = - \int d^2\theta \sum_i m_i \; \Phi \, \tilde{\Phi} + h.c \qquad (18.44b)$$

is the mass term of the SQCD Lagrangian, one deduces[278]:

$$m \simeq \frac{M_{\tilde{\pi}}}{\sqrt{2}} \quad \text{and} \quad \langle \phi \, \tilde{\phi} \rangle \simeq \frac{1}{\sqrt{2}} F_{\tilde{\pi}}^2 \quad , \qquad (18.45)$$

which is an expected behaviour in SQCD [283]. One can also use this behaviour in the relation among SQCD condensates :

$$m^2 \langle \phi^+ \phi + \tilde{\phi}^+ \tilde{\phi} \rangle = - m \left\langle \bar{\psi} \psi \right\rangle = \frac{1}{2} M_{\tilde{\pi}}^2 F_{\tilde{\pi}}^2 \quad , \qquad (18.46)$$

where the latter is the pion PCAC. It is easy to deduce the m behaviour of the condensates :

$$\langle \phi \, \tilde{\phi} \rangle \simeq \frac{1}{\sqrt{2}} \langle \phi^+ \phi + \tilde{\phi}^+ \tilde{\phi} \rangle \simeq \frac{1}{\sqrt{2}} F_{\tilde{\pi}}^2 \quad , \qquad (18.47)$$

while from (18.46), the $\left\langle \bar{\psi} \psi \right\rangle$ goes smoothly like m in the chiral limit*). This is a novel feature of SQCD.

---

*) A general study of the m behaviour of the condensates within an instanton dilute gas parametrization can be seen in Ref. 284).

Finally, one can solve Eqs (18.40) and (18.43) which leads with the help of (18.46) to :

$$\sqrt{\frac{N.}{2}} \cdot \frac{1}{\pi} m \, M^2_{\tilde{\pi}} \, F_{\tilde{\pi}} \simeq \mathcal{O}\left(M^3_{\tilde{\pi}}\right) \quad . \tag{18.48}$$

This condition is indeed realized owing to (18.45) and shows the consistency of the different results deduced from various constraints.

4. EXCITED QUARKS AND LEPTONS

Properties of the excited quarks and leptons have also been analyzed in Ref. 274) from the composite propagator :

$$S_F(q) = i \int d^4x \, e^{iqx} \left\langle 0 \left| T \, J^L_f(x) \, \bar{J}^L_f(0) \right| 0 \right\rangle \, , \tag{18.49a}$$

where the quark current is :

$$J^L_f(x) = \phi^+ \frac{1}{2} (1 - \gamma_5) \psi \quad . \tag{18.49b}$$

In a non-supersymmetric theory, the QHD expression of the propagator is :

$$S_F(q) = \frac{1}{2} (1 - \gamma_5) \hat{q} \left\{ - \frac{N}{32\pi^2} \log \frac{Q^2}{\nu^2} + \frac{\langle \phi^+ \phi \rangle}{Q^2} - \frac{\left\langle \alpha_H F^2 \right\rangle}{64\pi \, Q^4} + \ldots \right\} \, . \tag{18.50}$$

The phenomenological parametrization of the propagator is done as :

$$S_F(q) = \frac{1}{2} (1 - \gamma_5) \hat{q} \sum_{f, f^*} \lambda^2_i \frac{1}{\left(Q^2 + m^2_i\right)} \quad . \tag{18.51}$$

It is easy to derive the FESR constraints :

$$\sum_{f,f^*} \lambda_i^2 - \frac{N}{32\pi^2} t_c \simeq \langle \phi^+ \phi \rangle$$

$$\sum_{f,f^*} \lambda_i^2 m_i^2 - \left(\frac{N}{64\pi^2}\right) t_c^2 \simeq \frac{\left\langle \alpha_H F^2 \right\rangle}{64\pi} . \tag{18.52}$$

Positivity of λ_f^2 and neglecting of the m_f^2 term leads to the inequality :

$$m^2_{f^*} \geq \frac{1}{2} t_c \frac{1 + \pi \left\langle \alpha_H F^2 \right\rangle / 2 t_c^2}{1 + 16 \pi^2 \langle \phi^+ \phi \rangle / t_c} . \tag{18.53}$$

A light f^* would correspond to a large value of the scalar condensate :

$$\langle \phi^+ \phi \rangle \geq (0.8 \text{ TeV})^2 , \tag{18.54}$$

which looks unlikely for models with composite W bosons (Eq. 18.20). An analogous conclusion is also obtained for sypersymmetric theories. For completeness, we give the expression of the propagator in this case[274]:

$$S_F(q) = \frac{1}{2} (1-\gamma_5) \hat{q} \left\{ - \frac{N}{32\pi^2} \log \frac{Q^2}{\nu^2} + \frac{1}{2} \langle \phi^+ \phi + \tilde{\phi}^+ \tilde{\phi} \rangle / Q^2 \right.$$

$$\left. - \frac{3}{8} \sqrt{2}\, g_H \langle \phi^+ \lambda \psi + \tilde{\phi} \lambda \tilde{\tilde{\psi}} \rangle / Q^4 - \frac{\left\langle \alpha_H F^2 \right\rangle}{64\pi Q^4} \right.$$

$$- \frac{3}{4} \left(\frac{N^2 - 1}{N^2} \right) \pi \, \alpha_H \, (\phi^+ \phi + \tilde{\phi}^+ \tilde{\phi})^2 / Q^4$$

$$\left. - \frac{1}{4} \frac{(N^2-1)}{N^2} \pi \, \alpha_H \, (\phi \tilde{\phi} + \phi^+ \tilde{\phi}^+)^2 \Big/ Q^4 \right\} \quad . \qquad (18.55)$$

5. SPIN-TWO COMPOSITE BOSONS

Let us finally mention that the masses and couplings of a spin--two boson have been studied along the same lines as for spin 1/2 preons[285]. The current is the usual tensor current :

$$J_{\mu\nu} = \bar{\alpha} \, \gamma_\mu \, \overleftrightarrow{D}_\nu \, \beta \qquad (18.56)$$

and the two-point correlator has been studied in QCD (see Chapter 1 and Ref. 48b). A usual manipulation leads to the FESR constraint for the spin-two mass in the chiral limit :

$$M_2^2 \simeq \frac{3}{4} t_c \left\{ 1 + \frac{80}{27} \pi \frac{\left\langle \alpha_H \, F^2 \right\rangle}{t_c^2} \right\} \quad . \qquad (18.57)$$

A comparison with the W-mass relation in (18.14) indicates that although the W can be light owing to the possible compensation implied by the negative effect of the hypergluon condensate in (18.14), the spin-two boson is expected to be around the TeV scale. Findings of this exotic boson would provide a valuable test of the QHD sum rules approach to composite models as well as on the idea behind compositeness.

6. THREE-BOSON COUPLINGS

Three-boson couplings can also be estimated using vertex sum rules discussed in Chapter 14. However, within the accuracy of the approach, it would be worthwhile limiting ourselves to a qualitative result. In Ref. 279a), the $\tilde{\pi}Z\gamma$ coupling has been estimated as :

$$g_{\tilde{\pi}Z\gamma} \simeq m \left\langle \bar{\psi}\, \psi \right\rangle \Big/ M^2_{\tilde{\pi}}\, F_{\tilde{\pi}}\, M_Z\, F_Z \quad , \tag{18.58}$$

where it is interesting to notice that the coupling vanishes in a chiral theory. The $WW\tilde{\pi}$ coupling is dominated by the mixed $\left\langle \bar{\psi}\, \sigma^{\mu\nu} \lambda^a/2\, \psi\, F^a_{\mu\nu} \right\rangle$ condensate and also vanishes in a chiral theory. These behaviours of the bosonic couplings are very similar to the W-fermion-fermion one, evaluated within the same framework and may indicate that these bosonic couplings cannot be strong but must be of a weak-interaction type. An attempt to evaluate the ZWW coupling indicates that it can be strong and its structure can easily deviate from the usual gauge structure. An accurate measurement of this ZWW or some other three-vector boson couplings can provide an alternative test of the compositeness structure of vector bosons.

REFERENCES

Despite our efforts, the following list of references is far from being complete.

1) Gell-Mann,M., Acta Phys. Aust. Suppl. 9, 733 (1972) ;
Fritzsch,H. and Gell-Mann,M., Proc of the XVI Intern. Conf. on High Energy Phys. Chicago Vol 2, 135 (1972) ;
Fritzsch,H., Gell-Mann, M. and Leutwyler, H., Phys. Lett. 47B, 365 (1973).

2) Politzer, H.D., Phys. Rev. Lett. 30, 1346 (1973) ;
Gross, D.J. and Wilczek, F., Phys. Rev. Lett. 30, 1343 (1973).

3) Peterman, A., Phys. Reports 53, 157 (1979) ;
Mueller, A.H., Phys. Reports 73, 237 (1981) ;
Buras, A.J., Rev. Mod. Phys. 52, 199 (1980) ;
Altarelli, G., Phys. Reports 81, 1 (1982) ;
Yndurain, F.J., Quantum Chromodynamics, Springer-Verlag, New York (1982).

4) Yang, C.N. and Mills, R.L., Phys. Rev. 96, 191 (1954).

5) Faddeev, L.D. and Popov, Y.N., Phys. Lett. 25B, 29 (1967).

6) Becchi, C., Rouet, A. and Stora, R., Phys. Lett. 52B, 344 (1974); Commun. Math. Phys. 42, 127 (1975) ;

7) Belavin, A.A. et al, Phys. Lett. 59B, 15 (1975) ;
Hooft, G.'t., Phys. Rev. Lett. 37, 8 (1976) ; Phys. Rev. D14, 3432 (1976).

8) For reviews on current algebra, see e.g. Adler, S.L. and

Dashen, R.F. ; Current algebra and applications to particle physics (Benjamin, New-York N.Y. 1968 ; Pagels, H., Phys. Reports 16, 221, (1975) ; Alfaro, V.D., Fubini, S., Furlan, G. and Rosseti C. Currents in Hadron Physics (North Holland), Amsterdam (1973) ;

Reya, E., Rev. Mod. Phys. 46, 545 (1974).

9) Nambu, Y., Phys. Rev. 117, 648 (1960) ; Nambu, Y. and Iona-Lasinio G. Phys. Rev. 122, 345 (1961) ; Goldstone, J., Nuov. Cimento 19, 154 (1961).

10) Wigner, E., Group Theory, Princeton Univ. Press. N.J. (1939) ; Weyl, H., the classical group, Princeton Univ. Press. N.J. (1939).

11) See e.g. Das, T., et al, Phys. Rev. lett. 18, 759 (1967).

12) Gell-Mann, M., Oakes, R.J. and Renner, B., Phys. Rev. 175, 2195 (1968).

13) Langacker, P., and Pagels, H., Phys. Rev. D19, 2070 (1979) ; Langacker, P., Phys. Rev. D20, 2983 (1979) ; Gasser, J., Ann. Phys. (N.Y.) 136, 62 (1981).

14) For a recent review, see : Gasser, J., and Leutwyler, H., Phys. Reports 87, 79 (1982).

15) Weinberg, S., Phys. Rev. Lett. 18, 507 (1967).

16) Kawarabayashi, K., and Suzuki, M., Phys. Rev. Lett. 16, 255 (1966) ; Riazuddin and Fayazuddin, Phys; Rev. 147, 1071 (1966).

17) Das, T., Mathur, V.S. and Okubo, S., Phys. Rev. Lett. 19, 470 (1967).

18) Shifman, M.A., Vainshtein, A.I. and Zakharov, V.I., (hereafter referred as SVZ), Nucl. Phys. B147, 385, 448 (1979).

19) Bell, J.S. and Bertlmann, R.A., Nucl. Phys. B177, 218 (1981), Nucl. Phys. B187, 285 (1981) ; Phys. Lett. B137, 107 (1984). Bertlmann, R.A., Acta Phys. Austr. 53, 305 (1981) ; Proc. Non Perturbative Methods (Montpellier 1985) ed. S. Narison and Quarks 1988 Tbilisi, Preprint Wien UW Th Ph-1988-26

20) Radiative corrections to the Laplace transform sum rule have been first discussed in S. Narison and E. de Rafael, Phys. Lett. 103B,

57 (1981).

21) Logunov, A.A., Soloviev, L.D., and Tavkhelidze, A.N., Phys. Lett. 24B, 181 (1967) ; Sakurai, J.J., Phys. Lett. 46B, 207 (1973) : Bramon, A., Etim, E. and Greco, M., Phys. Lett. 41B, 609 (1972).

22) Shankar, R., Phys. Rev. D15, 755 (1977) ; Floratos, E.G., Narison S. and de Rafael, E., Nucl. Phys. B155, 155 (1979) ; Narison, S. and de Rafael, E., Nucl. Phys. B169, 253 (1980).

23) Bertlmann, R.A., Launer, G. and de Rafael, E., Nucl. Phys. B250, 61 (1985).

24) Chetyrkin, G., Krasnikov, N.V., and Tavkhelidze, A.N., Phys. Lett 76B, 83 (1978) ; Krasnikov, N.V., Pivovarov, A.A. and Tavkhelidze, A.N., Z. Phys. C19, 301 (1983), Gorishny, S.G., Kataev, A.L. and Larin, S.A., Phys. Lett. 135B, 457 (1984).

25) Bertlmann, R.A., et al, Z Phys. C39 , 231 (1988).

26) Kremer, M. et al, Phys. Rev. D34, 2127 (1986) ; Caprini, I. and Verzegnassi, C., Nuovo Cimento A75, 275 (1983) ; Ciulli, M. et al , J. Math. Phys. 25, 3194 (1984) ; Auberson, G. and Mennessier, G., Montpellier preprint PM 87/29;Causse, B. Montpellier Université Thesis (1989) unpublished

27) Yndurain, F.J., Phys. Lett. 63B, 211 (1976).

28) Novikov, V.A. et al., Phys. Reports 41, 3 (1978) ;

29) Reinders, L.J., Rubinstein H. and Yazaki, S., Phys. Reports 127, 1 (1985) and references therein.

30) Feynmann, R.P., Rev. Mod. Phys. 20, 367 (1948) ; Phys. Rev. 74, 939, 1430 (1948) ; 76, 749, 769 (1949) ; 80, 440 (1950) ; Schwinger, J., Phys. Rev. 73, 146 (1948), 74, 1439 (1948) ; 75, 651 (1949) ; 76, 790 (1949) ; 82, 664,914 (1951) ; 91, 713 (1953) ; Proc. Nat. Acad. Sci. USA 37, 452 (1951) ; Tomonaga, S., Prog. Theor. Phys. 1, 27 (1946) ; Koba, Z. and Tomonaga, S., ibid 3, 276 (1948) ; Koba, Z., Tati, T. and Tomonaga, S., ibid 2, 101 (1947) ; Kanazawa, S. and Tomonaga, S., ibid 3, 276 (1948) ; Dyson, F.J., Phys. Rev. 75, 486, 1736 (1949).

31) Pauli, N. and Villars, F., Rev. Mod. Phys. 21, 434 (1949).

32) Hooft, G.'t, and Veltman, M., Nucl. Phys. B144, 189 (1972) ;

Bollini, C.G. and Giambiagi, J.J., Nuov. Cimento 12B, 20 (1972) ; Ashmore, J., Nuov. Cim. Lett. 4, 289 (1972).

For reviews see e.g Refs 33 to 35 :

33) de Rafael, E., Lectures given at the Gif summer Institute (1978).

34) Narison, S., Phys. Reports 84, 263 (1982).

35) Leibbrandt, G., Rev. Mod. Phys. 47, 849 (1975) ; Hooft, G.'t and Veltmann, M., Diagrammar CERN Yellow report (1973).

36) Slavnov, A.A., Sov. Journ. Part. Nucl. 5, 303 (1975) ; Taylor, J.C., Nucl. Phys. B33, 436 (1971).

37) Stueckelberg, E.C.G. and Peterman, A., Helv. Phys. Acta 26, 499 (1953).

38) Gorishny, S.G., Kataev, A.L. and Larin, S.A., Yad. Fiz. 40, 517 (1984), erratum 42, 1312 (1985) ; Nuov. Cim. 92A, 119 (1986); Phys. Lett. B212, 238 (1988).

39) Abbott, L.F., Nucl. Phys. B185, 189 (1981). Kluberg-Stern, H. and Zuber, J.B., Phys. Rev. D12, 3159 (1975).

40) Tarrach, R., Nucl. Phys. B196, 45 (1982).

41) Espriu, D. and Tarrach, R., Z. Phys. C16, 77 (1982).

42) Narison, S. and Tarrach, R., Phys. Lett. 125B, 217 (1983).

43) Becchi, C. et al, Z. Phys. C8, 335 (1981).

44) Broadhurst, D.J., Phys. Lett. 101B, 423 (1981).

45) Gorishny S.G., Kataev A.L. and Larin, S.A., Phys. Lett. 135B, 457 (1984).

46) Broadhurst, D.J. and Generalis, S.C., OUT 4102-8 (1982). Open University preprint. (unpublished).

47) Broadhurst, D.J. and Generalis, S.C., OUT 4102-12 (1982) ; Generalis, S.C., Ph. D. Thesis 4102-13 (1983) (unpublished).

48) Bagan, E., Bramon, A. and Narison, S., Phys. Lett. 196B, 203 (1987) ; Bagan, E. and Narison S., Phys. Lett. 214B, 451 (1988).

49) Källen, G. and Sabry, A., Dan. Mat. Phys. Medd 29, n°17 (1955).

50) Schwinger, J., Particles, sources and fields, Vol.II (Addison Wesley, Reading, MA, 1973) p.407.

51) see e.g de Rafael, E., Lectures in QED, GIFT Summer School, UAB-FI-DI (1976) Barcelona Report.

52) Narison, S., Phys. Lett. 197B, 405 (1987).

53) For reviews see e.g:
Hasenfratz, P., Proc. Berkeley conference HEP (1986).
di Giacomo, A., Proc. Non-Perturbative Methods Montpellier (1985) ed. S. Narison ; Teper, M., ibid ; Gonzalez-Arroyo, A., ibid ; Rossi, G., ibid ; Engels, J., ibid ; Petronzio, R., Proc. of the LEAR Workshop Villars sur Ollon (1987); Parisi, G., Les Houches 1984, eds. K. Osterwalder and R. Stora (Elsevier 1986) J. Kogut, Les Houches 1982 eds. J.B. Zuber and R. Stora ; Creutz, M., Jacobs, L. and Rebbi, C., Phys. Reports 95, 201 (1983) ; Drouffe, J.M. and Zuber, J.B., Phys. Reports 102, 1 (1983) ; Martinelli, G., Munich Conf in HEP and Ringberg Conf, (1988); Maiani, L., Roma preprint (1988).

54) Hooft, G't., Nucl. Phys. 72B, 461 (1974) ; 75B, 461 (1974) ; Witten, E., Nucl. Phys. 160B, 57 (1979) ; Veneziano, G., Nucl. Phys. 117B, 519 (1976) ; Chew, G.G. and Rosenzweig, C., Phys. Reports 41C, 5 (1978).

55) De Angelis, G., de Falco, D. and Guerra, F., Lett. Nuov. Cim. 19, 55 (1977) ; Nambu, Y., Phys. Lett. 80B, 372 (1978) ; Polyakov, A.M., Phys. Lett. 82B, 247 (1978) ; Gervais, J.L. and Neveu, A., Phys. Lett. 80B, 255 (1979) ; Corrigan, E. and Hasslacher, H., Phys. Lett. 81B, 181 (1979).

56) Wilson, K.G., Phys. Rev. D10, 2445 (1974).

57) Makeenko, Yu.M. and Migdal, A.A., Phys. Lett. 88B, 135 (1979) ; Nucl. Phys. 188B, 269 (1981).

58) Tomboulis, E.T., Phys. Rev. Lett. 50, 885 (1983).

59) For reviews, see e.g : Crewther, R.J., Rev. Nuovo. Cim 2, 63 (1979) ; Christos, G.A., Phys. Reports 116, 251 (1984) ; Hooft, G.'t., Phys. Reports, 142, 357 (1986).

60) Witten, E., Nucl. Phys. 156B, 269 (1979) ; Veneziano, G., Nucl. Phys. 159B, 213 (1979) ; Di Vecchia, P. and Veneziano, G., Nucl. Phys. 171B, 253 (1980) ; Nath, P. and Arnowitt, R., Phys. Rev. D23, 473 (1981) ; Rosenzweig, C., Schechter, J. and Trahern, G., Phys. Rev. D21, 3388 (1980) ; Milton, K.A., Palmer, W.F. and

Pinsky, S.S., Phys. Rev. D22, 1647 (1980) ; Williams, P.G., Phys. Rev. D29, 1032 (1984) ; Bagchi, B. and Debnath, S., Z. Phys. C28, 597 (1985).

61) Skyrme, T.H.R., Proc. Royal, Soc. London A260, 127 (1961) ; Pak, N.K. and Tze, H.C., Ann. Phys. (NY), 117, 164 (1979).

62) Witten, E., Nucl. Phys. 223B, 433 (1983).

63) Wess, J. and Zumino, B., Phys. Lett. 37B, 95 (1971).

64) Adkins, G., Nappi, C. and Witten, E., Nucl. Phys. B228, 552 (1983).

65) Rubakov, V.A., Phys. Lett. 149B, 201 (1984).

66) Balog, J., Phys. Lett. 149B, 197 (1984) ; Adrianov, A.A., Phys. Lett. 157B, 452 (1985) ; Ebert, D. and Reinhardt, H., Nucl. Phys. B271, 188 (1986).

67) Donoghue, J.F., Golowich, E. and Hostein, B.R., Phys. Rev. Lett. 53, 747 (1984) ; Adrianov, A.A., Adrianov, V.A. and Novozhilov, V.Y., Phys. Rev. lett. 56, 1882 (1986) ; Pham, T.N. and Truong, T.N., Phys. Rev. D31, 3027 (1985).

68) Mattis, M.P. and Karliner, M., Phys. Rev. D31, 2833 (1985) ; Mattis, M.P. and Peskin, M., Phys. Rev D32, 58 (1985)

69) Narison, S. and Dosch, H.G., Phys. Lett. 180B, 390 (1986) ; 184B, 78 (1987).

70) Kaymakcalan, O., Rajeev, S. and Schechter, J., Phys. Rev. D30, 594,2345 (1984) ; D31, 1109 (1985) ; Pak, N.K. and Rossi, P., Nucl. Phys. 250B, 279 (1985) ; Brihaye, Y., Pak, N.K. and Rossi, P., Phys. lett. 164B, 111 (1985) ; Bando, M. et al, Phys. Rev. Lett. 54, 1215 (1985) ; Nucl. Phys. 259B, 493 (1985) ; Lacombe, M. et al, Phys. Rev. Lett. 57, 170 (1986) ; Abud, M., et al., Phys. Lett. B159, 155 (1985). For an alternative approach see however, Ecker, G., Gasser, J., Pich, A., and de Rafael, E., CERN TH 5185/88 (1988).

71) For reviews see e.g : Hansson, T.H., Proc. of Non-Perturbative Methods (1985) Montpellier ed. S. Narison ; Hasenfratz, P. et al, Phys. Lett. 95B, 299 (1980) ; Chodos, A. et al, Phys. Rev. D9, 3471 (1974) ; Thomas, A.W., Advances in Nucl. Phys. 13, 1 (1983).

72) Degrand, T. et al, Phys. Rev. D12, 2060 (1975).

73) Chanowitz, M.S. and Sharpe, S.R., Nucl. Phys. B222, 211 (1983) ; De Tar, C.E. and Donoghue, J.F., Ann. Rev. Nucl. Sci. 33 (1983).

74) Callan, C.G., Dashen, R. and Gross, D.J., Phys. Rev. D17, 2717 (1978) ; D19, 1826 (1979) ; D20, 3279 (1979).

75) Adler, S.L., Phys. Rev. D23, 205 (1981).

76) For reviews, see e.g: Richard, J.M., Gif Summer Institute (1986); Buchmüller, W., CERN TH 3938/84, Erice Lectures (1984) ; Gottfried, K., Proc. HEP 83, Brighton (UK) eds. J. Guy and C. Castain (1983) ; Grosse, H. and Martin, A., Phys. Rep. 60C, 341 (1980) ; Quigg, C., and Rosner, J.L., Phys. Rep. 56C, 169 (1979); Gromes, D., Z. Phys. C11, 147 (1981).

77) Richardson, J.L., Phys. Lett. 82B, 272 (1979).

78) Martin, A., Phys. Lett. 115B, 323 (1982).

79) For a review, see e.g. Martin, A., CERN TH4676/87 (1987).

80) Voloshin, M.B., Nucl. Phys. 187B, 365 (1981) ; Leutwyler, H., Phys. Lett. 98B, 447 (1981) ; Eichten, E., and Feinberg, F., Phys. Rev. D23, 2724 (1981) ; Gromes, D., phys. Lett. 115B, 483 (1982); see however, Dosch, H.G., and Marquard, U., Phys. Rev. D35, 2238 (1987) ; Dosch, H.G., Krämer, A., and Bertlmann, R.A., Phys. Lett.223B, 105 (1989) ; Campostrini, M., di Giacomo, A. and Olejnik,S Z. Phys. C31 (1986) 577.

81) Brodsky, S., and Pauli, H.C., Phys. Rev. D32, 1933 (1985), ibid 2001 (1985) and references therein.

82) Wilson, K.G., Phys. Rev. 179, 1499 (1969) ; 3D, 1818 (1971).

83) For a review, see e.g.: Shuryak, E.V., Phys. Reports, 115, 158 (1984). see also : Shuryak, E.V., Nucl. Phys. B214, 237 (1983) ; Geshkein, B.V., and Ioffe, B.L., Nucl. Phys. B166, 340 (1980).

84) Shifman, M.A., Vainshtein, A.I. and Zakharov, V.I., Phys. Lett. 76B, 471 (1978).

85) Analysis including renormalization effects is in Novikov, V.A. et al, Nucl. Phys. 249B, 445 (1985) ; Phys. Reports 116, 105 (1984).

86) Quinn H.R., and Gupta, S., Phys. Rev. D26, 499 (1982) ; Taylor, C., and Mc Clain, B., Phys. Rev. D28, 1364 (1983).

87) For a review, see e.g.: David, F., Proc. Non-perturbative Methods, Montpellier, 1985 ed. S. Narison.
88) Soldate, M., Ann. Phys. 158, 433 (1984).
Tavkkhelidze, A.N. and Tokarev, V.F., Sov. J. Part and Nucl. 16, 431 (1985) ;
Danilov, G.S., and Dyatolov, I.T., Yad Fiz 14, 1298 (1985).
89) David, F., Phys. Lett. 138B, 139 (1984).
90) David, F., and Hamber, H., Nucl. Phys. 248B, 381 (1984).
91) David, F., Nucl. Phys. 209B, 433 (1982) ; 234B, 237 (1984).
92) Parisi, G., Nucl. Phys. 150B, 163 (1979).
93) Mueller, A.H., Nucl. Phys. 250B, 327 (1985)
94) Pich, A., and de Rafael, E., Phys. Lett. 158B, 477 (1985).
95) Broadhurst, D.J., and Generalis, S.C., Phys. Lett. 139B, 85 (1984) ; Bagan, E. et al, Nucl. Phys. B254, 55 (1985).
96) Novikov, V.A. et al, Fortsch Phys. 32 , 585 (1984).
97) Nikolaev, S.N., and Radyushkin, A.V., Phys. Lett. 124B, 243 (1983) ; Nucl. Phys. B213, 285 (1983).
98) Kremer, M., Papadopoulos, N.A., and Schilcher, K., Phys. Lett. 143B, 476 (1984)
99) Chung, Y. et al, Z. Phys. C25, 151 (1984) ; Dosch, H.G., Jamin, M., and Narison, S., Phys. Lett.220B ,251 (1989).
100) Launer, G. Narison, S. and Tarrach, R., Z. Phys. C26, 433 (1984).
101) Bordes, J., Gimenez, V. and Penarrocha, J.A., Phys. Lett. 201B, 365 (1988)
102) Novikov, V.A. et al, Neutrino 78, Purdue Univ. Lafayette (1978).
103) Ioffe, B.L., Nucl. Phys. B188, 317 (1981) ; B191, 591 (1981).
104) Narison, S., Phys. Lett. B210, 238 (1988) and references therein
105) Kremer, M. and Schierholz, G., Phys. Lett. 194B, 283 (1987).
106) Espriu, D., Gross, M. and Wheater, J.F., Phys. Lett. 146B, 67 (1984).
107) Jamin M. and Kremer, M., Nucl. Phys. B277, 349 (1986).
108) Smilga, A.V., Sov. J. Nucl. Phys. 35, 271 (1982).
109) Pascual, P. and Tarrach, R., Lectures Notes in Physics Springer-Verlag 194 (1984).

110)Narison, S., Gif Summer Institute (1986) CERN TH 4624/86 (1986).
111)Dubovikov, M.S., Smilga, A.V., Nucl. Phys. 185B, 109 (1981).
112)Shuryak, E.V. and Vainshtein, A.I., Nucl. Phys. 199B, 451 (1982) ; 201B, 143 (1982).
113)Shifman, M., Nucl. Phys. 173B, 13 (1980).
114)Nikolaev, S.N. and Radyushkin, A.V., Nucl. Phys. 213B, 285 (1983).
115)Hubschmid, W. and Mallik, A., Nucl. Phys. 207B, 29 (1982).
116)Fock, V.A., Sov. J. Phys. 12, 404 (1937) and Works on Quantum Field Theory, Leningrad Univ. Press. Leningrad (1957).
117)Schwinger, J., Phys. Rev 82, 664 (1951) and Ref 50 Vols.I and II.
118)Pascual, P. and De Rafael, E., Z. Phys. C12, 127 (1982).
119)Chetyrkin, K.G., Gorishny, S.G. and Spiridinov, V.P., Phys. Lett. 160B, 149 (1985) ; Loladze, G.T., Surguladze, L.R. and Tkachov, F.V., Phys. Lett. 162B, 363 (1985).
120)Veltman, M. submitted by Strubbe, H., Comp. Phys. Comm. 18, 1 (1979).
121)Hearn, A.C., Reduce User's Manual, Version 3.2, CP78 (Rev 4/85) Santa Monica.
122)Nikolaev, S.N. and Radyushkin, A.V., Phys. lett. 124B, 243 (1983).
123)Broadhurst, D.J. and Generalis, S.C., Phys. Lett. 142B, 75 (1984) ; see also : Bagan, E., Latorre, J.I. and Pascual, P., Z. Phys. C32, 43 (1986) ; Gorishny, S.G., Larin, S.A. and Tkachov, F.V., Phys. Lett. 124B, 217 (1983).
124)Chetyrkin, K.G., Kataev, K.L. and Tkachov, F.V., Phys. Lett. 85B, 277 (1979) ; Dine, M. and Sapirstein, J., Phys. Rev. Lett. 43, 668 (1979) ; Celsmaster, W. and Gonsalves, R., Phys. Rev. Lett. 44, 560 (1980).
125)Broadhurst, D. and Generalis, S.C., Phys. Lett. 165B, 175 (1985); see also : Grozin, A.G. and Yu.F.Pinelis,Phys. Lett. 166B, 429 (1986).
126)Eidelman, S.I., Kurdadze, L.M., Vainshtein, A.I., Phys. Lett. 82B, 278 (1979).
127)Ioffe, B.L., Talk given at the High Energy Phys. Conf. Leipzig (1985) and references therein.
128)Floratos, E.G., Narison, S. and de Rafael, E., Nucl. Phys. B155,

115 (1979).
129)Narison, S., Z. Phys. C14, 263 (1982).
130)Trueman, T.L., Phys. Lett. 88B, 331 (1979). See also : Antoniadis, I., Phys. Lett. 48B, 223 (1979).
131)Peccei, R.D. and Solà, J., Nucl. Phys. B281, 1 (1987).
132)Pich, A. and de Rafael, E., Montpellier Workshop (1985) ed. S. Narison (World Scientific Co).
133)Machet, B. and Natale, A., Ann. Phys. (NY) 160, 114 (1985).
134)Novikov, V.A. et al, Nucl. Phys. B237, 525 (1984).
135)Narison, S., Paver, N. and Treleani, D., Nuov. Cim. 74A, 347 (1983)
136)Dominguez C.A., Z. Phys. C26, 269 (1984).
137)Dominguez, C.A. and de Rafael, E., Annals. Phys. 174, 372 (1987).
138)Pagels, H. and Zepeda, A., Phys. Rev. D5, 3262, (1972).
139)Narison, S., Rev. Nuov. Cim. Vol. 10, 1 (1987).
140)Narison, S., Phys. Lett. 104B, 485 (1981).
141)Dominguez, C.A. et al, Z. Phys. C27, 481 (1985).
142)For recent reviews on chiral perturbation theory, see e.g. : Gasser, J. and Leutwyler, H., Ann. Phys. (N.Y.) 158, 142 (1984) ; Nucl. Phys. 250B, 465 (1985).
143)Narison, S. et al, Nucl. Phys. B212, 365 (1983).
144)Socolow, R.H., Phys. Rev. 137, 211 (1965) ; Fuchs, N. and Scadron, M.D., Tucson preprint (1981) ; Gasser, J. and Leutwyler, H., Nucl. Phys. B250, 539 (1985) and references therein.
145)Narison, S., Contribution to the LEAR Workshop (1985) Tigne, Haute-Savoie, edited by U. Gastaldi et al.
146)Narison, S., Phys. Lett. 175B, 88 (1986).
147)Pich, A., Phys. Lett. 206B (1988) 322.
148)Bagan, E., et al, Phys. Lett. 135B, 463 (1984).
149)Dominguez, C.A. and Loewe, M., Phys. Rev. D31, 2930 (1985).
150)Particle Data group, Phys. lett. 170B (1986) 1 ; Rev. Mod. Phys. 56, S1 (1984).
151)Benaksas, D. et al, Phys. Lett. 39B, 289 (1972).
152)Reinders, L.J. and Rubinstein, H.R., Phys. Lett. 145B, 108 (1984).

153)Narison, S., Z. Phys. C22, 161 (1984) ; see also Ref 47).
154)Penso, G. and Verzegnassi, C., Nuov. Cim. A72, 113 (1982).
155)Esprui, D., Pascual, P. and Tarrach, R., Nucl. Phys. B214, 285 (1983). (note added in proof) ; Ioffe, B.L., Lectures given at the XX Cracow school, Zakopane (1983) ITEP-150 (1984).
156)Barducci, A., et al, Phys. Lett. B193, 305 (1987) ; Phys. Rev. D38, 238 (1988)
157)Ovchinnikov, A.A. and Pivovarov, A.A., Phys. Lett. 163B, 231 (1985).
158)Ayala, C., Bagan, E. and Bramon, A., Phys. Lett. 189B, 347 (1987).
159)Krasnikov, A.V. and Pivovarov, A.A., Nuov. Cim. 81A, 680 (1984).
160)Jamin, M., Narison, S. and Dosch, H.G. in Ref. 99).
161)Private communication from J. Gasser (Marseille 1984).
162)Reinders, J.L. and Rubinstein, H.R., Nucl. Phys. B196, 125 (1982).
163)Aliev, T.M. and Shifman, M.A., Phys. Lett. 112B, 401 (1982).
164)Guberina, B., et al, Nucl. Phys. B184, 476 (1982).
165)Narison, S. and Pich, A., Phys. Lett. B211, 183 (1988).
Marciano, W.J. and Sirlin, A., Phys. Rev. Lett. 61, 1815 (1988).
Braaten, A., Phys. Rev. Lett. 60, 1606 (1988).
166)Narison, S., Phys. Lett. 216B , 191 (1988)
167)Bertlmann, R.A., Nucl. Phys. 204B, 387 (1982) ;
Bertlmann, R.A., and Neufeld, H., Z. Phys. C27, 437(1985).
168)Marrow, J., Parker, J. and Shaw, G., Z. Phys. C37, 103 (1987).
169)Shifman, M.A., Vainshtein, A.I., Voloshin, M.B., and Zakharov, V.I. Phys. Lett. 77B, 80 (1978).
170)Narison, S., in preparation.
171)Narison, S., Phys. Lett. 198B, 104 (1987), and references therein.
172)Wasserbaech, S.R., Mark III experiment SLAC-PUB 4289 (1987).
173)Dominguez, C.A. and Paver, N., Phys. Lett. B197, 423 (1987).
174)Shuryak, E.V., Nucl. Phys. 198B, 83 (1982) ; Zhitnitsky, A.R., Zhitnitsky, I.R. and Chernyak, V.L., Sov, J. Nucl. Phys. 38, 773 (1983).
175)Private communication from A. Martin.
176)Narison, S., Phys. Lett. 218B , 238 (1989).
177)Shifman, M.A. and Voloshin, M.B., Sov. Nucl. Phys. 45 (2), 292

(1987) ; Politzer, H.D. and Wise, M.B., Phys. Lett. 208B, 504 (1988).
178)Gavela, M.B., Maiani, L., Petrarca, S., Martinelli, G. and Pène, O., Phys. Lett. 206B, 113 (1988) ;
Martinelli, G., Ringberg Conf. (1988), Munich Conf. (1988);
Maiani, L. , Roma preprint (1988).
179)Altarelli, G., HEP Conf. Uppsala (1987) ;
Altarelli, G., and Franzini, P., Z. Phys. C37, 271 (1988).
180)Argus collaboration, Albrecht, H. et al, Phys. Lett. B192, 245 (1987).
181)Chung, Y., Dosch, H.G., Kremer, M., and Schall, D., Phys. Lett. 102B, 175 (1981) ; Nucl. Phys. B197 (1982) 55 ; Z. Phys. C25, 151 (1984).
182)Espriu, D., Pascual, P., and Tarrach, R., Nucl. Phys. B124, 285 (1983).
183)Reinders, L.J., Rubinstein, H.R., and Yazaki, S., Phys. Lett. 120B, 209 (1983).
184)Ioffe, B.L. and Shifman, M.A, Nucl. Phys. B202 (1982).
185)Fritzsch, H. and Minkowski, P., Nuov. Cim. 30A, 393 (1975).
186)Sharpe, S.R., BNL Workshop on Glueballs...(1988) ; Michael, C. and Teper, M. , Nucl. Phys. B314, 367 (1989).
187)Novikov, V.A. Shifman, M.A., Vainshtein, A.I. and Zakharov, V.I., Nucl. Phys. B165, 67 (1980) ; B191, 301 (1981) ; Phys. Lett. 86B, 347 (1979).
188)Krasnikov, N.V., Pivovarov, A.A. and Tavkhelidze, N.N., Z. Phys. C19, 301 (1983) ; Pascual, P., and Tarrach, R., Phys. Lett. 113B, 495 (1982).
189)Narison, S., Z. Phys. C26, 209 (1984).
190)Narison, S., and Veneziano, G., Int. J. Mod. Phys A4, 2751 (1988); for reviews see Narison, S. , BNL Workshop on Glueballs ...(1988) and Moriond Conf on Hadronic Interactions (1989)
191)Dominguez, C.A., and Paver, N., Z. Phys. C31, 591 (1986) ; C32, 391 (1986).
192)Binon, F.G., et al, Nuov. Cim. 78A, 13 (1983) ;
Alde, D., et al., Nucl. Phys. B269, 485 (1988) ;

Alde, D., et al., Phys. Lett. B201, 160 (1988).

193)Alde, D., et al., Phys. Lett. B198 ,286 (1988).

194)Au, K.L., Morgan, D., and Pennington, R., Phys. Lett. 167B, 229 (1986) ; Phys. Rev. D35, 1633 (1987);
Au, K.L., Thesis RALT032 (1986) unpublished;
Mennessier, G. , Z. Phys. C16, 241 (1983).

195)For a review of the DM2 data, see, e.g. :
Jean Marie, B., Talk given at the XXIII International Conf. in HEP, Berkeley (1986).

196)Di Vecchia, P., and Veneziano, G., Nucl. Phys. B171, 253 (1980).

197)Montanet, L., Contribution at the Montpellier Workshop on Non-perturbative Methods, ed. S. Narison (1985) ;
Gilman, F.J., and Kauffman, R., Phys. Rev. D36, 2761 (1987).

198)Gershtein, S.S., Likhoded, A.A., and Prokoshkin, Y.D., Z. Phys. C24, 305 (1984), use the data of $J/\psi \to \gamma\eta,\eta'$ in order to predict r.

199)Crewther, R.K., Phys. Rev. Lett. 28, 1421 (1972).
Ellis, J. and Chanowitz, M.S., Phys. Lett. 40B, 397 (1972) ;
Phys. Rev. D7, 2490 (1973).

200)Ellis, J., and Lanik, J., Phys. Lett. 150B, 289 (1985) ; 175B, 83 (1986);
Chanowitz, M.S., LBL Berkeley preprint LBL-18701 (1984) : VI International Workshop on $\gamma\gamma$ Colission (Lake Tahoe, 1984) ;
Sharpe, S.R., Harvard preprint HUTP 84/A021 (1984) : Symposium on High Energy e^+e^- Interactions (Vanderbilt Univ., 1984).

201)Bramon, A. and Narison, S., Montpellier preprint PM 88/51 (1988).

202)Bienlein, J.K., BNL Workshop on Exotic-Glueballs-Hybrids (Sept 1988) G. Gidal, Ibid.

203)Latorre, J.I., Paban, S., and Narison, S., Phys. Lett. 191B, 437 (1987).

204)Hitlin, D., BNL Workshop on Glueballs-Hybrids and Exotics (1988) and references therein ; Montpellier Workshop on Non-Perturbative Methods (1985).

205)Narison, S., Munich Conf. in HEP (1988).

206)Ishikawa, K. et al, Phys. Rev. D37, 3216 (1988).

207)Kataev, A.L., Krasnikov N.V., and Pivovarov, A.A., Nucl. Phys. B198, 508 (1982).

208)Pak, N., Narison, S., and Paver, N., Phys. Lett. 147B, 162 (1984).

209)Balitsky, I.I., D'Yakonov, D.I., and Yung A.V., Phys. Lett. 112B, 71 (1982) ; Sov. J. Nucl. Phys. 35, 781 (1982).

210)Govaerts, J., de Viron, F., Gusbin, D., and Weyers, J., Phys. Lett. 128B, 262 (1983) and Nucl. Phys. B248, 1 (1984) ;
Latorre, J.I., Narison, S., Pascual, P., and Tarrach, R., Phys. Lett. 147B, 169 (1984).

211)Latorre, J.I., Narison, S., and Pascual, P., Z. Phys. C34, 347 (1987).

212)Govaerts, J., and de Viron, F., Phys. Rev. Lett. 53, 2207 (1984).

213)For a review, see Zilienski, M., BNL Workshop on Glueballs... Sept (1988).

214)Frere, J.M., and Titard, S., Phys. Lett. 214B, 463 (1988).

215)Alde, D., et al., Phys. Lett. 205B, 397 (1988).

216)Gastaldi, U., Contribution at the LEAR Workshop Tignes (1985), edit. U. Gastaldi et al and CERN preprint EP/88-71 (1988) Erice Lecture (May 1987).

217)Govaerts, J., Reinders, L.J., Rubinstein, H., and Weyers, J., Nucl. Phys. B258, 215 (1985) ;
Govaerts, J., Reinders, L.J., and Weyers, J., Nucl. Phys. B262, 575 (1985).

218)Jaffe, R.L., Phys. Rev. D15, 267 (1977) ;
Wong, D., and Liu, K.F., Phys. Rev. D21, 2039 (1980) ;
Weinstein, J., and Isgur, N., Phys. Rev. D27, 588 (1983).

219)Latorre, J.I., and Pascual, P., Journ. Phys. G11, 231 (1985).

220)Achasov, N.N., Devyanin, S.A., and Shestakov, G.N., Phys. Lett. 96B, 168 (1980) ; Z. Phys. C16, 55 (1982).

221)Craigie, N.S., and Stern, J., Nucl. Phys. 216B, 204 (1983).

222)a) Chernyak, V.I. and Zhitnitsky, R., Phys. Rep. 112, 173 (1984) ; b) Nesterenko, V.A., and Radyushkin, A.V., Phys. Lett. 115B, 410 (1982) ; c) Ioffe, B.L., and Smilga, A.V., Phys. Lett. 114B, 353 (1982) ; d) Gorskii, A.S., Sov. J. Nucl. Phys. 41, 1008 (1985) ; e) Lavelle, M.J., Z. Phys. C29, 203 (1985) ; f) King, I., and Sachrajda, C.T., Nucl. Phys. B279, 785 (1987) ;g) Narison, S., PhyS. Lett.224B, 184 (1989).

223)Reinders, L.J., Rubinstein, H.R., and Yazaki, S., Nucl. Phys. B213, 109 (1983).

224)Narison, S., and Paver, N., Z. Phys. C22, 69 (1984) ; Phys. Lett. 135B, 159 (1984).

225)Eletsky, V.L., Ioffe, B.L., and Kogan, Y.I., yad. Fiz 39, 138 (1984) ; Margvelashvili, M.V. and Shaposhnikov, M.E., Z. Phys. C38, 467 (1988).

226)Nakanishi, N., Progr. Theor. Phys. 25, 361 (1961) ; Theor. Phys. Suppl. 18, 1 (1961) ; Källen, G., and Wightman, A.S., Kgl. Danske Vidensk. Selsk. Mat. Phys. Skr 1 n°6 (1958) ; private communication from R. Stora.

227)*Symanzik, K., in Springer Tracts in Mod. Phys.* Vol 57 p.222, Berlin-Heidelberg-New York (1971).

228)A. Yu, Khodjamirian, Phys. Lett. 90B, 460 (1980) ;
Shifman, M.A., Z. Phys. C4, 345 (1980) ;
Shifman, M.A., and Vysotsky, M.I., Z. Phys. C10, 131 (1981) ;
Bailin, V.A., and Radyushkin, A.V., Yad. Fiz. 39, 1270 (1984).

229)Menessier, G., Narison, S., and Paver, N., Phys. Lett. 158B, 153 (1985).

230)Ayala, C., Bramon, A. and Cornet, F., presented at the Workshop on Non-Pert. Methods. Montpellier (1985) ed. S. Narison.

231)Koniuk, R. and Tarrach, R., Z. Phys. C18, 179 (1983) ;
Chetyrkin, K.G., et al., INR preprint (1984) P-0337.

232)Brodsky, S.J. and Lepage, G.P., Phys. Rev. D22, 2157 (1980).

233)Guberina, B., Machet, B. and de Rafael, E., Phys. Lett. 128B, 269 (1983).

234)For reviews, see e.g : Buras, A.J., Bari conference in HEP (1985) eds. L. Nitti and G. Preparata ;
de Rafael, E., Munich Lectures MPI-PAE/PTh 72/84 (1984).

235)Cabibbo, N., Phys. Rev. Lett. 10, 531 (1963) ;
Kobayashi, M., and Maskawa, K., Prog. Theor. Phys. 49, 653 (1973) ;
Maiani, L., Cargèse Lectures (1976).

236)Gaillard, M.K., and Lee, B.W., Phys. Rev. D10, 897 (1974).
237)Pich, A. and Narison, S., (unpublished).
238)Chetyrkin, K.G., et al., Phys. Lett. 174B, 104 (1986) ; Decker, R., Nucl. Phys. B277, 661 (1986) ; Reinders, L.J. and Yazaki, S., Nucl. Phys. B288, 789 (1987).
239)Bilic, N., Dominguez, C.A., and Guberina, B., Z. Phys. C39, 351 (1988).
240)Narison, S., (unpublished).
241)Donoghue, J.F., Golowich, E., and Holstein, B.R., Phys. Lett. 119B, 412 (1982).
242)Bardeen, W.A., Buras, AJ., and Gerard, J.M., Nucl. Phys. 293B, 787 (1987).
243)Shifman, M.A., Vainshtein, A.I., and Zakharov, V.I., JETP. Lett. 22, 55 (1975) ; Nucl. Phys. 120B, 316 (1977).
244)Guberina, B., Pich, A. and de Rafael, E., Phys. Lett. 163B, 198 (1985).
245)Guberina, B., Pich, A. and de Rafael, E., Nucl. Phys. 277B, 197 (1986).
246)Pich, A, CERN TH 5187/88, Munich Conf. in HEP (1988) ; CERN TH 5102/88, Ringberg Workshop on Weak decays (1988), Pich, A. and de Rafael, E., (to appear).
247)Gilman, F.J. and Wise, M.B., Phys. Rev. D20, 2392 (1979).
248)Blok, B. and Shifman, M.A, Yad. Fiz. 45, 211,478,841 (1987).
249)Dosch, H.G., Jamin, M. and Stech, B., Z. Phys. C42, 167 (1989).
250)Stech, B., Phys. Rev. D36, 975 (1987) and HD-THEP-88-11 (to appear in C. Jarlskog's book WSC Singapore).
251)Gorishny, S.G., Kataev, A.V. and Larin, S.A, Phys. Lett. B212, 238 (1988).
252)Gan, K.K. and Perl, M.L., Int. Journ. Mod. Phys. A3, 531 (1988).
253)Berend, H.J. et al (CELLO) Z. Phys. C23, 103 (1984) ; Gan, K.K., et al., (MARK II) Phys. Rev. Lett. 59, 411 (1987).
254)Bylsma, B.G. et al., (HRS), Phys. Rev. D35, 2269 (1987).
255)Gilman, F.J. and Rhie, S.H., Phys Rev. D31, 1066 (1985).
256)Mills, G.B., et al. Phys. Rev. Lett. 54, 624 (1985).

257)Bramon, A., Narison, S. and Pich, A., Phys. Lett. 196B, 543 (1987) ;
Pich, A., Phys. Lett. 196B, 561 (1987).

258)Band, H.R., et al. (MAC), Phys. Lett. 198B, 297 (1987).

259)Kogut, J., et al. Phys. Rev. Lett. 51, 869 (1983) ;
Celik, T., Engels, J., Satz, H., Phys. Lett. 125B, 411 (1983) ;
Gottlieb, S.A., Kuti, J., Toussaint, D., Kennedy, A.D., Meyer, S., Pendleton, B.J., Sugar, L., Phys. Rev. Lett. 55, 1958 (1985).

260)Hasenfratz, P., Karsch, F., Stamatescu, I.O., Phys. Lett. 133B, 221 (1983) ;
Celik, T., Engels, J., Satz, H., Phys. Lett. 133B, 427 (1983).
Polonyi, J., Wyld, H.W, Kogut, J.B., Shigemitsu, I., Sinclair, D.K., Phys. Lett. 53, 644 (1984) ;
Fucito, F., Rebbi, C., Solomon, S., Nucl. Phys. B252, 727 (1985) ;
Kovacs, E.V.E., Sinclair, D.K., Kogut, J.B., Phys. Rev. Lett. 58, 751 (1987) ;
Engels, J., in Non-Perturbative Methods, Montpellier Workshop 1985, S. Narison, Ed.

261)Bochkarev, A.I., Shaposhnikov, E., Nucl. Phys. B268, 220 (1986) ;
Z. Phys. C36, 267 (1987) ; Shaposhinikov, M., Lectures given at the CERN-INR school, CERN yellow report 86-03 (1986).

262)Landau, L.D., Zh ETF 37, 805 (1958).

263)Matzubara, I., Progr. Theor. Phys. 14, 351 (1955).

264)Dosch, H.G., and Narison, S., Phys. Lett. B203, 155 (1988).

265)Gasser, P., and Leutwyler, H., Phys. Lett. 188B, 477 (1987) ;
Kogut, J.B., Phys. Lett. 161B, 367 (1986).

266)Bagan, E., Latorre, J.I., Tarrach, R., Phys. Lett. 152B, 113 (1985) ; Phys. Lett. 158B, 145 (1985) ;
Bagan, E., and Latorre, J.I., Phys. Lett. 176B, 449 (1986) ;
Bagan, E., Phys. Lett. (1987).

267)Bagan, E., and Latorre, J.I., Nucl. Phys. B298, 613 (1988).

268)Craigie, N.S., Katznelson, E., and Rebbi, C., Nucl. Phys. B247, 360 (1984).

269)Ditsas, P., and Shaw, G., Nucl. Phys. B229, 29 (1983) and references therein ;

Broadhurst, D.J., Phys. Lett. 123B, 251 (1983).
270)Hung, P.Q., and Sakurai, J.J., Nucl. Phys. 143B, 81 (1978).
271)For reviews see e.g : Fritzsch, H., MPI/PAE/PTh 76/83 (1983) ;
Harari, H., WIS 85/2 (SLAC Summer School) ;
Peccei, R.D., MPI/PAE/PTh 35/84 ;
Buchmüller, W., Acta Phys. Austr. Suppl. 27, 517 (1985).
Lyons, L., Prog. Part and Nucl. Phys. 10, 227 (1983).
272)See e.g :Kneur, J.L., Larbi, S., and Narison, S., Phys. Lett. 194B, 147 (1987) and Trieste preprint IC/88/14 (1988) (sub to Z. Phys. C);
Chiapetta, P., Kneur, J.L., Larbi, S., and Narison, S., Phys. Lett. 193B, 346 (1987) ;
Chiapetta, P. and Narison, S., Phys. Lett. 198B, 412 (1987).
Narison, S., Phys. Lett. 194B, 420 (1987).
273)Narison, S., Phys. Lett. 122B, 171 (1983).
274)Narison, S., and Pascual, P., Phys. Lett. 150B, 363 (1985).
275)Dosch, H.G., Kremer, M., and Schmidt, M.G., Z. Phys. C26, 569 (1985) ;
Matveev, V.A, Shaposhnikov, M.E. and Tavkhelidze, A.N., Trieste preprint IC/83/214 (1983).
276)Buchmüller, W., and Schmidt, M.G., Nucl. Phys. B258, 230 (1985).
277)Devyanin, S.A., and Jaffe, R.L., Phys. Rev. D D33, 2615 (1986).
278)Narison, S., Phys. Lett. 142B, 168 (1984).
279)Girardi, G., Narison, S., and Perrottet, M., Phys. Lett. 133B, 234 (1983) ;
Narison, S., and Perrottet, M., Nuov. Cim. 90A, 49 (1985).
280)Narison, S., Z. Phys. C28, 591 (1985) and Contribution at the Bari Conf. in HEP (1985) eds by L. Nitti and G. Preparata .
281)Fritzsch, H., and Mandelbaum, G., Phys. Lett. 102B, 319 (1981).
282)Abbott, L., and Farhi, E., Phys. Lett. 101B, 69 (1981) ;
Claudson, M., Farhi, E. and Jaffe, R.L., Phys. Rev. D34, 873 (1986).
283)Veneziano, G., Phys. Lett. 124B, 357 (1983) ; 128B, 199 (1983).
284)Amati, D., et al., Phys. Rep. 162, 169 (1988) and references

therein

285)Narison, S., Unpublished notes and Phys. Lett. 194B, 420 (1987).

286)Dwight, H.B., Tables of integrals and other mathematical data, the Macmillian Company, New-York (1961) ; Gradshtein, I.S. and Ryzhik I.M., Table of Integrals, Series and Products, Academic Press, New-York (1980)

287)Lewin, L., Dilogarithms and associated functions, Mac Donald London (1958); Bateman, H., Higher Transcendental Functions Vol.1 *McGraw-Hill book company (New-York, Toronto, London) 1953 edited* by Erdelyi.

288)Maison, D. and Petrman, A., Comput. Phys. Comm 7, 121 (1974); Kolbig, K.S., J. Symbolic Comput. 1, 109 (1985).

289)Kolbig, K.S., J. Comput. and Applied Maths 18, 369 (1987).

EPILOGUE

There is little to be said in conclusion. My wishes and hopes are that after a careful reading of this book, you have :

1) learnt new things and techniques which will improve your knowledge in this field and will also help you to tackle new problems ;

2) observed the progress achieved within QSSR over the last ten years of activity.

Suffice to say :

"Ataovy sotro ranon'akoho : be azo andrandraina, kely azo andrandraina"

which means :
Do like the chicken drinking water. Whether they get a lot or a little , they always raise their heads in gratitude.

Before concluding this book, allow me to express my *"fisaorana lehibe"* (great gratitude) to :

i) my collaborators for their invaluable contributions in the different original works reviewed in this book ;

ii) the sponsors of my doctoral fellowship (European Economic Communites), of the long-term invitations (from ICTP Trieste, LAPP Annecy and CERN Geneva) and of the international exchange programme (DRCI of the CNRS) for their financial support.

iii) the institutes and conference organisers for giving me the opportunity to present the subjects discussed here.

I have also benefitted from the typing of Josette Cellier (University of Montpellier), the drawings skills of Arlette Coudert (CERN - Geneva), the advice of Jonathan Wilkinson of the CERN Translation and Minutes Service on English usage and the help of Claude Razanajao (University of Montpellier) in the preparation of the cover-design. I should not also forget some indirect encouragement from my family and friends. However, despite all our efforts, some misprints and errors may remain, for which I crave the reader's indulgence.
For, as we say in Malagasy :

"Ny homana aza misy latsaka"

which means : Not everything you put on your spoon reaches your mouth.

As I introduced this book in the same way as my report in Ref. 34), let me conclude it analogously by expressing my vision on the future of physics à la Oppenheimer (Reith Lectures, BBC, 1953) :

"Physics will change even more ... If it is radical and unfamiliar ... We think that the future will be only more radical and not less, only more strange and not more familiar, and that it will have its own new insights for the inquiring human spirit".

A M E N

APPENDIX

Late Prof Richard P. Feynman who has introduced the diagram techniques for perturbation theories

A. FEYNMAN RULES IN QCD

The following rules can be obtained from the effective action :

$$S_{eff} = \int d^4x \, \mathcal{L}_{QCD}(x) \quad .$$

We shall present these rules in the class α_G of covariant gauges :

Incoming and outgoing lines

quarks : in 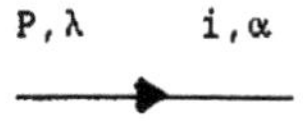

$u^i_\alpha\left(\vec{P}, \lambda\right)$

out 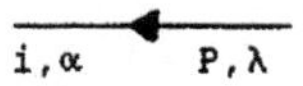

$\bar{u}^i_\alpha\left(\vec{P}, \lambda\right)$

antiquarks : in 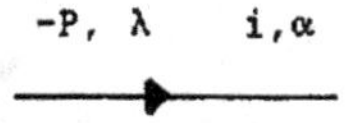

$\bar{v}^i_\alpha\left(\vec{P}, \lambda\right)$

out 

$v^i_\alpha\left(\vec{P}, \lambda\right)$

gluons : in 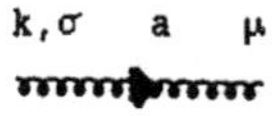

$\epsilon^a_\mu\left(\vec{k}, \sigma\right)$

out 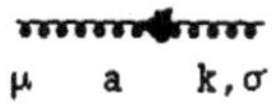

$\left(\epsilon^a_\mu\right)^* \left(\vec{k}, \sigma\right)$ (A.1)

with :

$$(\hat{p} - m)\, u\left(\vec{p}\right) = 0 \qquad (\hat{p} + m)\, v\left(\vec{p}\right) = 0$$

$$\bar{u}\left(\vec{p},\ \lambda\right)\ u\left(\vec{p},\ \lambda\right) = -\ \bar{v}\left(\vec{p},\ \lambda\right)\ v\left(\vec{p},\ \lambda\right) = 2\ E\ \left(\vec{p}\right)$$

$$\sum_{\sigma=1}^{3} \epsilon_\mu\left(\vec{k},\sigma\right)\ \epsilon^*_\nu\left(\vec{k},\sigma\right) = -\ g_{\mu\nu} + \frac{k_\mu k_\nu}{m^2}$$

$$\epsilon_\mu\ \left(\vec{k},0\right)\ \epsilon^*_\nu\left(\vec{k},0\right) = \frac{k_\mu k_\nu}{m^2}$$

<u>Propagators</u>

quark : 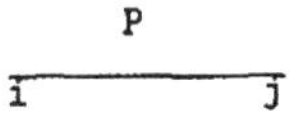

$$= \frac{i}{\hat{p} - m + i\epsilon'}\ \delta_{ij}$$

gluon : 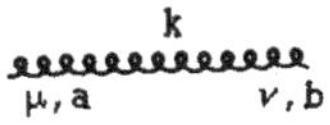

$$= (-i)\ \frac{\delta_{ab}}{k^2+i\epsilon'}\left\{g^{\mu\nu} - (1-\alpha_G)\ \frac{k_\mu k_\nu}{k^2}\right\}$$

ghost : 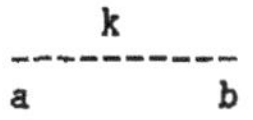

$$= (-i)\ \delta_{ab}\ \frac{1}{k^2+i\epsilon'} \qquad \text{(A.2)}$$

<u>Vertex</u>

quark-gluon-quark : 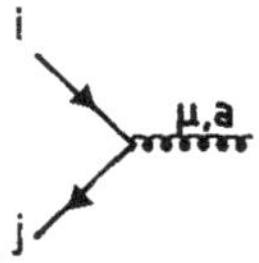

$$=(+ig)\ \gamma^\mu\ T^{(a)}_{ij}$$

3-gluon : 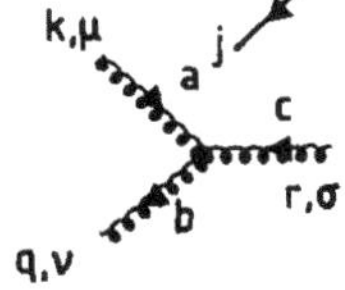

$$=(g)\,f^{abc}\left[g_{\mu\nu}(k+q)_\sigma - g_{\nu\sigma}(q+r)_\mu + g_{\sigma\mu}(r-k)_\nu\right]$$

4-gluon : 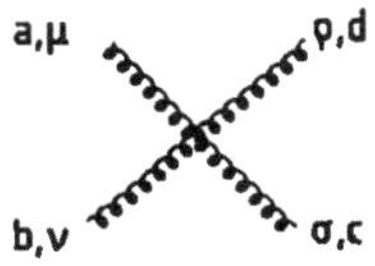

$$= -ig^2\ \left\{f^{abe}f^{cde}(g^{\mu\sigma}g^{\nu\rho}-g^{\mu\rho}g^{\nu\sigma}) + f^{ace}f^{bde}(g^{\mu\nu}g^{\sigma\rho}-g^{\mu\rho}g^{\nu\sigma}) + f^{ade}f^{cde}(g^{\mu\sigma}g^{\nu\rho}-g^{\mu\nu}g^{\sigma\rho})\right\}$$

ghost-gluon-ghost : 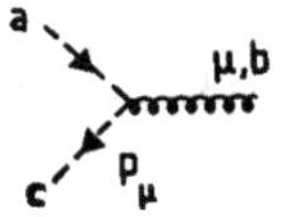

$= (+\,g)f^{abc}\;p_\mu .$ (A.3)

Factors

- $\displaystyle\int \frac{d^n p}{(2\pi)^n}$: for each loop integration and extract a δ function expressing the conservation of total energy momentum

- (-1) : *for each closed fermion or ghost loop*

- $\dfrac{1}{2!}$ for :  or  or 

- $\dfrac{1}{3!}$ for :  (A.4)

- (-1) for diagrams which differ by exchange of two identical fermion lines

- $i(2\pi)^4\;\delta\left(\sum P_{in} - \sum P_{out}\right)$ for each vertex

- arrows along the quark and/or ghost lines <u>all</u> point in the same direction

- the resulting S and T matrices are :

$$S = 1 - i(2\pi)^4\;\delta\left(\sum P_{in} - \sum P_{out}\right)T\quad .$$

B. WEIGHT FACTORS OF $SU(N)_c$

The generators T_a of the $SU(N)_c$ Lie algebra obey the properties :

$$[T_a, T_b] = i f_{abc} T_c \tag{B.1}$$

$$\mathrm{Tr}\, T_a = 0$$

where the structure constant f_{abc} is a real and totally antisymmetric tensor.

a) <u>In the adjoint representation</u> of the gluon fields :

$$(T_a)_{bc} = -i f_{abc} \tag{B.2a}$$

with the properties :

$$f_{abc} f_{dbc} = N\delta_{ad},$$

$$f_{abe}f_{cde} = \frac{2}{N}[\delta_{ac}\delta_{bd} - \delta_{ad}\delta_{bc}] + d_{ace}d_{dbe} - d_{ade}d_{bce} \quad ,$$

$$f_{abe}d_{cde} + f_{ace}d_{dbe} + f_{ade}d_{bce} = 0 \quad . \tag{B.2b}$$

where d_{abc} is a real and totally symmetric tensor :

$$d_{abb} = 0 \quad ,$$
$$d_{abc}d_{dbc} = (N - 4/N)\,\delta_{ad} \quad , \tag{B.3}$$

In this representation, one has the trace :

$$\mathrm{Tr}\, T^{(a)}T^{(b)} = N\delta_{ab}$$

$$\mathrm{Tr}\ T^{(a)}T^{(b)}T^{(c)} = \frac{i}{2} N f_{abc}$$

$$\mathrm{Tr}\ T^{(a)}T^{(b)}T^{(c)}R^{(d)} = \delta_{ab}\delta_{cd} + \delta_{ad}\delta_{bc} + \frac{N}{4}(d_{abe}d_{cde} - d_{ace}d_{dbe} + d_{ade}d_{bce}). \quad (B.4)$$

b) <u>In the fundamental representation</u> of the quark fields, one has :

$$T_a = \frac{1}{2}\lambda_a \quad (B.5)$$

with the properties :

$$[\lambda_a, \lambda_b] = 2i\ f_{abc}\ \lambda_c$$

$$\{\lambda_a, \lambda_b\} = \frac{4}{N}\delta_{ab} + 2\ d_{abc}\lambda_c$$

$$\lambda_a\ \lambda_b = \frac{2}{N}\delta_{ab} + d_{abc}\lambda_c + i\ f_{abc}\ \lambda_c\ . \quad (B.6)$$

The trace properties are :

$$\mathrm{Tr}\ \lambda_a\lambda_b = 2\ \delta_{ab}$$
$$\mathrm{Tr}\ \lambda_a\lambda_b\lambda_c = 2(d_{abc} + i\ f_{abc})$$

$$\begin{aligned}\mathrm{Tr}\ \lambda_a\lambda_b\lambda_c\mu_d = & \frac{4}{N}(\delta_{ab}\delta_{cd} - \delta_{ac}\delta_{bd} + \delta_{ad}\ \delta_{bc}) \\ & + 2(d_{abe}d_{cde} - d_{ace}d_{abe} + d_{ade}d_{bce}) \\ & + 2i(d_{abe}f_{cde} - d_{ace}f_{abe} + d_{ade}f_{bce})\ . \quad (B.7)\end{aligned}$$

Some other useful relations are :

$$(\lambda_a)_{\alpha\beta}\ (\lambda_a)_{\beta\gamma} = 4\left(\frac{N^2-1}{2N} \equiv C_2(R)\right)\delta_{\alpha\gamma}$$

$$(\lambda_b \lambda_a \lambda_b)_{\alpha\beta} = - \frac{2}{N} (\lambda_a)_{\alpha\beta}$$

$$(\lambda_a \lambda_b)_{\alpha\beta} \, (T_b)_{ca} = N(\lambda_c)_{\alpha\beta} \qquad (B.8)$$

and in the adjoint :

$$(T_a)_{bc} \, (T_a)_{cd} = (N \equiv C_2(G)) \, \delta_{bd} \quad . \qquad (B.9)$$

C. DIRAC ALGEBRA IN N DIMENSIONS

We shall use the basic property of Dirac matrices :

$$\{\gamma_\mu, \gamma_\nu\} = 2 \, g_{\mu\nu} \quad ; \quad \sigma^{\mu\nu} \equiv \frac{i}{2} [\gamma_\mu, \gamma_\nu] \quad . \qquad (C.1)$$

The algebra in n dimensions reads :

$$\gamma_\mu \gamma^\mu = n\mathbb{1} \quad \text{and} \quad g^{\mu\nu} g_{\mu\nu} = n\mathbb{1} \quad ,$$

$$\gamma_\mu \gamma_\alpha \gamma^\mu = (2-n) \, \gamma_\alpha \quad ,$$

$$\gamma_\mu \gamma_\alpha \gamma_\beta \gamma^\mu = 4 \, g_{\alpha\beta}\mathbb{1} + (n-4) \, \gamma_\alpha \gamma_\beta \quad ,$$

$$\gamma_\mu \gamma_\alpha \gamma_\beta \gamma_\gamma \gamma^\mu = - 2 \, \gamma_\gamma \gamma_\beta \gamma_\alpha - (n-4) \, \gamma_\alpha \gamma_\beta \gamma_\gamma \quad . \qquad (C.2)$$

The trace properties are :

$$\mathrm{Tr}\left(\gamma^{\mu_1} \ldots \gamma^{\mu_m}\right) = 0 \quad \text{for } m \text{ odd}$$

$$T\left(\gamma^{\mu_1} \ldots \gamma^{\mu_m}\right) = - \mathrm{Tr}\left(\gamma^{\mu_m} \gamma^{\mu_1} \ldots \gamma^{\mu_{m+1}}\right) + 2 \sum_{i=1}^{m-1} (-1)^{i+1} \quad .$$

$$\cdot \quad \mathrm{Tr}\left(\gamma^{\mu_1}\ldots\gamma^{\mu_{i-1}}\gamma^{\mu_{i+1}}\ldots\gamma^{\mu_{m-1}}\right) g_{\mu_i\mu_m}$$

$$\mathrm{Tr}\ \mathbb{1} = 4 \quad . \tag{C.3}$$

Then

$$\mathrm{Tr}\ \gamma_\mu\gamma_\nu = 4\ g_{\mu\nu} \quad ,$$
$$\mathrm{Tr}\ \gamma_\mu\gamma_\nu\gamma_\rho\gamma_\sigma = 4\ \{g_{\mu\nu}g_{\rho\sigma} - g_{\mu\rho}g_{\nu\sigma} + g_{\mu\sigma}g_{\nu\rho}\} \quad ,$$

$$\mathrm{Tr}\ \gamma_{\lambda\mu\nu\rho\sigma\tau} = g_{\lambda\mu}T_{\nu\rho\sigma\tau} - g_{\nu\lambda}T_{\mu\rho\sigma\tau} + g_{\lambda\rho}T_{\mu\nu\sigma\tau} - g_{\lambda\sigma}T_{\mu\nu\rho\tau} + g_{\lambda\tau}T_{\mu\nu\rho\sigma} \quad , \tag{C.4}$$

with

$$\gamma_{\lambda\mu\nu\rho\sigma\tau} \equiv \gamma_\lambda\gamma_\mu\gamma_\nu\gamma_\rho\gamma_\sigma\gamma_\tau$$

$$T_{\mu\nu\rho\sigma} \equiv \mathrm{Tr}\ \gamma_\mu\gamma_\nu\gamma_\rho\psi_\sigma . \tag{C.5}$$

The γ_5 matrix is unambiguously defined in four dimensions :

$$\gamma_5 = i\ \gamma_0\gamma_1\gamma_2\gamma_3 = \gamma^5 \tag{C.6}$$

with :

$$(\gamma_5)^2 = \mathbb{1} \qquad \gamma_5\gamma_\mu = -\ \gamma_\mu\gamma_5$$

and :

$$\left[\gamma_5, \sigma^{\mu\nu}\right] = 0 \qquad \gamma_5\sigma^{\mu\nu} = \frac{i}{2}\ \epsilon^{\mu\nu\rho\sigma}\ \sigma_{\rho\sigma} \tag{C.7}$$

In n dimensions some other possible definitions have been proposed which have been compared in Ref. 34). Among these, the one satisfying the properties in (C.7) is practically convenient in n dimensions. The trace properties of γ_5 matrices are :

$$\mathrm{Tr}\ \gamma_5 = 0\ ,$$
$$\mathrm{Tr}\ \gamma_5\gamma_\mu\gamma_\nu = 0\ ,$$
$$\mathrm{Tr}\ \gamma_5\gamma_\mu\gamma_\nu\gamma_\rho\gamma_\sigma = 4 \times i\epsilon_{\mu\nu\rho\sigma}\ ,$$
$$\mathrm{Tr}\ \gamma^5\gamma^\lambda\gamma^\mu\gamma^\nu\gamma^\rho\gamma^\sigma\gamma^\tau = 4i\ \Big\{g^{\mu\lambda}\epsilon^{\nu\rho\sigma\tau} - g^{\lambda\nu}\epsilon^{\mu\rho\sigma\tau} + g^{\mu\nu}\epsilon^{\lambda\rho\sigma\tau} + g^{\sigma\tau}\epsilon^{\lambda\mu\nu\rho} - g^{\rho\tau}\epsilon^{\lambda\mu\nu\sigma} + g^{\rho\sigma}\epsilon^{\lambda\mu\nu\tau}\Big\}$$
$$\mathrm{Tr}\ \gamma_5\gamma_{\mu_1}\ldots\gamma_{\mu_m} = 0 \text{ for m odd}\ , \tag{C.8}$$

where in n dimensions the totally antisymmetric tensor has the properties :

$$\epsilon_{\mu\nu\alpha\beta}\ \epsilon^{\rho\nu\alpha\beta} = -(n-3)(n-2)(n-1)\ g^\rho_\mu$$
$$\epsilon_{\mu\nu\alpha\beta}\epsilon^{\rho\sigma\alpha\beta} = -(n-3)(n-2)\ \left(g^\rho_\mu g^\sigma_\nu - g^\rho_\nu g^\sigma_\mu\right)$$
$$\epsilon_{\mu\nu\alpha\beta}\ \epsilon^{\rho\sigma\gamma\beta} = (3-n)\ \begin{vmatrix} g^\rho_\mu & g^\sigma_\mu & g^\gamma_\mu \\ g^\rho_\nu & g^\sigma_\nu & g^\gamma_\nu \\ g^\rho_\alpha & g^\sigma_\alpha & g^\gamma_\alpha \end{vmatrix} \tag{C.9}$$

Some other useful properties of the γ-matrices are the hermiticity :

$$\gamma^0\gamma^\mu\gamma^0 = (\gamma^\mu)^+ \qquad \gamma^0\gamma_5\gamma_0 = -\ \gamma_5^+ = -\ \gamma_5 \tag{C.10}$$

and the parity :

$$C\gamma_\mu C^{-1} = -\ \gamma_\mu^T \qquad C\gamma_5 C^{-1} = \gamma_5^T$$
$$C\sigma_{\mu\nu}C^{-1} = -\ \sigma^T_{\mu\nu} \qquad C(\gamma_5\gamma_\mu)\ C^{-1} = (\gamma_5\gamma_\mu)^T \tag{C.11}$$

D. FEYNMAN INTEGRALS

1. The Feynman parametrization

The Feynman parametrization is needed to recombine the product of dominators appearing in the momentum integral. Two alternative ways of parametrization are available. The first is obtained after the exponentiation of the propagator denominators and leads to

$$\frac{1}{a_1 \ldots a_n} = \int_0^\infty dz_1 \ldots \int_0^\infty dz_n \exp\left(- \sum_{i=1}^{n} a_i z_i \right) . \qquad (D.1)$$

The second alternative is obtained from the original Feynman parametrization :

$$\frac{1}{a_1 \ldots a_n} = (n-1)! \int_0^1 dz_1 \ldots \int_0^1 dz_n \frac{\delta(1 - \Sigma_i z_i)}{(\Sigma_i a_i z_i)^n} . \qquad (D.2)$$

After a suitable change of variables, one can eliminate the δ function and one ends up with :

$$\frac{1}{a_1 \ldots a_n} = (n-1)! \int_0^1 dx_1 \int_0^{x_1} dx_2 \ldots \int_0^{x_{n-2}} dx_{n-1}$$

$$[a_1 x_{n-1} + a_2 (x_{n-2} - x_{n-1}) + \ldots a_n (1 - x_1)]^{-1} \qquad (D.3)$$

or :

$$\frac{1}{a_1 \ldots a_n} = (n-1)! \int_0^1 u_1^{n-2}\, du_1 \int_0^1 u_2^{n-3}\, du_2 \ldots \int_0^1 du_{n-1}$$

$$\times [(a_1 - a_2) u_1 \ldots u_{n-1} + (a_2 - a_3) u_1 \ldots u_{n-2} + \ldots + a_n]^{-n} . \qquad (D.4)$$

This last parametrization is very convenient. In fact, it allows possible cancellations of the terms of two different propagators and offers the advantage of defining constant bounds of integration. This last property especially concerns the numerical calculation (see e.g. various QED calculations). In the following we shall adopt the parametrization in Eq. (D.4). A particular useful case of Eq. (D.4) is :

$$\frac{1}{a^\alpha b^\beta} = \frac{\Gamma(\alpha+\beta)}{\Gamma(\alpha)\ \Gamma(\beta)} \int_0^1 dx\ \frac{x^{\alpha-1}\ (1-x)^{\beta-1}}{[(a-b)x+b]^{\alpha+\beta}} \ ,$$

$$\frac{1}{a^n b^m c^r} = \frac{\Gamma(n+m+r)}{\Gamma(n)\Gamma(m)\Gamma(r)} \int_0^1 dx\ x^{m+n-1}(1-x)^{r-1} \int_0^1 dy\ \frac{(1-y)^{m-1} y^{n-1}}{[(a-b)xy+(b-c)x+c]^{n+m+r}} . \qquad (D.5)$$

In the case where a_i is $\log k^2$, the following representation integral is useful :

$$\frac{1}{(\log\ k^2)^{n+1}} = \frac{1}{\Gamma(n+1)} \int_0^\infty dx\ x^n\ (k^2)^{-x} \ . \qquad (D.6)$$

2. **Preliminary treatment of the one-loop Feynman integral**

After a Feynman parametrization of the propagators and a shift of the momentum variable, the momentum integrals for the one-loop diagram reduce usually to the integral of the type :

$$I(m,r) = \int \frac{d^n\tilde{k}}{(2\pi)^n} \frac{(\tilde{k}^2)^r}{[\tilde{k}^2 - R^2]^m} \ . \qquad (D.7)$$

For the evaluation of such integrals, it is convenient to rotate the path of integration in the complex $\tilde{k}_0$ plane $\left[\tilde{k} \equiv \left(\tilde{k}_0, \vec{\tilde{k}}\right)\right]$, by $+\pi/2$ without crossing the two poles :

$$\tilde{k}_o = \pm \sqrt{|\tilde{k}|^2 + \mathbb{R}^2} \quad . \tag{D.8}$$

Then, the k_o integration has the limits $- i\infty$ to $+ i\infty$. Let us define (Euclidean space)

$$\tilde{k}_o \equiv ik_o , \quad \tilde{k} \equiv k \quad \text{and} \quad k \equiv (k_o , k) \quad , \tag{D.9}$$

so that the k_o integral goes from $-\infty$ to $+\infty$. Then, we find

$$I(m,r) = (-1)^{r-m}\, i \int \frac{d^n k}{(2\pi)^n} \frac{(k^2)^r}{[k^2 - \mathbb{R}^2]^m} \quad . \tag{D.10}$$

Going over polar co-ordinates, we have

$$\int d^n k = \int_0^\infty \rho^{n-1}\, d\rho \int_0^\pi d\theta_{n-1}\, (\sin\theta_{n-1})^{n-2} \ldots \int_0^\pi d\theta_2 \sin\theta_2 \int_0^{2\pi} d\theta_1 , \tag{D.11}$$

where ρ is the length of the vector k. Given that the integrand of $I(m,r)$ depends only on ρ, one can perform the angular integration using

$$\int_0^\pi d\theta\, (\sin\theta)^m = \sqrt{\pi}\, \frac{\Gamma\left(\frac{1}{2}(m+1)\right)}{\Gamma\left(\frac{1}{2}(m+2)\right)} \quad , \tag{D.12}$$

where Γ is the gamma function defined as :

$$\Gamma(z) = \int_0^\infty dt\, t^{z-1} e^{-t} \quad \text{for complex } z \quad . \tag{D.13}$$

Then, one gets :

$$I(m,r) = (-1)^{r-m}\, i\, \frac{2(\pi)^{n/2}}{\Gamma(n/2)} \int_0^\infty d\rho\, \rho^{n-1} \frac{(\rho^2)^r}{[\rho^2 - \mathbb{R}^2]^m} \,, \qquad (D.14)$$

which leads to the basic formula :

$$I(m,r) \equiv \int \frac{d^n \tilde{k}}{(2\pi)^n} \frac{(\tilde{k}^2)^r}{[\tilde{k}^2 - \mathbb{R}^2]^m}$$

$$= \frac{i}{(16\pi^2)^{n/4}} (-1)^{r-m} (\mathbb{R}^2)^{r-m+n/2} \frac{\Gamma(r + n/2)\, \Gamma(m - r - n/2)}{\Gamma(n/2)\, \Gamma(m)} \,. \qquad (D.15)$$

Note that by a symmetric integration, it is easy to show that

$$\int \frac{d^n \tilde{k}}{(2\pi)^n} \frac{\tilde{k}_\mu\, \tilde{k}_\nu}{[\tilde{k}^2 - \mathbb{R}^2]^m} = \frac{1}{n} g_{\mu\nu} \int \frac{d^n \tilde{k}}{(2\pi)^n} \frac{\tilde{k}^2}{[\tilde{k}^2 - \mathbb{R}^2]^m} \,, \qquad (D.16)$$

and in the case where r is odd :

$$\int \frac{d^n \tilde{k}}{(2\pi)^n} \frac{\tilde{k}_{\mu_1} \ldots \tilde{k}_{\mu_r}}{[\tilde{k}^2 - \mathbb{R}^2]^m} = 0 \,. \qquad (D.17)$$

Note also that within dimensional regularization, the typical tadpole integrals : $\int (d^n \tilde{k}/(2\pi)^n)(\tilde{k}^2)^{\beta-1}$ vanish identically for $\beta = 0,1,2,\ldots$ We shall also see that the divergence of $I(m,r)$ for $r-m+n/2 > 0$ can be transformed into $\epsilon \equiv 4-n$ poles thanks to the properties of the Γ function.

3. **Useful properties of the Γ and B functions**

a) The Γ function defined in Eq. (D.13) has the properties

$$\Gamma(1+z) = z\,\Gamma(z) \quad , \tag{D.18}$$

$$\Gamma(1+z) = \exp\left\{-\gamma z + \sum_{n=2}^{\infty} (-1)^n \frac{z^n}{n}\,\xi(n)\right\} \quad , \tag{D.19}$$

where $\xi(n)$ is the Riemann function :

$$\xi(n) = \sum_{k=1}^{\infty} \frac{1}{k^n} : \quad \xi(2) = \frac{\pi^2}{6}, \quad \xi(3) = 1.202\ldots, \quad \xi(4) = \frac{\pi^4}{90} \quad , \tag{D.20}$$

and γ is the Euler constant

$$\gamma = \lim_{n\to\infty} \left\{1 + \frac{1}{2} + \ldots + \frac{1}{n} - \log n\right\} = 0.5772 \quad . \tag{D.21}$$

The following expansion is particularly interesting for the ϵ-regularization

$$\lim_{\epsilon\to 0} \Gamma(1+\epsilon) = 1-\gamma\epsilon + \frac{\epsilon^2}{2}\left(\gamma^2 + \frac{\pi^2}{6}\right) - \frac{\epsilon^3}{3}\left[\frac{\gamma^3}{2} + \frac{\gamma\pi^2}{4} + \xi(3)\right] + O(\epsilon^4) \quad , \tag{D.22}$$

from which one can deduce $\Gamma(\epsilon)$ with the help of (D.18). The following properties are also useful :

$$\Gamma(n) = (n-1)! \quad \text{for integer } n,$$

$$\Gamma(x)\,\Gamma(1-x) = \frac{\pi}{\sin \pi x} \quad ,$$

$$\Gamma\left(\frac{1}{2}\right) = \sqrt{\pi} \quad . \tag{D.23}$$

b) The beta function B(x,y) :

$$B(x,y) = \int_0^1 dt\ t^{x-1}\ (1-t)^{y-1} = \frac{\Gamma(x)\ \Gamma(y)}{\Gamma(x+y)} \quad , \tag{D.24}$$

has the useful properties :

$$B(x+1,\ y) = \left(\frac{x}{x+y}\right) B(x,y) \quad ,$$

$$B(x,\ 1+y) = \left(\frac{y}{x+y}\right) B(x,y) \quad ,$$

$$\lim_{\epsilon\to 0} B(1+\epsilon a, 1+\epsilon b) = 1-\epsilon(a+b)+\epsilon^2\left[(a+b)^2 - ab\ \frac{\pi^2}{6}\right]$$

$$+\ \epsilon^3(a+b)[-(a+b)^2 + ab\ \xi(2) + ab\ \xi(3)] + \ldots$$

$$\lim_{\epsilon\to 0} B\left(n - \frac{\epsilon}{2},\ 1 - \frac{\epsilon}{2}\right) = \frac{1}{n}\left\{1 + \frac{\epsilon}{2}\left[\frac{2}{n} + \sum_{j=1}^{n-1} \frac{1}{j}\right]\right\} + O(\epsilon^2) \quad ,$$

$$\lim_{\epsilon\to 0} B\left(n - \frac{\epsilon}{2},\ 2 - \frac{\epsilon}{2}\right) = \frac{1}{n(n+1)} \left\{1 - \frac{\epsilon}{2} \quad .\right.$$

$$.\quad \left[1 - \frac{2}{n} - \frac{2}{n+1} - \sum_{j=1}^{n-1} \frac{1}{j}\right\} + O(\epsilon^2) \quad . \tag{D.25}$$

c) The incomplete beta function $B_a(x,y)$:

$$B_a(x,y) = \int_0^a dt\ t^{x-1}\ (1-t)^{y-1} \quad . \tag{D.26}$$

has the properties :

$$a\ I_a(x,y) - I_a(x+1,y) + (1-a)\ I_a(x+1,\ y-1) = 0,$$

$$(x+y - xa)\ I_a(x,y) - y\ I_a(x,\ y+1) - x(1-a)\ I_a(x+1,\ y-1) = 0,$$

$$y\ I_a(x,\ y+1) + x\ I_a(x+1,\ y) - (x+y)\ I_a(x,y) = 0, \tag{D.27}$$

with :

$$I_a(x,y) \equiv B_a(x,y)/B_1(x,y) \quad . \tag{D.28}$$

d) The hypergeometric function ${}_2F_1(a,b,c,z)$:

$${}_2F_1(a,b,c,z) = \frac{\Gamma(c)}{\Gamma(b)\Gamma(c-b)} \int_0^1 dt\ t^{b-1}(1-t)^{c-b-1}(1-tz)^{-a}, \tag{D.29}$$

for Re c > Re b > 0, $|\arg(1-z)| < \pi$, with

$${}_2F_1(a,b,c,z) = 1 + \frac{ab}{1.c}\ z + \frac{a(a+1)\ b(b+1)}{1.2.c(c+1)}\ z^2 + \ldots$$

$$= \frac{\Gamma(c)}{\Gamma(b)\ \Gamma(a)} \sum_{n=0}^{\infty} \frac{\Gamma(a+n)\ \Gamma(b+n)}{\Gamma(c+n)} \frac{z^n}{n!} \quad , \tag{D.30}$$

is also useful for the treatment of multiloop integrals within Gegenbauer polynomial techniques.

4. Tables of one-loop integrals

a) Massless integrals

The most useful integral is :

$$I(\alpha,\beta) \equiv \int \frac{d^n k}{(2\pi)^n} \frac{1}{(k^2+i\epsilon')^\alpha} \frac{1}{(k-q+i\epsilon')^{2\beta}}$$

$$= \frac{i}{(16\pi^2)^{n/4}} (-1)^{-\alpha-\beta}(-q^2)^{-\alpha-\beta+n/2} \frac{\Gamma(\alpha+\beta-n/2)}{\Gamma(\alpha)\Gamma(\beta)} B\left(\frac{n}{2} - \beta, \frac{n}{2} -\alpha\right) , \quad \text{(D.31)}$$

from which one can derive in 4-ϵ dimensions (see e.g. Refs 35b and 109)

$$I^\mu(\alpha,\beta) \equiv \int \frac{d^n k}{(2\pi)^n} \frac{k^\mu}{(k^2+i\epsilon')^\alpha((k-q)^2+i\epsilon')^\beta}$$

$$= \nu^{-\epsilon} \left(\frac{i}{16\pi^2}\right) \left(\frac{-q^2}{4\pi\nu^2}\right)^{-\epsilon/2} (q^2)^{2-\alpha-\beta} q^\mu .$$

$$\cdot \frac{\Gamma(3-\alpha-\epsilon/2)\ \Gamma(2-\beta-\epsilon/2)\ \Gamma(\alpha+\beta-2+\epsilon/2)}{\Gamma(\alpha)\ \Gamma(\beta)\ \Gamma(5-\alpha-\beta-\epsilon)} , \quad \text{(D.32)}$$

$$I^{\mu\nu}(\alpha,\beta) \equiv \int \frac{d^n k}{(2\pi)^n} \frac{k^\mu k^\nu}{(k^2+i\epsilon')^\alpha\ ((k-q)^2 + i\epsilon')^\beta}$$

$$= \nu^{-\epsilon} \left(\frac{i}{16\pi^2}\right) \left(\frac{-q^2}{4\pi\nu^2}\right)^{-\epsilon/2} (q^2)^{2-\alpha-\beta} .$$

$$\cdot \left\{g^{\mu\nu}q^2 \frac{\Gamma(3-\alpha-\epsilon/2)\ \Gamma(3-\beta-\epsilon/2)\ \Gamma(\alpha+\beta+\epsilon/2)}{2\ \Gamma(\alpha)\ \Gamma(\beta)\ \Gamma(6-\alpha-\beta-\epsilon)}\right.$$

$$\left. + q^\mu q^\nu \frac{\Gamma(4-\alpha-\epsilon/2)\ \Gamma(2-\beta-\epsilon/2)\ \Gamma(\alpha+\beta-2+\epsilon/2)}{\Gamma(\alpha)\ \Gamma(\beta)\ \Gamma(6-\alpha-\beta-\epsilon)}\right\} . \quad \text{(D.33)}$$

For small values of α and β, these integrals are given in the following tables where :

$$\frac{2}{\bar{\epsilon}} \equiv \frac{2}{\epsilon} - 2\left[\gamma_E - \log\,(-q^2/4\pi\nu^2\;)\right] \tag{D.34}$$

Table D.1 :

α	β	$I(\alpha,\beta)\;.\;\nu^{\epsilon}.\left(\frac{16\pi^2}{i}\right)\;.\;(q^2)^{\alpha+\beta-2}$
1	1	$\frac{2}{\epsilon} + 2$
2	1	$-\frac{2}{\epsilon} + 0$
3	1	-1
2	2	$-\frac{4}{\epsilon} - 2$
4	1	$-1/3$
3	2	$-\frac{4}{\epsilon} - 5$

Table D.2 :

α	β	$I^{\mu}\;(\alpha,\beta)\;\nu^{\epsilon}\left(\frac{16\pi^2}{i}\right)\;q^{\mu}(q^2)^{\alpha+\beta-2}$
1	1	$\frac{1}{\epsilon} + 1$

2	1	$+ 1$
1	2	$-\frac{2}{\epsilon} - 1$
3	1	$-\frac{1}{\epsilon} - \frac{1}{2}$
2	2	$-\frac{2}{\epsilon} - 1$
1	3	$+\frac{1}{\epsilon} - \frac{1}{2}$

Table D.3 :

$$\nu^{-\epsilon} I^{\mu\nu} \equiv \frac{i}{16\pi^2} (q^2)^{(2-\alpha-\beta)} [Aq^2 g^{\mu\nu} + Bq^\mu q^\nu]$$

α	β	A	B
1	1	$-\frac{1}{6\epsilon} - \frac{2}{9}$	$+\frac{2}{3\epsilon} + \frac{13}{18}$
2	1	$+\frac{1}{2\epsilon} + \frac{1}{2}$	$+\frac{1}{2}$
1	2	$+\frac{1}{2\epsilon} + \frac{1}{2}$	$-\frac{2}{\epsilon} - \frac{3}{2}$
3	1	$-\frac{1}{2\epsilon} - \frac{1}{4}$	$+\frac{1}{2}$
2	2	$+\frac{1}{2}$	$-\frac{2}{\epsilon} - 2$
1	3	$-\frac{1}{2\epsilon} - \frac{1}{4}$	$+\frac{2}{\epsilon} + \frac{1}{2}$

b) Massive integrals in 4-ϵ dimensions :

$$I(\alpha,\beta,m^2) \equiv \int \frac{d^n k}{(2\pi)^n} \frac{1}{(k^2-m^2-i\epsilon')^\beta}$$

$$= \frac{i}{16\pi^2} (-m^2)^{2-\alpha-\beta} \nu^{-\epsilon} \left(\frac{m^2}{4\pi\nu^2}\right)^{-\epsilon/2} \frac{\Gamma(2-\alpha-\epsilon/2)\ \Gamma(\beta+\alpha-2+\epsilon/2)}{\Gamma(\beta)\ \Gamma(2-\epsilon/2)} \tag{D.35}$$

$$I(\alpha,\beta,q^2,m^2) \equiv \int \frac{d^n k}{(2\pi)^n} \frac{1}{[(k-q)^2-m^2+i\epsilon']^\alpha\ (k^2+i\epsilon')^\beta}$$

$$= \frac{i}{16\pi^2} \frac{\nu^{-\epsilon}}{(q^2)^{\alpha+\beta-2}} \left(\frac{-q^2}{4\pi\nu^2}\right)^{-\epsilon/2} \frac{\Gamma(\alpha+\beta-2+\epsilon/2)\ \Gamma(2-\beta-\epsilon/2)}{\Gamma(\alpha)\ \Gamma(2-\epsilon/2)} \cdot$$

$$\cdot \left(1-\frac{m^2}{q^2}\right)^{2-\alpha-\beta-\epsilon/2} {}_2F_1\left(\alpha+\beta-2+\epsilon/2, 2-\beta-\epsilon/2, 2-\epsilon/2\ ;\ \frac{1}{1-\frac{m^2}{q^2}}\right) \tag{D.36}$$

$$I^\mu(\alpha,\beta,q^2,m^2) \equiv \int \frac{d^n k}{(2\pi)^n} \frac{k^\mu}{[(k-q)^2-m^2+i\epsilon']^\alpha\ (k^2+i\epsilon')^\beta}$$

$$= \left(\frac{i}{16\pi^2}\right) \frac{\nu^{-\epsilon}}{(q^2)^{\alpha+\beta-2}} \cdot \left(\frac{-q^2}{4\pi\nu^2}\right)^{-\epsilon/2} \cdot q^\mu \cdot \frac{\Gamma(\alpha+\beta-2+\epsilon/2)\ \Gamma(3-\beta-\epsilon/2)}{\Gamma(\alpha)\ \Gamma(3-\epsilon/2)}$$

$$\cdot \left(1-\frac{m^2}{q^2}\right)^{2-\alpha-\beta-\epsilon/2} {}_2F_1\left(\alpha+\beta-2+\epsilon/2,\ 3-\beta-\epsilon/2,\ 3-\epsilon/2,\ \frac{1}{1-\frac{m^2}{q^2}}\right) \tag{D.37}$$

$$\tilde{I}(\alpha,\beta,q^2,m^2) \equiv \int \frac{d^n k}{(2\pi)^n} \frac{1}{[(k-q)^2-m^2-i\epsilon']^\alpha\ [k^2-m^2-i\epsilon']^\beta}$$

$$= \frac{i}{16\pi^2} \frac{\nu^{-\epsilon}}{(q^2)^{\alpha+\beta-2}} \left(\frac{-q^2}{4\pi\nu^2}\right)^{-\epsilon/2} \cdot \frac{\Gamma(\alpha+\beta-2+\epsilon/2)}{\Gamma(\alpha)\ \Gamma(\beta)}$$

$$\cdot \int_0^1 dx . x^{\alpha-1} (1-x)^{\beta-1} \left[x(1-x) - \frac{m^2}{q^2}\right]^{2-\alpha-\beta-\epsilon/2} \qquad (D.38)$$

where :

$$\tilde{I}(\alpha,\beta,q^2,m^2) = I(\beta,\alpha,q^2,m^2) \quad . \qquad (D39)$$

For $\alpha+\beta>2$, one can rewrite the x-integral by letting $\epsilon \to 0$.
For some particular values of α and β, one has :

$$I(1,1,q^2,m^2) = \frac{i}{(16\pi^2)} \nu^{-\epsilon} \left\{\frac{2}{\epsilon} - \log\left(\frac{-q^2}{4\pi\nu^2}\right) - \gamma_E - \frac{m^2}{q^2} \log \frac{m^2}{-q^2} - \left(1 - \frac{m^2}{q^2}\right) \log\left(1 - \frac{m^2}{q^2}\right) + 2\right\} \qquad (D.40)$$

$$I(1,2,q^2,m^2) = \frac{i}{(16\pi^2)} \nu^{-\epsilon} \frac{1}{q^2-m^2} \left\{- \frac{2}{\epsilon} + \log \frac{-q^2}{4\pi\nu^2} + \gamma_E - \frac{m^2}{q^2} \log \frac{m^2}{-q^2} + \left(1 + \frac{m^2}{q^2}\right) \log\left(1 - \frac{m^2}{q^2}\right)\right\} \qquad (D.41)$$

$$I^\mu(1,1,q^2,m^2) = \frac{i}{(16\pi^2)} \nu^{-\epsilon} \frac{q^\mu}{2} \left\{\frac{2}{\epsilon} - \log \frac{-q^2}{4\pi\nu^2} - \gamma_E - \frac{m^2}{q^2} \left(2 - \frac{m^2}{q^2}\right) \log \frac{m^2}{-q^2}\right.$$

$$- \left(1 - \frac{m^2}{q^2}\right)^2 \log \left(1 - \frac{m^2}{q^2}\right) - \frac{m^2}{q^2} + 2 \Bigg\} \ . \tag{D.42}$$

$$I^\mu(1,2,q^2,m^2) = \frac{i}{16\pi^2} \nu^{-\epsilon} \frac{q^\mu}{q^2} \left\{ - \frac{m^2}{q^2} \log \left(- \frac{m^2}{q^2}\right) + \frac{m^2}{q^2} \log \left(1 - \frac{m^2}{q^2}\right) + 1 \right\} \tag{D.43}$$

$$\tilde{I}(1,1,q^2,m^2) = \frac{i}{(16\pi^2)} \nu^{-\epsilon} \Bigg\{ \frac{2}{\epsilon} - \log \frac{m^2}{4\pi\nu^2} - \gamma_E$$

$$- \sqrt{1 - \frac{4m^2}{q^2}} \log \frac{\sqrt{1 - \frac{4m^2}{q^2}} + 1}{\sqrt{1 - \frac{4m^2}{q^2}} - 1} + 2 \Bigg\} \tag{D.44}$$

$$I(2,1,q^2,m^2) = \frac{i}{16\pi^2} \nu^{-\epsilon} \left\{ \frac{1}{q^2 \sqrt{1 - \frac{4m^2}{q^2}}} \log \frac{\sqrt{1 - \frac{4m^2}{q^2}} + 1}{\sqrt{1 - \frac{4m^2}{q^2}} - 1} \right\} . \tag{D.45}$$

5. Two- and three-loop massless integrals

These integrals come from Ref. 34) :

$$I_2 \equiv \int \frac{d^n k_1}{(2\pi)^n} \int \frac{d^n k_2}{(2\pi)^n} \frac{1}{k_2^2 (k_1 - k_2)^2 \ (q - k_1)^2}$$

$$= \frac{-1}{(16\pi^2)^{n/2}} (-1)^{-3+n} (-q^2)^{-3+n} \Gamma(3-n) \; B\left(\frac{n}{2} - 1, \frac{n}{2} - 1\right) B\left(\frac{n}{2} - 1, n-2\right)$$

$$= (-1)^{-\epsilon} \frac{1}{(4\pi)^{4-\epsilon}} (q^2)(-q^2)^{-\epsilon} \frac{1}{2}\left(\frac{1}{\epsilon} + \frac{13}{4} - \gamma\right) \quad \text{for } n = 4-\epsilon \;,$$

(D.46)

$$I_3 \equiv \left\{ \int \frac{d^n k_1}{(2\pi)^n} \frac{1}{k_1^2 (k_1 - q)^2} \right\}^2$$

$$= \frac{(-1)}{(16\pi^2)^{n+2}} (-1)^{-4} (-q^2)^{-4+n} \left\{ \Gamma\left(2 - \frac{n}{2}\right) B\left(\frac{n}{2} - 1, \frac{n}{2} - 2\right)\right\}^2$$

$$= \frac{(-1)}{(4\pi)^{4-\epsilon}} (-q^2)^{-\epsilon} \; 4\left\{\frac{1}{\epsilon^2} + \frac{1}{\epsilon}(2-\gamma) + 3 - 2\gamma + \frac{\gamma^2}{2} - \frac{\pi^2}{24}\right\} \quad \text{for } n = 4-\epsilon \;,$$

(D.47)

$$I_4 \equiv \int \frac{d^n k_1}{(2\pi)^n} \int \frac{d^n k_2}{(2\pi)^n} \frac{1}{k_1^2 (k_1 - q)^2 (k_1 - k_2)^2 (k_2 - q)^2}$$

$$= \frac{(-1)}{(16\pi^2)^{n/2}} (-1)^{-4+n} (-q^2)^{-4+n} \frac{\Gamma(2-n/2)\Gamma(4-n)}{\Gamma(3-n/2)} B\left(\frac{n}{2} -1, \frac{n}{2} -1\right) \quad B\left(\frac{n}{2} -1, n-3\right)$$

$$= \frac{(-1)}{(4\pi)^{4-\epsilon}} (-1)^{-\epsilon} (-q^2)^{-\epsilon} \left\{\frac{2}{\epsilon^2} + \frac{5-2\gamma}{\epsilon} + \frac{19}{2} - 5\gamma + \frac{1}{4}\left(4\gamma^2 - \frac{\pi^2}{3}\right)\right\} \quad \text{for } n=4-\epsilon$$

(D.48)

$$I_5 \equiv \int \frac{d^n k_1}{(2\pi)^n} \int \frac{d^n k_2}{(2\pi)^n} \frac{k_1^2}{k_2^2 (k_1 - k_2)^2 \; (q - k_1)^2 (q - k_2)^2}$$

$$= \frac{4}{(4\pi)^{4-\epsilon}} (-q^2)^{1-\epsilon} \frac{\Gamma(\epsilon)}{\epsilon} B\left(1 - \frac{\epsilon}{2}, 2 - \frac{\epsilon}{2}\right) B\left(2 - \frac{\epsilon}{2}, 1 - \epsilon\right)$$

$$= \frac{1}{(4\pi)^{4-\epsilon}} (-q^2)^{1-\epsilon} \left\{ \frac{1}{\epsilon^2} + \frac{1}{\epsilon} \left(\frac{11}{4} - \gamma \right) + \frac{1}{2} \left(\frac{89}{8} - \frac{11\gamma}{2} + \gamma^2 - \frac{\pi^2}{12} \right) \right\}. \quad \text{for } n=4-\epsilon.$$

(D.49)

Most of the integrals encountered in the evaluation of loop diagrams can be reduced to the above integrals, e.g. by means of the formula

$$kq = \frac{1}{2} \{k^2 + q^2 - (k-q)^2\} .$$

(D.50)

A more complicated type of integral needed for the three-loop calculation has been done analytically by the authors in Ref. 86) using the Gegenbauer polynomial techniques. Within our convention, the integral is :

$$I(\alpha, \beta, \lambda) \equiv \int \frac{d^n k_1}{(2\pi)^n} \int \frac{d^n k_2}{(2\pi)^n} \frac{1}{k_1^{2\alpha} (k_1 - q)^{2\beta} (k_1 - k_2)^2 \, k_2^2 (k_2 - q)^2}$$

$$= \frac{(-1)}{(16\pi^2)^{\lambda+1}} (-1)^{2\lambda-1-\alpha-\beta} (-q^2)^{2\lambda-1-\alpha-\beta} F(\alpha,\beta,\lambda) \quad \text{for } \alpha,\beta \geq 1, \ \lambda \equiv n/2-1,$$

(D.51)

with

$$F(1,1,1) = 6 \sum_{k=0}^{\infty} \frac{1}{(k+1)^3} = 6\xi(3) ,$$

(D.52)

$$F(\alpha > 1, \ \beta > 1, \lambda) = \frac{\Gamma(1-2\lambda)\ \Gamma(\lambda-\alpha)\ \Gamma(\lambda-\beta)\ \Gamma(\lambda)\ \Gamma(\alpha+\beta-2\lambda)}{\Gamma(\alpha)\ \Gamma(\beta)\ \Gamma(3\lambda-\alpha-\beta)}$$

$$\times \left\{ \frac{\Gamma(3\lambda-\alpha-\beta)}{\Gamma(1+\lambda-\alpha-\beta)} - \frac{\Gamma(\alpha+\beta-\lambda)}{\Gamma(\alpha+\beta+1-3\lambda)} + \frac{\Gamma(\alpha)}{\Gamma(1+\alpha-2\lambda)} - \frac{\Gamma(2\lambda-\alpha)}{\Gamma(1-\alpha)}\right.$$

$$+ \frac{\Gamma(\beta)}{\Gamma(1+\beta-2\lambda)} - \frac{\Gamma(2\lambda-\beta)}{\Gamma(1-\beta)}\Bigg\} \quad . \tag{D.53}$$

Another type of two-loop integral which is useful for deep inelastic analysis is :

$$I_7 \equiv \int \frac{d^n k_1}{(2\pi)^n} \int \frac{d^n k_2}{(2\pi)^n} \frac{(\Delta k_1)^i \ (\Delta k_2)^j}{k_1^2 (k_1-k_2)^2 \ k_2^2 (k_2-q)^2}$$

$$= \frac{(-1)}{(4\pi)^{4-\epsilon}} (-1)^{-\epsilon} \ (-q^2)^{-\epsilon} \ \frac{2}{\epsilon} \Gamma(\epsilon) \ \frac{\Gamma(i+1-\epsilon/2) \ \Gamma(1-\epsilon/2)}{\Gamma(i+2-\epsilon)} \ \frac{\Gamma(i+j-\epsilon) \ \Gamma(1-\epsilon/2)}{\Gamma(i+j+2 - \frac{3}{2}\epsilon)} \tag{D.54}$$

for $n = 4-\epsilon$ and for $\Delta^2 = 0$.

6. Some useful logarithmic integrals

a) The following integrals can be found in many textbooks (see in particular Ref. 286)) :

$$\int dx \ x^n \ \log(ax-b) = \frac{1}{n+1} \left\{ x^{n+1} - \left(\frac{b}{a}\right)^{n+1} \log(ax-b) - \sum_{i=1}^{n+1} \left(\frac{b}{a}\right)^{n+1-i} \cdot \frac{x^i}{i} \right\} \tag{D.55}$$

$$\int_0^1 dx \ x^n \log x = - \frac{1}{(n+1)^2} \tag{D.56}$$

$$I_n \equiv \int_0^1 dx \ x^n \ \log\left[a-x(1-x)\right] \tag{D.57}$$

with :

$$I_0 = -2 + \log a + \sqrt{1-4a}\,\log \frac{\sqrt{1-4a}+1}{\sqrt{1-4a}-1} \quad ,$$

(D.58)

$$I_1 = \frac{1}{2} I_0 \quad ,$$

(D.59)

$$I_2 = \frac{1}{3}\left[-\frac{13}{6} + 2a + \log a + (1-a)\sqrt{1-4a}\,\log\frac{\sqrt{1-4a}+1}{\sqrt{1-4a}-1}\right] \quad ,$$

(D.60)

$$I_3 = \frac{1}{2}\,[I_0 - 3\,I_1 + 3\,I_2] \quad ,$$

(D.61)

where one takes advantage of the invariance under $x \leftrightarrow (1-x)$ allowing us to deduce the integrals odd in n from the even ones.

b) In addition to the properties of dilogarithms (see e.g. Ref. 287)), one can use the following integrals from Refs 88), 286) and 109) :

$$\int_0^x \frac{dt}{t}\log(1+t^2) = -\frac{1}{2}\,Li_2(-x^2)$$

(D.62)

$$\int_0^{x>0} \frac{dt}{t}\log(1-t+t^2) = -\frac{1}{3}\,Li_2(-x^3) + Li_2\,(-x)$$

(D.63)

$$\int_0^{x>0} \frac{dt}{t}\log(1+t+t^2) = -\frac{1}{3}\,Li_2\,(+x^3) + Li_2\,(+x)$$

(D.64)

$$\int_0^1 \frac{dx}{1+x}\log(1+x)\,\log^2 x = -\frac{3}{2}\,\xi^2(2) + 4S_4 + \frac{7}{2}\,\xi(3)\log 2 -$$

$$- \zeta(2) \log^2 2 + \frac{1}{6} \log^4 2$$

(D.65)

$$\int_0^1 \frac{dx}{1+x} \log x \log^2(1+x) = - \frac{4}{5} \zeta^2(2) + 2 S_4 + \frac{7}{4} \zeta(3) \log 2$$

$$- \frac{1}{2} \zeta(2) \log^2 2 + \frac{1}{12} \log^4 2$$

(D.66)

$$\int_0^1 \frac{dx}{x} \log^2 x \log(1+x) = \frac{7}{10} \zeta^2(2) = - \frac{1}{3} \int_0^1 \frac{dx}{1+x} \log^3 x$$

(D.67)

$$\int_0^1 \frac{dx}{x} \log (1+x) \, Li_2 (-x) = \frac{1}{8} \zeta^2(2)$$

(D.68)

$$\int_0^1 \frac{dx}{1+x} \log(1+x) \, Li_2(-x) = \frac{6}{5} \zeta^2(2) - 3 S_4 - \frac{21}{8} \zeta(3) \log 2 + \frac{1}{2} \zeta(2) \log^2 2$$

$$- \frac{1}{8} \log^4 2$$

(D.69)

$$\int_0^1 \frac{dx}{1+x} \log x \, Li_2(-x) = \frac{13}{8} \zeta^2(2) - 4S_4 - \frac{7}{2} \zeta(3) \log 2 + \zeta(2) \log^2 2 - \frac{1}{6} \log^4 2$$

(D.70)

$$\int_0^1 \frac{dx}{1+x} \log^2(1+x) \log(1-x) = - \frac{4}{5} \zeta^2(2) + 2 S_4 + 2 \zeta(3) \log 2$$

$$- \zeta(2) \log^2 2 + \frac{1}{3} \log^4 2$$

(D.71)

with

$$S_4 \equiv \sum_{n=1}^{\infty} \frac{1}{2^n n^4} = 0.5174790616\ldots$$

(D.72)

$$\int_0^1 dx \, x^{\alpha-1} (1-x)^{\beta-1} \log x = \frac{\Gamma(\alpha) \Gamma(\beta)}{\Gamma(\alpha+\beta)} [s_1(\alpha-1) - s_1(\alpha+\beta-1)]$$

(D.73)

$$\int_0^1 dx \, x^{\alpha-1} (1-x)^{\beta-1} \log^2 x = \frac{\Gamma(\alpha) \Gamma(\beta)}{\Gamma(\alpha+\beta)} \Big\{ (s_1(\alpha-1) - s_1(\alpha+\beta-1))^2$$

$$+ s_2(\alpha+\beta-1) - s_2(\alpha-1) \Bigg\} \quad ,$$

(D.74)

where :

$$s_1(n) \equiv \sum_1^n \frac{1}{k^1} \quad .$$

(D.75)

Differentiating with respect to β one gets :

$$\int_0^1 dx \; x^{\alpha-1}(1-x)^{\beta-1} \log x \log(1-x) = \frac{\Gamma(\alpha)\;\Gamma(\beta)}{\Gamma(\alpha+\beta)} \quad .$$

$$. \{s_2(\alpha+\beta-1) - \xi(2) + (s_1(\alpha-1) - s_1(\alpha+\beta-1)(s_1(\beta-1)-s_1(\alpha+\beta-1))\} \quad .$$

(D.76)

c) A general expression as well as particular forms of :

$$\int_0^1 dx \; x^{\nu-1} \; (1-x)^{-\lambda} \log^m x$$

(D.77)

are given in Ref. 289) for integer and half-integer values of β and λ. For instance :

$$\int_0^1 dx \; \frac{x^2 \; \log^2 x}{(1-x)^2} = -4\xi(3) + \frac{\pi^2}{3} + 2$$

(D.78)

$$\int_0^1 dx \; \frac{x^{3/2} \; \log^2 x}{(1-x)^2} = -21 \; \xi(3) + \pi^2 + 16 \quad .$$

(D.79)

E. SOME USEFUL FORMULA FOR QSSR

Let $\hat{\mathcal{L}}$ the Laplace transform operator in Eq. (1.38). Then, we have[20,23] :

$$\hat{\mathbb{L}}\ \frac{1}{(Q^2+m^2)^{\alpha+1}} = \frac{1}{\Gamma(\alpha+1)}\ \tau^{\alpha+1}\ e^{-m^2\tau} \qquad \text{(E.1)}$$

$$\hat{\mathbb{L}}\ \frac{1}{(Q^2)^{\alpha+1}} \log \frac{Q^2}{\nu^2} = \frac{\tau^{\alpha+1}}{\Gamma(\alpha+1)} [- \log \tau\nu^2 + \psi(\alpha+1)] \qquad \text{(E.2)}$$

$$\hat{\mathbb{L}}\ \frac{1}{(Q^2)^{\alpha+1}} \log^2 \frac{Q^2}{\nu^2} = \frac{\tau^{\alpha+1}}{\Gamma(\alpha+1)} \Big[\log^2 \tau\nu^2 - 2\psi(\alpha+1).\log \tau\nu^2 + \psi^2(\alpha+1) - \psi'(\alpha+1) \Big] \qquad \text{(E.3)}$$

$$\hat{\mathbb{L}}\ \frac{1}{x^{\alpha+1}} \frac{1}{(\log x)^{\beta+1}} = y\ \mu(y,\beta,\alpha)$$

$$\tau\Lambda^2 \equiv y \underset{\to}{\simeq} 0\ \frac{1}{\Gamma(\alpha+1)}\ y^{\alpha+1} \frac{1}{(-\log y)^{\beta+1}} \left[1 + (\beta+1)\ \psi(\alpha+1) \frac{1}{\log y} + \mathcal{O}\left(\frac{1}{\log^2 y}\right)\right] \qquad \text{(E.4)}$$

$$\hat{\mathbb{L}}\ \frac{\log \log x}{x^{\alpha+1} (\log x)^{\beta+1}} \simeq \frac{1}{\Gamma(\alpha+1)} \frac{\log (\log \alpha)}{(- \log y)^{\beta+1}}\ y^{\alpha+1}\ . \cdot \left[1 + (\beta+1)\ \psi(\alpha+1) \frac{1}{\log y} + \mathcal{O}\left(\frac{1}{\log^2 y}\right)\right] \qquad \text{(E.5)}$$

where :

$$\mu(y,\beta,\alpha) = \int_0^\infty dx\ \frac{y^{\alpha+x}\ x^\beta}{\Gamma(\beta+1)\ \Gamma(\alpha+x+1)}$$

$$\mu(y,\ -m,\ \alpha)\ =\ (-1)^{m-1}\left[\frac{d^{m-1}}{(dx)^{m-1}}\left[\frac{y^{\alpha-x}}{\Gamma(\alpha+x+1)}\right]\right]_{x=0} \qquad m = 1,2,\ldots \tag{E.6}$$

$$\psi(z)\ \equiv\ \frac{d}{dz}\ \log\ \Gamma(z) \tag{E.7}$$

are defined by Erdelyi et al in Ref. 287) with :

$$\psi(1)\ =\ -\ \gamma_E$$

$$\mu(y,\ -1,\ \alpha)\ =\ \frac{y^{\alpha}}{\Gamma(\alpha+1)} \tag{E.8}$$

$$\mu(y,\ -2,\ \alpha)\ =\ \frac{y^{\alpha}}{\Gamma(\alpha+1)}\ [-\ \log\ y\ +\ \psi(\alpha+1)]$$

$$\mu(y,\ -3,\ \alpha)\ =\ \frac{y^{\alpha}}{\Gamma(\alpha+1)}\ [\log^2 y\ -\ 2\ \psi(\alpha+1)\ \log\ y\ +\ \psi^2(\alpha+1)\ -\ \psi'(\alpha+1)]$$

For the treatment of the continuum, we also have :

$$\int_0^{t_c} dt\ t^n\ e^{-t\tau}\ =\ (n-1)\ !\ \tau^{-n}\ [1\ -\ \rho^n] \tag{E.9}$$

with :

$$\rho_n\ =\ e^{-t_c\tau}\left(1\ +\ t_c\tau\ +\ \ldots\ +\ \frac{(t_c\tau)^n}{n\ !}\right)\ .$$

Finally, let us notice that for the FESR, the integral :

$$\int_0^{t_c} dt\ t^n\ \log\ \frac{t}{\nu^2} \tag{E.10a}$$

induces the extra-term :

$$\frac{t_c^{n+1}}{(n+1)} \left(- \frac{1}{n}\right) \tag{E.10b}$$

after a RGI of the QCD series.

www.ingramcontent.com/pod-product-compliance
Lightning Source LLC
Chambersburg PA
CBHW050632190326
41458CB00008B/2243
* 9 7 8 9 9 7 1 5 0 6 5 3 7 *